12th International Conference on Turbochargers and Turbocharging 2016

London, United Kingdom
17 - 18 May 2016

ISBN: 978-1-5108-5620-2

Printed from e-media with permission by:

Curran Associates, Inc.
57 Morehouse Lane
Red Hook, NY 12571

Some format issues inherent in the e-media version may also appear in this print version.

Printed by Curran Associates, Inc. (2018)

For permission requests, please contact Institution of Mechanical Engineers (IMechE) at the address below.

Institution of Mechanical Engineers (IMechE)
One Birdcage Walk
London, SW1H 9JJ
United Kingdom

library@Imeche.org
www.imeche.org

Additional copies of this publication are available from:

Curran Associates, Inc.
57 Morehouse Lane
Red Hook, NY 12571 USA
Phone: 845-758-0400
Fax: 845-758-2633
Email: curran@proceedings.com
Web: www.proceedings.com

CONTENTS

The 12th International Conference on Turbochargers and Turbocharging

Combustion Engines & Fuels Group Organising Committee

Dr Kian Banisoleiman *(Chairman)*	Lloyd's Register EMEA
Prof Ricardo Martinez-Botas	Imperial College
Prof Joerg Seume	Leibniz Universitaet Hannover
Dr Colin Copeland	University of Bath
Prof Roland Baar	Technische Universität Berlin
Prof Stephen Spence	Queen's University Belfast
Dr Jan-Christoph Haag	MAN Diesel & Turbo
Dr Ennio Codan	ABB
Michael Dolton	Cummins Turbo Technologies
Steve Birnie	BorgWarner
Peter Davies	Honeywell Garrett
Dr Dietmar Filsinger	IHI
Dr Seiichi Ibaraki	Mitsubishi Heavy Industries Limited
Prabhu Ramasamy	Caterpillar
Takashi Otobe	Honda R&D
Dr Geoff Capon	Ford
Per-Inge Larsson	SCANIA CV
Dr Les Smith	Jaguar Land Rover
Dr Elias Chebli	Porsche Motorsport
Bernd Wietholt	Volkswagen AG
Andrew Banks	Ricardo

The 12th International Conference on Turbochargers and Turbocharging

Organised by

Institution of MECHANICAL ENGINEERS

17–18 May 2016

The Kia Oval, Surrey County Cricket Club, London

Conference Proceedings sponsored by

MITSUBISHI HEAVY INDUSTRIES, LTD.

Our Technologies, Your Tomorrow

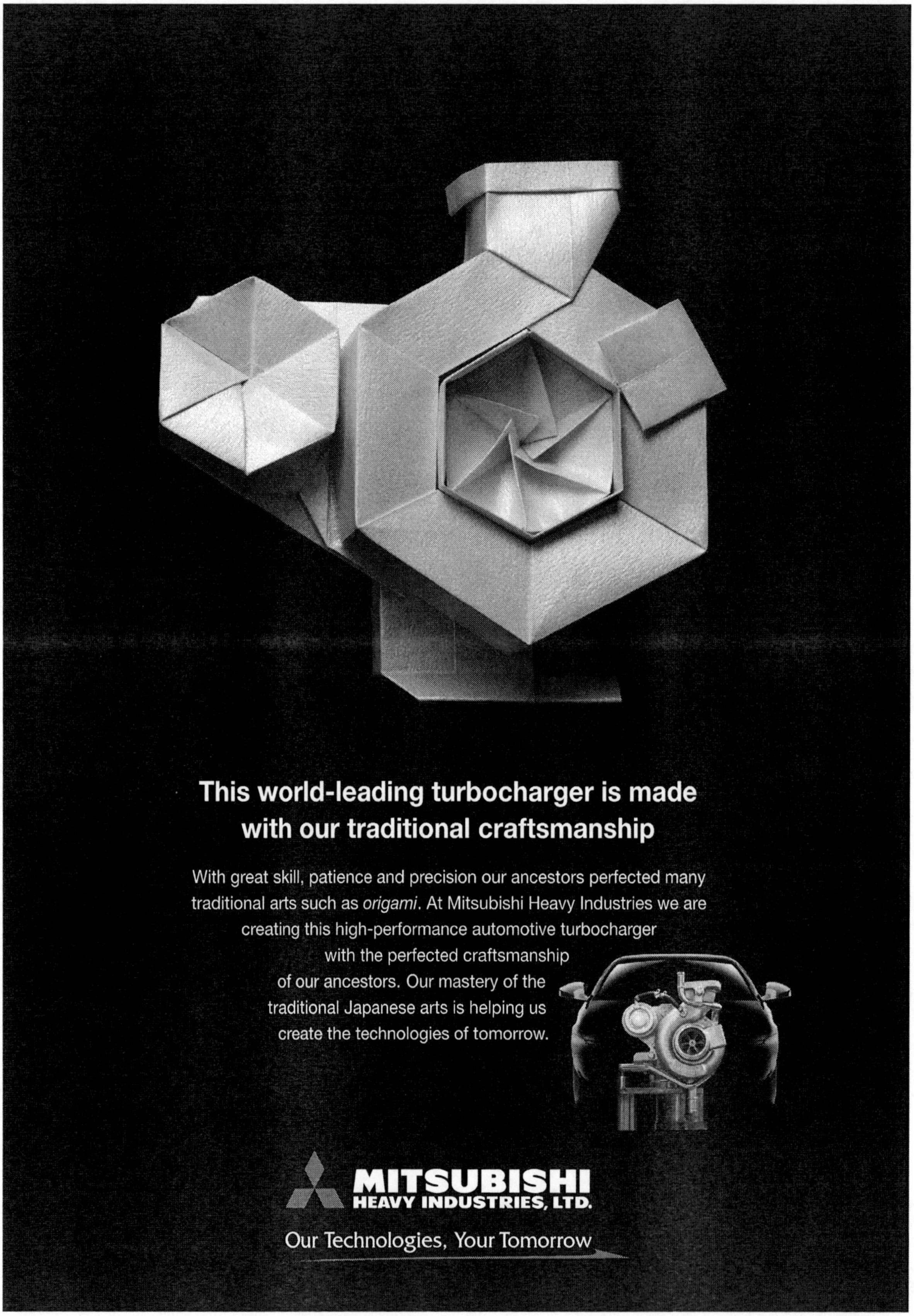
This world-leading turbocharger is made with our traditional craftsmanship
With great skill, patience and precision our ancestors perfected many traditional arts such as origami. At Mitsubishi Heavy Industries we are creating this high-performance automotive turbocharger with the perfected craftsmanship of our ancestors. Our mastery of the traditional Japanese arts is helping us create the technologies of tomorrow.
MITSUBISHI
HEAVY INDUSTRIES, LTD.
Our Technologies, Your Tomorrow

INNOVATION
comes as standard
Our sophisticated turbocharger powering the all new 2016 Nissan TITAN XD
M^2 Two-Stage System with Rotary Turbine Control (RTC)

SESSION 1A

COMPRESSOR:
FLOW RANGE ENHANCEMENT
DESIGN, APPLICATIONS
AND MATCHING

Investigation of performance and flow mechanism of a non-axisymmetric casing treatment in centrifugal compressor for turbochargers

B T Wang, R Numakura, H Tamaki
IHI Corporation, Japan

ABSTRACT

In a turbocharger compressor, the scroll inevitably causes non-axisymmetric flow field to affect the performance and flow stability near surge condition. This paper discusses the feasibility of introducing a non-axisymmetric casing treatment (naCT) to improve the asymmetrical flow field in the original design with counter-swirl casing treatment (csCT) and enhance the performance. A naCT design was tried by using four guide vanes in circumferential range of 90 degrees to partially impact the counter-swirl flow at compressor inlet. The CFD results show improvement of static pressure distortion and flow structure in diffuser by setting the new design in certain circumferential position.

NOMENCLATURE

Q	: Volume Flow Rate[m^3/s]
D	: Diameter of Positions[mm]
p	: Pressure[Pa]
T	: Temperature[K]
Cp	: Specific Heat Ratio at Constant Pressure [J/kg·K]
U	: Peripheral Speed [m/s]

Abbreviation

woCT	: Without Casing Treatment
CT	: Conventional Casing Treatment Without Guide Vanes
csCT	: Counter-swirl Casing Treatment
naCT	: Non-axisymmetric Casing Treatment
CFD	: Computational Fluid Dynamics

Subscript

1	: Impeller Inlet
2	: Impeller Exit
c	: Value at Choke Condition

Definition of non-dimensional parameter

μ_0	: Work Input Coefficient, $\mu_0 = Cp(T_{2t} - T_{1t})/U_2{}^2$

1 INTRODUCTION

Currently, the turbocharging technology is widely utilized in the internal combustion engine for automobile. Compared with the conventional engine with naturally aspiration, a turbocharged engine is beneficial to increase output power, reduce fuel consumption and improve emission [1].

High-pressure turbocharging technology is regarded as one of the developing trends due to its significant downsizing effect to lower the fuel consumption and emissions for satisfying the future regulation [2]. Therefore, a compressor with high pressure ratio and wide operating range becomes indispensable for development of a future turbocharger. However, the map width tends to be narrow for a high-pressure-ratio compressor, in which the stall can be easily induced by the transonic flow and increased relative Mach number. Hence, the technology to enhance the map width is always considered in the design process of a compressor with high pressure ratio.

Casing treatment is widely adopted to improve the surge characteristics in a turbocharger compressor. Generally, the casing treatment consists of a bleed slot, an annular cavity and an upstream slot fabricated into the compressor housing. When the compressor is operated near stall condition, the reverse flow with low momentum near the shroud of compressor impeller is extracted through the bleed slot and annular cavity to the upstream of compressor inlet. Hence, the stall will be prevented to shift the surge to lower flow rate. This idea was first proposed by Fisher [3] in 1989. In 2002, Yamaguchi [4] proposed a "counter-swirl casing treatment" (csCT) by putting the equally pitched guide vanes in the annular cavity and upstream slot to impact counter-swirl flow at the compressor inlet. Thus, the loading in each blade was greatly increased to further postpone the surge to lower flow rate, as shown in Figure 1. Recently, Harley et al. [5] proposed a casing treatment by using equispaced axial grooves in the compressor housing to link the high pressure region near shroud to the upstream inflow, and the test results shown in Figure 2 present an effective surge margin improvement within a smaller package space than conventional casing treatment.

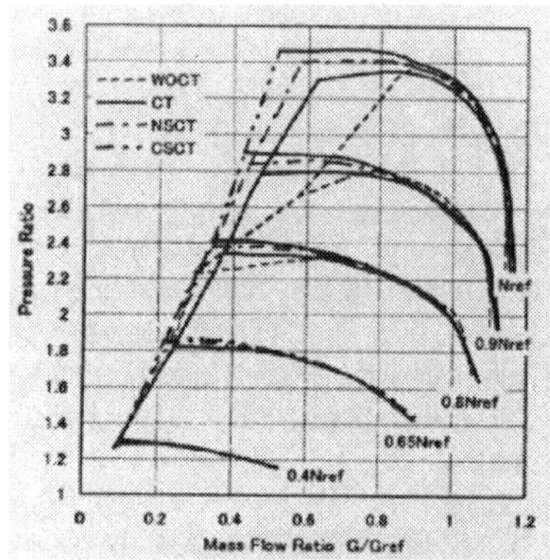

Figure 1: Test results of compressor with csCT [4]

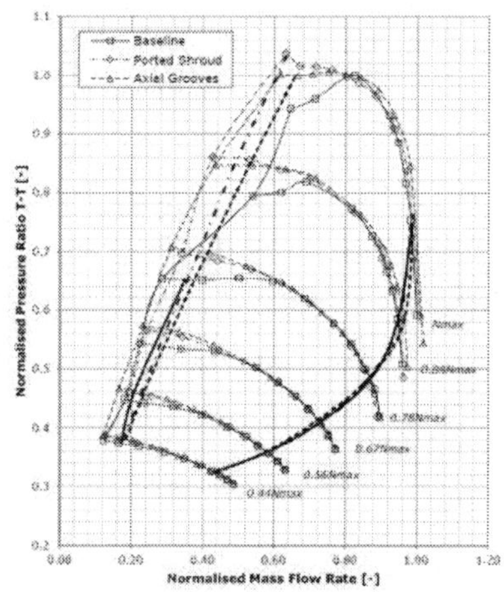

Figure 2: Test results of compressor with axial grooves [5]

Most of the casing treatments have symmetrical geometries along the circumferential direction. However, the scroll with asymmetrical geometry is always equipped to collect the fluid after diffuser. Therefore, the scroll works as a diffuser at the low flow rate to cause the static pressure distortion along the circumferential direction. As a result, the flow field in the upstream diffuser and impeller is inevitably affected, which has been confirmed by previous researches [6-7]. Also, the asymmetrical flow field induced by the scroll is considered as the main reason to cause the stall in the compressor [8]. To deal with this problem, Zheng [9] suggested a non-axisymmetric casing treatment on the basis of investigating the asymmetrical flow field in a centrifugal compressor [10]. By utilizing the design with ununiform axial positions of bleed slot, the occurrence of surge was shifted to lower flow rate by 10%, compared to conventional symmetrical design. Chen [11] summarized recent development of casing treatment and pointed out that casing treatment with asymmetrical geometry should be a promising direction to enhance its effect.

In the present work, the design of casing treatment is discussed by considering the non-axisymmetric flow structure in a compressor. Firstly, the numerical and experimental studies are conducted to investigate flow field in a high-pressure-ratio compressor with csCT. Based on the results, a non-axisymmetric casing treatment (naCT) design is made, and the effect of new design on the performance and flow field is investigated by CFD.

2 METHODOLOGY

2.1 Compressor configuration

Figure 3(a) shows the cross-sectional view of the compressor used as a research object in this work. The compressor is equipped in a large turbocharger for diesel truck. The compressor impeller has 7 full blades and 7 splitters, and the outlet diameter is 92mm. After the compressor impeller, the vaneless diffuser is employed to recover the dynamic pressure before the asymmetrical scroll for fluid collection. The csCT is adopted to widen the stable working range by utilizing an insert ring with 16 inclined guide vanes, and the recirculation flow is directed to impact counter swirl at compressor inlet near surge. The model of the insert ring is shown in Figure 4, and this technology proposed by Yamaguchi [4] is used to gain more surge margin than conventional casing treatment without guide vanes.

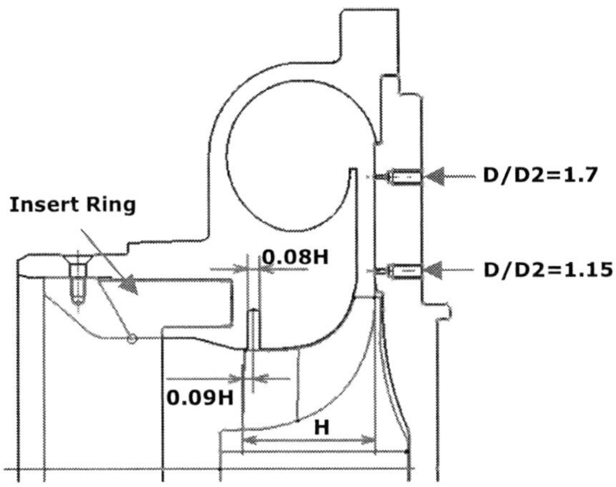

(a) Cross-sectional view

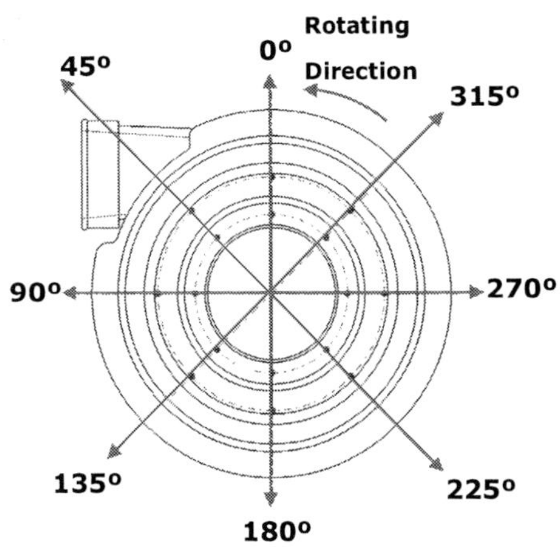

(b) Measurement positions of static pressure

Figure 3: Compressor for investigation

Figure 4: Schematic view of insert ring

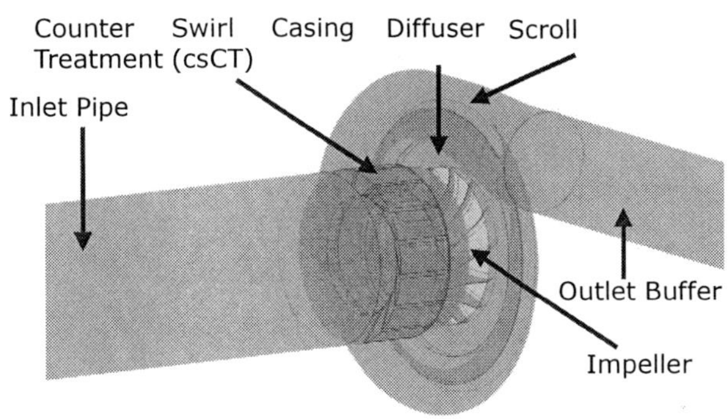

Figure 5: Simulated zone for numerical study

2.2 Experimental procedure

The experiment was done to investigate the performance and flow field in the compressor. The measurement was conducted at the compressor inlet and outlet to obtain the total pressure and total temperature for calculating the stage performance. Moreover, in order to get a view of the asymmetrical flow field along the circumferential direction, 16 measurement points, as shown in Figure3 (a) and Figure 3(b), were set at different circumferential positions to take the static pressure at diffuser inlet and outlet. For the sake of checking the effect of different types of casing treatment, the case of conventional casing treatment (CT) was also measured by removing the guide vanes in the insert ring. Also, the measurement of compressor without casing treatment was conducted to make a comparison.

2.3 Numerical procedure

In this study, the commercial solver ANSYS-CFX 15.0 was employed to solve three-dimensional Reynolds-averaged compressible Navier-Stokes equations in the simulated zone, which consists of inlet pipe, impeller, counter-swirl casing treatment, diffuser, scroll and outlet buffer. The simulated zone shown in Figure 5 was divided into several blocks to generate mesh respectively. The finite volume method was used to discretize each zone with unstructured grid mesh, and the whole mesh number is around 9,600,000, as shown in Figure 6. In term of the turbulent model, the SST (shear stress transportation) model was used for turbulence closure.

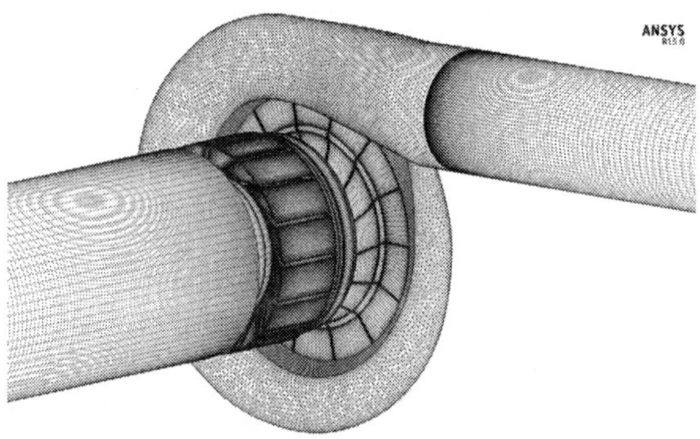

Figure 6: Mesh for numerical study

As for the boundary conditions, total pressure, total temperature and flow direction were specified at the stage inlet, and the mass flow rate was set at the stage outlet except for the conditions near choke, in which the static pressure was imposed to reach a good convergence. The wall was assumed to be non-slip and adiabatic. The calculation was conducted at the flow rates between choke and surge, obtained from the test results. The residual of each equation lower than 10^{-5} was taken as a criterion to judge if the convergence has been achieved.

The interfaces were set between each block to transfer the flow information. To treat the interface between rotor and stationary zones, the Frozen-Rotor method was used to locally transfer data. In order to check the effect of relative position between impeller and scroll on the performance and flow field, the calculation with changed relative position between impeller and scroll was also made. The differences of efficiency and pressure ratio between varied relative positions are less than 1% and 1.5%, respectively. Also, for the static pressure distribution along the circumferential direction at diffuser inlet and outlet, similar profile can be noticed between varied positions. Therefore, the treatment is considered to be reasonable for the discussion in this work.

Corresponding to the experimental study, the CFD was also conducted for the cases of woCT and CT to make a comparison with experimental results.

3 RESULTS AND DISCUSSION FOR ORIGINAL DESIGN

3.1 Compressor performance

Figure 7 shows the compressor performance at 90krpm from both CFD prediction and experimental measurement. The volume flow rate is normalized by the measured choke flow rate for the case of woCT. Also, the normalized efficiency and work input coefficient are adopted by the tested efficiency and work input coefficient at design point (Q/Qc=0.76) for the case of woCT.

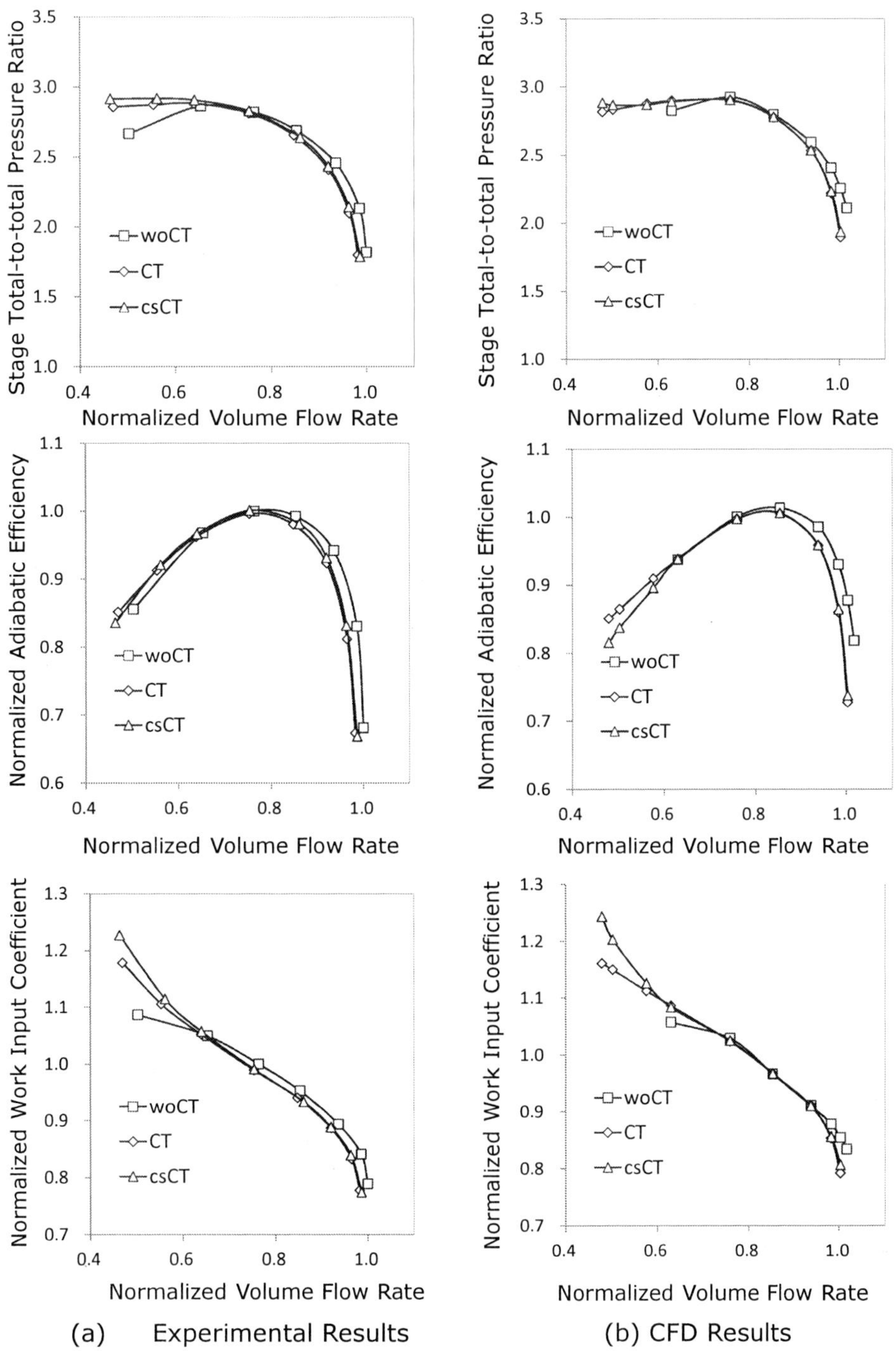

Figure 7: Comparison of CFD and experimental results for compressor performance at 90krpm

Compared with experimental results (Figure 7(a)), the CFD results shown in Figure 7(b) have a fairly well prediction of the performance for different cases. At the high flow rate, both results show the drop of pressure ratio and efficiency caused by the cases with casing treatment, compared with that of woCT. In the present design, the position of bleed slot close to impeller lead edge is at the upstream position of impeller throat, as shown in Figure 3(a). Thus, the choke flow rate is not enhanced by using casing treatment. On the other hand, when the calculation approaches low flow rate, the convergence cannot be achieved in the case of woCT at the normalized volume flow rate lower than 0.63. Moreover, the pressure ratio starts to drop in the normalized flow rate around 0.65, which is also shown in the test results. At the normalized flow rate lower than 0.6, the pressure ratio of the case with csCT presents weak tendency to increase with the decrease of flow rate. On the contrary, the pressure ratio of case with CT slightly tends to drop towards surge, which is captured by both CFD and experimental results. As for the efficiency near surge, the case of CT has better efficiency than csCT in the test results, which qualitatively agrees with CFD prediction, even though the CFD results have an overestimation of efficiency drop for the case of csCT. With regards to the work input coefficient, the CFD results agrees well with test results to show the strong work input increase at the low flow rate owing to the effect of counter-swirl inlet flow induced by guide vanes. On the basis of this comparison, it is concluded that the present CFD calculation is reasonable to discuss the performance of compressor in this work.

3.2 Static pressure distribution

In order to investigate the non-axisymmetric flow field in the compressor, the static pressures were measured at different circumferential locations of diffuser inlet and outlet, as shown in Figure 3. The measurement results against circumferential angle are plotted in Figure 8 at the flow conditions near choke, peak and surge. The measured static pressure is normalized by the total pressure at the stage inlet. Also, the CFD results are shown by taking the static pressures at the same locations with the test to make a comparison.

Generally, the scroll is designed to keep uniform static pressure along the circumferential direction at the design flow rate. However, at the off-design conditions, the scroll works as a nozzle at the high flow rates, and acts as a diffuser at the low flow rates to cause the asymmetrical static pressure distribution. This distribution inevitably affects the flow field in the diffuser. As shown in Figure 8(a) and Figure 8(c), the clear ununiform static pressure distribution can be observed near the surge and choke conditions, compared to that near peak showing almost flat profile along the circumferential direction. Also, by comparing the distribution between the diffuser inlet and outlet, they show quite similar shapes with each other, which implies that the effect of asymmetrical flow field caused by scroll has reached diffuser inlet. Moreover, it should be noticed that the test results show the obvious static pressure distortion at the circumferential position from 45° to 90° in the diffuser at the working condition near surge.

Compared with the test results, the CFD results successfully capture the asymmetrical flow field in the diffuser at the off-design conditions, even though the amplitude of static pressure distortion near surge shows a little difference from the test results.

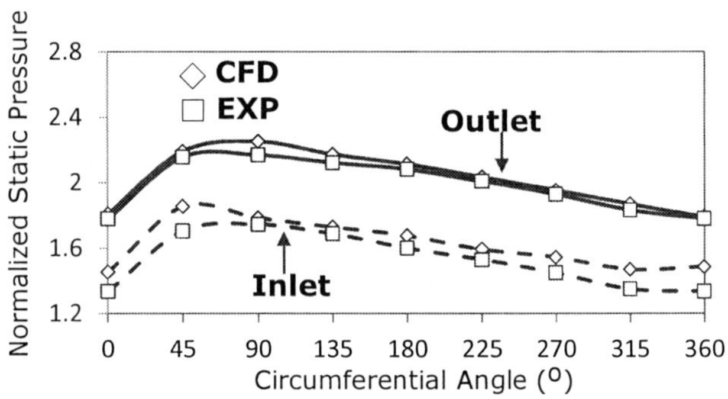

(a)Near Choke (Q/Qc=0.98)

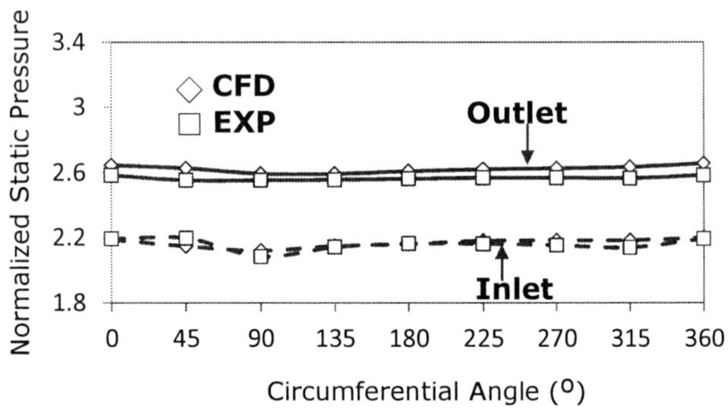

(b)Near Peak (Q/Qc=0.76)

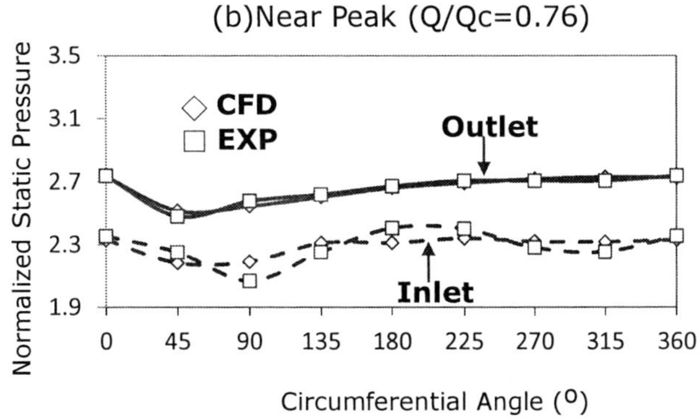

(c) Near Surge (Q/Qc=0.57)

Figure 8: Static pressure distribution at diffuser inlet and outlet for the case with csCT

4 NON-AXISYMMETRIC CASING TREATMENT

4.1 Design concept

On the basis of static pressure distribution near surge shown in Figure 8(c), a non-axisymmetric casing treatment (naCT) design is to be discussed in this section. In the past researches [9-10], a naCT design has been proposed by using a sinusoidal distribution of axial positions for the bleed slot along the circumferential direction. Then, the surge was shifted to lower flow rate with improved uniformity of flow field along circumferential direction. In the present work, for the sake of improving the static pressure distortion in diffuser, a naCT design was tried by placing only four guide vanes in certain circumferential range to impact the counter-swirl flow at the corresponding positions of compressor inlet, as shown in Figure 9. The range of guide vanes was chosen to be 90 degrees, which corresponds to the range of static pressures distortion in Figure 8(c). In Figure 9(a), the guide vanes are placed from the circumferential angle 0°to 90°, where the static pressure distortion is observed in the diffuser. This case was named "naCT_0" for the further discussion. Also, another case named "naCT_180", as shown in Figure 9(b), is investigated to set the guide vanes from 180° to 270° for checking the effect of circumferential position on the flow field.

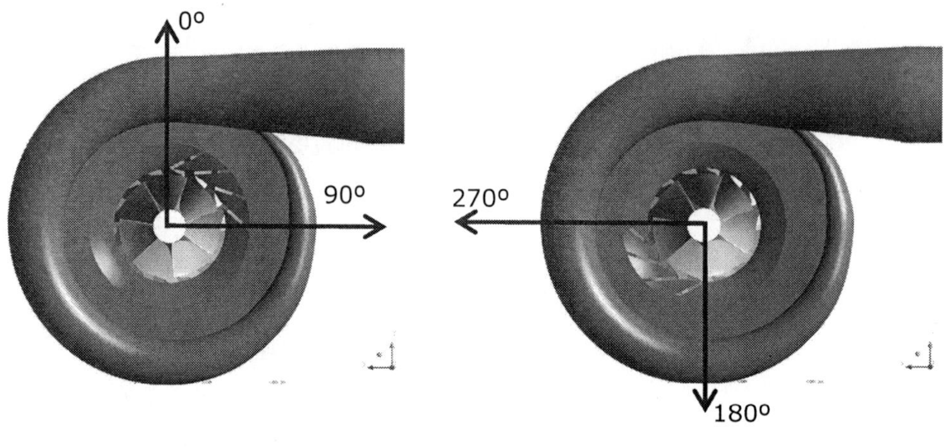

(a) naCT_0 (b) naCT_180

Figure 9: Schematic view of new naCT design

4.2 Performance predicted by CFD

Figure 10 presents the comparison of performance for the cases with csCT and naCT designs. The volume flow rate and adiabatic efficiency are normalized by using the same way with those in Figure 7. As shown in Figure 10, the difference can hardly be found at the conditions of peak and high flow rates. On the contrary, obvious difference is observed at the low flow rate. The case of naCT_0 shows stronger tendency of pressure ratio increase, as well as higher efficiency, compared with original csCT design. As for the cases with naCT_180, no obvious performance difference can be observed, except for slightly lower pressure ratio and higher efficiency near surge than case of csCT.

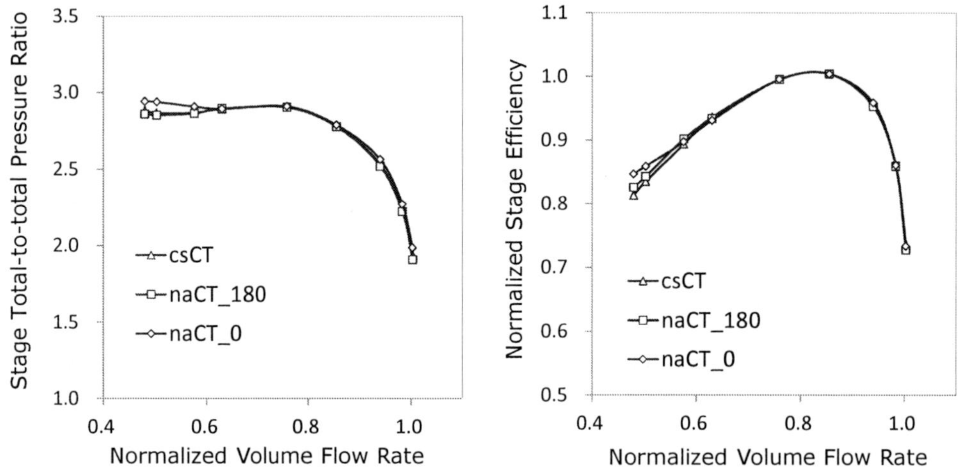

(a) Stage pressure ratio (b) Normalized stage efficiency

Figure 10: Performance of naCT predicted by CFD

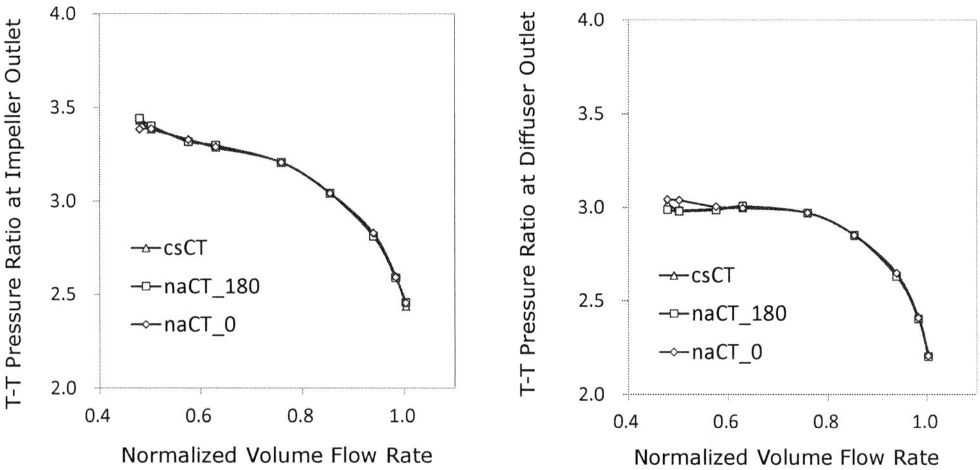

(a) Pressure ratio at impeller outlet (b) Pressure ratio at diffuser outlet

Figure 11: Pressure ratios at impeller and diffuser outlet predicted by CFD

In order to make a further discussion, the total-to-total pressure ratios at impeller and diffuser outlets are plotted in Figure 11(a) and Figure 11(b). The mass-averaged total pressures at the radial sections with D/D_2 equal to 1.05 and 1.7 are used to calculate the pressure ratio at impeller and diffuser outlets, respectively. It can be observed that the pressure ratio tends to increase towards surge for the cases with csCT and naCT_180 at the impeller outlet. On the other hand, the tendency of pressure ratio increase for the case of naCT_0 becomes weaker near surge, compared with other two cases. As for the pressure ratio at diffuser outlet, a different trend is shown compared with that at impeller outlet. The case of naCT_0 presents tendency of pressure ratio increase towards surge. With regards to the cases of csCT and naCT_180, the curve of pressure ratio becomes almost flat when approaching surge, which implies that serious pressure loss are produced in the diffuser for both cases. By comparing the results of cases with naCT_0 and naCT_180 in Figure 10 and Figure 11, it can be concluded that the circumferential position to set the guide vanes for new naCT design has a clear effect on the performance near surge.

4.3 Effect on internal flow field

For the sake of discussing the effect of new naCT design on the flow field of impeller, the contours of relative Mach number at 90% span are portrayed in Figure 12. In addition to the cases with csCT and naCT, the cases of woCT and CT are also presented for comparison.

As shown in Figure 12, for the case of woCT at the normalized flow rate 0.63, the reverse flow with low momentum reaches the leading edge of full blades to form a large stall region. On the contrary, in the case of CT, the low momentum fluid is extracted by the bleed slot. Therefore, the reverse flow hardly affects the flow field near leading edge of full blades, even at a lower flow rate ($Q/Qc=0.5$) than the case of woCT. When the counter swirl flow is impacted at compressor inlet by inclined guide vanes in the case of csCT, the flow is accelerated near the suction side of each blade. Hence, blade loading is clearly increased at the leading edge, compared to the case of CT. With regards to the case of naCT, the flow field near the leading edge of each full blade is quite dependent on their circumferential positions. For the case of naCT_0, the high loading blade at leading edge lies at the positions with circumferential angle from 0° to 240°. In contrast, the loadings of other blades at leading edge are quite low, which can explain the weaker tendency of pressure ratio increase shown in Figure 11(a) than case of csCT towards surge. As for the low flow momentum region, the area is clearly decreased, especially for the flow passages with circumferential angle from 0º to 180º. On the other hand, for the case of naCT_180, the blade with high loading corresponds to the region with circumferential angle from 180ºto 300º. In term of the low momentum flow region, the area is almost the same with that of csCT.

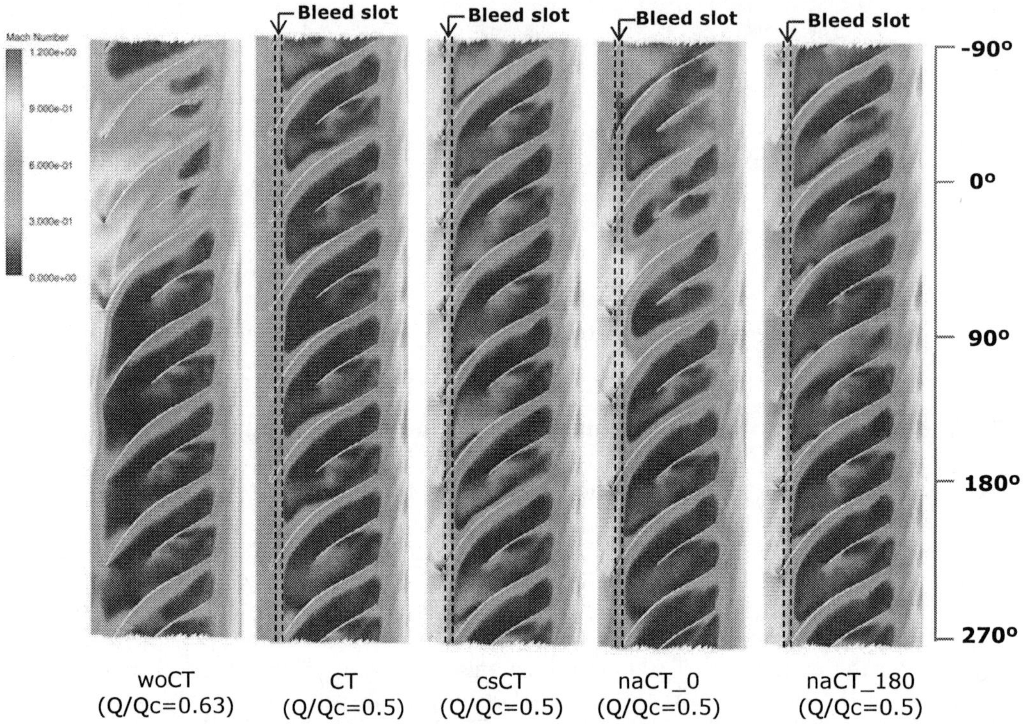

Figure 12: Relative Mach number contours at 90% span in impeller

In order to check the effect of naCT on the static pressure distortion, Figure 13 presents the static pressure distribution at the diffuser inlet for the cases with csCT and naCT near surge. Compared with the case of csCT, the static pressure distortion in the circumferential range from 45ºto 90ºis improved at the diffuser inlet for the case of naCT_0 owing to the varied inlet flow condition by asymmetrical placement of guide vanes. As for the case of naCT_180, the static pressure distortion becomes a little worse than original csCT, which implies that the circumferential position of present naCT design has a great effect on the static pressure distortion in the diffuser.

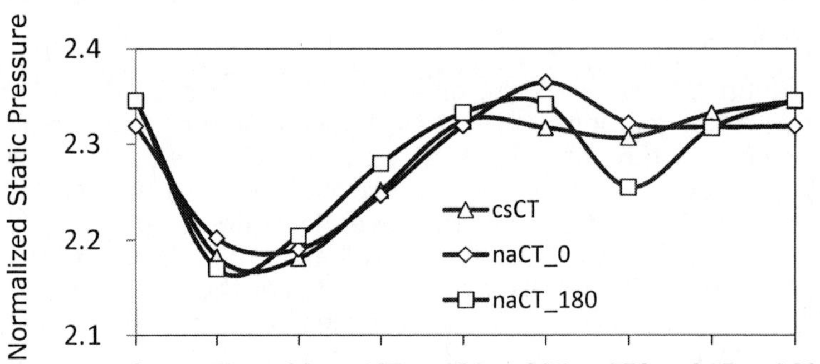

Figure 13: Static pressure distribution at diffuser inlet for the cases of csCT and naCT from CFD results (Q/Qc=0.5)

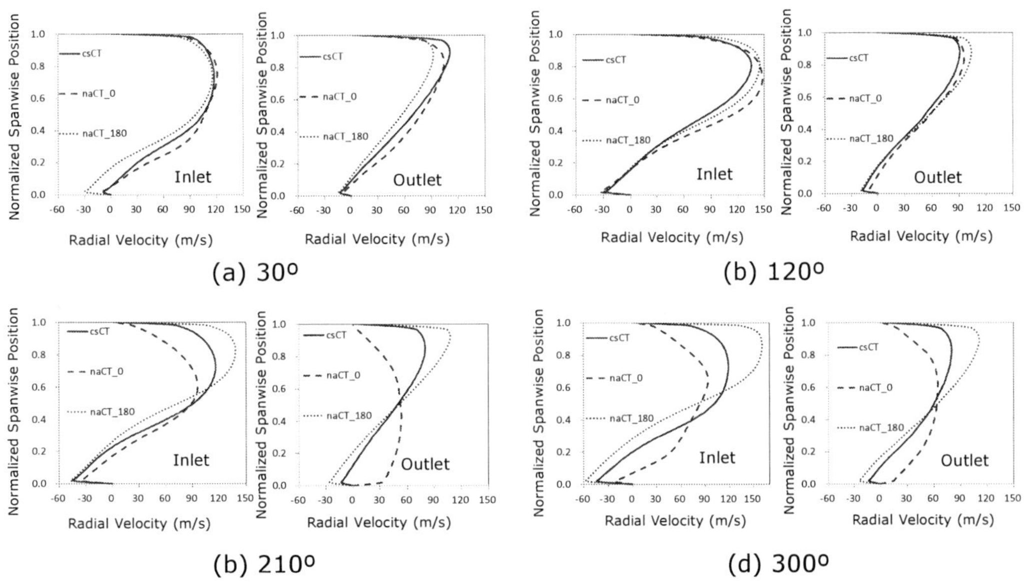

Figure 14: Radial velocity distribution along the span direction in different circumferential sections near diffuser inlet and outlet (Q/Qc=0.5)

In Figure 12, the stall region in the passage of impeller is not observed for case of csCT, even near the surge condition. Therefore, the flow field in the diffuser is to be discussed as the next step. It was reported that the separation near the diffuser hub side is the main reason to cause the diffuser stall for a compressor with vaneless diffuser [12]. Figure 14 plots the radial velocity against the normalized spanwise position from hub to shroud near the diffuser inlet (D/D_2=1.3) and outlet (D/D_2=1.7) at four circumferential locations. It is clear that the separation happens near diffuser hub side for the case of csCT. Especially at the circumferential locations of 210⁰ and 300⁰, the separation region almost reaches higher than 20% span. On the other hand, the small separation region can still be found near the diffuser outlet, even though the region becomes much smaller than that near the diffuser inlet. In contrast, the case of naCT_0 shows the separation improvement near the diffuser inlet, especially at the location of 210⁰ and 300⁰. Moreover, the separation region cannot be observed at the circumferential locations of 210⁰ and 300⁰ near the diffuser outlet, while only quite small regions can be observed in other two locations. As for the cases with naCT_180, the separation becomes even worse at each location, compared with case of csCT.

In Figure 11, the tendency of pressure ratio is greatly changed from impeller outlet to diffuser outlet. On the basis of above discussion, the separation near hub side is considered to be mainly responsible for the pressure loss in the cases of csCT and naCT_180. Moreover, the separation near the hub side is also closely related to the diffuser stall [12]. Therefore, the flow stability in the diffuser is expected to be enhanced for the case of naCT_0, compared to other two cases.

5 CONCLUSIONS

In this work, the flow field in a compressor with csCT was investigated by CFD and experiment. On the basis of non-axisymmetric flow structure, a naCT design was tried to place four guide vanes with circumferential range of 90 degrees in annular cavity and upstream slot, and CFD was conducted to clarify the effect on the performance and flow field. The conclusions could be drawn as follows:

(1) In the original csCT design with guide vanes in the whole circumference, the clear static pressure distortion along the circumferential direction was shown at the diffuser inlet and outlet for both CFD and experimental results. Also, the CFD

results show clear flow separation near the hub side of diffuser to produce significant pressure loss.

(2) According to CFD results, the naCT design has a clear effect on the flow field and performance towards surge. In the case of naCT_0, the static pressure distortion in diffuser is reduced by impacting counter swirl at certain circumferential range of the compressor inlet. Moreover, the flow separation in the diffuser is improved to enhance the performance near surge.

(3) By changing the circumferential position of new naCT design, the performance and flow field are obviously affected. The case of naCT_180 shows increased static pressure distortion and deteriorated flow field in the diffuser to produce more loss than case of naCT_0.

REFERENCES

[1] Baines, N. C, 2005, "Fundamentals of Turbochargers", Concept ETI, Inc., Vermont.

[2] Clenci, A. C., Descombes, G., Podevin, P., and Hara, V., 2007, "Some Aspects Concerning the Combination of Downsizing with Turbochargering, Variable Compression Ratio, and Variable Intake Valve Lift", Proc. Inst. Mech. Eng., Part D J. Automob. Eng., 221(10),pp. 1287-1294.

[3] Fisher, F. B., 1989, "Application of Map Width Enhancement Devices to Turbocharger Compressor Stages", SAE Trans., 97(6), pp.1303-1310.

[4] Yamaguchi, S., Yamaguchi, H., Goto, S., Nakano, H., and Nakamura, F., 2002, "The Development of Effective Casing Treatment for Turbocharger Compressors", IMechE 7th Intl. Conf. on Turbochargers and Turbocharging, London, UK. C602/008/2002.

[5] Harley, P., Starke, A., Bamba, T., and Filsinger, D., 2016, "Axial Groove Casing Treatment in An Automotive Turbocharger Centrifugal Compressor", IMechE 12th Intl. Conf. on Turbochargers and Turbocharging, London, UK. (currently under review).

[6] Hillewaert, K. and Van den Braembussche, R.A, T., 1999, "Numerical Simulation of Impeller-Volute Interaction in Centrifugal Compressor", ASME J. Turbomach., 121(3) pp. 603-608.

[7] Reunanen, A., Pitkanen, H., Siikonen, T., Heiska, H., Larjola, J., Esa, H., and Sallinen P., 2000, "Computational and Experimental Comparison of Different Volute Geometries in a Radial Compressor", ASME Paper No. 2000-GT-469.

[8] Sorokes, J., Borer, C., and Koch, J., 1997, "Investigation of the Circumferential Static Pressure Non-uniformity Caused by a Centrifugal Compressor Discharge Volute", ASME Paper No. 98-GT-326.

[9] Zheng, X. Q., Zhang, Y. J., Yang, M. Y., Bamba, T., and Tamaki, H., 2013, "Stability Improvement of High-Pressure-Ratio Turbocharger Centrifugal Compressor by Asymmetrical Flow Control-Part II: Non-axisymmetrical Self-Recirculation Casing Treatment", ASME J. Turbomach., 135(2), 021007.

[10] Yang, M. Y., Zheng, X. Q., Zhang, Y. J., Bamba, T., Tamaki, H., Joern, H., and Li, Z. G., 2013, "Stability Improvement of High-Pressure-Ratio Turbocharger Centrifugal Compressor by Asymmetrical Flow Control - Part I: Non-axisymmetrical Flow Field in Centrifugal Compressor", ASME J. Turbomach., 135(2) 021006.

[11] Chen, H., and Lei, V., 2012, "Casing Treatment and Inlet Swirl of Centrifugal Compressors", ASME J. Turbomach., 135(4), 041010.

[12] Ishida, M., Sakaguchi, D., and Ueki, H., 2001, "Suppression of Rotating Stall by Wall Roughness Control in Vaneless Diffusers of Centrifugal Blowers", ASME J. Turbomach., 123(1), 63-72.

Turbocharger compressor map enhancement for highly efficient combustion engines

T Czapka [(a)], D Hagelstein [(a)], J Theobald [(a)], J Seume [(b)]

[(a)] Volkswagen AG, Germany

[(b)] Leibniz Universität Hannover, Institute for Turbomachinery and Fluid-Dynamics, Germany

ABSTRACT

An optimal matching between turbocharger and engine aims for the best possible utilization of the provided turbine power on the compressor side in order to increase manifold pressure. To cover the engine´s demand for combustion air especially at high loads over a wide operating range, a compressor with a wide and stable compressor map is needed. This paper examines a new methodology for turbocharger development processes as well as different methods for the enhancement of the map characteristics of a turbocharger centrifugal compressor for automotive engine applications.

To analyze the potential of methods for compressor map enhancement, both numerical and experimental investigation methods are used. An evaluation of the compressor aerodynamics with respect to the engine characteristics is performed to this end.

The results show that enhancement of the compressor map is beneficial for the engine, especially by shifting the surge line towards lower mass flow rates. Through stabilizing the compressor map and improving the efficiency in the low flow area of the compressor map, both the stationary and dynamic engine behavior is improved. The methodology introduced into the turbocharger development process proved to be very effective regarding the process of examining the enhancement of the compressor map. From all concepts examined, the combination of variable trim compressor (VTC) upstream and low solidity vaned diffuser (LSVD) downstream of the compressor impeller offers the best performance results.

1 INTRODUCTION

In automotive transportation, the combustion engine is still the globally dominating propulsion concept due to its high level of development. In order to be able to claim this key position in the contested field of mobility in the near future, the focus of engine development is on the continuous technological enhancement of all components. The requirements are increasingly tightened due to pressure from forces such as politics, economy and the environment (emission legislation, cost, shortage of resources) as well as the increased customer requirements for higher performance, driving dynamics and efficiency need to be dealt with. The aim of development therefore is reducing the fuel consumption through an increase of the effective driving power and the effective efficiency and improving the dynamic behavior simultaneously.

In the last years the downsizing concept – reducing displacement and simultaneously increasing power density – has emerged as the most successful strategy for the reduction of fuel consumption in the gasoline engine [1], [2]. Through the shifting of the operating points to higher mean effective pressure levels, the combustion engine can be operated in the design range more often with higher efficiency [3].

Through the correct sizing of the compressor and turbine, as well as corresponding control measures (e.g. waste gate) on the turbine side, an appropriate supply of charge pressure to cover of the engine´s demand for combustion air can be guaranteed over the entire engine map. In the course of further increasing charging and consequently higher mass flow rates, increasing demands are placed upon the turbocharger design, i.e. the demanded low-end-torque (LET) and the rated power at engine full load along with a simultaneously acceptable dynamic behavior in case of step changes in loading (see figure 1).

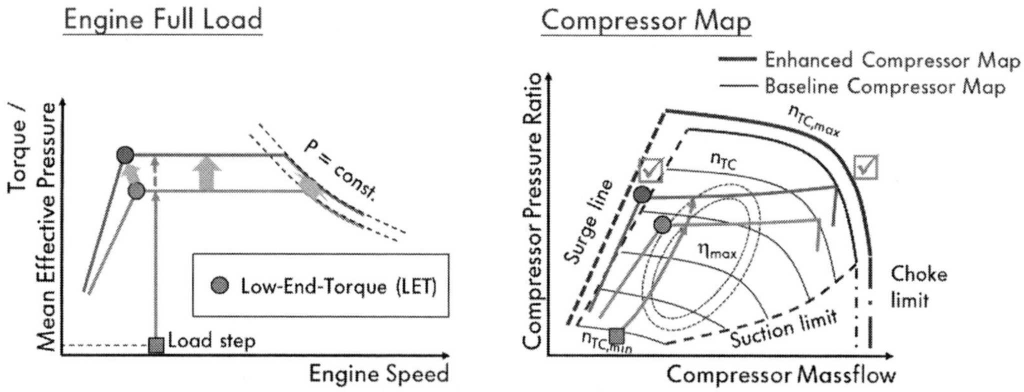

Figure 1 Full load and load step characteristics of engine and compressor

In order to be able to cover the increasing engine requirements with a single-stage turbocharger-system in the future, the available turbine power needs to be used as efficiently as possible on the compressor side for the supply of the charge pressure in all operating points. Independently of the compressor sizing there is the demand for the enhancement of the stable compressor operating range, in particular towards the lower mass flow rate along with a simultaneously improved charge pressure level as well as high efficiency in the compressor map.

2 COMPRESSOR MAP ENHANCEMENT

Within the thermodynamic interaction of combustion engine and turbocharger the compressor is mostly the limiting component due to its limited operating range. From the literature different systems for compressor map enhancement are known. These systems either act geometrically variable (active) or rigid (passive). Regarding the position to the compressor impeller they can be divided into upstream or downstream devices with respect to the fact that a combination of systems is possible.

Ported shrouds with a shroud-sided bleeding slots and variable inlet guide vanes are proper devices for compressor map enhancement both at the surge limit and the choke limit. With both systems there is an opportunity for reducing inlet incidence as well as inlet losses. [4], [5], [6]

By decreasing of the effective flow area at the compressor impeller inlet, a variable trim compressor (VTC) also has a great potential for shifting the surge line towards lower mass flow rates while simultaneously greatly increasing efficiency. A partial blocking of the flow field, preferably near the shroud, leads to a reduction of the inflow incidence at the inlet of the compressor blades. In particular this is due to higher flow velocities in the free flow area near the hub. A shielding of the shroud-sided impeller inlet additionally raises the compressor efficiency as incidence and therefore flow losses are avoided in this range. However, the reduction of the compressor impeller inflow area also leads to a decrease of the usable compressor map width, as the compressor tends to choke at lower mass flow rates. To operate the compressor properly at high flow rates, the VTC device needs to be shifted such that it opens up the whole inlet flow area. [7], [8]

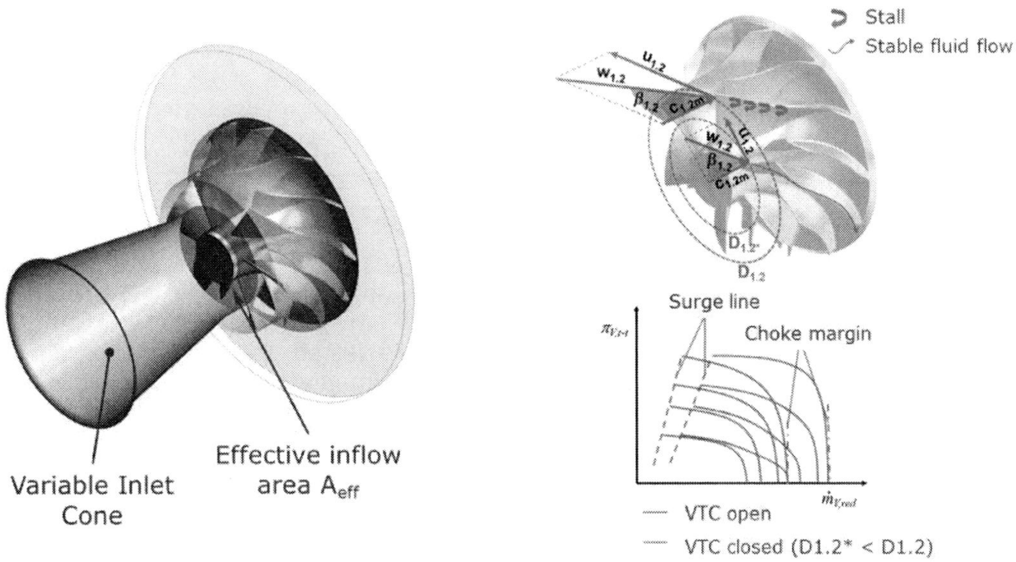

Figure 2 VTC design and methodology

Having a pinch on the diffuser inlet (taper of the diffuser inlet width), preferably both on the shroud and the hub side, contributes to stabilizing and homogenizing of the impeller outflow [9], [10]. By the implementation of guiding vanes into the compressor diffuser the charge pressure level and the efficiency can be further increased within the compressor map due to the improved flow conditions in the diffuser [11]. Nevertheless the flow field of the volute is influenced by the flow conditions at the diffuser outlet and it depends on the setting angle of the diffuser vanes. Low solidity vaned diffusers (LSVD) offer a wider effective operating range and better flow conditions in the off-design flow region due to reduced blockage of the cross-section flow area compared to conventional diffusers with flat vanes. [12], [13], [14]

A variability of the diffuser can be achieved in different ways:

In case of full variability, the LSVD is closed by rotating the vanes towards the surge line and opened for operation towards choke while the radial vane position remains constant. This is disadvantageous at the maximum choke flow rate because it causes higher flow losses compared to a vaneless diffuser. [15]

Vane variability can alternatively be achieved by switching the LSVD in and out of the diffuser as illustrated in Figure 3. This provides the possibility of optimizing several compressor map regions without reducing the choke flow rate [16].

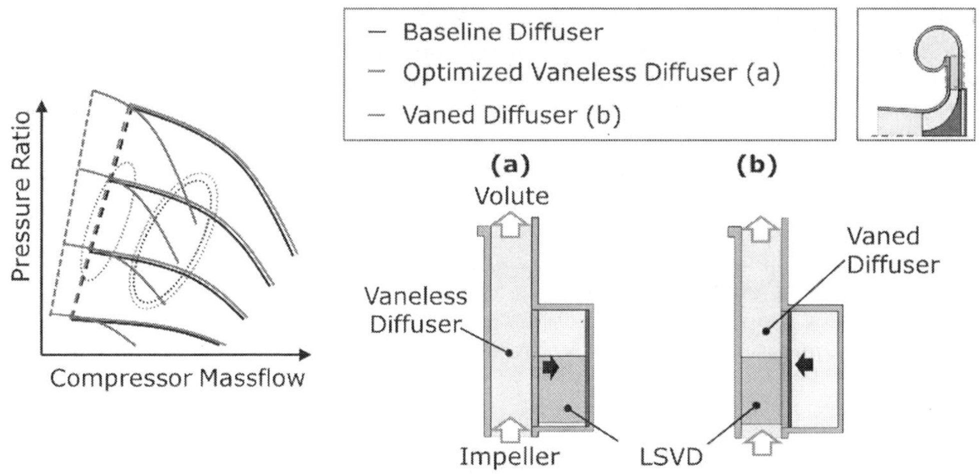

Figure 3 LSVD design and methodology – designed for high flow (a) and surge flow improvement (b)

3 METHODOLOGY OF INVESTIGATION

The design of the compressor map enhancement and the analysis of the resulting aerodynamics are done numerically by 3D-CFD-simulation. The most promising configurations are manufactured as prototypes and tested under steady conditions on the turbocharger hot gas test bench. The experimental results are used for the engine cycle simulation in order to evaluate the potential impact of the compressor map enhancement on the engine characteristics. Finally, the dynamic response of the turbocharger with the improved compressor is analyzed experimentally on a hardware-in-the-loop (HiL) turbocharger dynamic test bench controlled by a new real time capable virtual engine environment.

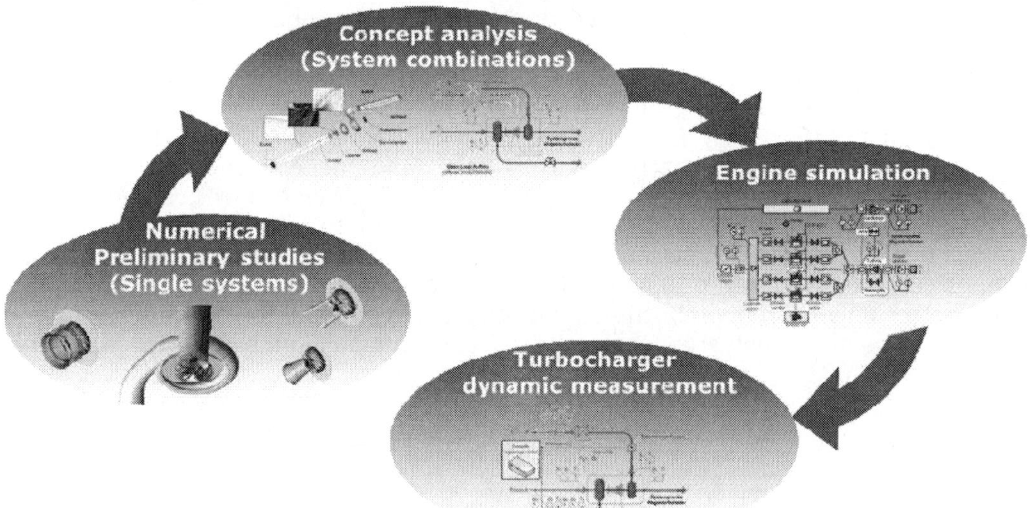

Figure 4 Methodology of investigation

4 TURBOCHARGER BASELINE COMPRESSOR

All configurations analyzed were measured on the same baseline turbocharger. The compressor impeller has a trim of T = 75% and consists of six main blades and six splitter blades. For the increase of the maximum flow capacity as well as for the stabilization of the flow towards the surge line, the impeller blade design includes backsweep. The compressor diffuser is designed as a parallel-walled radial diffuser with linear shroud pinch at the diffuser inlet. The external volute is designed with a logarithmically widening flow area.

5 TEST BENCH FACILITIES

5.1 Turbocharger hot gas test bench

Different turbocharger configurations are compared on the basis of steady measurements of the performance map on the turbocharger hot gas test bench. Measurements of the turbocharger are performed both with an open-loop and a closed-loop compressor configuration on the hot gas test bench [17], [18], [19].

5.2 Hardware-in-the-loop (HiL) turbocharger dynamic test bench

Due to the thermodynamic coupling of turbocharger and combustion engine as well as the great number of resulting factors, a measurement without the engine focusing on the turbocharger-specific properties is not easily possible. A new test method with which steady and dynamic investigations can be carried under engine-realistic conditions without having a combustion engine was developed by [19]. This hardware-in-the-loop ([HiL) test bench is also used in the present work in order to measure the turbocharger transient behavior. The model environment and methodology was enhanced for this purpose [17].

Unlike a conventional turbocharger test bench, the turbine receives compressed air from a central supply which is preheated by an electrical heater. A sufficiently fast and precise control of the required flow conditions for dynamic operation is achieved by control valves (butterfly flaps) upstream of the turbine and downstream of the compressor.

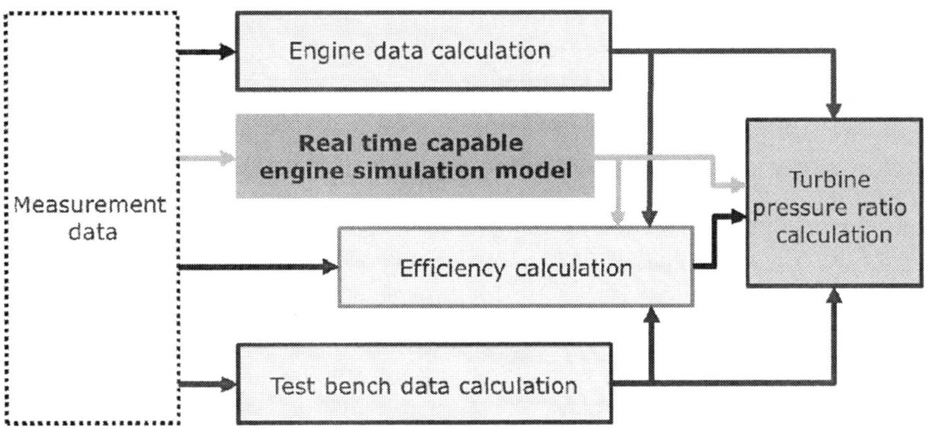

Figure 5 Load step control model

The integration and coupling of a new real time capable engine simulation model for setting the engine exhaust temperature is the core element of the load-step method (see figures 5 and 6). By this a quasi-physical compensation calculation of the exhaust gas temperature is enabled considering gas-dynamic effects in the model during the load step. The control is based on a required value calculation of the turbine pressure ratio at the test bench, necessary to control the compressor and turbine flap.

The control methodology is based on a virtual mass flow coupling between compressor and turbine. The mass flow provided by the compressor to the combustion chamber and the fuel add up to the exhaust gas mass flow to the turbine with the resulting exhaust gas energy after the combustion. For this purpose the compressor needs to be driven on the respective engine-specific operation line. A calculation based on the principle of the virtual mass flow coupling ensures that the turbine power at the dynamic test bench agrees with the turbine power on the engine at all times.

6 NUMERICAL ANALYSIS

6.1 3D-CFD-Compressor simulation model

The preliminary numeric studies of the baseline compressor and all further configurations are carried out with the three dimensional CFD simulation code Ansys CFX. Every model is established as a full 360° model. Except for the compressor volute all model domains are meshed with a structured highly resolved hexahedral mesh with a prismatic wall junction region. The discretization of the volute is carried out through an unstructured tetrahedron mesh, with the wall junction region also prismatically highly resolved.

To describe the rotating compressor impeller, the quasi-steady frozen rotor interface is used. This is a tried-and-tested approach for the steady-state model of flow interactions in the compressor. At the inlet boundary, total temperature and pressure are set while air mass flow is specified at the outlet boundary and so is the shaft speed of the rotor. With this configuration, every stable operating point of the compressor map can be calculated. The air is modeled as an ideal gas. Wall friction is modeled while heat transfer is not. As a turbulence model the two-equation model k-ω SST with automatic wall functions is used.

6.2 1D-Engine simulation model

The cycle simulation to analyze the engine operating characteristics as well as the implemented turbocharger configurations is carried out with GT-Power. The simulation model used represents a four-cylinder direct-injection gasoline engine with mono-turbocharger and waste gate.

The thermodynamic properties of compressor and turbine are implemented via operating maps into the given templates in the model. The maps for compressor and turbine are extrapolated. This is necessary especially for the dynamic simulation because map regions at low flow rate cannot be measured properly. These regions are normally crossed by the engine´s air flow operation line at the beginning of the load step.

6.3 1D-Real time engine simulation model

For the calculation of the engine´s exhaust gas temperature at turbine inlet, a new real time capable simulation model was set up which is connected to the load step model via provided interfaces [17]. This provides the following advantages for calculation of the load step on the turbocharger dynamic test bench:

- transient engine performance calculation considering gas-dynamic effects
- calculation and monitoring of engine-specific properties during the load step
- quasi-physical calculation of the engine´s exhaust gas temperature at turbine inlet depending on the turbine backpressure characteristics

In the present work, the Mean Value (MV) modelling strategy is used for deriving a real time capable engine simulation model from the detailed baseline engine model [20]. This approach is particularly characterized by a high numerical stability and a very fast calculation. At the same time, a high accuracy is achieved by the integration of neural networks into the model structure. The one-dimensional (1D) real time engine simulation model as well as the data structure for the coupling to the zero-dimensional (0D) load step model is schematically presented in figure 6.

Data between the 0D-model and the 1D-model are exchanged through an interface which provides measurement variables from the test bench directly to the real time engine simulation model as constraints at the model boundaries. In turn, engine specifications from the 1D-simulation are transferred through the interface into the 0D-environment.

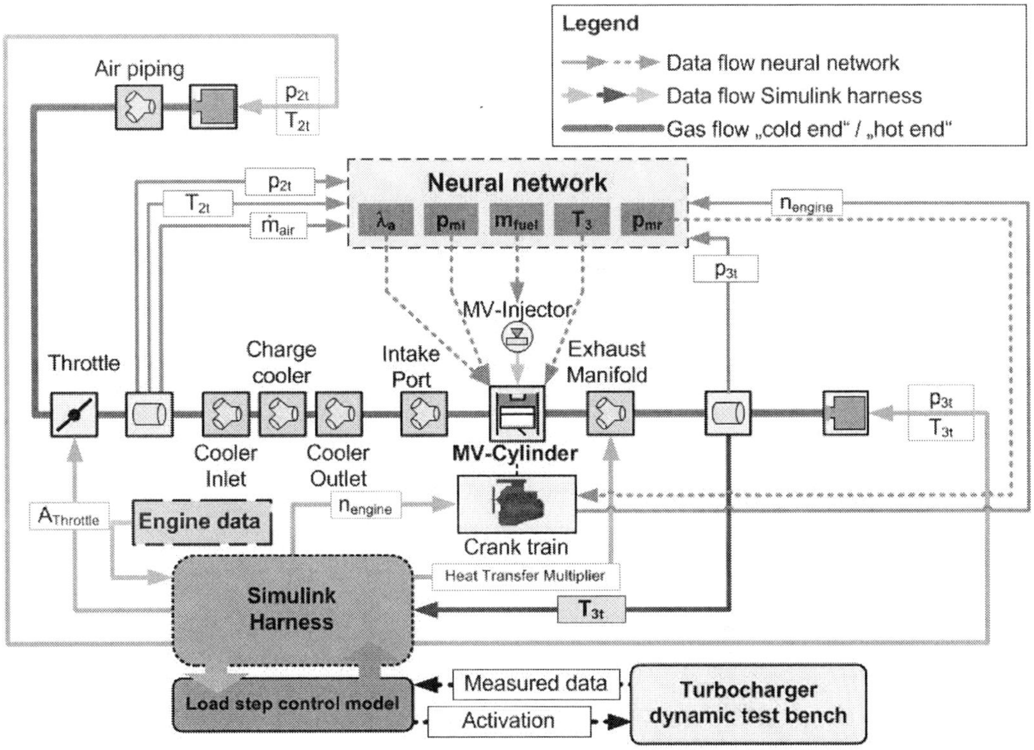

Figure 6 1D-Real time engine simulation model

7 RESULTS

This chapter first describes the results of the preliminary numerical studies which were carried out ahead of the concept definition. Afterwards, the chosen concept is introduced. After the discussion of the simulated compressor map and the measured prototypes, the concept is evaluated for the gasoline engine via 1D-engine simulation and the turbocharger dynamic test bench.

7.1 Numerical studies of variable geometry options

7.1.1 Investigation of the variable trim compressor (VTC)

In this study a number of fixed inlet cones with reduced effective inflow area are considered. Through the reduction of the effective inflow area at the compressor impeller inlet, a significant shift of the surge line is achieved up to 65.8% in the direction towards lower mass flow rates (see figure 7).

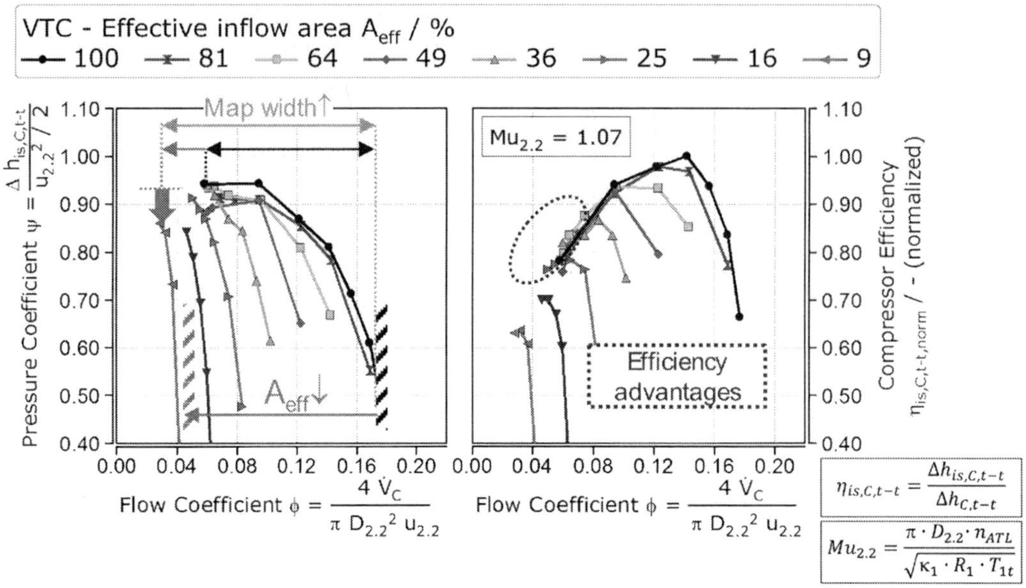

Figure 7 VTC simulation results

In particular at low mass flow rate as well as low charge pressure level a clear increase of compressor efficiency is achieved. Through the reduction of effective inflow area at the impeller inlet flow losses near the shroud wall as well as incidence near the rotor hub are reduced effectively (see figure 8).

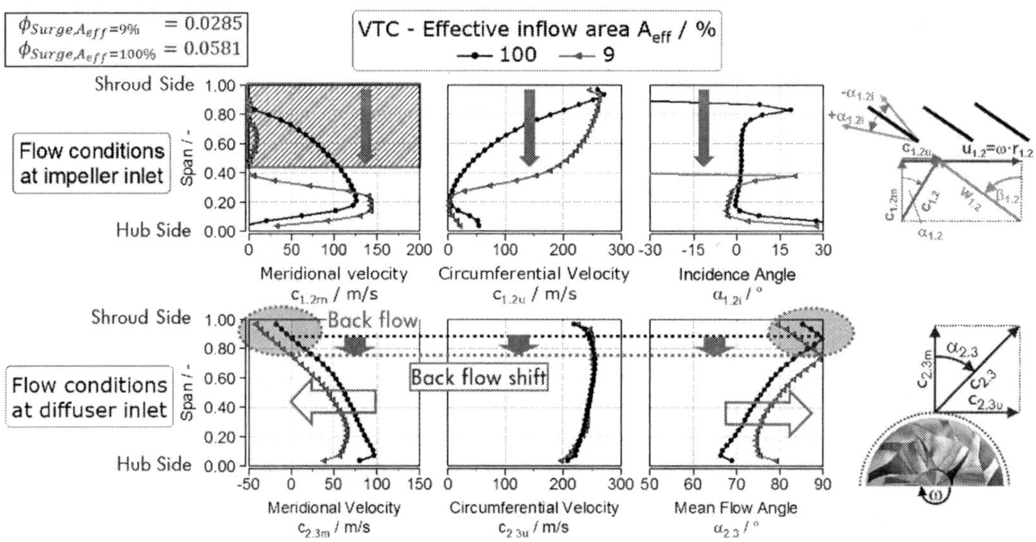

Figure 8 VTC flow field simulation results ($Mu_{2.2}$ = 1.07)

With reduced effective inflow area at low mass flow rates, high flow velocities are achieved at the compressor impeller inlet near the hub. This in turn stabilizes the flow in the compressor impeller because flow separation at the blade leading edge is avoided. Correspondingly the effective compressor map width is narrowed significantly due to a reduced effective inflow area because the compressor chokes at lower mass flow rates due to higher relative Mach numbers at the impeller inlet.

The reduction of the effective inflow area is reasonable only to a certain level because an excessive reduction leads to an enlargement of the backflow at the shroud of the diffuser inlet. Thus the amount of fluid flowing back into the compressor impeller and therefore dissipation would increase. Consequently this leads to a clear increase of the compressor exit temperature and therefore to drop in the compressor efficiency.

Of all concepts examined, the VTC device shows the greatest potential for shifting the surge line to lower mass flow rates. On the other hand, shifting the choke margin to higher mass flow rates is not possible with the shown device since it can only reduce the effective inflow area. [17]

7.1.2 Investigation of VTC in combination with LSVD

The LSVD showed a great potential for the enhancement of the compressor map at lower mass flow rates. The LSVD consists of N = 8 vanes with a solidity of σ = 0.51 [17]. Therefore the potential for the enhancement of the compressor map through a suitable combination of VTC and LSVD is explored numerically. Fundamentally two scenarios are considered:

- VTC closed in combination with vaneless diffuser (LSVD shifted outside of the diffuser flow area, as shown in figure 3a)
- VTC closed in combination with LSVD (see figure 3b)

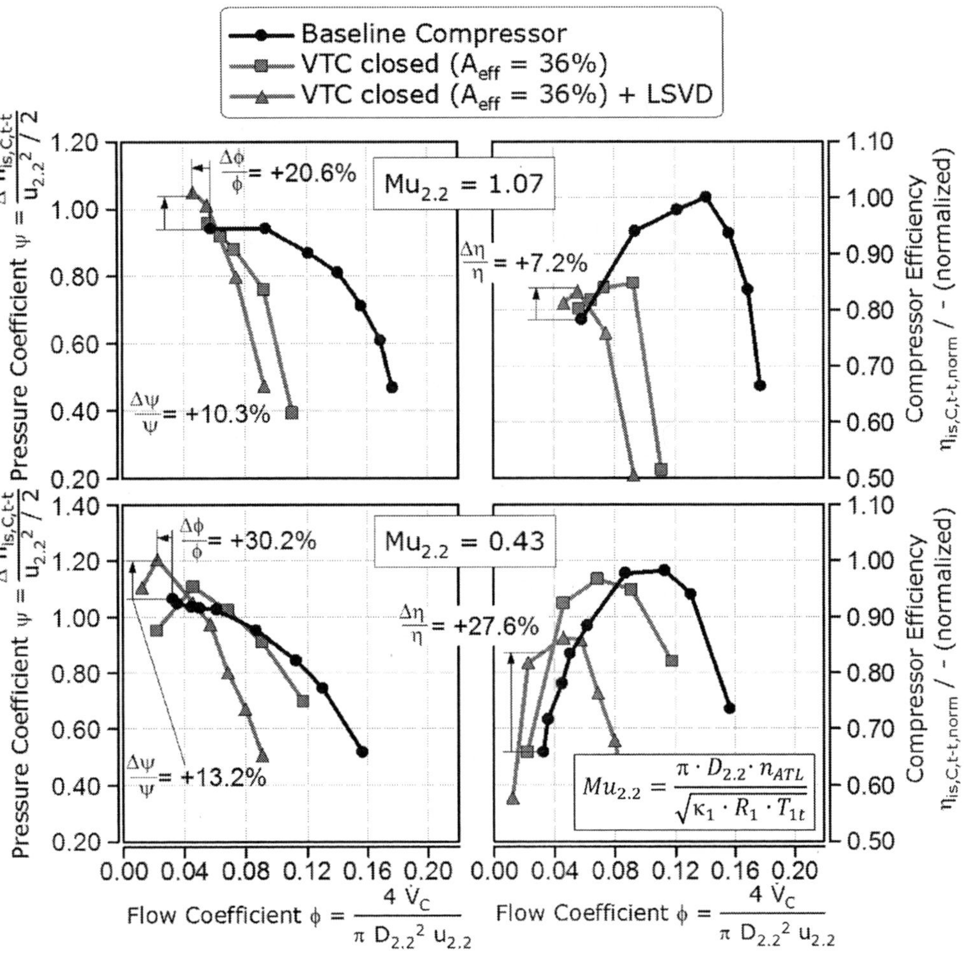

Figure 9 VTC and LSVD simulation results

In order to utilize the enlarged compressor map range effectively, in particular at lower mass flow rates, an increase in compressor efficiency is necessary. Only then the offered turbine power can be utilized efficiently by the compressor to augment the charge pressure. At that case only one fixed VTC position with relative inflow area reduced to A_{eff} = 36% of the initial inflow area is considered. The influence of the configurations on the pressure ratio and the efficiency are presented in figure 9 for the circumferential Mach numbers of the compressor of $Mu_{2.2}$ = 0.43 and 1.07.

The VTC with vaneless diffuser causes a moderate improvement of the surge line at both speeds but the width of the compressor map is simultaneously reduced due to the narrower inflow area. In the low speed range significant efficiency advantages compared to the baseline compressor are gained but this gain decreases as the circumferential Mach number of the compressor is increased. The cause for the efficiency improvement is the reduction of the entrance losses especially near the shroud in compressor impeller and diffuser.

Shifting the LSVD into the diffuser and having the VTC closed leads to a further shift of the surge line in the direction of lower mass flow rates. Simultaneously clear advantages are achieved with regard to the gained charge pressure level and the efficiency benefit when compared to the VTC with vaneless diffuser. Nevertheless the usable compressor map range turns out to be reduced. The LSVD stabilizes the flow field at the diffuser inlet. The combination of VTC and LSVD is thus very effective in improving the compressor map especially near the surge line.

7.2 Turbocharger hot gas test bench results

The numerically examined variants were manufactured as prototype modules and tested on the turbocharger hot gas test bench. They were produced as discrete and inflexible devices, thus avoiding geometrical variances.

The LSVD is implemented as a fixed cartridge, the VTC is represented by a cone. Compressor map measurement results are shown in figure 10.

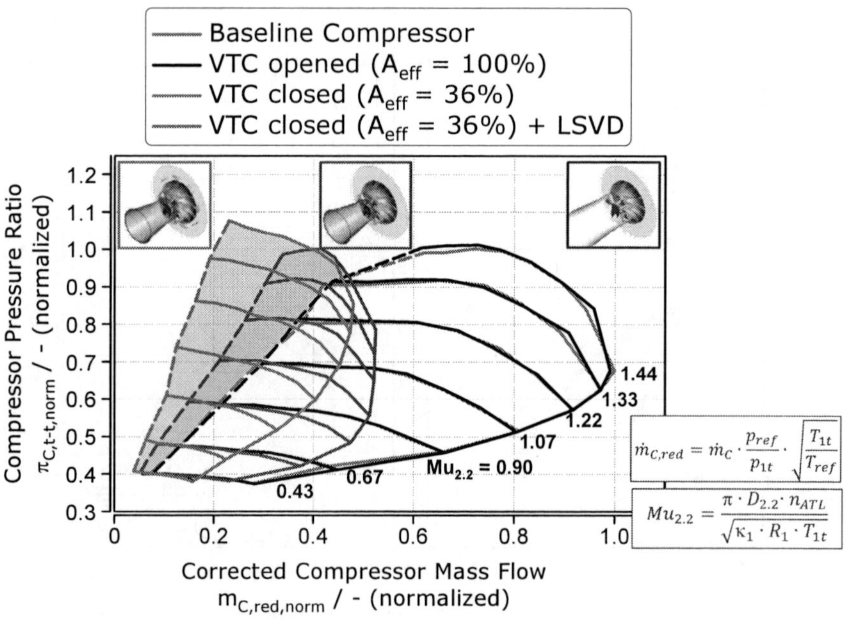

Figure 10 VTC and LSVD measurement results I

With an optimized vaneless diffuser and the VTC open, the compressor map width of the baseline compressor is achieved with some benefit in efficiency. With the VTC in closed position, the surge line is shifted by about 45.2% in the direction of lower mass flow rates.

By bringing in the LSVD vanes into the diffuser, the surge line is shifted further towards lower mass flow rates, thus the surge line is shifted between 50.0% and 67.9% compared to the baseline compressor. Figure 11 shows that additionally the pressure ratio is raised by up to 8.8% due to the increased circumferential Mach number of the compressor compared to the baseline compressor.

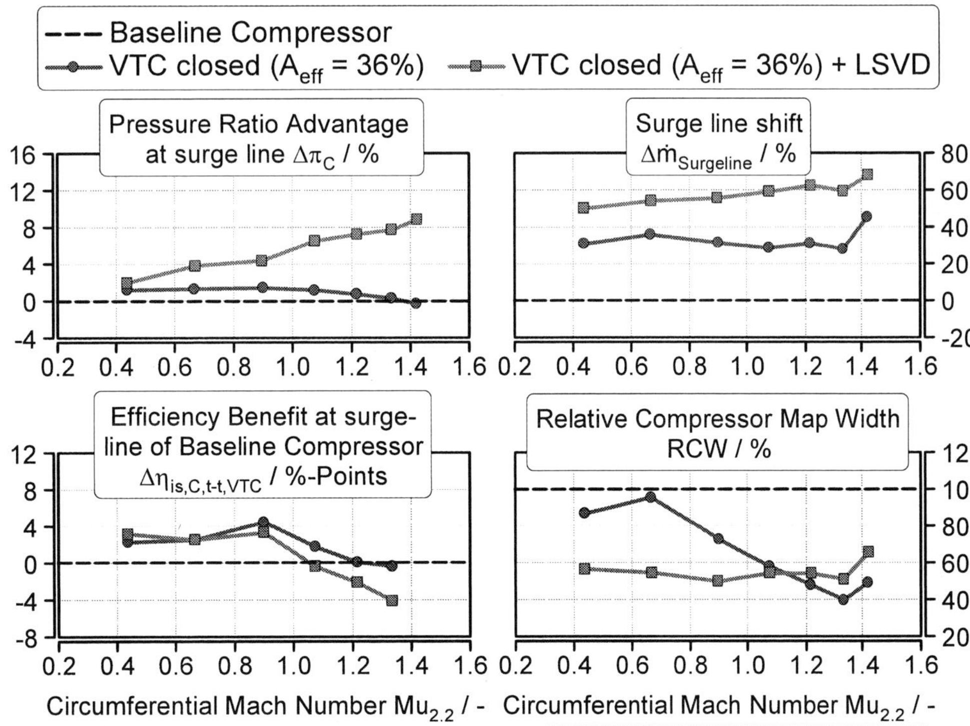

$$Mu_{2.2} = \frac{\pi \cdot D_{2.2} \cdot n_{ATL}}{\sqrt{\kappa_1 \cdot R_1 \cdot T_{1t}}} \qquad RCW = \frac{(\dot{m}_{C,red,Choke} - \dot{m}_{C,red,Surge})_{VTC}}{(\dot{m}_{C,red,Choke} - \dot{m}_{C,red,Surge})_{Baseline}}$$

Figure 11 VTC and LSVD measurement results II

It is furthermore striking that both the charge pressure and the surge line displacement benefit from the stabilizing effect of the LSVD as the circumferential Mach number of the compressor is increased. In the lower speed range, the compressor map width is reduced by implementation of the LSVD. Compressor speed lines decrease with a smaller gradient towards choke so that there is only little difference regarding the map width compared to the configuration with the VTC closed and a vaneless diffuser. The advantages in pressure ratio and efficiency are to be attributed particularly to the improved flow condition in the LSVD especially at low flow rate. The LSVD stabilizes the compressor impeller outflow and the flow downstream of the diffuser.

With the VTC device closed, efficiency advantages of up to 4.5 percentage points at the surge line of the baseline compressor and up to 8 percentage points in the low flow region are achieved compared to the baseline compressor. By switching in the LSVD into the diffuser, the map area with efficiency improvement is shifted further towards lower mass flow rates. With implemented LSVD, the peak efficiency decreases slightly. But due to the effect of flow stabilization in the diffuser the efficiency is improved all over the resulting compressor map.

The results of compressor measurements show that the combination of VTC device and LSVD are advantageous compared to the VTC with a vaneless diffuser with regard to compressor operating range as well as pressure ratio and efficiency in particular in the low flow area. For an effective utilization of this potential, an active variability of both devices is essential in order to adapt the compressor to the engine´s demands.

7.3 Engine simulation results

The compressor maps as well as the turbine map steadily measured on the turbocharger hot gas test bench are used to evaluate the potential of the engine´s performance.

7.3.1 Full load simulation

Full load simulation results are shown in figure 12. With VTC closed and vaneless diffuser for all engine speeds in the LET-range of n = 1000 min^{-1} up to 2000 min^{-1} significant improvements of the LET characteristics of the engine is achieved at full load compared to the baseline application. The LET level of maximum torque is already reached at n = 1550 min^{-1}.

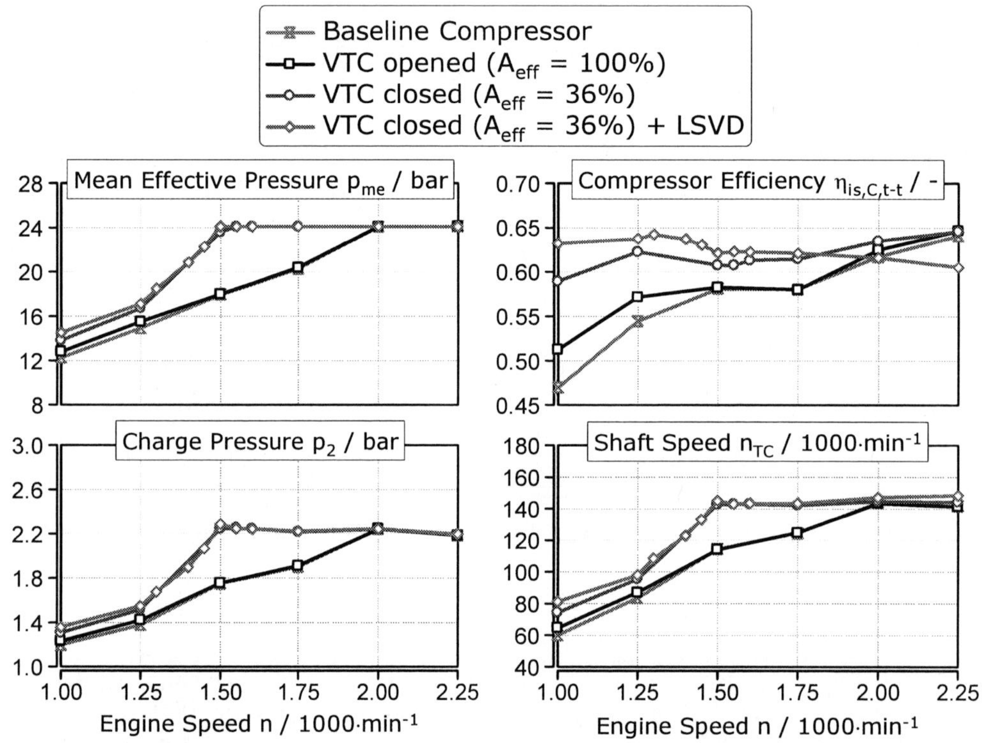

Figure 12 Engine full load simulation results

In the low speed range (n < 1250 min^{-1}) the mean effective pressure is somewhat higher compared to the reference. Due to significant advantages in compressor efficiency, the turbocharger efficiency increases up to six percentage points. Furthermore, the turbine is driven at higher blade speed ratio and therefore at higher efficiency. The results show that in the case of reduction of the effective inflow area at the compressor impeller inlet due to a VTC device significant improvement in the engine´s LET performance can already be achieved without adaptation of the VTC inflow area until the maximum torque level is reached. With further reduction of the effective inflow area below 36% of the initial inflow area no more advantages regarding compressor map characteristics are gained. Consequently also no further advantages regarding the engine´s full load behavior in the low speed range are achieved. The results correlate with the estimates made in section 7.1.1 concerning the reduction of the effective inflow area with closed VTC.

With the implementation of the LSVD into the compressor diffuser and the VTC in closed position the engine's full load characteristics are improved further, so that the LET is reached already at n = 1500 min^{-1}. At full load the optimal compressor map areas with respect to the efficiency are ran through. So the operation points can be driven continuously in a high efficiency plateau. With VTC and LSVD the compressor efficiency is increased again by about four percentage points. The advantages in compressor efficiency also propagate to the turbine side due to the higher scavenging rate and pressure ratio so that the turbocharger efficiency turns out to be higher by three percentage points compared to the VTC with vaneless diffuser.

The simulation results show that in particular through the compressor map enhancement towards lower mass flow rates there is great potential for the improvement of the full load characteristic of the engine. Nevertheless there needs to be sufficient amount of turbine power in order to be able to cover the power requirements of the compressor in order to generate the required charge pressure level. High compressor efficiency particularly at low engine speeds is required for the effective utilization of the extended compressor map width. Furthermore disadvantages regarding the blade speed ratio of the turbine can be compensated.

7.3.2 Load step simulation

In the next step the compressor configurations are analyzed numerically with regard to the engine's transient performance. The ramp up of charge pressure and therefore torque from part load to full load is defined as load step. Load step simulation results are shown in figure 13.

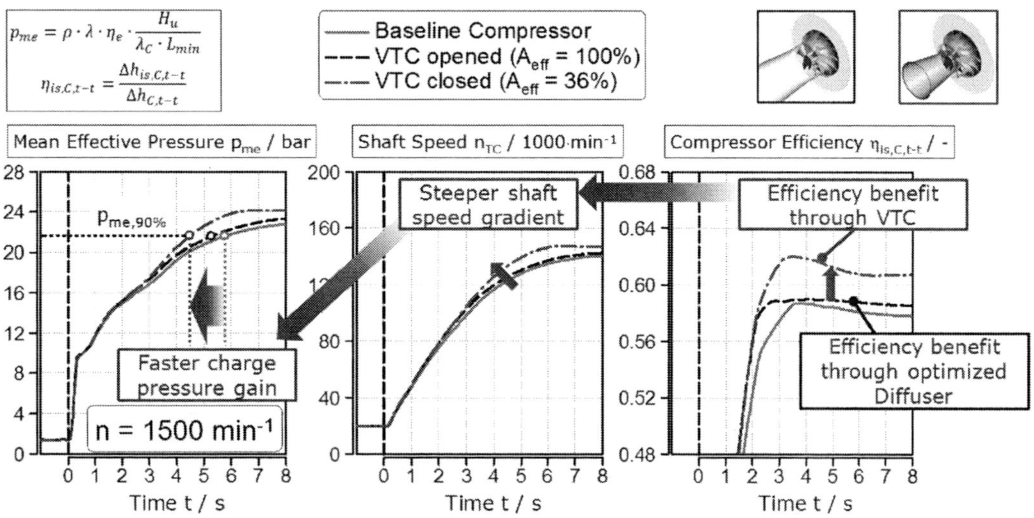

Figure 13 Engine load step simulation results

In this case the load step for constant engine speed at n = 1500 min^{-1} is calculated. The load step is started at part load level (M_d = 20 Nm). The beginning of the load step is introduced through opening the throttle valve and closing of the waste gate. With attaining the full load mean effective pressure level of p_{me} = 24 bar the load step is completed. The comparison of the load step times is carried out in each case at the moment of attaining the 90% of mean effective pressure full load level. Due to the compressor map characteristics and the range of the engine operation lines, only the configuration with VTC and vaneless diffuser is considered.

With closed VTC and vaneless diffuser efficiency advantages up to 4.5% points are achieved in the lower turbocharger shaft speed range and furthermore about 2.5%

points at the end of the load step in comparison to the reference. This leads to load step advantages of up to ${}^{\Delta t}/_{t}$ = 29.1%. Due to the improved diffuser design the efficiency the compressor configuration with opened VTC with vaneless diffuser gains load step advantages up to one percentage point in comparison to the baseline compressor.

The increased efficiency of the compressor causes a faster gain of charge pressure and therefore mean effective pressure because more fuel can be injected in the combustion chamber. At the end of the power stroke also the turbine enthalpy is increased. Due to the higher turbine power output and simultaneously reduced compressor power because of higher efficiency there is more power left for the acceleration of the turbocharger shaft. With increasing shaft speed, the turbine runs at higher blade speed ratios which benefits efficiency. Consequently, the load step is finished earlier. In the case of the VTC suitable control ensures that the engine´s operation lines run through a compressor map range with high efficiency following the load step.

The results show that not only the efficiency of compressor and turbine, but rather the thermodynamic interaction of both along the engine operation lines lead to an improved response.

7.4 HiL-Turbocharger dynamic test bench results

Finally, an experimental dynamic measurement of the turbocharger compressor concepts is carried out on the turbocharger dynamic test bench with the help of the new load step method. Experimental load step results are shown in figure 14.

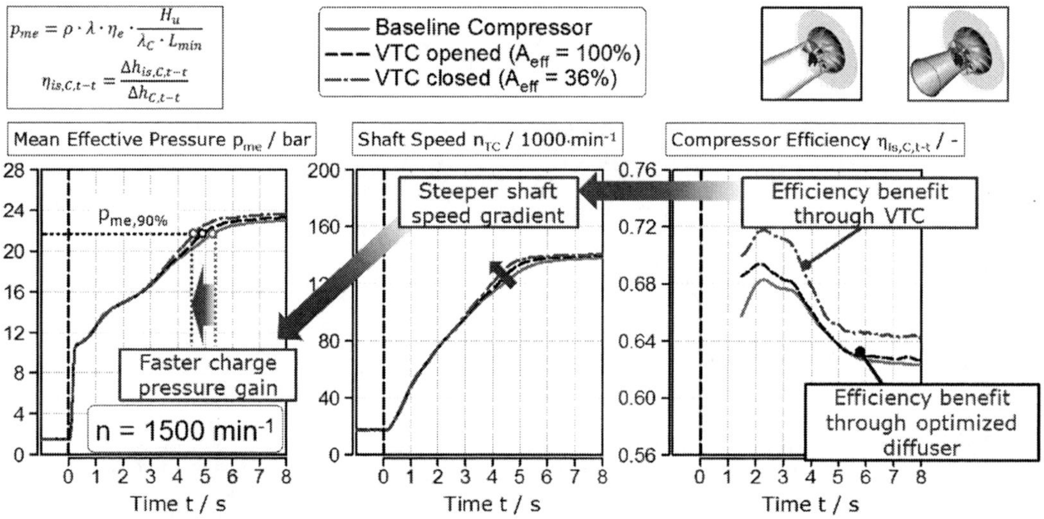

Figure 14 Turbocharger load step measurement results

The compressor configuration with VTC and optimized vaneless diffuser is generally superior compared to the reference due to the improved efficiency. With closed VTC there is a load step time advantage of ${}^{\Delta t}/_{t}$ = -13.3% at constant engine speed of n = 1500 min^{-1} compared to the reference. Reason for that is that the engine operation line runs through the compressor map range with high efficiency. Due to the efficient charge pressure gain the load step takes place with a steeper turbocharger shaft speed gradient. Since the turbine is also driven earlier at higher efficiencies, the load step and thus also the rise in charge pressure take place even faster.

Due to the advantages in efficiency, the power demanded by the compressor is lowered. The performance surplus directly results in a steeper turbocharger shaft speed gradient in particular by the end of the load step. This in turn leads to an improved rise in charge pressure so that the load step proceeds more quickly.

The trends from the preliminary numerical studies are generally confirmed by the turbocharger load step measurements. Results show that already apparently small advantages in compressor efficiency have a positive effect on the load step behavior of the turbocharger and engine. The load step measurements prove that compressor map enhancement devices not only lead to improvement of the steady but also the dynamic behavior of the engine performance.

8 CONCLUSIONS

This paper presents a new design methodology for devices to enhance the turbocharger compressor map and for the analysis of the benefit of the compressor map enhancement on the steady and transient operating characteristic of a charged vehicle combustion engine. Improvements to the engine performance are achieved by the implementation and combination of suitable enhancement devices on the compressor of the turbocharger, without modifications having to be carried out on the design of the compressor impeller or the turbine.

The maximum potential for enhancing the compressor operating range is achieved using a combination of a variable trim compressor (VTC) in combination with a variable low solidity vaned diffuser (LSVD), ahead of every other system analyzed. The functionality of enhancement devices upstream of the compressor impeller is generally preserved when implementing a vaned diffuser,

The results of the engine cycle simulation show that in particular the improvement of the compressor near the surge line improves the steady full load performance of a vehicle combustion engine especially in the low-end-torque (LET) region. It is further proven that the enhancement by VTC in combination with a vaneless as well as the vaned diffuser has the greatest potential to improve the LET characteristics of the engine. It is remarkable that a binary switching in and out of the VTC is sufficient to improve the engine´s LET.

In the transient calculation, all examined compressor enhancement devices achieve a faster response to the load step, as long as the engine operation lines run through the compressor map range with improved efficiency.

The experimental transient measurements of the turbocharger configurations were carried out on a hardware-in-the-loop (HiL) turbocharger dynamic test bench. A new load step method was conceived, in which an engine model capable of real time simulation is coupled to the controls of the dynamic test bench. Thus a detailed analysis of the engine´s response to load steps is possible under same constraints as the steady performance tests. The results of the load step simulation are confirmed by the experimental measurements on the turbocharger dynamic test bench.

The results show that there is great potential to improve the steady as well as transient engine characteristics with the help of suitable devices for the enhancement of the turbocharger compressor map.

REFERENCES

[1] Golloch, R. (2005): Downsizing bei Verbrennungsmotoren – Ein wirksames Konzept zur Kraftstoffverbrauchssenkung (Downsizing the combustion engine – a useful concept for the reduction of fuel consumpion), Springer-Verlag, Berlin Heidelberg, 2005

[2] Szengel, R., Middendorf, H., Voeltz, S., Laumann, A., Tilcher, L., Theobald, J., Etzrodt, T., Krebs, R. (2007): Der TSI-Motor mit 90 kW – Erweiterung der verbrauchsgünstigen Ottomotoren-Baureihe von Volkswagen (The TSI-engine with 90kW – Enhancement of the efficient gasoline engine family of Volkswagen), Motorentechnische Zeitschrift, Nr. 68, Ausgabe 07-08, Vieweg-Verlag, Wiesbaden, 2007

[3] Zinner, K., Pucher, H. (2012): Aufladung von Verbrennungsmotoren – Grundlagen, Berechnung, Ausführungen (Charging of the combustion engine), 4. Auflage, Springer-Verlag, Berlin, 2012

[4] Sivagnanasundaram, S., Early, J., Spence, S., Nikpour, B. (2011a): An impact of various shroud bleed slot configurations an cavity vanes on compressor map width and the inducer flow field, Proceedings of ASME Turbo Expo, GT2011-45629, Vancouver, British Columbia, Canada, 6-10 June 2011

[5] Hagelstein, D. (2001): Experimentelle Untersuchungen an einer Radialverdichterstufe mit verschiedenen Vorleitgittern variabler Geometrie, Dissertation (Experimental analysis of a radial compressor with different variable inlet guide vanes), Universität Hannover, 2001

[6] Mohseni, A., Goldhahn, E., Van den Braembusche, R., Seume, J. (2010): Novel IGV designs for centrifugal compressors and their interaction with the impeller, Proceedings of ASME Turbo Expo, GT2010-23048, Glasgow, UK, 14-18 June 2010

[7] Grigoriadis, P., Benz, A., Sens, M. (2011): Variable Trim Compressor – Ein neuer Ansatz für eine variable Verdichtergeometrie (Variable Trim Compressor – A new approach for compressor variability), Paper, 16. Aufladetechnische Konferenz (ATK), Dresden, 29.-30.09.2011

[8] Grigoriadis, P., Böttcher, L., Binder, E., Sens, M. (2014): Advanced boosting technologies for future SI engine concepts, 11th International Conference on Turbochargers and Turbocharging, IMechE, C1384/010, London, 13-14 May 2014

[9] Jaatinen, A. (2009): Performance improvement of centrifugal compressor stage with pinched geometry or vaned diffuser, Dissertation, Lappeenranta University of Technology, 2009

[10] Turunen-Saaresti, T. (2004): Computational and experimental analysis of flow field in the diffusers of centrifugal compressors, Dissertation, Lappeenranta University of Technology, 2004

[11] Eynon, P. A., Whitfield, A. (1998): Performance of a turbocharger compressor with a range of low solidity vaned diffusers, SAE International, International Congress and Exposition, 980772, Detroit, Michigan, USA, 23-26 February 1998

[12] Senoo, Y. (1978): Japanese Patent Disclosure, 119411/78, 1978

[13] Hohlweg, W. C., Direnzi, G. L., Aungier, R. H. (1993): Comparison of conventional and low solidity vaned diffusers, ASME-Paper, 93-GT-98, Cincinnati, Ohio, USA, 24-27 May 1993

[14] Engeda, A. (2001): The design and performance results of simple flat plate low solidity vaned diffusers, Proc IMechE, Vol. 215 Part A, pages 109-118, 2001

[15] Japikse, D. (1996): Centrifugal compressor design and performance, Concepts ETI, Wilder, Vermont, USA,1996

[16] Tange, H., Ikeya, N., Takanashi, M., Hokari, T. (2003): Variable geometry diffuser of turbocharger compressor for passenger vehicles, SAE International, 2003-01-0051, 2003

[17] Czapka, T. (2015): Erweiterung des Verdichterkennfeldes eines Abgasturboladers für verbrauchsoptimierte Verbrennungsmotoren (Turbocharger compressor map enhancement for highly efficient combustion engines), Dissertation, Leibniz Universität Hannover, 2015

[18] Uhlmann, T., Lückmann, D., Pischinger, S., Schorn, N., Teetz, C. (2012): Erweiterte Turbinenkennfeldmessung - Erweiterung des messbaren Turbinenkennfeldbereichs (Enhanced measurement of the turbine operation map), Abschlussbericht, FVV-Vorhaben Nr. 1038, Heft 981, 2013

[19] Scheller, P. (2012): Eine neue Methode zur Bewertung des transienten Turboladerverhaltens bei aufgeladenen Ottomotoren (A new method to analyze the dynamic behavior of turbochargers for charged gasoline engines), Dissertation, Technische Universität Dresden, 2012

[20] Gamma Technologies (2011): GT-Suite - Engine Performance Tutorials, Version 7.3, Westmond, Illinois, USA, 2011

Aerodynamic design of a high flow and high pressure ratio centrifugal compressor for a marine-use turbocharger

C Mikami, K Yamada, T Kawakubo
Turbo Machinery and Engine Technology Dept., Products Development Center, IHI Corporation, Japan

ABSTRACT

High flow and high pressure ratio compressor is demanded to marine-use turbocharger for Miller cycle engines. To meet the above requirement, the new centrifugal compressor is developed which pressure ratio is 5.7 and the flow coefficient is about 10% higher than the previous one. In order to achieve these performance targets, the multi-splitter impeller is applied to the new compressor and circumferential spacing of splitter blades is optimized to maximize adiabatic efficiency. And also, in order to investigate the structural issues, which are blade vibration and compressor wheel's metal temperature rise, blade vibration and disk temperature measurements are conducted for newly developed compressor.

NOMENCLATURE

D_2	Impeller tip diameter (m)
T_{1t}	Inlet total temperature (K)
M_{u2}	Impeller tip Mach number
n	Rotor speed (rps)
N_S	Specific speed ($=nQ^{1/2}/\Delta h_{is}^{3/4}$)
p_S	Statistic pressure (Pa)
Q	Volume flow rate (m^3/s)
ϕ	Flow coefficient ($=Q/\pi D_2^3 n$)
η_{ad}	Adiabatic efficiency
μ_y	Pressure coefficient
π	Total pressure ratio

1 INTRODUCTION

Recently, the regulation of NOx emission becomes stringent to marine vehicles and the Miller cycle has been applied to marine-use diesel engines for the NOx reduction. In order to recover the engine power output, a marine-use turbocharger has to charge more amount of air in a shorter period, therefore a centrifugal compressor with high flow rate and high pressure ratio is required for the marine-use turbocharger. To meet the above requirement, studies of high flow and high pressure ratio centrifugal compressors have been carried out in recent years [1-3].

The authors have conducted the new compressor which of a stage pressure ratio is more than 5 and wheel diameter is smaller than the existing one. Due to the smaller wheel diameter, the flow coefficient is approximately 10% higher and the inlet relative Mach number should be substantially higher, which exceeds 1.5 at the tip.

This paper describes the aerodynamic design of a new compressor. Under the high Mach number condition, the strong shock wave occurred at the inducer shock-boundary layer interaction and this interaction increases the aerodynamic loss and/or the reduction of the operating range. In order to overcome these aerodynamic difficulties, the authors adopted double-splitter impeller design. Moreover, a recirculation passage which is known as operating range enhancement device of compressor[4, 5] applied to the shroud. The aerodynamic design of the compressor stage including a compressor wheel, diffuser vanes and recirculation passage in the shroud is evaluated carefully by three dimensional steady-state computational fluid dynamics simulations, and also the structural performances, which are centrifugal stress and eigen-frequency, were estimated by finite element method (FEM) analysis. Finally, the actual compressor performance test is conducted at the turbocharger test facility in IHI, in which not only the aerodynamic performance but also blade vibration and metal temperature measurements of the compressor wheel are evaluated.

2 METHODOLOGY OF COMPRESSOR DESIGN

2.1 Design Target

Table 1 shows the design specification of each compressor stage in the past and the present developments (PR5.7-1 and PR5.7-2). Fig.1 shows the experimentally acquired performance curves of PR5.7-1, and the target performance curve of PR5.7-2. Design specifications of PR5.7-2 are almost kept the same as PR5.7-1 but flow coefficient is approximately 10% higher. The high flow coefficient is preferred for the cost and installation reasons. The increase of the flow coefficient is achieved by making the impeller shroud diameter smaller than the original. The higher through-flow velocity results in a stronger shock wave and an increased concern on the shock-boundary layer interaction. The higher wall curvature incurs a higher flow turning and a possibility of the boundary layer separation. All of these phenomena may cause a reduction of the efficiency and/or the stable operating range. Fig.2 shows an efficiency trend of pressure ratio 5.7 for the specific speed of a centrifugal compressor. Even the original PR5.7 stage is far away from the optimum as shown in Fig.2 and a substantial amount of the aerodynamic efforts have already been made for this original stage. Thus the development of the new PR5.7 stage with a higher flow coefficient must be an even more difficult technical challenge than the previous one.

Table.1 Compressor specifications

	PR5.7-1 (Previous Design)	PR5.7-2 (Design Target)
π	5.7	
M_{u2}	1.61	
ϕ	-	9% UP
μ_y	-	same value
N_s	0.149	0.156

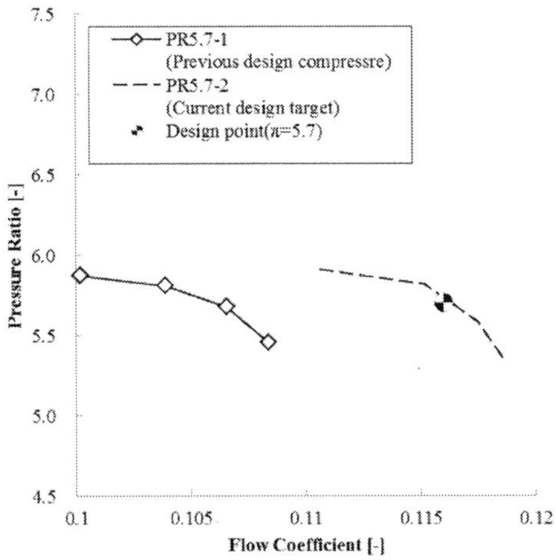

Fig.1 Compressor specifications

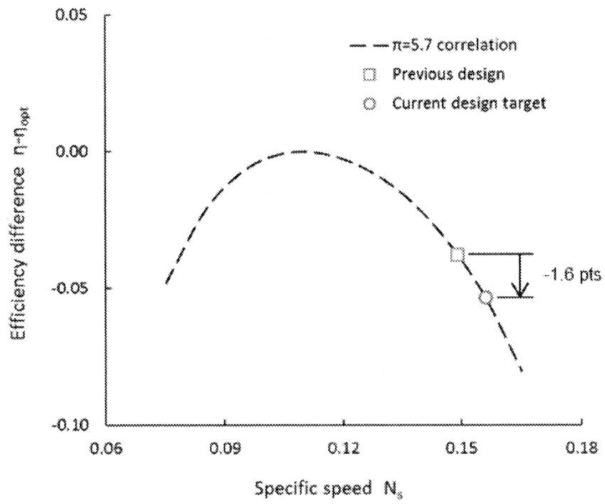

Fig.2 Estimated efficiency against specific speed

2.2 Basic Design

The main dimensions of impeller and diffuser are selected based on the one-dimensional aerodynamic design analysis and the supporting design criteria or guidelines, which are based on the past design and experiment experiences.

The inlet diameter is selected so that relative Mach number will be the lowest possible value. The design incidence angle is set to be substantially zero due to the high inlet Mach number. But the off-design incidence angle is much closer to the possible inducer stall occurrence at the specified minimum flow condition.

The most important geometric parameters for the impeller exit region are the tip diameter, which has already been specified as a design constraint, tip blade height and backward angle. The number of blades is another important parameter and is set to be 18, the same number as the previous design.

The increase of the design difficulty seems unavoidable at inlet. In contrast, it is possible to set the assessment parameters to the similar level to the previous design at exit by selecting the geometric parameters similar to those used in the previous design except for the blade height. The blade height must be increased because the flow coefficient has been increased. This can be done without any increase of design severity. This means that by selecting the same level of the degree of reaction, the flow capacity can be increased without the increased difficulty downstream of the impeller. By selecting the similar degree of reaction, the designer can resort on the past experiences on the stages with smaller flow coefficients and concentrate only on the inlet region. In the present case, even though some of the design parameters violate the design criteria, such a violation has already been verified to be acceptable in the past experiments, so hereafter we can ignore these violations.

2.3 Numerical Procedure

Numerical simulations have been performed using the in-house CFD code developed in IHI. This code solves the three dimensional Reynolds-averaged compressible Navier-Stokes equations. For the convective fluxes, the MUSCL-ROE TVD scheme is used. High order accuracy can be achieved by the MUSCL scheme and shock wave can be captured by the TVD scheme. For turbulence closure, one-equation Spalart-Allmaras model is used.

Computational domain includes an impeller, a vaned diffuse and a casing treatment passage. These passages are meshed with H-type structural grids. The minimum cell size at wall is set as y+ parameter approximately less than 3. The total number of grids became 6.8 million.

Finite Element Code is ANSYS Mechanical and MSC Nastran. The modal analysis calculated by ANSYS and stress analysis calculated MSC. The impeller has three different type blades and they are connecting on the disk. But the result of eigen-frequencies test by hammering indicates that the influence of the blades and disk coupling effect is very small. Therefore the domain is each single blade with fillet for the modal analysis. On the other hand, the domain for the stress analysis is one segment which consist of three blades and part of the disk with cyclic symmetric boundary condition.

2.4 Test method

The performance test of compressor stage is conducted in a turbocharger aerodynamic component test cell in IHI Corporation. The compressor is installed into an existing production turbocharger and its turbine is driven by the compressed air or the combustion gas. In the experiment, total pressures and temperatures are measured at the inlet and the exit of the stage to acquire the stage pressure ratio and the efficiency, while the compressor mass flow rate is measured by the orifice plate installed far downstream of the stage. The compressor housing is insulated thermally to isolate the compressor from the environment.

For the blades vibration measurement, the strain gauges are attached on the blade surfaces where the vibration stress can be measured enough to identify the vibration modes. These strain-gage locations are determined by FEM analysis as shown in Fig.3. And also, due to the high pressure ratio, the air temperature at compressor exit becomes high. The compressor wheel needs careful design to avoid the creep failure. Therefore, in order to investigate the structural performance to the high temperature operation, the impeller disk temperature by thermos-couples set on the impeller disk surface shown in Fig.4. The signals of strain-gages and thermos-couples are transmitted by telemetry-system in which the transmitter is installed at the end of the rotor shaft.

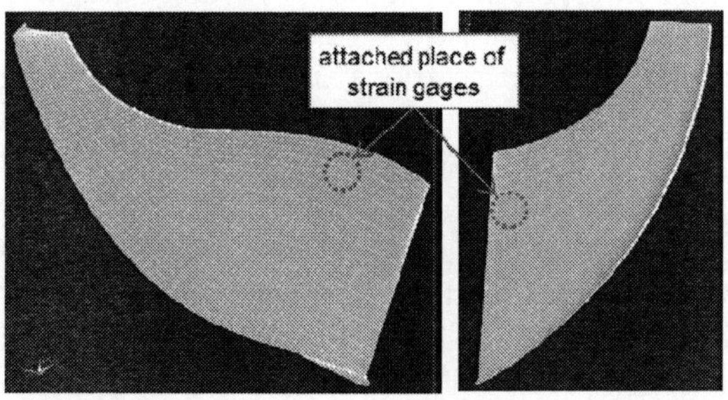

Fig.3 Strain-gage measurement points

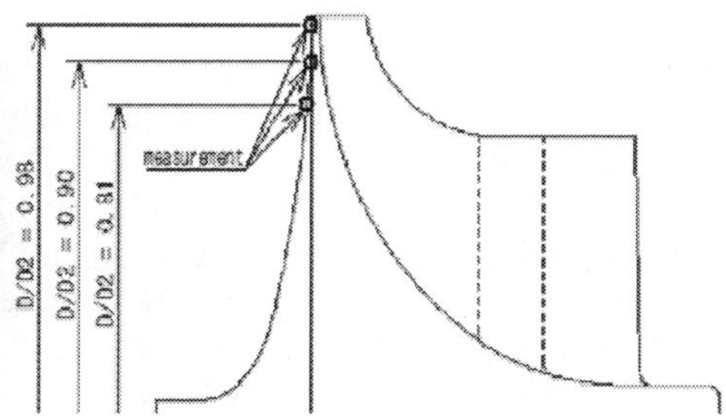

Fig.4 Disk temperature measurement points

3 RESULTS AND DISCUSSIONS

3.1 Impeller blade layout

The impeller has three difference blades, long blade and two splitter blades. Two splitter blades also have different length, middle and short against the long blade. The splitter blades layout of impeller of previous design compressor PR5.7-1 was adopted the equal distance from the other blade elements. Therefore preliminary design of new one also adopted the equal distance layout. Fig.5 shows the blade surface Mach number distribution at 90% span of the design point. CFD result indicates that blade loading of each blade is not the same and blade loading balance seems not appropriate. The short blade loading is clearly smaller than the other blades, thus it is suspected that the flow separation might be occurred on the short blade surface.

To investigate the influence of blade loading on a compressor performance, some parametric studies are conducted in which the splitter blade layout is changed. In this study, four types of blade layout shown in Fig.6 are evaluated by the steady-state numerical simulation. At first, in order to evaluate the effect of the short blade on the compressor performance, only the short blade layout is changed. In Case 1, the short blade is moved to the middle blade, and in Case 2, moved to the long blade. Next, the middle blade layout is changed in which the middle blade is close to the long blade in Case3 and close to the short blade in Case4.

Fig.7 shows the impeller performance for different blade layout calculated by CFD analysis. This result indicates that the splitter blade layout has a very large effect on the compressor performance and the efficiency difference between the best case, Case1, and the worst case, Case2, exceeds 1% which value is not negligible.

In order to clarify the reason why such performance difference occurs, the relation between the blade loading and the loss generation within the impeller is investigated. Fig.8 shows the averaged entropy distribution along the meridional passage which are expressed as the difference from the base type, i.e. equally spaced configuration. Loss evaluation is done for the three divided regions, which are near the middle blade leading edge defined as Region I, around the short blade leading edge as Region II and around the trailing edge as the Region III. And also, Fig.9 shows the comparison of entropy contours between the base and the unequal spacing types at 90% span.

The blade surface Mach number distributions at 90% span are compared between the base and Case1-2 in Fig.10 and Case3-4 in Fig.11 for each blade, respectively.

In the cases of different short blade layout, which are Case1 and Case2, the blade loadings both of middle and short blades are varied in Fig.10 (b) and (d). In detail, the middle blade loading is suppressed and the short one is enhanced in Case1 and the opposite trend can be seen in Case2. The short blade L/E is located in a high Mach number flow region but by moving it close to the middle one the inlet Mach number can be reduced. As a result, the blade loading of each blade becomes more balanced in Case1. That leads the entropy decrease at region II in Case1 within the passage between the short and middle blades as shown in Fig.9 (b). Case 2 shows opposite trend in Fig.9 (c) by moving the short blade distance to the long blade. That adjustment of Case1 reduces the loss of the region II and the total loss is the least, even though the loss of the region III increases.

In the cases of different middle blade layout, which are Case3 and Case4, the blade loadings both of long and middle blades are varied. In detail, the long blade loading is suppressed and the middle one is enhanced in Case3 and the opposite trend can be seen in Case4 as shown in Fig.10 (a) and (b). The balances of each blade loading of Case3 and Case4 are worth than Case1. Fig.9 (d) and (e) show the entropy of the region II and III increase cause by interaction of shock wave and boundary layer.

Case1 which has the most balanced blade loading shows the best compressor performance and it is very important to balance the blade loading of each blade to minimize the loss generation within the impeller.

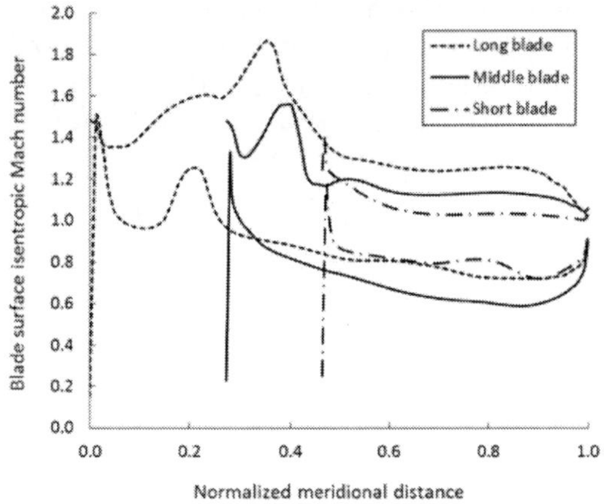

Fig.5 Mach number distribution at 90% span

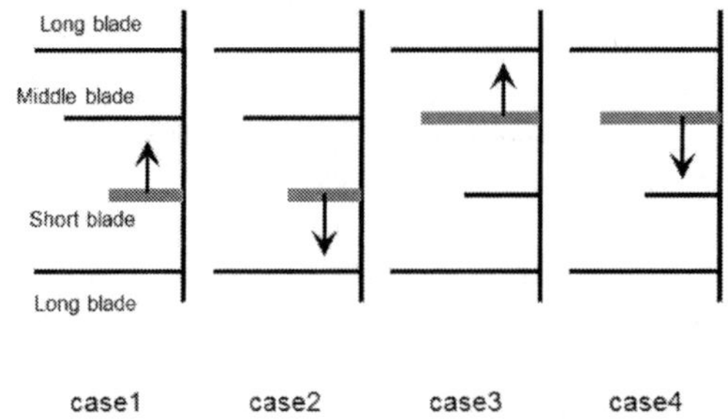

Fig.6 Schematic of splitter blade's layout

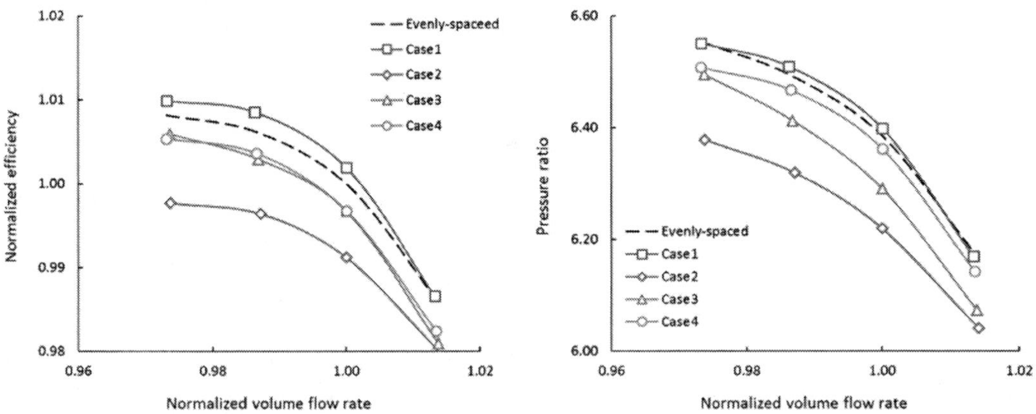

(a) Efficiency (b) Pressure ratio

Fig.7 Impeller performance for difference blade layouts

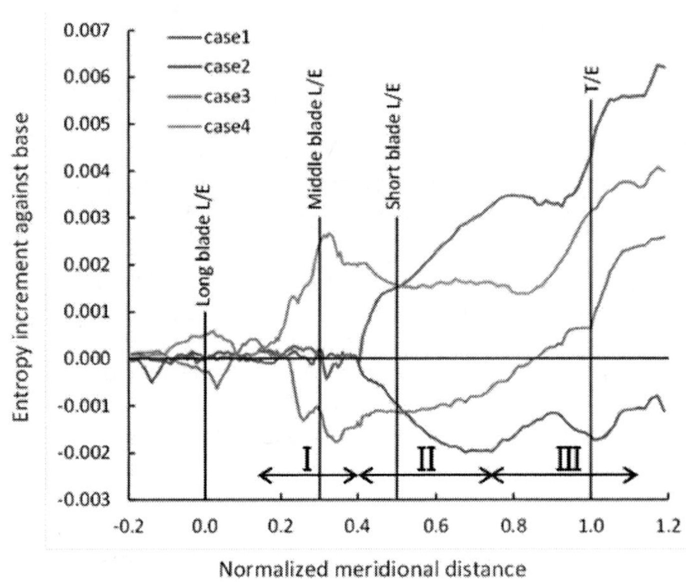

Fig.8 Entropy distribution of mean line average at impeller

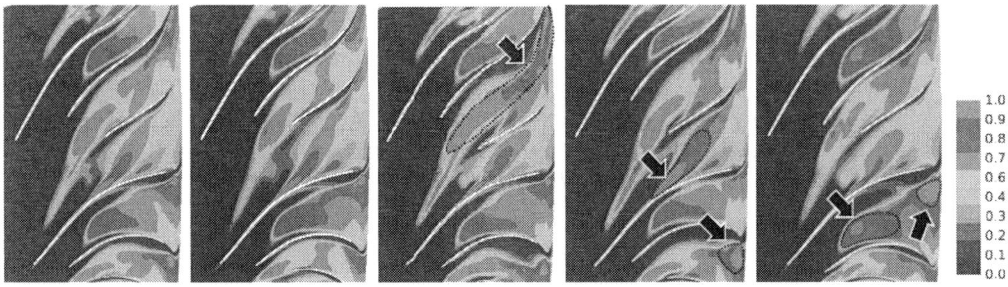

(a) base (b) Case1 (c) Case2 (d) Case3 (e) Case4

Fig.9 Entropy contour at 90%Span

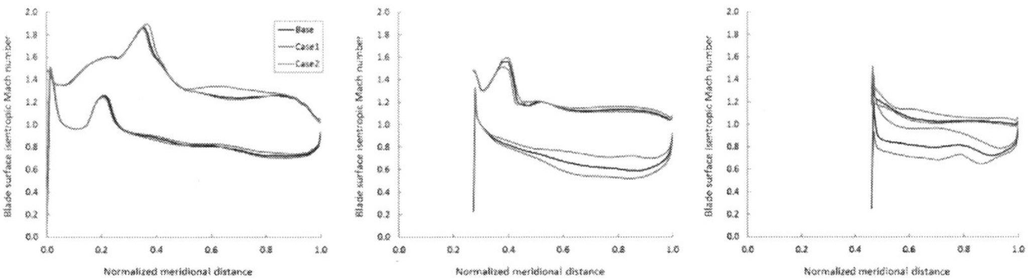

(a) Long blade (b) Middle blade (c) Short blade

Fig.10 Mach number distribution at 90% span blade surface for short blade arrangement

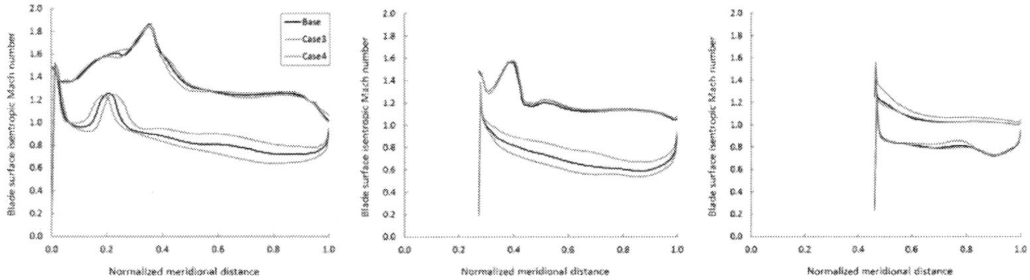

(a) Long blade (b) Middle blade (c) Short blade

Fig.11 Mach number distribution at 90% span blade surface for middle blade arrangement

3.2 Effect of Casing treatment

In order to improve the surge margin, the casing treatment are adopted in a compressor casing which has an annular cavity and connected main flow region by upstream slot located upstream from impeller leading edge and impeller shroud slot. Fig.12 shows a schematic shape of casing treatment. Casing treatment has a possibility of working as circulation and also recirculation device and which depends on the slot design. Furthermore, casing treatment may

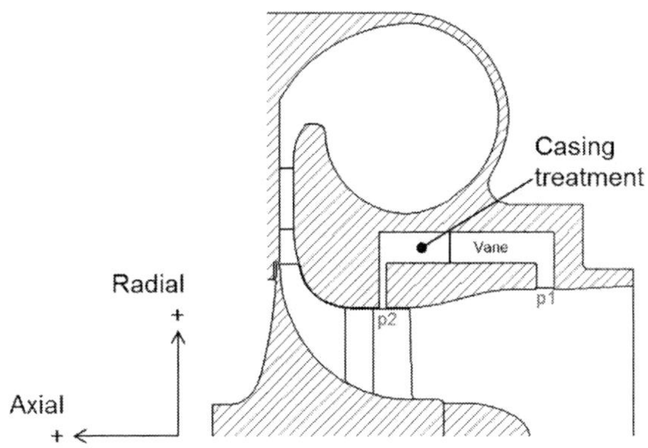

Fig.12 Schematic of casing treatment

possibly reduce impeller efficiency by corner loss, wall loss, mixing loss and so on. Therefore it is important to design the casing treatment geometry optimally. Whether the flow in the cavity circulates or recirculates depends on the pressure balance between upstream and downstream slots. When upstream slot static pressure p1 is higher than downstream slot p2, the flow runs to downstream. Contrariwise p2 is higher than p1, the flow recirculates to upstream. On the other hand, the amount of the air through the cavity is controlled by p2 and the width of the downstream slot.

The recirculating flow angle can be controlled by installed vanes at upstream slot. Counter swirl, the swirl momentum of the opposite direction to the impeller rotation, can enhance the effect of improve surge margin, according to the previous studies by Yamaguchi[6] and Tamaki[7]. The casing treatment with the counter swirl vanes is called as CSCT and without CT is called as NOCT in this paper.

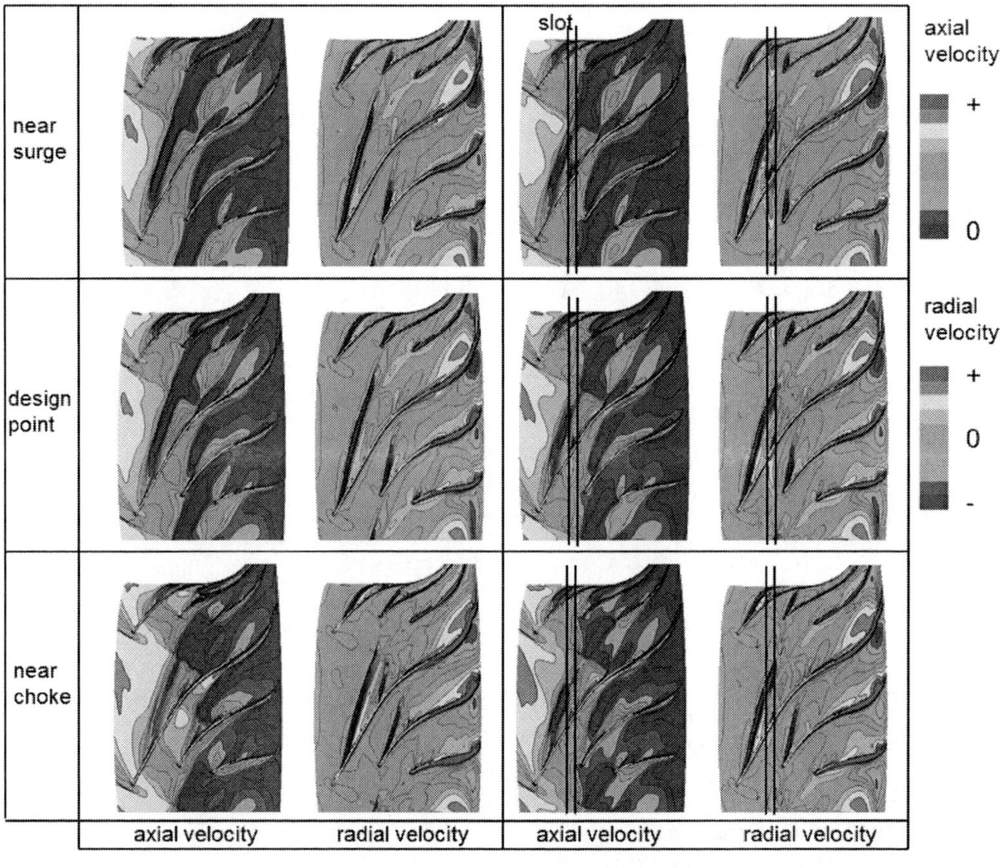

(a) Without casing treatment (NOCT)

(b) With counter swirl casing treatment (CSCT)

Fig.13 Axial and radial velocity contours at 98%Span

In this CT study, the slot location was set the pressure position which is p1<p2 at near surge identified from the CFD result of NOCT analysis and the width was adjusted not to sacrifice the efficiency at design point. The new compressor adopts the vanes within the casing treatment to achieve the wide range operation. Fig.13 shows the CFD results of NOCT (a) and with designed CSCT (b), which shows axial velocity contour at 98% span (left) and radial velocity contour (right). The axially reversed flow near inlet of NOCT is getting bigger from the choke to the near surge. With CSCT, the inlet axially reversed flow is not varied between the near surge and the choke and the

region is more reduced than that of NOCT. Focusing on the radial flow of CSCT, there is a recirculating flow at pressure side of long blade in the near surge only, and there is little recirculation flow at the design flow rate. That means the slot location, width and counter swirl vanes are designed appropriately.

3.3 Performance testing of the newly designed compressor

On the basis of the above studies, the new compressor PR5.7-2 is designed. As a vaned diffuser combined with the new impeller, the same vane profile as the previous one is adopted because the impeller outlet flow is similar to the previous design. The leading edge metal angle is optimized the incidence characteristic by CFD estimating and trailing edge metal angle is adjusted so as to give the appropriate throat area and the best matching with the scroll.

The appearance of the fabricated compressor stage is shown in Fig.15. Fig.16 shows the experimental results of the newly designed stage in comparison with the deign prediction. The performance prediction is based on the CFD prediction for the impeller passage coupled with the scroll performance estimated by one-dimensional analysis tool. The experimental result confirms that the current stage satisfies the performance target. Accurate prediction of compressor performance at highly pressure condition is usually difficult because there are the unsteadiness conditions by strong shock waves, leakage flows, rotor-stator interaction and so on. But the experimental result agrees quite well with the design prediction not only at the design speed but also at part speed. Near the design speed, CFD overestimate the width of the stable operating range.

Fig.16 compares the performance of the previous (PR5.7-1) with the new design (PR5.7-2). The efficiency levels of the two stages are almost the same and 5.7-2 achieved the volume flow ratio of PR5.7-1, i.e., the increasing of flow coefficient succeeded. But the operation range more than 95% rotation speed is small. Thus, the design approach taken in the current design can consider to be valid and retain the preferable performance of the previous stage successfully even though the inlet Mach number is substantially raised in the PR5.7-2. Focusing on CT effect, the surge margin at 85% rotation speed is greater than the NOCT type. 50% and more than 95% rotation speed, the surge margin is not improved from NOCT type and becomes a narrow range compared to PR5.7-1. This result is different from CFD prediction and indicates the performance prediction of high Mach number region is very difficult by the current steady-state CFD prediction.

Fig.14 Newly designed compressor stage (PR5.7-2)

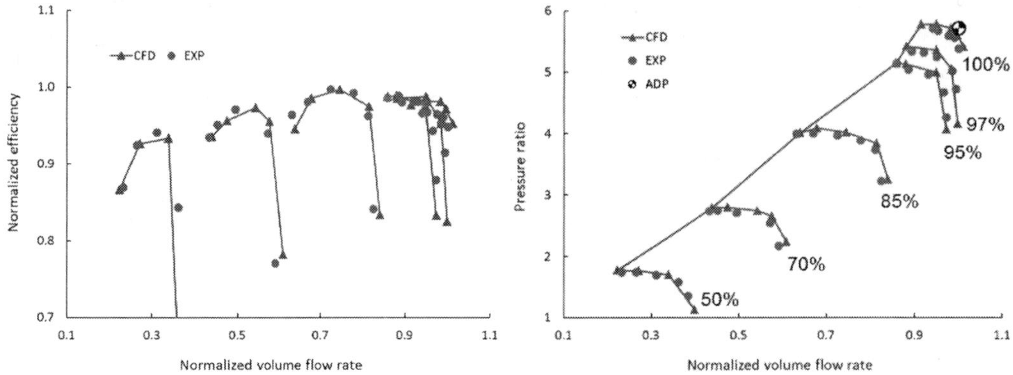

(a) Stage efficiency (b) Stage pressure ratio

Fig.15 Performance of Newly designed compressor stage (CFD and EXP.)

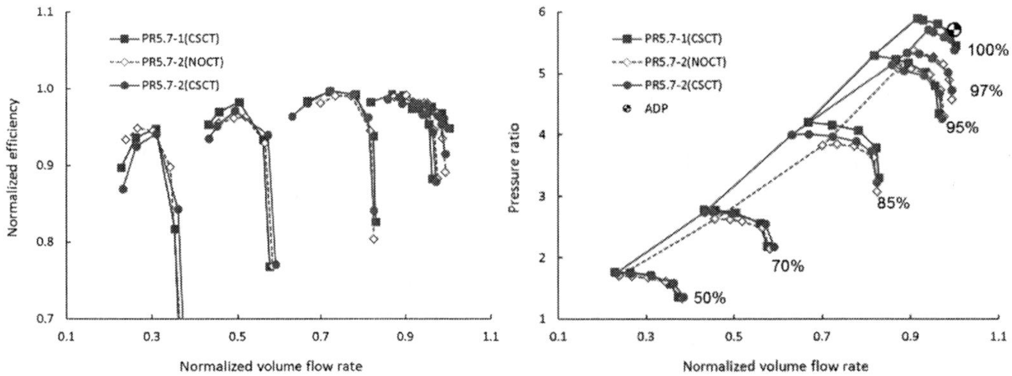

(a) Stage efficiency (b) Stage pressure ratio

Fig.16 Performance of Previous and Newly designed compressor stage (PR5.7-1 and PR5.7-2)

3.4 Characteristics of impeller blade vibration and disk temperature

Due to the high pressure ratio, the rotor-stator interaction between impeller and diffuser will be strong and increase the possibility of failure due to high cycle fatigue. And also, high pressure leads high temperature rise, which reduces the strength of compressor wheel's material. In order to evaluate these structural performances of newly designed compressor PR5.7-2, the impeller blade vibration and the impeller disk temperature are measured in this study. Fig.17 shows the Campbell diagram of this impeller. Broken lines show eigen-frequencies calculated by FEM analysis and circles show the magnitude of vibration stress obtained by strain-gage measurement. 15EO is same as number of diffuser vanes. The Strain-gages are set on long and short blades only, not for a middle one, because the distance between a short blade and a middle one is very small due to its unequal spacing design described in the precious section, thus the middle blade's vibration stress could not be evaluated in this study. From Fig.17 (a) and (c), long and short blades eigen-frequencies of FEM analysis show a good agreement with the blade vibration measurement. The stresses of long blade are responded to 15EO and 12EO but the the vibration stress values are smaller than the criteria stress. The stresses of short blade are responded not only to 15EO and 12EO but also 11EO. However, the vibration stress values are also smaller than the criteria stress. That is confirmed that the magnitude of the vibration stress is within allowable stress limit of the compressor wheel's material. As a remained issue, the middle blade vibration must be evaluated, but it is very difficult to achieve by strain-gage measurement, thus non-intrusive blade vibration measurement is desirable in the near future.

Fig.18 shows the result of impeller disk temperature measurement. ΔT_{flow} is difference between inlet flow temperature and exit one and ΔT_{metal} is difference between inlet temperature and the impeller disk one. The measurement result shows that the metal temperature increases by the near disk outlet or the high rotational speed. That relation of metal and flow temperature is like linear variation. It means that impeller disk temperature can predict by impeller outlet one.

Fig. 19 shows a principal stress distribution at maximum rotational speed obtained by FEM analysis. High centrifugal stress is observed near the blade root and that value compered the material characteristic at prediction temperature of same disk position. It can be judged allowable or not against the material strength, i.e. 0.2% proof stress. In this impeller, it was satisfied enough margin of safety.

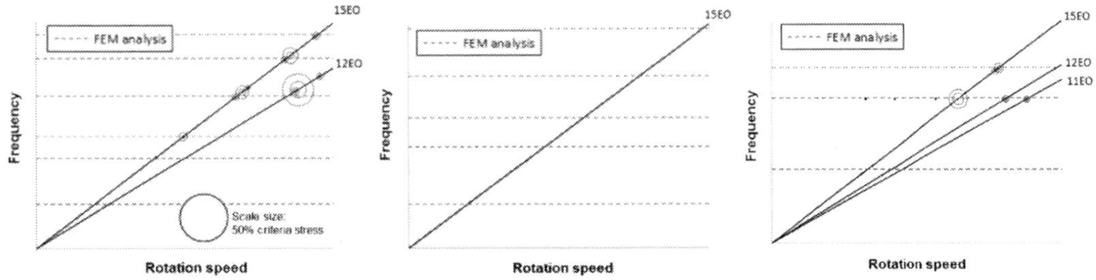

(a) Long blade (b) Middle blade (c) Short blade

Fig.17 Campbell diagram of PR5.7-2

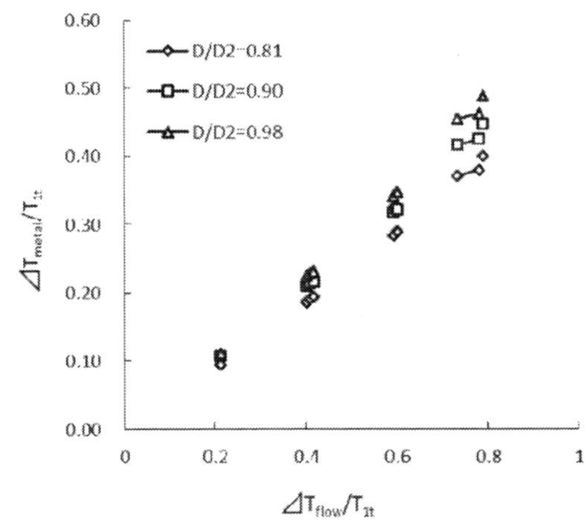

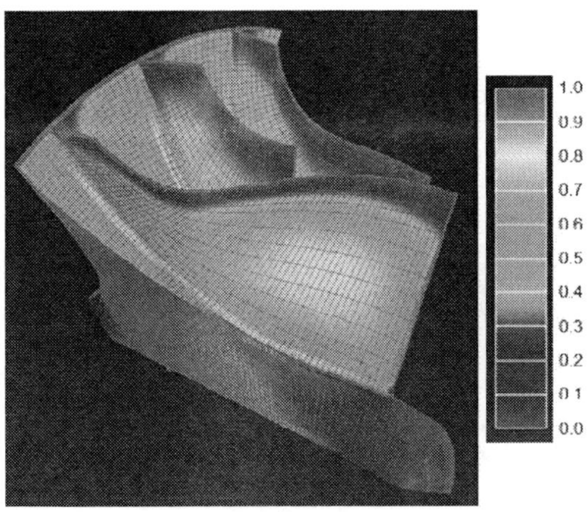

Fig.18 Measurement temperature of impeller disk

Fig.19 Normalized stress contour by 0.22% yield stress

4 CONCLUSIONS

A high pressure ratio centrifugal compressor stage with a pressure ratio 5.7 is designed with a higher flow coefficient than the previous one. Aerodynamic design and performance prediction for the newly developed compressor are carried out by means of a combination of the one-dimensional analysis and steady-state CFD.

From the experimental results of the performance, it is confirmed that the new compressor stage satisfies the target performance. The experimental and the design prediction results agree very well, thus it confirms the validity of the design and performance prediction methods for the high pressure ratio stage with a high flow coefficient. And also, regarding the structural performances, it confirms that both blade vibration and disk metal temperature are within allowable design limit.

REFERENCES

[1] Ibaraki S., Matsuo T., et al., 2004, "Design Optimization of Turbocharger Compressor for High Pressure Turbocharged Diesel Engine", CIMAC Congresses, paper no. 59, Kyoto.

[2] Marconcini M., Rubechini F., et al., 2010, "Numerical Analysis of the Vaned Diffuser of a Transonic Centrifugal Compressor", ASME, Journal of Turbomachinery, Vol. 132(4).

[3] Tamaki, H., Unno M., et al., 2009, "Aerodynamic Design to Increase Pressure Ratio of Centrifugal Compressors for Turbochargers", ASME Turbo Expo, GT2009-59160, Orland.

[4] Fisher F. B., 1989, "Application of Map Width Enhancement Devices to Turbocharger Compressor Stages", SAE paper, no. 880794.

[5] Hunziker R., Dickmann H-P., et al. 2001, "Numerical and experimental investigation of a centrifugal compressor with an inducer casing bleed system", IMechE, Part A: Journal of Power and Energy, Vol. 215(6), pp. 783-791.

[6] Yamaguchi, S., 2002, "The development of effective casing treatment for turbocharger compressors", Seventh International Conference on Turbochargers and Turbocharging, C602/008/2002, London.

[7] Tamaki, H., 2012, "Effect of Recirculation Device With Counter Swirl Vane on Performance of High Pressure Ratio Centrifugal Compressor", ASME, Journal of Turbomachinery, Vol. 134(5).

SESSION 1B

COMPRESSOR: SURGE LINE MEASUREMENT AND INTERACTION

Turbo charger compressor inlet and outlet pipe length and volume, and the effects on the characteristics and location of surge

M Whittlesea, E Roberts
Cummins Turbo Technologies, UK

ABSTRACT

To understand some of the influences of the onset of surge, with a view to understand potential variation between testing facilities a literature review was completed. This highlighted the Riemann characteristic which links the outlet volume, effective length and area of axial and radial compressors to a non-dimensional number $B = \frac{U}{2a}\left(\frac{V_b}{A_c L_c}\right)^{\frac{1}{2}}$. The Riemann characteristic can be used to categorise what type of surge may occur and deduce the frequency of pressure oscillations that result from instability. This is essentially an approximation of the ratio of pressure to inertia forces in the system where the compressor discharges into a length of pipe (relating to inertia) then into a plenum (relating to the pressure term). On a gas stand the boost section essentially combines the two, a pipe and a plenum with no distinction between the two. The purpose of the experiment was to try and establish a link between the Riemann characteristic by transferring the term that described the compressor effective discharge length to the inlet. It was discovered that inertia and pressure are both very important properties within the boost section, and the relationships are complex and nonlinear. It was deemed necessary that further running points are required to better model the relationship with many more repeats to better define the means for each test condition to establish them as different populations. The results showed there was minimal practical impact on the units surge margin within the design constraints of our test cells, but the concept of changing the diameter to alter the volume in the boost section should be explored further. Additionally it may be useful to experimentally derive a lumped parameter for the inlet, and boost sections inertia and volume to make the Riemann characteristic more functional.

INDEX TERMS

DOE, Design of experiment
U, Wheel speed
a, Speed of sound
V_b, Boost volume
A_c, compressor effective cross sectional area
L_c, Compressor effective length
CTT, Cummins Turbo Technologies
Cx/U, Flow velocity/Tip speed
NDMF, Non dimensional flow units (Kg/s$\frac{\sqrt{K}}{Mpa}$)
Statistically valid, To 95% confidence limits

I. INTRODUCTION

Turbo charging has become a necessity to reduce engine emissions, increase fuel economy and generate more power. This has led to pushing the design limits to achieve higher compression ratios and mass flows and also fulfilling industrial demand for high pressures and low mass flows, a high risk condition for surge. However to achieve the two with one turbo would enable a boost of low end torque whilst having the capability to supply the maximum flow rate for a given engine [1].

What is surge

A compressor map is a plot of flow vs pressure ratio for fixed speed lines. The right hand limit is governed by choke, which CTT defined at the point compressor efficiency drops below 59.5%. The left hand limit is described as the surge limit. A physical limitation beyond which stable operation is not achievable. Tracking along the constant speed line from choke to surge, as the flow is reduced due to a restriction downstream, pressure builds in the boost section to the stall limit, the flow becomes unstable leading to surge [3].

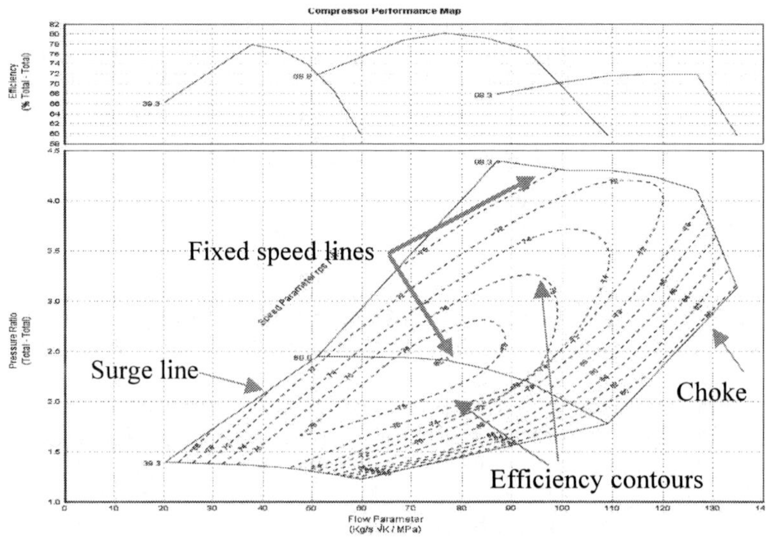

The instability can either manifest itself as a large scale oscillation, or a steady reduced pressure or flow (rotating stall) [2]. Papers generally refer to surge as a large oscillations in pressure and flow, averaged through an annulus, while rotating stall is a cell of stagnated flow revolving around the circumference at approximately 25-30% the rotational speed of the rotor depending on conditions. Once the pattern is fully developed stable mass flow is a possibility [3] [4].

Rotating stall and surge are two different characteristics. When the unit is surging there are strong pulsations in compressor outlet pressure, mass flow, and the compressor inlet and outlet temperatures will rise, associated with the lower efficiency operation and some back flow warming the inlet. This may be accompanied by a strong vibration and noise from the turbo. Static or total pressure analysis can determine if we have stable operation of the unit [2].

There are various stages of surge, it may be possible to determine what type or at what stage in the surge cycle we are, whether we are in a rotating stall, classic surge or deep surge condition by reviewing the transient response. In Fig 1 clock wise from the top left, Rotating stall shows a fast low amplitude oscillation in pressure and flow but essentially non-damaging and 'stable'. When the flow is reduced further the same 'stable behaviour continues at a significantly lower flow and pressure ratio. Top right shows rotating stall progressing to classic surge.

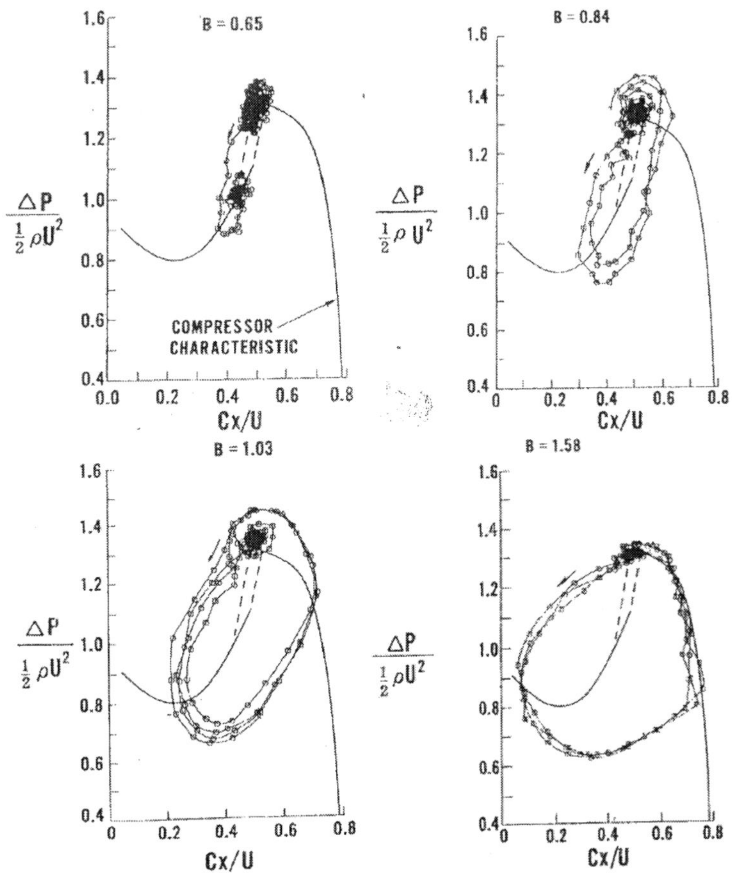

Figure 1 Illustration of different of different surge characteristics [4]

Bottom left shows classic surge, and bottom right is indicative of deep surge, where there is no period of rotating stall before largescale oscillations begin. The trace around the map draws much larger circles than classic surge or rotating stall.

The importance of understanding the location of surge

From a thermodynamic efficiency point of view, around surge the compressor efficiency can be as low as 20% which although isn't damaging is undesirable for an engine running condition especially as the focus is moving to try and minimise large particle carbon emissions or clog a DPF.

The more important consideration is from a safety point of view and that surge has been seen to severely reduce turbocharger life and induce failure by the application of significant repetitive stresses on the bearing system [7]. The author has seen the destructive power of surge, sheering a wheel from a shaft while Galindo et al [1], had also broken a turbocharger shaft whiles testing negative flows. This is due to the forceful pressure oscillations caused by operating in deep surge [8]. High internal temperatures can cause blade weaknesses, this could result in a blade off or the blades creeping. Either leading to a wheel touch and a fatigue failure due to the blade tips rubbing and breaking away, or binding with the housing and causing a seizure. One could cause an imbalance and the other extreme torsional stress sheering the shaft, both pose a containment risk.

Furthermore, additional on application risks include damage to the intake system through over pressure because of reverse flow. For gas turbine applications running at low flows the danger of high temperatures in the burner and turbine [3].

Due to the complexity and lack of knowledge about the theory governing compressor instability a physical model cannot meet the requirements, therefore we need a way to measure and map the surge location [2].

What has been discussed highlights the importance and the key problem of being able to detect the onset of surge automatically, accurately and fast while mapping to protect the validity of the compressor performance test. This is because the decay of rotating stall to potentially damaging surge is very sharp [4]. This will then allow for safe operating limits to be prescribed.

Previous work
In all reviewed literature, surge was always attained by conditioning the back pressure and thus flow for a temperature corrected speed line [2].

Many authors have tried various techniques to mitigate the onset of surge or understand the relationships involved. W C Oakes [9], drew the conclusion that surge behaviour was a function not only of wheel design but influenced by geometry around the compressor.

The techniques used to manipulate the conditions at the inlet to the compressor to expand map width that have shown to have positive results are the use of variable geometry compressors to account for low and high flow conditions. Compressor inlet vanes to induce pre-swirl, not only can these double as an intake throttle but can significantly increase the map width, although the pressure and enthalpy drop across a stator can reduce the achievable pressure ratio and efficiency for given engine condition. Other options include recirculation so that although the net flow through the unit to the engine is below the surge line for the compressor, the compressor is able to avoid stall due to bleeding of boost pressure back to the inlet. The most complex method to achieve low flow and avoid compressor surge is to use a multiple stage turbo. This requires a large and a small turbo with closed coupled valves to direct flow through the appropriate stage. This also has the benefits of reducing turbo lag at low flows by using the smaller stage, increasing low down torque, and with the larger stage facilitating larger flows at highway speeds [5].

Theory and related previous work
It has been found that poor inlet geometry of the intake can have the effect of inducing surge, this may be due to an upstream restriction, possibly induced turbulence, both increasing the NDMF at which critical loading occurs [10].

Other research completed has linked the outlet volume and effective length and area of axial and radial compressors to a non-dimensional number. The Riemann characteristic which can be used to predict what type of surge may occur and deduce the frequency of pressure oscillations that result from instability [4].

$$B = \frac{U}{2a}\left(\frac{V_b}{A_c L_c}\right)^{\frac{1}{2}} \qquad 1.$$

In previous experiments by Grietzer [3], the axial compressor was set up, discharging directly into a section of pipe (the compressor effective length (AcLc)) which then fed into a plenum of a volume (Vb). Galindo [1], also used this set up for turbo charger testing on an engine. Both were able to demonstrate that as they modify the Riemann characteristic they were able to characterise the expected resulting surge.

When throttling the compressor exit to increase pressure and reduce flow. It was found for large values of B firstly classic surge would occur progressing to deep surge as the throttling is increased. For values of 0.8>1 we expect to see rotating stall, then classic surge progressing to deep surge as flow is further reduced. For values of B below 0.8 only rotating stall is expected, but eventually escalating to deep surge [4].

As well as the type of surge, the amplitude and frequency can be modelled and predicted. With the governing equations for unsteady, compressible non-homentropic gas flow using finite difference, and approximating the flow using Taylor series it is possible to model the behaviour based on the system parameters, essentially this is a 1D gas model that can account for wave affects [1]. Most of the work is based on work by Grietzer who in 1976 created a lumped parameter from axial compressors. Accounting for the filling and discharging during surge and fluid flow decay [11].

The surge cycle is dependent upon the work being done on the system by damping forces, where the oscillations are associated with the dynamic instability [3]. This can be likened to a second order mass spring damper with a nonlinear restoring force and damping, the result being the oscillations can be modelled as a change in mechanical energy with time [3].

It was found for an impeller between 80mm and 179mm that the frequency of surge is 5-8 Hz [2], which is an order of magnitude less than the passage of rotating stall cells [3]. Although Grietzer [4] found rotating stall to occur at ~10hz instead of 250hz by Galindo, which had good agreement with his models [1]. The reason for the vast difference in frequencies is likely linked to the use of different Riemann parameters in there set up.

Frequency response depends on whether B is below or above a critical value. Up to 0.7, the frequency of the surge event which is likely to be rotating stall (and near constant pressure) and based on the Helmholtz resonating frequency of the arrangement. Where the kinetic energy of the oscillation is linked to the motion of fluid with the potential energy linked to the compression of the fluid [3]. At larger values of B then the frequency becomes a function of the physical mechanics of the system to facilitate the filling and discharging of air (large oscillations at reduced pressure and flow).

Where Galindo's [1] models ability to maintain good agreement with the experimental data was eroded was in linking the variables of the mechanical system with the acoustic properties. For example, the models predicted the largest amplitude as the compressor out duct is decreased and B is increased. In reality the largest amplitude was due to reflected wave phenomena in the duct. Where we have resonance, this might indicate that Helmholtz resonators could be used to dampen the oscillations and stabilize the system as well as disrupt it [1]. The models also lacked the detail that rotating stall happens over substantial parts of the classic surge cycle, and before rotating stall becomes classic surge, FFT evaluations of the transient logs indicate a clear signal at the expected frequency of surge emerges from the noise [2]. In the cases where classic surge is expected this can provide an early warning of the onset of surge and this information on expected frequencies can provide useful insight into how the channel should be set up to remove unwanted noise for example with the use of an appropriate low pass filter.

As with any experiment there are some pitfalls to be mindful of, firstly measurement of flow at this region of the compressor mapping can be tricky. Instantaneous mass flow cannot be measured with precision unless somehow you are able to have sensors covering the entire annulus to create an averaged flow [4]. If the relying on the skill of the operator, or designing an algorithm to run a test program it is important to know which mode of surge may be expected. Deep surge has a large amount of hysteresis with the onset and cessation. Compressor response is not governed by speed or the boost valve setting but coupled to the flow, turbulence and boundary layer development on the blades. Why this may be an issue is that the boost valve may be closed to the brink of surge but upon opening it by a pre-defined value surge would persist and it may not be possible to bring the unit back to an un-stalled condition without more severe action [3].

Purpose

Inertia forces are equal to $\rho\omega U L_c A_c$. Looking at similar terms in the Riemann parameter we can see this is an approximation of pressure and inertial forces. So if B is increased by a fixed percent then this will result in inertial forces that are relatively smaller compared to the driving force, the pressure differential [4]. Therefore the relative amplitude of axial velocity oscillations cannot remain constant as B increases, but intensify to maintain balance. This leads to the more circular transient traces around a flow vs ΔP Graph shown in figure 1.

We also know that if the downstream volume of the compressor is sufficiently small enough, linked by the Riemann parameter that relates Helmholtz frequency to surge dynamics that deep surge can be avoided [11].

On application a turbocharger will have a significant amount of inlet pipework which was not reflected in the previous work. Additionally the turbocharger does not discharge into an effective length of pipe and then into a plenum, but directly into a single volume, albeit convoluting.

The purpose of this experiment is to build on the knowledge of the Riemann Characteristic and to look to create a relationship that links the application of the ratio of pressure to inertia forces to that of that of the outlet and inlet respectively. Assigning the inlet geometry as the compressor effective length to describe the inertia force and pressure term to the boost section geometry. If this relationship is transferable for compressor effective length from the boost side to the inlet side then it may be possible to widen the range at which stable points can be achieved.

The results may include moving the surge line or modifying the behaviour to be non-damaging in a predictable way. Ultimately a relationship may be formed which link results from different mapping conditions, between testing facilities as well as to operating conditions. This may allow us to better describe to customers what performance they may expect with application pipework.

The null-hypothesis: It is not possible to directly translate parameters from the Grietzer parameter to other structures within the system boundary that in theory share the same properties. I.e. that the inlet pipework can be linked to the inertia, and the outlet can be linked to the pressure term and demonstrate the same trends as found in the literature by modifying the map position at which the unit moves from a stable rotating stalled condition to a classic or deep surge event.

II. EXPERIMENT DESIGN

Apparatus

A Holset HX40 unit is set up as shown in figure 2. The turbine side is solely used as a drive unit for the purposes of this testing. A main air valve downstream of a 6 bar main air supply shown in figure 2 is used to control the mass flow through the turbine and thus the speed of the turbo while the boost valve is used to control the flow through the compressor stage.

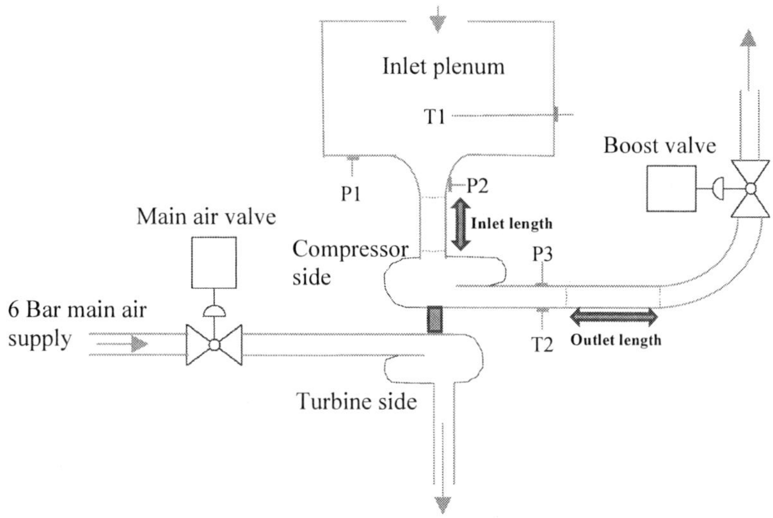

Figure 2 Diagram of apparatus

Flow is calculated by taking a differential pressure measurement from the total pressure in the plenum to the static pressure measurement at the throat of a calibrated nozzle (P1-P2), P2 is reported as Cmp_srg_p on any following plots.

Pressure ratio is calculated by looking at the ratio of total pressures P3÷P1. It should be noted that in the cases where there is a long inlet then there may be a significant pressure drop across the pipe at higher flows leading to a lower pressure ratio being represented which is in fact is an error.

Cell control and test script procedure

The test cell is controlled via closed loop PID controllers, tuned specifically for the characteristics of the range of units that are compatible with the cell hardware. The control system is based around proprietary software designed to run on a real time operating system, Linux. This means tasks are run exactly when they are scheduled, and needed, which is important with the rapidly changing dynamics of turbo charger testing for safety and data acquisition.

To induce surge at the desired speed the boost valve is slowly closed to drive the unit towards surge. As long as critical parameters are stable and within tolerance this continues, otherwise we pause to maintain test validity. The test cell continues to log points as it steps the unit towards surge, this means that we will always have a record of the last stable operating point governed by whether the following conditions are within tolerance; speed, compressor in, out and wall convergence temperatures, and compressor out pressure.

Using a proprietary method of surge detection as soon as the test cell detects a surge event the boost valve is opened to a predefined position within 0.3s to prevent damage to the unit. The data is reviewed to ensure stability criteria was met at the point of surge, if it hasn't then the test is repeated. This is repeated for several speed lines to generate data for high and low pressures and flows.

Surge determination

It is essential the test script has the right logic to drive the unit to achieve full surge since allowing the system to run the unit provides several key benefits over the subjectivity of an operator. The unit receives the same ramp rate of flow reduction each time, the system constantly monitor parameters at 25 Hz so can react much faster than an operator reducing the exposure of the unit to the damaging effects of surge when it occurs.

When the compressor surges there are certain characteristics that will change that if monitored can allow us to statistically determine via an algorithm surge has occurred.

It is also important that the system can determine the point of instability if a classic or deep surge event does not occur, for example at very low speeds, where rotating stall events are more likely and the behvaior is considerably different to that of a hard surge event.

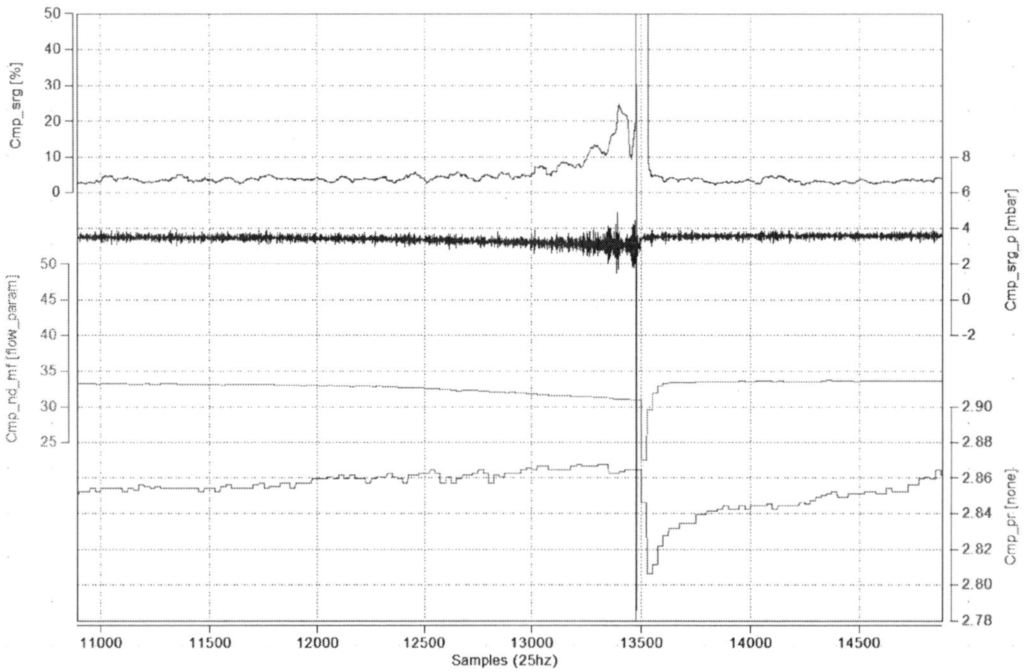

Figure 3 100krpm line with 1.5m inlet and 0 outlet length

In Figure 3, as flow is slowly reduced we can see an increase in variation indicating rotating stall. Mass flow and compressor ratio are then both shown to drop indicating complete compressor stall and reverse flow. This is not always completely apparent since in this experiment to maintain accuracy these channels are still heavly filtered to remove any noise to make average logs for compressor performance measures more repeatable.

Preliminary work

Taking the information from the literature some preliminary work was undertaken in order to set up the channel monitoring the flow and compressor intake pressure differential to gain the most insight from the data. The Helmholtz calculations where completed for the standard set up and showed we would need to be targeting a frequency of up to 13 Hz considering harmonics.

Fig 4 shows the benefits of applying low pass filtering to a noisy channel. The purple trace is the raw data sampled at 1000 Hz, the red trace is the same data, offset for visual purposes but with a 16 Hz low pass filter applied.

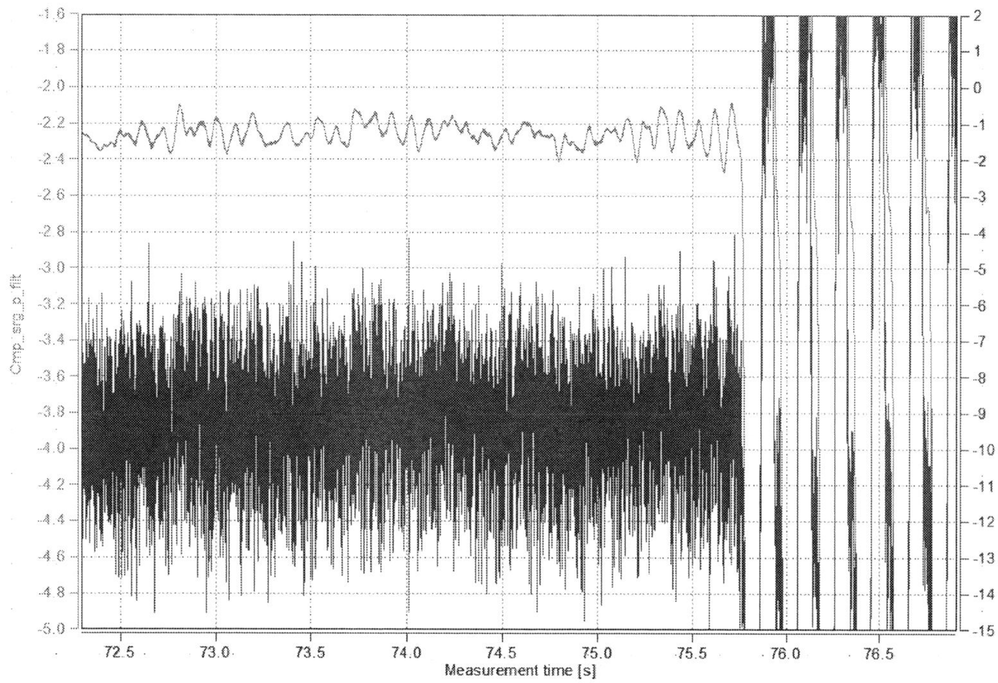

Figure 4 Compressor inlet data with and without proposed filter applied

We can clearly see that this filter removes a significant amount of the noise associated with the rotational frequency of rotating stall cells. What we can see that as we approach the large oscillations associated with classic surge we can see a 5 Hz signal emerge. This seems to support the results from Hongmin Chen [2]. We can also now determine from the filtered data a frequency for the flow instability leading up to classic surge as well as surge itself.

Limitations of the data acquisition to be used on test, also to avoid collecting an unnecessary amount of data meant that although the 16 Hz low pass filter was applied to the raw signal it could only be sampled and the surge percentage calculated at 25 Hz. One drawback being we lose the associated frequency with the rotation of the stalled cells although this wasn't deemed important for this test.

Test matrix

Minitab 17 was used to design a DOE, and to evaluate the results by the production factorial plots as well as presenting statistical values to gauge validity and then make a judgement on whether to accept or reject the null hypothesis.

Repeats	Inlet length	outlet extension
2	1.5	0
2	0.5	2
3	1	1
2	1.5	2
2	0.5	0
2	0	0

The repeats refer to full repeats of the mapping test script which incorperate running 3 speeds, 60krpm, 100krpm and 125krpm. The limits of extending the boost section by 2m (base length was 8m) and the inlet by 1.5m were simply due to the practicalities of fitting the pipework into the test cell. Repeats of the tests were important to try and discern part to part variation. Expecially important when trying to work with a phenomenon as coupled and compelex as compressor surge.

III. RESULTS

The results were processed through minitab with the response being the surge NDMF. The factors being inlet length and outlet extension. This allowed factorial plots to be created to evaultate the influence of the parameters by speed.In the following interaction plots show the performance for different setups. With 0-2m outlet extensions and 0.5-1.5m inlet extensions.

Table 1 Results table including averaged surge NDMF for each set up

Inlet length	outlet extension	Speed (Krpm)	B parameter	Surge NDMF
1.5	0	60	0.55	14.03
1.5	2	60	0.61	14.25
1	1	60	0.71	14.48
0.5	0	60	0.95	13.47
0.5	2	60	1.06	13.58
1.5	0	100	0.92	30.73
1.5	2	100	1.02	31.42
1	1	100	1.19	31.27
0.5	0	100	1.59	30.50
0.5	2	100	1.77	30.87
1.5	0	125	1.15	42.77
1.5	2	125	1.28	42.51
1	1	125	1.49	42.65
0.5	0	125	1.98	42.90
0.5	2	125	2.22	43.36

Surge NDMF

The interaction plot for the 125krpm in figure 5 line shows a negative slope for increasing the inlet length at both extreme outlet conditions, this is what would have been expected by adding more inertia to the system.

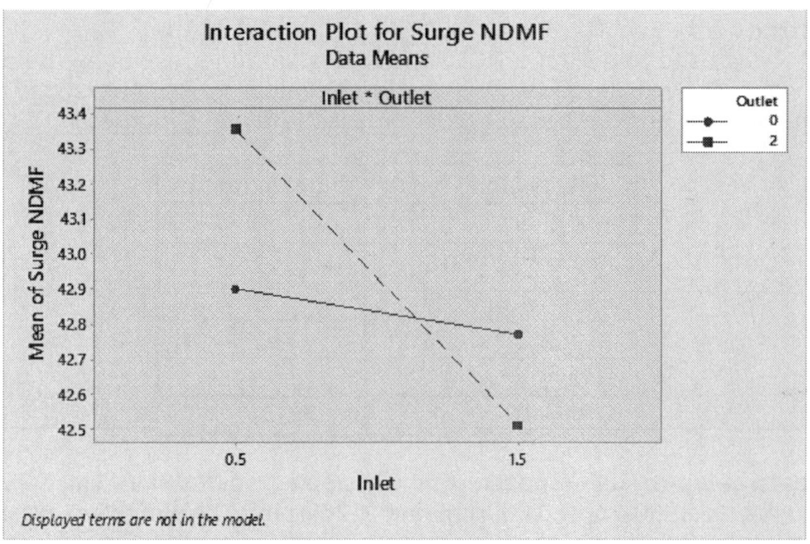

Figure 5 125krpm interaction plot. P values; inlet 0.02, outlet 0.567

The relationship is only statisctically valid for the affect of the inlet when working to 95% confidence limits. It is hypothesised that as the inlet length increases and so does the inertia in the system and that this reduces the B value to one where the characteristic that will develop is rotating stall at the surge line and then to classic surge.

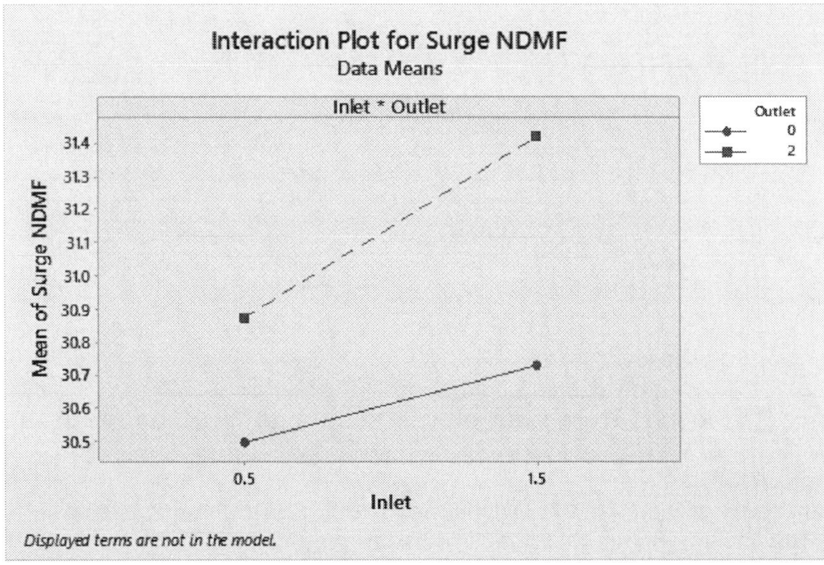

Figure 6 100kprm interaction plot. P values; inlet 0.092, outlet 0.031

If this is the case then the automated surge routine would consider rotating stall as stable at the surge line due to the much smaller ratio of variation to the absolute differential pressure from the inlet to the plenum. The effect of the boost volume and length in this case does not have the effect of creating an offset but a pivot in the relationship between inlet length and surge NDMF. The outlet appears to have little impact on modifying the surge line NDMF apart from potentially modifying the influence of the inlet length. From table 1, it appears that changing the boost volume has a minimal impact on B compared to the inlet within the constraints of what was possible to test.

Figure 6 shows the interaction plot for the surge NDMF for the 100krpm speed line. This highlights two interactions, as boost volume and inlet length increases so does the surge NDMF, as Boost volume increases as does the affect of the inlet on NDMF, the opposite of the 125krpm result. The B parameter increases much more significantly again with the inlet variable however on this occasion only the interatction for the boost volume is statistically valid. By increasing the B parameter from the literature we expect to go from stable flow directly to classic or deep surge without the period of rotating stall. So an increase in volume has had the expected result of increasing the NDMF at which "stable" flow can be achieved.

On the lowest speed line it could be said the automatic surge detection algorithm and data acquisition logged stable flow at a lower NDMF for shorter inlet pipework if 90% confidence limits were to be accepted. There was no discernible or statistical difference based on boost volume changes. This was unexpected that more inertia in the system was linked to instability at a larger mass flow, the opposite would have been expected.

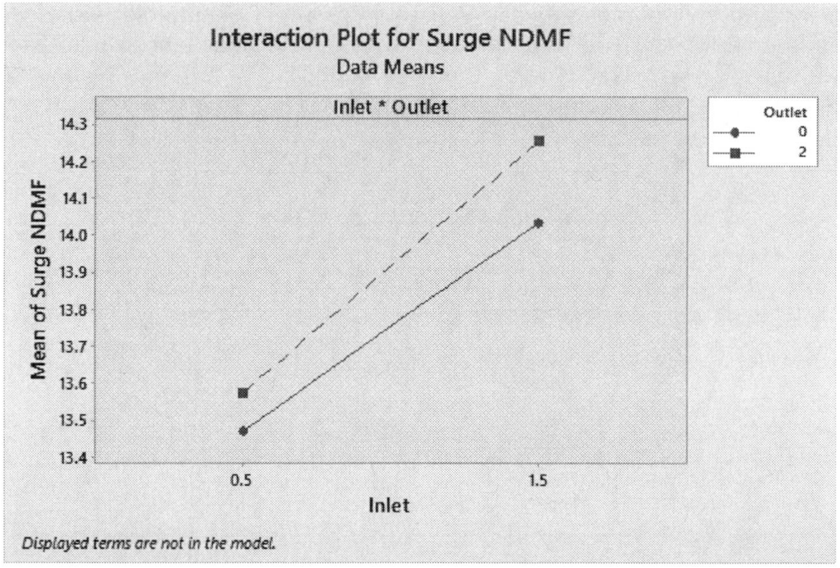

Figure 7 60krpm interaction plot. P values; inlet 0.106, outlet 0.648

Surge frequency

In contrast to what was seen for NDMF, for the 125krpm line, it is the boost volume that has the more significant impact on frequency shown in figure 8. As the boost volume is increased, B changes from an average of 1.215 to 2.1 the surge frequency is decreased 25%. This was expected since, if the system is likened to a mass spring system then adding additional air mass into boost section will decrease the frequency since we are now relying on the physical dynamics of oscillating more air. This is also an expected outcome from calculating the Helmholtz frequency response for the system.

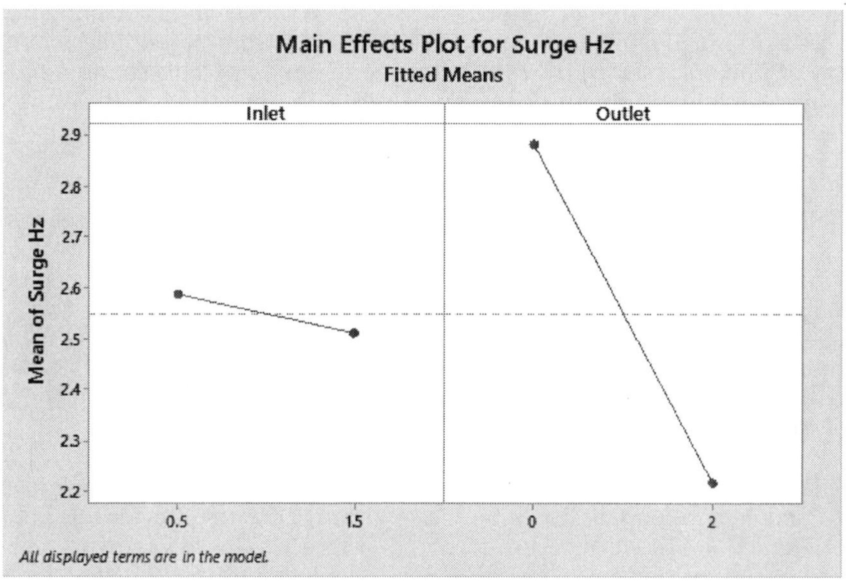

Figure 8 125krpm main effects plot. P values; inlet 0.834, outlet 0.097

At 100krpm the inlet geometry governs the frequency of instability shown in figure 9. As the inlet pipe is increased so does the frequency of surge by 16% as B changes from an average of 1.68 to 0.97. This is more complex to explain, at this value of B the oscillations where supposed to be determined by the physical limitations of the geometry. Following the same trend as the Helmholtz frequency calculation, this should generate lower frequencies for larger volumes so this cannot be the lead factor here.

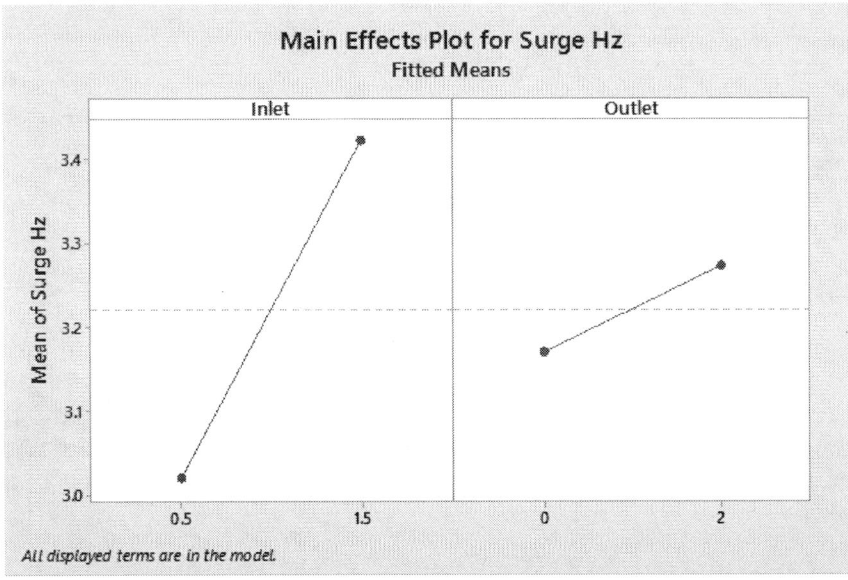

Figure 9 100krpm main effects plot. P values; inlet 0.081, outlet, 0.619

Instead it is believed this result is coupled to two properties of the oscillatory system, one, that there is less pressure in the boost section with less mass flow, lowering the potential energy stored here and so the system is less mass dependent even with the relatively small increase in boost volume. This reduces the time it would take following critical loading and the subsequent pressure drop for the compressor to recover. Second, that the hypothesis was that increasing the inlet increases the inertia in the system and therefor the kinetic energy, and this would support the compressors ability to increase the pressure rise rate. In this case the system is more sensitive to inertia to determine how quickly it can respond.

IV. DISCUSSION

As inlet length was increased, on the 125krpm line, we saw a statistically valid but small improvement in surge margin, but only when the 2m boost extension was used. By that it is meant it surged at a lower flow providing a wider map. This improvement was not shown when the boost section was setup without the 2m extension. Evaluating the B parameter to establish if there is a larger swing in B values with varying inlet length depending if a 0m or 2m boost extension was used, as a possible reason why this may be a more dominant parameter, it is evident that with the 2m boost extension a greater range of B is covered, 0.94 compared to 0.83 without. The performance may have been the same if the inlet could have been proportionally extended with the boost length to keep this value the same.

Altering the boost volume by adding 2m, may not have had a significant or the expected outcome since the relative change in boost volume is small and perhaps we are trying to detect a change in surge NDMF that is also very small, and where the cell variation is large enough to prevent us from seeing the true trends. Looking at the centre point values we can say the relationship is nonlinear and although we would be able to quote an equation that describes a model to fit the current data we may be unable to determine the true relationship for two reasons. Not running enough repeats to account for cell variation as well as enough combinations to put together a full picture of the interaction between the variables.

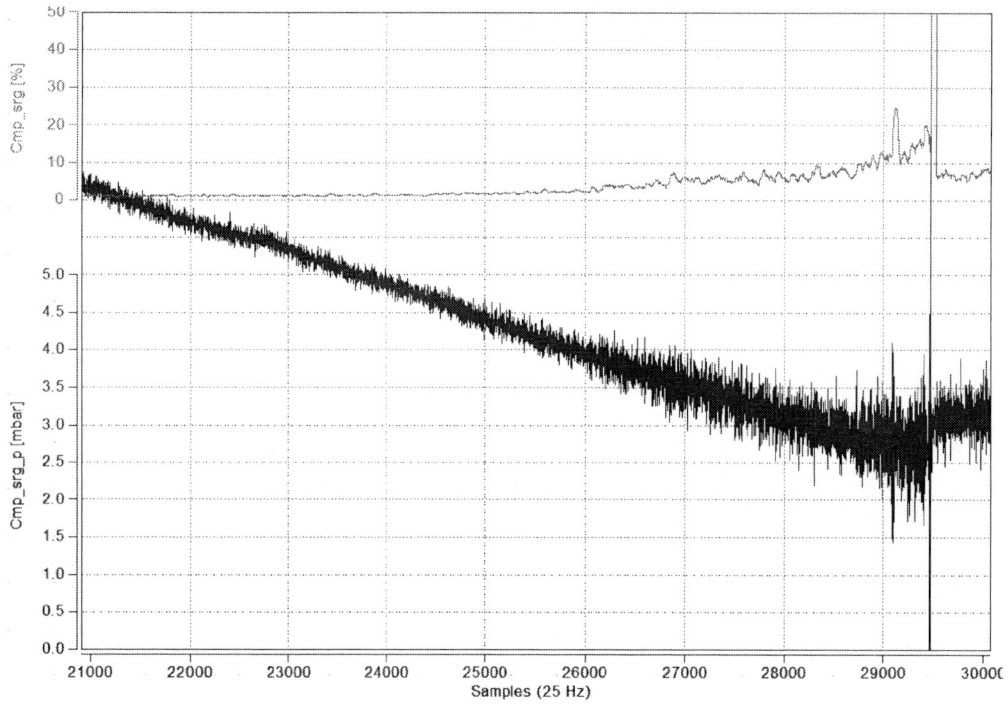

Figure 10 100krpm 1.5m inlet 0 outlet, Compressor inlet pressure and coefficient of variance

Similarly another unexpected result was when the speed is reduced to 100krpm, the dominant factor switches to that of the boost volume instead of the inlet length. This could be because there is significantly lower flow in the inlet and the outlet pipework also contains much less pressure and therefore irrespective of the value of B, which does not fully take into account the ratio of inertia to pressure. That the energy is much more balanced between the inlet and outlet allowing a relatively small change of 2m in outlet volume to have more influence on the results.

The two boost conditions then may result in a situation where there is a clearer definition between either, the development of rotating stall at the surge line and a lower NDMF at which classic or deep surge occurs or where at the surge line we move swiftly into deep surge and therefore this occurs at a higher NDMF. Fig 10 shows the building of rotating stall, to the point where there is a classic surge event. Fig 12 on the other hand demonstrates a distinct lack of rotating stall before the oscillatory deep surge occurs.

The inlet may be relatively unresposive to changes in geomertry due to the lower inertia per meter due to the low speed. In which case the addition of 1.5m to the inlet has little to no effect in realterms of altering the inertia pressure balance between the two sides significantly.

Although NDMF results were presented for the 60krpm speed line. Investigating the transient logs of the surge events it was identified that classic or deep surge was not achieved. We can see in figure 11 that there is not a defined pressure rise and drop as shown in figure 10 and 12. On the graphs Cmp_srg_p is the differential between the plenum and the inlet nozzle which the coefficient of variance, cmp_srg is calculated from used for surge identification. The B parameter for the lowest speed line, for all runs was below 1.06 with most below the critical 0.7 value. So we would mainly expect to see rotating stall behaviour, this coupled to the sensitivity of the algorithm at low absolute inlet depression pressure values, a small amount of variation leads to a significant coefficient of variance. However we can see the compressor inlet pressure differential dipping below zero, indicating the pressure in the flow nozzle has increased above the plenum, shown in fig 11, this leads to the 200% spike and the algorithm logic falsely detecting it as a surge event. This highlights an easy improvement to achieve better data in the low flow, low pressure regions.

This indicates the use of this variable as a measure in this test maybe susceptible to erroneous results as the pressure fluctuations around rotating stall are inherently noisy. Since hard surge was not achieved it may be prudent to discount the use of the low speed data, however the increase in inlet pressure variation indicating an increasing instability in rotating stall may be a fair approximation that a surge event is very close. If this is the case the results are still unexpected. As more inertia was added we would have more likely expected to achieve stable flow at lower NDMFs not the opposite.

To present three possible explanations to why surge NDMF increased with the inlet length and theoretically inertia in the system. Firstly, it could be that the inlet pipework created the turbulence that at small inlet depression pressures combined with the additional variation from turbulence may lead to false positives on surge detections. Secondly is that the difference measured in NDMF of approximately 4% is within the test cell variation and is therefore not a significant result which is not evident because of the lack of repeat testing, however it should be considered within the confines of the test and the number of repeats we still achieved a P value of 0.1. Thirdly, as reported in the literature one of the weaknesses of analytical models was the inability to account for the phenomena of reflected pressure waves, it could be possible the additional volume added to the inlet, at these lower flows facilitated the false detection through resonation in the pipework.

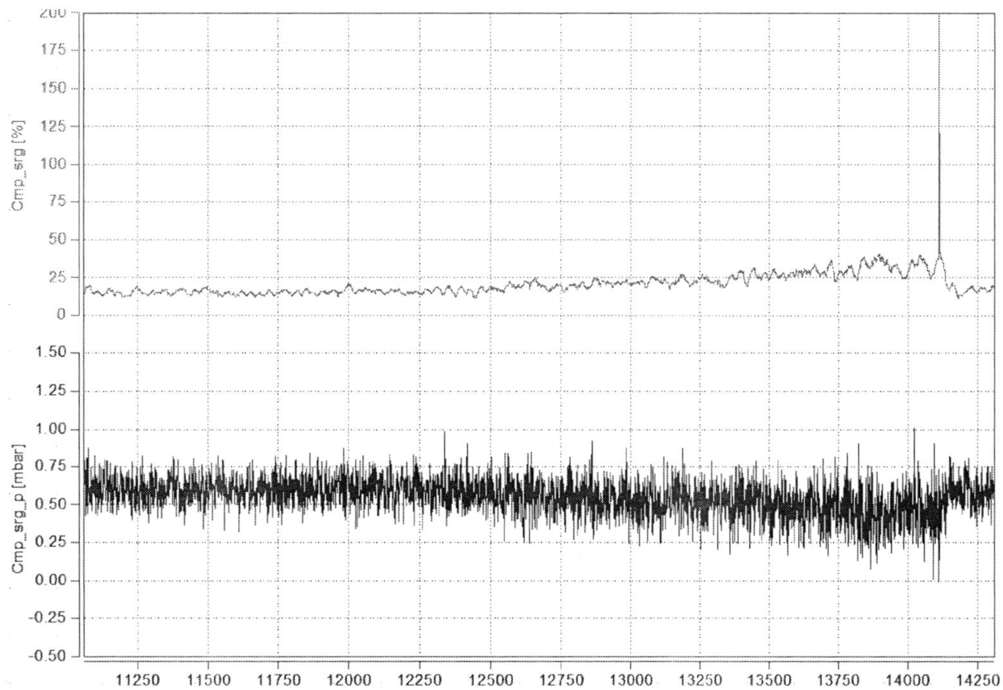

Figure 11 60krpm 1.5m inlet 0 outlet, Compressor inlet pressure and coefficient of variance

It may have been possible to drive beyond the rotating stall to a classic or deep surge event to log its position, however these are base units, and are used for test cell validation and approval so it is undesirable to compromise there integrity. Secondly the boost pressure was very low at the 60kprm speed, it may be that the effect of B could become irrelevant if the flow could have been reduced to the leak rate past the boost valve seat and all we may have experienced is larger amplitude rotating stall.

Similarly there were difficulties in identifying the frequency of surge at the low speed since only rotating stall triggered the Surge detection system. This looked to be around 13 Hz however only sampling at 25 Hz made it difficult to reliably state a frequency for each case. So this analysis was omitted for the 60krpm line. The significantly higher frequency of instability was deemed to be linked to the significantly reduced pressure ratio with much reduced air mass in the system facilitating a much higher oscillation rate.

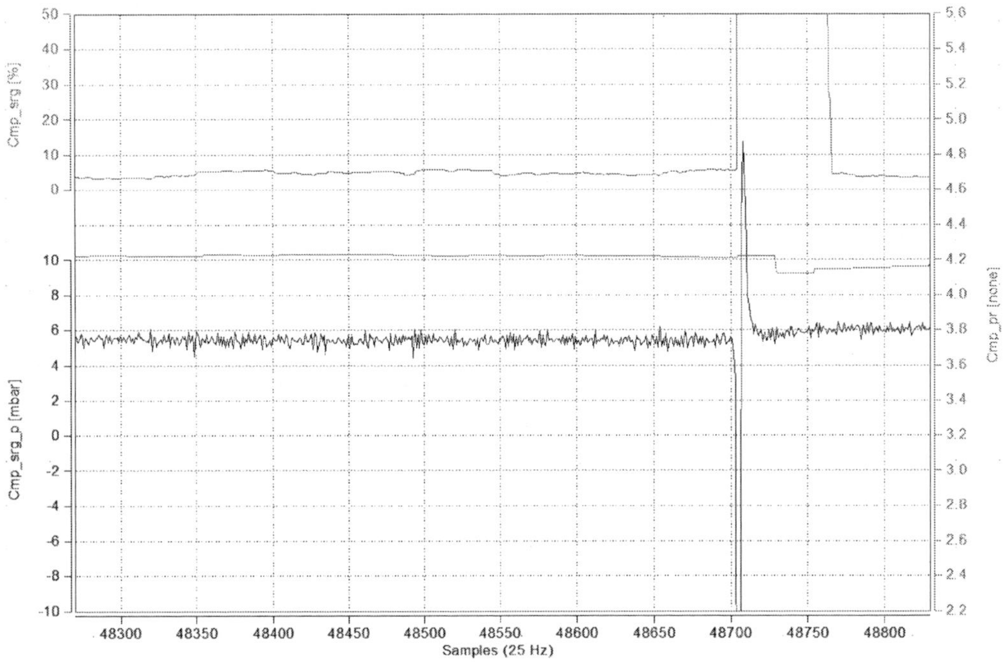

Figure 12 125krpm 0.5m inlet 2 outlet, Compressor PR, inlet pressure and coefficient of variance

The data rate as well as the filtering applied to the compressor out pressure channel and the frequency response of the sensors means that it was not possible to attain what type of surge was apparent by over plotting this on a compressor map. As we can see in fig 12 the Compressor pressure ratio does not respond in the same timeframe as the compressor inlet pressure differential measurement which is used to determine flow so attempting to over plot this on a compressor map was fruitless.

It was expected from the literature the pressure variation at surge would be the virtually the same for all speeds when normalised to absolute pressure value. This was found not to be the case with the coefficient of variance indicating significantly different values of around 4%, 22% and 38% for the 125krpm, 100krpm, and 60krpm line respectively at the onset of the oscillations. Then again Chen [2], who reported this did run fairly low speed tests over a small speed range of 30-50krpm so maybe under these circumstances this approximation is adequate.

V. CONCLUSIONS

The test depended on two repeats of each condition apart from the centre point, so being able to develop a predictive model experimentally was going to be difficult taking into consideration cell repeatability and capability. Looking at the variation between the entire data set and calculating two standard deviations from this. For all of the results this is equal to +-1.06 NDMF units, this is nearly double the same calculation when purely considering repeats of the same test setup which was is equal to +-0.39 NDMF units. The variation due to the cell measurement system is very small at approximately +-0.5 NDMF. This indicates for the most part that the variation we are seeing appears to be outside the normal limits from the test cell which is why the statistical P values are fairly strong, and also the relationships that are reported as statistically valid, are probably linked to the inlet and outlet volumes changing in some way.

It looks as if, although there is a relationship between the inlet and outlet volume and the effect on the surge NDMF and frequency, we should accept the null hypothesis. We cannot simply transfer the inertia term of the Riemann characteristic to inlet geometry when the boost plenum and effective length are the same structure. It is not straight forward and not demonstrated in this experiment.

What we can say we have learned is that the balance of forces concerning pressure and inertia are important between the boost section and the inlet, and the results are significant for the most part to 95% confidence limits. We can also say the results do not simply follow the hypothesis that reducing the pressure parameter and increasing the inertia term, by means of reducing the boost volume and increasing inlet length respectively results in creating wider operating points by extending the range at which rotating stall occurs in place of a classic or deep surge event.

The evidence instead shows that within the limitations of the physical changes we could make to the test cell and the sample sizes we have used that the dominant factor is dynamic between the two, either the inlet or outlet and is dependent on speed. Considering the high speed lines 125krp and 100krpm, the relationships are almost as expected, but on the high speed line the hypothesis is true for the inlet geometry and for the 100kprm line the outlet geometry.

This may be due to other complex relationships, perhaps linking blade velocity and angles of attack to the critical loading. In this case the use of the Riemann characteristic falls short of a fully characterising pressure and inertia in this experiment because the boost section to which the pressure term is applied is not a static volume of pressure, and will make up a significant amount of the inertia in the system.

A progression from this work is to build on the Riemann characteristic by adding in an additional term to factor in the inertia from the inlet pipework as well as a lumped parameter for the pressure and inertia that exists in the boost section. Then the parameter may be able to better describe expected behaviour from a typical turbocharger application or test setup.

We have a need to understand the variation in surge NDMF induced by different testing facilities due to cell design. We have learned and demonstrated statistically, that increasing the boost section volume at the 125kprm speed decreases the surge frequency by the addition of mass, and alters the effect of the inlet on NDMF. The inlet geometry was more important in terms of map potential map width, with increasing length achieved a statistically wider map.

While on the 100krpm line, increasing the boost section volume also statistically narrowed the map it had no significant affect on the surge frequency. The inlet was now not generally important to the surge NDMF but had significant effect on increasing the surge frequency.

On reflection, due to a higher level of uncertainty at low flows, and unexpected behaviour all we can summarise for low flows is that as the inlet is increased the map width appears to decrease. The reasons are unknown, and although statistically valid the change is very close to the test cell capability, i.e. it could be cell variation.

The centre points show highly nonlinear behaviour, and this could be that we are seeing that the rate of influence changes as the parameters do, which when comparing one speed line to another tends to support this. It is possible to state the experimentally derived equations of the models however these did have high error terms, this is probably due to the non-linearity and the lack of intermediary data points between the corners as this was a simple 4 corner and mid-point DOE design. The relationship changes with speed so ideally the data would be analysed again by inlet and outlet conditions individually against speed, with many more repeats and at different data points to generate a more thorough model.

Further work should be conducted to fully investigate the impact of increasing the volume by increasing the diameter rather than the length. Increasing the diameter would create slower moving gas with less inertia and possibly different characteristics. Although this may be more difficult to test it may be more relevant to correcting data when moving between test cells with different boost volumes due to different pipe diameters. It is also desirable to use what has been learned about modifying the frequency of the instability right before surge to target a Helmholtz resonator to dampen out application specific pressure frequencies with the aim to achieve a wider compressor map.

REFERENCES

[1] Galindo, J. Experiments and modelling of surge in small centrifugal compressor for automotive engines, 2007, Experimental thermal and fluid science 32 (2008) 818-826

[2] Chen, H. Study on algorithm for turbocharger surge detection and its application, 2008, Society of automotive engineers 2008-01-1695. 0148-7191

[3] Grietzer, E. M. Surge and rotating stall in axial flow compressors part 1: Theoretical compression system model, 1976, Gas turbine conference, Houston, Texas, March 26, 1975.190-199

[4] Grietzer, E. M. Surge and rotating stall in axial flow compressors part 2: Theoretical compression system model, 1976, Gas turbine conference, Houston, Texas, March 26, 1975.199-217

[5] J, Portalier, J.C. Conception of twin turbo boosting system for new generation of PSA 2,2 liter HDI, in: Proceedings of the Conference on thermos and fluid Dynamics processes in Diesel Engines Thiesel 2006, Valencia, September 2006, pp. 589-607

[6] J.E. Pinsley, Active stabaliszation of centrifugal compressor surge, Journal or turbomachinery-Transactions of the ASME 113 (1991) 113-732

[7] R, Kurz, R.C White, Surge avoidance in gas compression systems, Journal of Turbomachinery-Transactions of the ASME 126 (2004) 501-506

[8] H.P. Bloch, A Practical Guide to compressor Technology, McGraw Hill, 1996, ISBN 0-07-005937-3

[9] W.C. Oakes, P.B. Lawless, J.R. Fagan, S. Fleeter, High-speed centrifugal compressor surge initiation characterization, Journal of propulsion and Power 18 (2002) 1012-1018

[10] H, Schulte, H. Schorn, H. Kindle, Influence of various compressor inlet designs on compressor performance, in: J.h. Whitelaw, F. Payri, C. Arcoumanis, J.M. Desantes, Thermo- and fluid Dynamic Processes in Diesel Engines, Springer-Verlag, Berlin, 2004, PP.103-116

[11] D.A. Fink, Surge dynamics in free spool centrifugal compressor system, Journal of turbomachinery-Transactions of the ASME 114 (1992) 321-332

Experimental investigation of upstream installation effects on the turbocharger compressor map

B Kerres, A Cronhjort
KTH Royal Institute of Technology, Department of Machine Design, CCGEx, Sweden
M Mihaescu
KTH Royal Institute of Technology, Department of Mechanics, CCGEx, Sweden

ABSTRACT

This paper experimentally investigates the effects of an upstream bended pipe on the compressor speedline slopes and surge line. Different orientation angles for the incoming bended pipe relative to the compressor scroll are investigated. The tests were carried out on a cold gas stand on a passenger car sized turbocharger. A bended pipe upstream of the compressor leads to an increase of the surge margin. This effect does not depend on the orientation of the bend. Comparisons with a straight inlet with artificially generated pressure losses indicate that the increase in operating range is an effect of the pressure losses generated in the bend.

NOMENCLATURE

Latin letters

A	Cross section area
B	Greitzer B parameter
D	Diameter
f	Fanning friction factor
L	Length
N	Impeller speed [rpm]
MF	Mass flow
p	Pressure
P	Signal power
R	Radius
Re	Reynolds number
S	Power spectral density
T	Temperature
TPR	Total pressure ratio
TTR	Total temperature ratio
U	Impeller tip speed [m/s]

Greek letters

η	Isentropic efficiency
κ	Ratio of specific heats
ψ	Pressure parameter $\Delta p/(0.5\rho U^2)$
ϕ	Mass flow parameter $MF/(A\rho U)$
ρ	Density
ω	Helmholtz angular frequency

Indices

1	Before compressor
2	After compressor
C	Compressor
c	corrected
p	Pressure signal
T	Throttle
t-t	Total-to-total
ref	Reference value

1 INTRODUCTION AND BACKGROUND

The operating range of turbocharger compressors is an important factor in modern boosted internal combustion engines. The compressor must be able to deliver mass flow and boosting pressure as demanded by the engine operating point. The mass flow is approximately proportional to the engine speed for constant brake mean effective pressure (BMEP). A higher BMEP, on the other hand, requires both a higher boosting pressure and a higher mass flow for a given fuel-to-air ratio.

The range of a turbo-compressor is limited by three factors. At high mass flows, transonic flow conditions may be reached. The flow gets choked, and this limit is called the choke line. At high impeller speeds, the compressor operation is limited by mechanical stresses. Finally, at low mass flows, the compression system becomes unstable, with the flow periodically flowing from the high-pressure volume at the compressor outlet towards the low-pressure volume at the compressor inlet. This pumping behaviour is called surge. The surge line limits the engine operating range at low speeds and high boosting pressure demands, i.e. high BMEP.

The impeller and the diffuser of a centrifugal compressor are usually designed for a certain mass flow at a certain rotational speed, the so-called aerodynamic design point. Deviations from this operating point will alter the flow path, increasing the likelihood for wall adverse pressure gradients and flow separation to develop. Naturally, this decreases compressor's efficiency. For very low mass flows, the losses can become so large that the pressure ratio decreases for decreasing mass flows, and the speedline slope becomes positive. As shown by Greitzer for axial compressors in (1; 2) and confirmed by Fink et al for centrifugal compressors (3), a positive compressor slope can lead to an unstable compression system, depending on the resistance of other elements in the flow loop, e.g. the throttle valve, and the energy that can be stored in the compressed volume between compressor and throttle. The pressure ratio and mass flow start oscillating, because the volume is filled and emptied with a characteristic low frequency. This pumping behaviour is called surge. A theoretical limit for the stability of a compression system derived in (1), using a static (Equation (1)) and a dynamic (Equation (2)) stability criterion, is

$$\frac{\partial \psi_C}{\partial \phi_C} \leq \frac{\partial \psi_T}{\partial \phi_T} \tag{1}$$

$$\frac{\partial \psi_C}{\partial \phi_C} \leq \frac{1}{B^2}\left(\frac{\partial \psi_T}{\partial \phi_T}\right)^{-1} \tag{2}$$

where ψ is the dimensionless pressure parameter and ϕ is the dimensionless mass flow parameter. The indices C and T refer to the compressor and the throttle (or other flow resistance downstream the compressor). B is a parameter that relates the pressure to the inertia forces in a duct, see (4). It is defined as

$$B = \frac{U}{2\omega_H L_c} \tag{3}$$

where U is the impeller tip speed, ω_H is the Helmholtz frequency of a system with equivalent duct and plenum dimensions, and L_C is the length of the compressor duct. A larger value of B makes a system more likely to enter surge operation. This theoretical model has been incorporated into larger engine models e.g. by Theotokatos and Kyrtatos in (5) and by Galindo et al in (6).

The upstream geometry of the piping leading to the compressor has been shown to influence the stability of the compression system. Kim et al (7) showed experimentally that a pipe bend upstream the impeller reduces the compressor efficiency, and concluded that this happens mainly due to deterioration of the impeller flow. More specifically for the surge limit, Galindo et al (8) showed that a tapered inlet, a bent pipe, and a volume at the inlet all improve the surge margin, especially at higher pressure ratios, and Serrano et al (9) investigated the effect of different elbows, with and without guide vanes, on pressure ratio and the surge limit.

In this paper the surge line dependency on a bended inlet pipe is further investigated by also changing the relative orientation of the bend to the compressor volute outlet. The objective is to determine whether this orientation has any effects on the lower mass flow half of the compressor map. In section 2, the experimental setup and the inlet geometries tested are explained. In section 0, the compressor maps measured for a straight and a bended inlet are given. The effects of the bend orientation relative to the compressor are investigated in section 3.2. Based on these preliminary results, a more detailed investigation into the surge line at high speeds using theoretical and experimental criteria is carried out in section 3.3. The results are concluded in section 4.

2 METHOD

2.1 Experimental setup

The tests were run on a Garrett GT1752 turbocharger, which was installed on a cold gas stand. The compressor impeller has an exducer diameter of 52 mm and a TRIM of

$$TRIM = \left(\frac{D_{inducer}}{D_{exducer}}\right)^2 = 0.53. \tag{4}$$

Its design mass flow is 115 g/sec. The six main blades and six splitter blades are backwards swept. It is followed by a vaneless diffuser and a volute. The turbine is fed with compressed air, which is heated to 65 °C in order to prevent icing after the expansion.

The air is induced by the compressor from the test chamber through a funnel when using the straight inlet pipe, or through a bellmouth for the bended pipe configurations. A schematic and sensor locations are shown in Figure 1. The straight pipe has a length of L = 17 D. Total temperature and pressure were measured at the inlet. For the configurations with a bended inlet, static wall pressure was also measured downstream the bend at two other circumferential positions (p_{1a}, p_{1b}) using Kulite WCT-312M (0...1.7 bar abs) sensors.

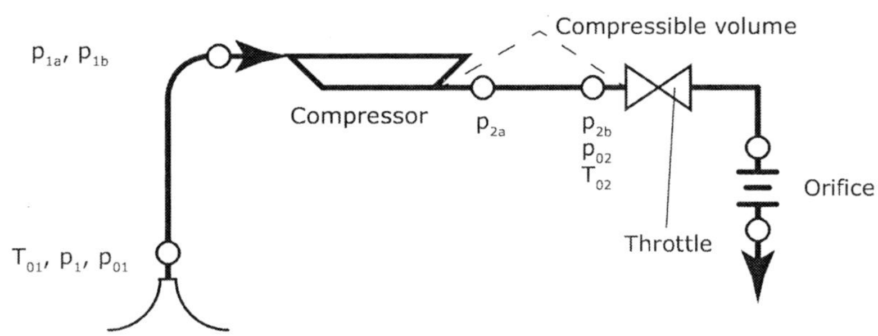

Figure 1: Schematic of the rest rig. Measurements are marked by white circles.

The compressor mass flow is throttled by a pneumatic valve located 16 D downstream the compressor. This valve position relatively close to the compressor was chosen in order to keep the volume small, $V \approx 1$ L, and limit the amplitude of surge cycle pressures. Total pressure and temperature are measured 10 D downstream the compressor; wall static pressure is measured both 3 and 10 D

downstream (p_{2a}, p_{2b}) using a piezoresistive Kistler 4045A5 (0...5 bar abs) sensor. The mass flow is measured with the help of an orifice downstream the pneumatic valve. An orifice was chosen to get very accurate mass flow measurements, at the cost of not being able to determine mass flow oscillations in surge.

The compressor maps were measured in accordance with the SAE testing standard for turbocharger maps (10), i.e. speedline wise from high to low mass flows. Since the current investigation is mainly concerned with surge, and a larger orifice would be needed for high mass flows, measurements were taken only at mass flows lower than the aerodynamic design mass flows. The last operating point displayed corresponds to the minimum mass flow before the compression system enters deep surge. Since the mass flow is not measured in a time-resolved manner, the common definition of deep surge as operation with an at times negative mass flow could not be used. A criterion based on the outlet pressure oscillation magnitude as described in section 0 is applied instead.

2.2 Inlet pipe geometries and orientation tested

The compressor inlet geometries tested are given in Figure 2.

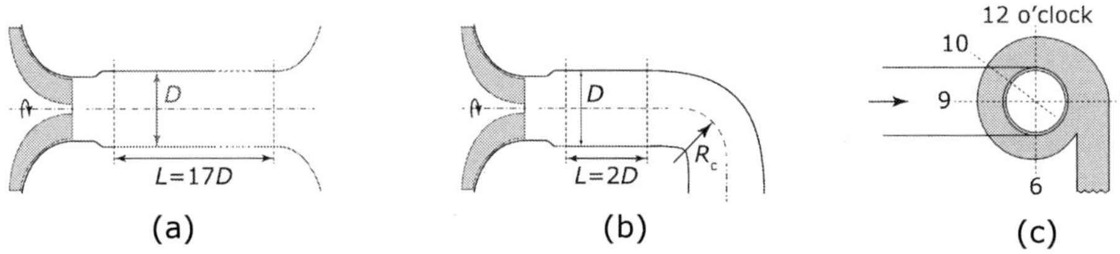

Figure 2: Inlet geometries that were tested. (a) shows the default straight inlet, while (b) and (c) show the bended inlet configuration and its clocking.

The baseline case shown in Figure 2 (a) consists of a straight inlet with an inner diameter of D = 44 mm, the same as the compressor inlet. For the next cases, a bent pipe is mounted 2D upstream the compressor inlet. The pipe has a ratio of inner radius to centreline radius of R_i/R_c = 1.04. It has the possibility to be rotated around the compressor axis, as shown in Figure 2 (c) for verifying "clocking" effects. In this paper, the results for the orientations 3, 6, 9, 10, and 12 o'clock were tested. Since the volute is not axisymmetric, with the tongue at a position of 4 o'clock, a non-uniform pressure distribution is expected for mass flows different from the design mass flow. The aim is to investigate whether any optimal positioning between inlet angle and volute outlet exists.

2.3 Repeatability

A compressor operating point was determined as steady state if the isentropic efficiency as calculated acc. to Equation (8) fluctuated with a maximum amplitude of 0.3 percentage points around a constant mean value. The sampling time for each point was two minutes. In order to estimate the repeatability of the measurements, some operating points were sampled at least five times over two days. The operating points were reached both by throttling from a higher mass flow, i.e. according to the SAE standard (10), and by de-throttling from lower mass flows. The mass flow standard deviation of the samples for one operating point was $s_{MFC} \approx 0.3...0.5\%$ of the mean (depending on the operating point), which is within the measurement accuracy of mass flow measurements using the pressure drop over an orifice. The total pressure ratio standard deviation of the samples was around $s_{TPR} \approx 0.15\%$, again within the measurement accuracy of the pressure transducer. There was no detectable difference between throttling and de-throttling towards a given operating point. The measurements are therefore deemed repeatable.

3 RESULTS

3.1 Compressor Maps

In this section, the compressor maps for the different inlet geometries are compared. The compressor maps are characterized by the corrected mass flow (MF_c), the corrected speed (N_c), the total pressure ratio from inlet to outlet (*TPR*), and the isentropic efficiency ($\eta_{is,t-t}$) defined by the relations (5), (6), (7), and (8), respectively.

$$\mathrm{MF}_c = \mathrm{MF}\cdot\frac{\sqrt{T_1/T_{ref}}}{p_1/p_{ref}} \tag{5}$$

$$N_c = \frac{N}{\sqrt{T_1/T_{ref}}} \tag{6}$$

$$\mathrm{TPR} = \frac{p_{02}}{p_{01}} \tag{7}$$

$$\eta_{is,t-t} = \frac{\mathrm{TPR}^{\frac{\kappa-1}{\kappa}} - 1}{\mathrm{TTR} - 1} \tag{8}$$

In the above relations, *MF* is the physical mass flow; T_1 and p_1 are the static inlet temperature and pressure; T_{ref} and p_{ref} are a reference pressure and temperature; *N* is the physical impeller rotational frequency; and *TTR* is the total temperature ratio T_{02}/T_{01}. The correction of mass flow and impeller speed is made to ensure Mach similarity, i.e. the same axial and circumferential Mach number irrespective of inlet conditions. The isentropic efficiency compares the actual enthalpy increase with that of an isentropic compression reaching the same pressure ratio.

Figure 3 shows a comparison between the map measured by the supplier on a hot gas stand (in blue), and the map measured on the cold gas stand, for a straight inlet with inner diameter D = 44 mm (in red). The speedlines for a configuration using a bended inlet pipe with a bend towards 9 o'clock, see also Figure 2 (c), are also drawn in black.

Comparing the hot and cold gas stand measurements using a straight inlet pipe, one can see that the *TPR* at high mass flows is higher on the cold gas stand than on the hot gas stand. This could be due to the lack of heat transfer from the turbine to the compressor, or even cooling of the compressor housing at higher speeds. Heat addition to a compressible flow reduces its total pressure, cp. a Rayleigh flow (11). The pressure difference is more pronounced for higher mass flows, which could be explained by the higher Nusselt numbers in both turbine and compressor, and thus higher heat transfer (12). The slope of the speedlines on the cold gas stand is close to zero or even positive, whereas the hot gas stand compressor map shows consistently negative slopes except at the very highest speedline. At speeds of $N_c \geq 160$ krpm, this leads to higher pressure ratios on the hot gas stand. A possible explanation could be that the acceleration of the flow due to the heat addition prevents local stall, since the flow is better aligned with the geometry designed for higher flow rates. The surge line given by the supplier is at higher mass flows than measured, though this could be purely an effect of a safety margin towards surge.

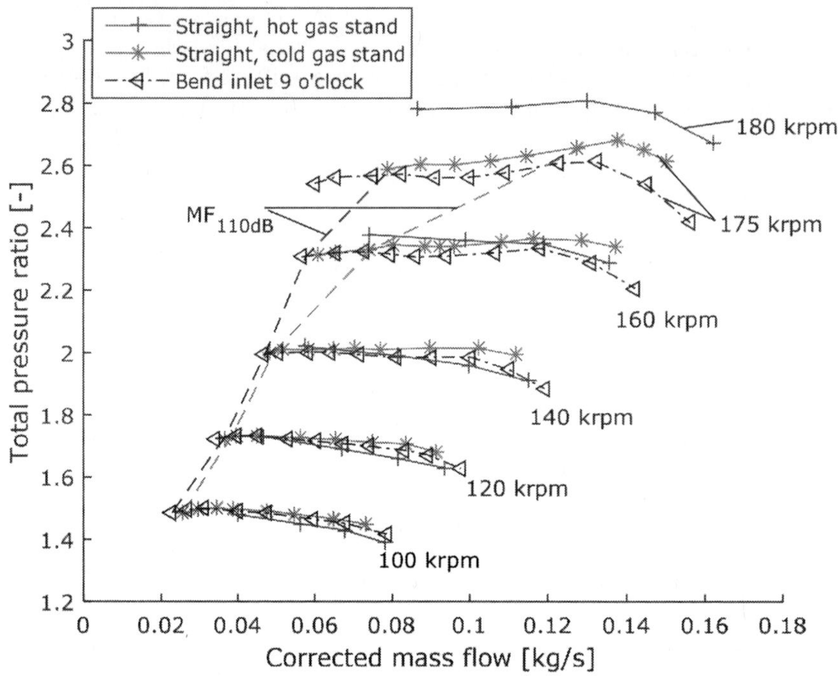

Figure 3: Left half of the compressor map measured on a hot gas stand (blue) and on the cold gas stand (red) with a straight inlet, and with a bend at the inlet on the cold gas stand (black).

As a surge criterion, the signal power of the static outlet pressure p_{2b} in the frequency interval [f_{surge} - 10 Hz; f_{surge} + 10 Hz] was calculated by integrating the power spectral density of the outlet pressure signal p_{2b}, S_{pp}:

$$P_{p2b} = \int_{f_{surge}-10\text{Hz}}^{f_{surge}+10\text{Hz}} S_{pp} \mathrm{d}f \tag{9}$$

This criterion thus is a measure of the outlet pressure oscillations in the frequency interval around the surge frequency. A reference pressure of 20 μPa was used to convert to decibel. A similar criterion was also used e.g. by Dehner in (13). The points shown in Figure 3 all have a value $P_{p2b} \leq 120$ dB, while operating points in deep surge are above this threshold. The dashed lines, coloured in red for the straight inlet and in black for the bended inlet, are the iso-lines $P_{p2cb} = 110$ dB. This value was chosen as an exemplary lower threshold. P_{p2b} is not monotonically increasing with reduced mass flows at some higher impeller speeds, which means that the compressor in some regions becomes more stable by decreasing the mass flow. In this case, the lines connect the largest mass flow where the value of 110 dB is reached. For low impeller speeds of 100 krpm, there is no significant difference in pressure ratio, but a slightly improved surge margin for the bended pipe configuration. At higher speeds and mass flows, the pressure ratio of the bended inlet configuration drops compared to the straight inlet configuration. The $P_{p2b} = 120$ dB surge line is moved towards lower mass flows by about 10% at low impeller speeds. For the highest impeller speed, the shift becomes much larger at about $\Delta MF_{120dB} = 20$ g/s, which amounts to 25% of the straight inlet surge line mass flow. Thus, the 110 dB surge line differs from the 120 dB line only at high impeller speeds. The difference between the straight and the bended inlet increases significantly from $N_c = 160$ krpm to $N_c = 175$ krpm. This confirms what was found by other authors as well (8; 9).

The drop in pressure ratio shown here is entirely the result of losses in the bend. If one corrects for those, the differences in pressure ratio disappear. Figure 4 shows the measured static wall pressures at 2 circumferential locations, 1 o'clock and 6 o'clock, for five different orientations of the bended pipe, namely 3, 6, 9, 10, and 12 o'clock (see also Figure 2 (c) for orientation purposes).

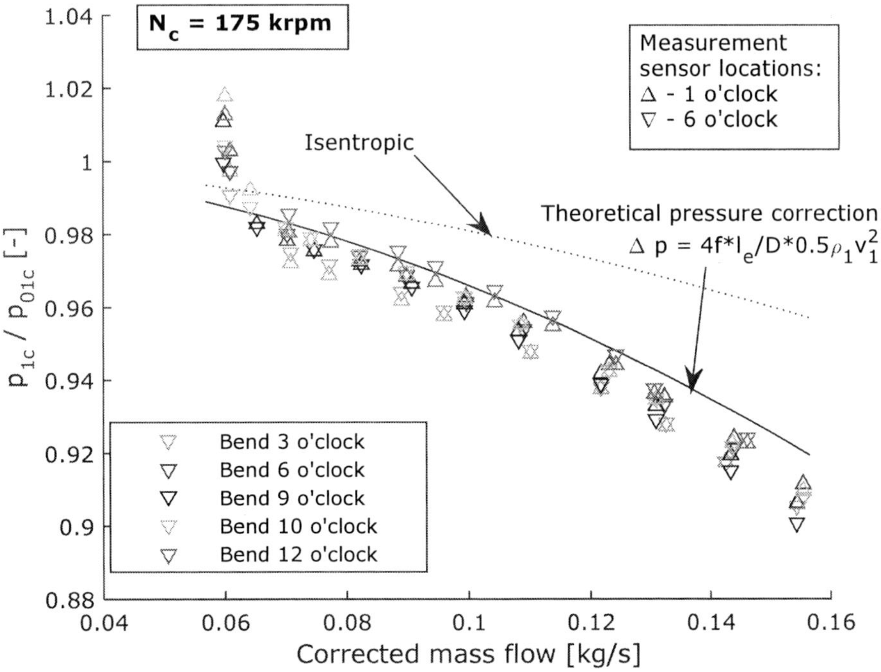

Figure 4: Static inlet pressure p_{1c} relative to test room stagnation pressure p_{01c} for N_c = 175 krpm.

Also plotted are the theoretical static pressure for an isentropic acceleration from stagnation to the bulk inlet velocity (which in turn is calculated from the measured mass flow), and a theoretical loss correction for bends using the equation proposed by Crawford et al in (14). For a given bend ratio R_i/R_c = 1.04, the equivalent pipe length may be estimated as

$$\frac{l_e}{D} = 1.25 \cdot \left(\frac{R_i}{R_c}\right)^{1.5} \cdot \mathrm{Re}_D{}^{0.35} \tag{10}$$

where Re_D is the Reynolds number based on pipe diameter. The pressure losses can then be calculated as

$$\Delta p = 4f \cdot \frac{l_e}{D} \cdot \frac{\rho}{2} v^2 \tag{11}$$

where ρ is the inlet density, v is the bulk velocity, and f is the Fanning friction factor, estimated here using the equation from Haaland in (15) for turbulent circular pipe flow:

$$\frac{1}{\sqrt{4f}} = -1.8 \cdot \log_{10}\left(\left(\frac{\varepsilon}{3.7D}\right)^{1.11} + \frac{6.9}{\mathrm{Re}_D}\right) \tag{12}$$

with a surface roughness ε. Figure 4 shows that this estimate gives a good fit to the measured pressures, although it slightly underestimates the losses for this bend at higher mass flows. Close to surge, the inlet static pressure is higher than the stagnation pressure in the test chamber. This is an indication of backflow from the impeller. Also noticeable is the fact that when this pressure increase is measured, the inlet pressure is higher at the circumferential position where the volute pressure is also expected to be higher. This points to a possible circumferentially uneven backflow from the compressor in operating regimes close to surge, since the backflow has a higher potential in the 1 o'clock position.

If the pressure ratio is corrected for the reduced inlet pressure using the equations above, the difference in *TPR* can be explained entirely.

3.2 Inlet bend orientation effects

Since the largest differences between a straight inlet and a bent inlet were detected at the higher impeller speeds, these speedlines were chosen to investigate the bend clocking (orientation) effects. Figure 5 shows the highest speedline N_c = 175 krpm for all five different bend orientations, 3, 6, 9, 10, and 12 o'clock. It shows that there is no difference outside the experimental uncertainty between the tested orientations, except at the peak *TPR*. Here, the orientation 9 o'clock results in a lower *TPR*, with differences up to $\Delta TPR = 0.02$, and a drop in efficiency by $\Delta\eta_{t\text{-}t} = 1\%$. These results could also be confirmed for the lower speedline of N_c = 160 krpm (not shown here). This drop in *TPR* could be related to the fact that in the 9 o'clock orientation, the highest velocities at the compressor inlet are at the same position where the highest pressures in the volute are expected. There is no discernible difference in the surge line as defined by the P_{p2b} = 120 dB threshold between the different orientations of the bended inlet.

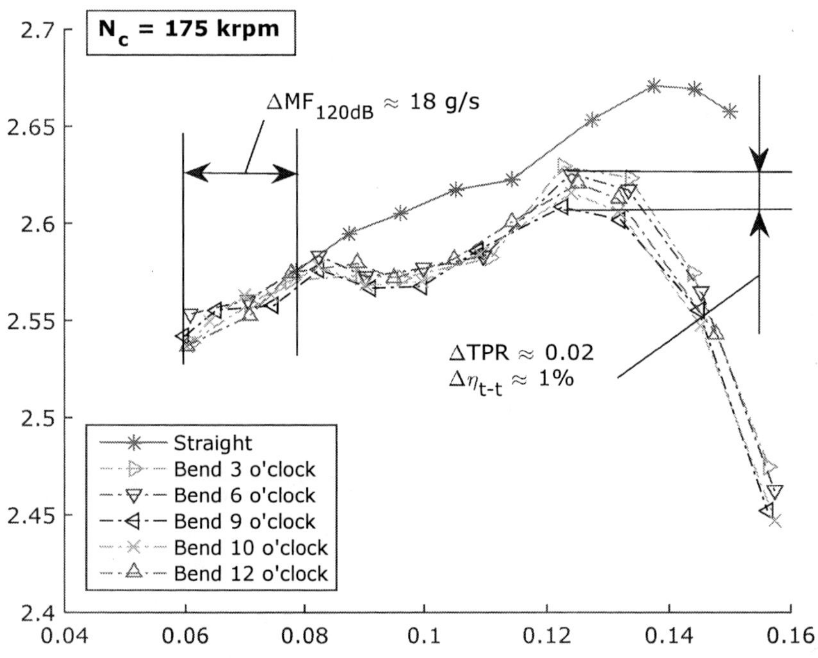

Figure 5: Speedline N_c = 175 krpm for the straight inlet and bend inlet configuration in different clocking positions.

This result suggests that axisymmetric effects of the bend on the inlet velocity field are at least partly responsible for the shift in surge margin. Those could be a static pressure drop, or a corresponding increase in the axial inlet velocity. The increase in the inlet velocity is almost completely accounted for, however, by the correction of the mass flow, which ensures Mach similarity. The pressure drop also leads to a small drop in the static temperature and thus lower speed of sound. If one assumes an adiabatic pressure drop in the bend, the difference in axial inlet velocity is of the order of magnitude of 1%, for the same corrected mass flow. This can be shown using Fanno flow relations. It is thus unlikely that the increase in axial velocity is the cause for this shift in surge margin.

Another possibility is the stabilizing effects of a pressure drop in the bend on the compression system as given by the static and dynamic stability criteria, Eq. (1) and (2). The pressure losses occur upstream of the compressor. They increase approximately quadratic with the inlet axial velocity, and thus the (physical) mass flow. Therefore, they decrease the positive slope of the combined bend and compressor system, and have a stabilizing effect on the compression system.

3.3 Compressor speedline slope effects

In order to test this hypothesis, the highest speedlines N_c = 160 krpm and N_c = 175 krpm were tested with the straight inlet where a pressure loss was created in the inlet by a diffuser with flow separation. The diffuser was located 12*D* upstream of the compressor, so that the inlet flow is fully developed. It was decided to design the diffuser to generate pressure losses that result in an approximately flat speedline near surge for the combination of pressure loss device and compressor, in order to test whether an upstream device can in fact dampen the oscillations that occur in a dynamically unstable system.

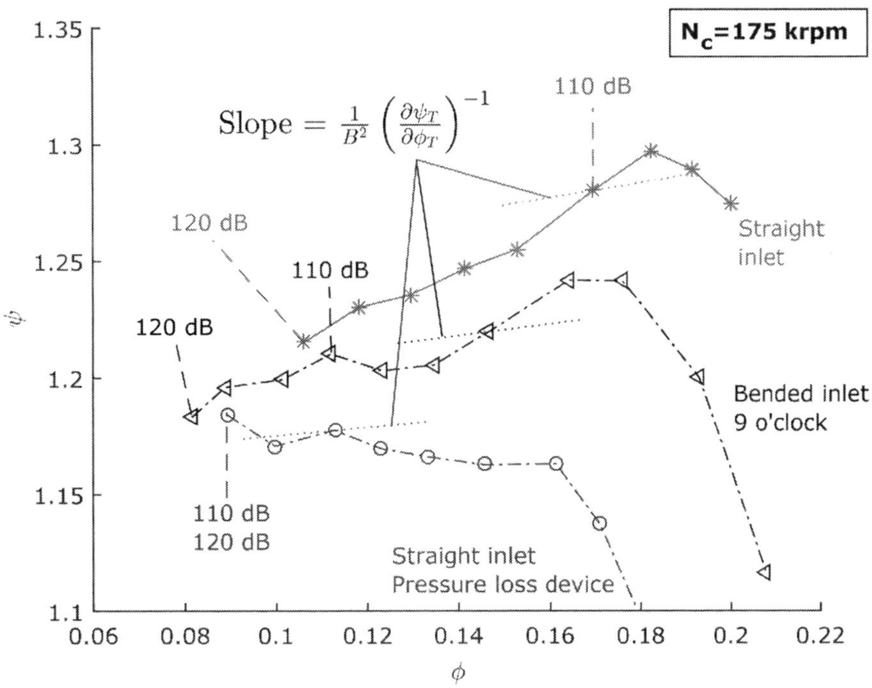

Figure 6: Surge criteria for the straight inlet, bended inlet 9 o'clock, and straight inlet with pressure loss device configurations.

Figure 6 shows the highest measured speedline N_c = 175 krpm in a non-dimensionalized ϕ-ψ compressor map for the straight inlet configuration (red stars), the bended inlet 9 o'clock configuration (black triangles), and the straight inlet configuration with the aforementioned pressure loss in the inlet pipe (blue circles). For each of these configurations, the limit where the outlet pressure fluctuation power P_{p2b} reaches 110 dB and 120 dB respectively is marked. For the straight inlet and the bended inlet, the results correspond to those shown in Figure

3. There are some pressure oscillations at surge frequencies for the positively sloped part of the speedline, and the leftmost shown point is the last point with $P_{p2b} \leq 120$ dB. For the straight inlet with a pressure loss device, the speedline is negatively sloped except at very low mass flows. This means that the system is theoretically stable for most parts of the measured range. The threshold of $P_{p2b} = 110$ dB is not reached at all, except for values of $\phi < 0.089$, i.e. to the left of the last shown point. The results for $N_c = 160$ krpm show similar behaviour and are therefore not shown here.

The static stability criterion, Equation (1), is fulfilled for all points shown in the figure, since the throttle slope is very steep. Figure 6 shows theoretical estimates of the surge line using the dynamical stability criterion from Equations (2) and (3). The Helmholtz frequency for the calculation of the B parameter was estimated using (see also Dehner et al (16))

$$\omega_H = a\sqrt{\frac{A}{L_c V_p}}, \qquad \frac{A}{L_c} = \sum_i \frac{1}{\frac{L_{c,i}}{A_i}} \tag{13}$$

where a is the speed of sound at the compressor outlet, A is the cross-sectional area and L_c the length in meridional direction of the component, and i counts through the inlet duct, the compressor and volute, and the outlet duct. The volume of the pipe between compressor outlet and throttle was used as plenum volume V_p. The frequencies estimated this way are within 10% of the frequencies measured in mild surge. The value of B at 175 krpm is 0.53 for the straight inlet configurations and 0.55 for the bended inlet configurations (because the inlet duct is slightly shorter). The throttle slope $\partial\psi_T/\partial\phi_T$ was estimated assuming that only the pressure after the throttle and the temperature before the throttle are constant close to the measured operating point, so that only the pressure before the throttle changes. In Figure 6, the critical slope of the dynamic stability criterion is drawn at the point where it is first breached for each speedline. This happens when the speedline slope turns positive in all configurations. Thus, using the strict model criteria, the surge limit at the highest speedline would be located at a mass flow with a local *TPR* peak. Due to the large losses created by the pressure loss device, the peak is shifted towards low mass flows, and the outlet pressure oscillations are much weaker.

Setting the threshold at $P_{p2b} = 120$ dB, one can see that while the pressure loss device in a straight inlet increases the operating range, the increase is less than that for the bended inlet. Therefore one can conclude that the speedline slope explains the milder oscillations of the compressor outlet pressure and a part of the shift of the onset of deep surge. It does not fully explain the increased operating range, however. The axis-asymmetric flow field following a bend is likely also partly responsible for this effect.

4 CONCLUSIONS

The left half of the compressor map was compared for two inlet configurations: A straight inlet pipe and a bended pipe. Bended pipe orientation effects relative to the compressor axis were investigated. The losses in the bend lead to a reduced total pressure ratio for high mass flows. The reduction in peak *TPR* is almost independent of the clocking position, only the 9 o'clock position consistently shows slightly lower *TPR* than other positions. Considering the surge line, the effects of the bend are strongly dependent on the definition of surge. With a theoretical criterion, the compression system becomes unstable as soon as the speedline slope turns positive. This gives an increase in the surge margin for a bended inlet as compared to the straight pipe inlet. When using a measure of the compressor outlet pressure signal power, P_{p2b}, the result is also an increase in the surge margin; the exact value of the increase depends on the chosen threshold.

A comparison with a straight inlet with a pressure drop device indicates that the increase in surge margin for the bended inlet is mainly a result of the stabilization of the compression system through the pressure losses in the bend. Since this also results in significantly lower boost pressures at high mass flows, however, it would likely not be beneficial to install such a system from an engine perspective. The results suggest that in SI engines with a throttle upstream the compressor, the throttle could be used to sacrifice boost pressure in order to suppress intake manifold pressure oscillations.

ACKNOWLEDGMENT

The authors would like to thank Dr. Nils Tillmark at the Department of Mechanics, KTH Royal Institute of Technology, for his able support in setting up and running the experiments. The support from the Swedish Energy Agency, Scania AB, Volvo Car Corporation, Volvo Group Trucks Technology and BorgWarner Turbo Systems is greatly appreciated.

REFERENCE LIST

1. Greitzer, E. *Surge and Rotating Stall in Axial Compressors I: Theoretical Compression System Model.* 1976, Journal of Engineering for Gas Turbines and Power, Vol. 98, pp. 190-198. DOI: 10.1115/1.3446138.
2. Greitzer, E. *Surge and Rotating Stall in Axial Compressors II: Experimental Results and Comparison with Theory.* 1976, Journal of Engineering for Gas Turbines and Power, Vol. 2, pp. 199-211. DOI: 10.1115/1.3446139.
3. Fink, D.A., Cumpsty, N.A. and Greitzer, E. *Surge Dynamics in a Free-Spool Centrifugal Compressor System.* 1992, Journal of Turbomachinery, Vol. 114, pp. 321-332. DOI: 10.1115/1.2929146.
4. Greitzer, E. *The Stability of Pumping Systems - The 1981 Freeman Scholar Lecture.* 1981, Journal of Fluids Engineering, Vol. 103, pp. 194-242.
5. Theotokatos, G. and Kyrtatos, N. P. *Diesel Engine Transient Operation with Turbocharger Compressor Surging.* 2001. SAE Technical Paper 2001-01-1241. DOI: 10.4271/2001-01-1241.
6. Galindo, J., Serrano, J. R., Margot, X, et al. *Experiments and modelling of surge in small centrifugal compressor for automotive engines.* 2008, Experimental Thermal and Fluid Science, Vol. 32, pp. 818-826. DOI: 10.1016/j.expthermflusci.2007.10.001.
7. Kim, Y., Engeda, A., Aungier, R. and Direnzi, G. *The influence of inlet flow distortion on the performance of a centrifugal compressor and the development of an improved inlet using numerical simulations.* 2001, Journal of Power and Energy, Vol. 215 (A), pp. 323-338. DOI: 10.1243/0957650011538550.
8. Galindo, J., Arnau, F. J., Tiseira, A, et al. *Measurement and Modeling of Compressor Surge on Engine Test Bench for Different Intake Line Configurations.* SAE Technical Paper 2011-01-0370. DOI: 10.4271/2011-01-0370.
9. Serrano, J. R., Margot, X., Tiseira, A. and Garcia-Cuevas, L. *Optimization of the inlet air line of an automotive turbocharger.* 2013, International Journal of Engine Research, Vol. 14, pp. 92-104. DOI: 10.1177/1468087412449085.

10. SAE. Turbocharger Gas Stand Test Code. *Turbocharger Gas Stand Test Code.* 1995.
11. Anderson, J.D. *Modern Compressible Flow: With Historical Perspective.* 3rd. Edition: McGraw-Hill, New York, 2004.
12. Sirakov, B. and Casey, M. *Evaluation of Heat Transfer Effects on Turbocharger Performance.* 2013, Journal of Turbomachinery, Vol. 135 (2). DOI: 10.1115/1.4006608.
13. Dehner, R., Figurella, N., Selamet, A., et al. *Instabilities at the Low-Flow Range of a Turbocharger Compressor.* 2013, SAE International Journal of Engines, Vol. 6. DOI: 10.4271/2013-01-1886.
14. Crawford, N.M., Cunningham, G. and Spence, S.W.T. *An experimental investigation into the pressure drop for turbulent flow in 90 deg elbow bends.* 2007, Proceedings of the IMechE, Part E: Journal of Process Mechanical Engineering, Vol. 221, pp. 77-88. DOI: 10.1243/0954408JPME84.
15. Haaland, S.E. *Simple and Explicit Formulas for the Friction Factor in Turbulent Pipe Flow.* 1983, Journal of Fluids Engineering, Vol. 105(1), pp. 89-90. DOI: 10.1115/1.3240948.
16. Dehner, R., Selamet, A., Keller, P. and Becker, M. *Prediction of Surge in a Turbocharger Compression System vs. Measurements.* 2011, SAE International Journal of Engines, Vol. 4 (2). DOI: 10.4271/2011-01-1527.

Influence of the hot gas test bench piping system on the surge line of an automotive turbocharger compressor

C Schäfer, I Sandor, M Klaus
Continental Automotive GmbH, Germany

R Baar
Fachgebiet Verbrennungskraftmaschinen, Technische Universität Berlin, Germany

ABSTRACT

Due to the demand for higher specific power and performance of modern gasoline internal combustion engines, turbocharger compressors are forced to be operated closer to the surge line. Hence, knowledge of the effects leading to this operating boundary becomes even more of interest.

Because literature on the influence of the inlet ducting on the surge line is rare, in this study the dependency of the surge line on the intake piping system is investigated based on carefully conducted hot gas test bench measurements. Furthermore, the combined influence of different pipe volumes downstream and upstream of the centrifugal compressor is shown. Possible reasons for the observed results are discussed and in addition, the feasibility of simulating the effect employing an adjusted Greitzer/Theotokatos model is assessed.

The goal is to contribute to the development of a model, which allows the quantitative comparison of surge lines measured at different test beds in order to ease the turbocharger matching for engine developers and to estimate the influence of engine packaging on the surge line of turbocharger compressors.

NOMENCLATURE

A	area	$\dot{m}$	mass flow
a	speed of sound	p	pressure
c	velocity	r	radius
d	diameter	R	gas constant
f	frequency	T	temperature
F	force	U	blade speed
L	length	V	volume

Greek Letters

κ	specific heat ratio	τ	time constant
Π	pressure ratio	σ	standard deviation
μ	expected value	ω	Helmholtz frequency

Subscripts

amb	ambient	φ	circumferential
ax	axial	p	pressure
C	compressor	r	radius
corr	corrected	ref	reference
ED	exhaust duct	rel	relative
h	throat	R	radial
ID	intake duct	s	static
imp	impeller	SL	surge line
ini	initial condition	SS	steady state
inl	inlet	t	total
Norm	normalized	T	Throttle

1 INTRODUCTION

Modern downsized and turbocharged gasoline internal combustion engines demand for higher specific power and performance. This means for the torque characteristic of these engines, that the so called low end torque is increased and shifted towards lower engine speeds. As a result, the respective turbocharger compressor is operated at higher pressure ratios and lower mass flow rates at the same time which means an operation closer to the surge limit [1]. Hence, for the selection of a turbocharger for a particular engine, the actual surge line in the given engine package is highly relevant. Therefore, understanding of the formation of this operating boundary as well as of the effects affecting it becomes even more crucial for automotive turbocharger compressor design.

The surge line of a centrifugal compressor is not just determined by the aerodynamic design of the compressor stage itself, but as well by the system it is operated in and furthermore by design details like the recirculation valve cavity or inlet elbows [2]. Further possibilities of affecting the surge line are given in [3]. Within this work it is outlined, that different compressor test bench pipings inhibit a quantitative comparison of diverse compressors, which is fundamental for turbocharger selection. But on the other hand, identical test stand piping for the complete industry and affiliated research institutions would be a very hard demand due to the variety of different test bed types and facility limitations. These can result on the inflow side from different air mass flow measurement techniques, possible inflow conditioning or noise control regulations and on the pressurized side mainly from test stand space availability. Often in industrial test centers, a high degree of test bench utilisation is of great interest. Therefore, turbines are mounted to the hot gas producer and the exhaust duct in a standardized way. On the compressor side piping is adjusted respectively employing standardized elements in order to not adjust the turbocharger assembly itself. Hence, a reliable model to convert the measurement results of various test benches into compressor maps according to a standard setup would be very helpful for engine developers in order to quantitatively compare them correctly. Furthermore, such a model would help assessing the influence of the actual engine packaging and to customize the turbocharger compressor best possible.

The first step for the development of a respective model is the understanding and modeling of the influence of the compressor ducting system. Much work has been done investigating the influence of the piping downstream of the compressor, mainly based on the well known Greitzer study ([4] and [5]). Hence, the effect of the pressurized side of the compression system on the stability limit is reasonably known. But in contrast to this, literature regarding the influence of the intake piping apart from inflow bent pipes or inlet restrictions is rare [6]. From the simulation point of view, Theotokatos and Kyrtatos [7] suggested a Greitzer-based model including a representation of the inlet ducting, but its influence has not explicitly been investigated. Nevertheless, this model is often implemented in engine simulation tools ([8] and [9]).

Hence, the influence of inflow and pressurized side piping on the surge line of a small centrifugal compressor for automotive applications will be shown employing hot gas measurements. Therefore, the experimental facility, investigated piping geometries and the measurement approach will be introduced at first. Then, the experimental results will be presented. Within this context, the precision of the measurement results and hence their reliability will be assessed and the influence of an employed inflow elbow as a possible source of error will be discussed. Afterwards, the feasibility of simulating the respective alterations using the common Greitzer/Theotokatos approach will be assessed. This part is divided into two sections. At first, the numerical model and the simulation approach will be presented. Secondly, the computational results will be shown and compared to the measurement results. Finally, a conclusion will be given and future work will be discussed.

2 EXPERIMENTAL FACILITY

The measurements for this study have been conducted at the hot gas test stand of the chair for internal combustion engines of TU Berlin (Berlin institute of technology). At this test bed hot gas is delivered from a burning chamber fired with diesel fuel to propel the turbine. The compressor mass flow is measured via an inflow nozzle. Static pressure is measured via four bores and averaged hydraulically in a measurement pipe. One of these is located upstream and downstream of the compressor each. They are employed for temperature measurement as well, averaging signals of three gauges. Compressor throttling is done via two parallel valves, one for coarse and one for fine adjustment of the operating point each. Details on the experimental facility can be found in [10].

A small centrifugal compressor for automotive applications was investigated. It consists of a milled aluminum compressor wheel and an aluminum high pressure die cast housing. The turbomachine has been designed for a modern three-cylinder gasoline engine with a displacement of one liter.

As surge line identification is crucial for the current study, this operating boundary has been detected carefully via a slow approach in single throttle valve position steps. Surge has been diagnosed acoustically and in addition via the temperature signal of the measurement pipe upstream of the compressor. Here, a very accurate distinction between instabilities and surge could be made, because in the event of surge a significant temperature increase could be recognized due to back flow of compressed air.

Table 1: Investigated piping geometries

geometry	P1 small	P1 large	P2 small	P2 large
Abbreviation	P1s	P1L	P2s	P2L
Volume [L]	0.8	17.3	1.8	13.2

For suction (P1) and pressurized side (P2) of the compression system a large (L) and small (s) geometry has been investigated each, details can be found in Table 1. Hence, four different combinations P1-P2 have been measured and a principle sketch of each can be found in Figure 1, employing colours which will be used to identify the respective setup further. The P2 volume was mainly determined by a long flexible hose connecting the compressor outlet and the throttle valve. In the case of the small volume, an additional throttle valve has been located upstream of that hose reducing the P2 volume. The small P1 volume has been formed by a measurement pipe and the intake nozzle for mass flow determination. In order to increase the ducting volume, additional pipes have been integrated. Because axial space upstream of the compressor inlet was limited by the test bed walls and due to a desired large volume, a 90° bend had to be employed, as shown in Figure 2.

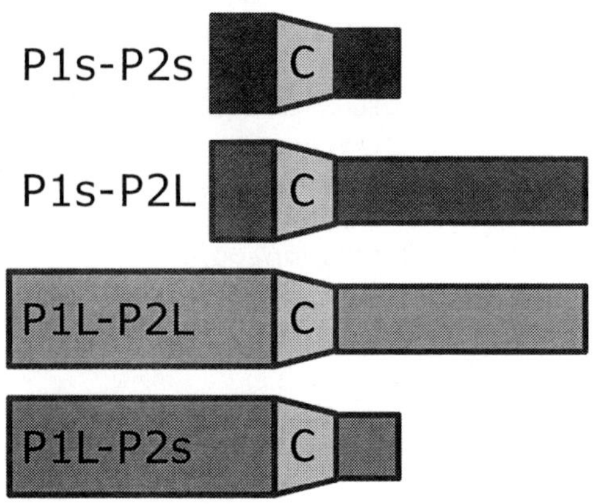

Figure 1: Schematic illustration of investigated geometry setups

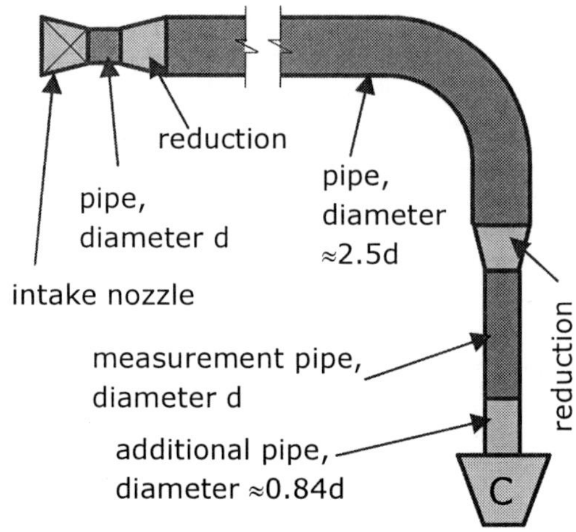

Figure 2: Sketch of large P1 volume (top view, no scale)

3 MEASUREMENT RESULTS

3.1 Precision of experimental surge line determination

An accurate surge line determination is limited due to several errors influencing the results. As for every measurement, uncertainties of the measurement equipment affect the determination of mass flow, pressure ratio and all other quantities of interest. But in addition to this, the accuracy and repeatability is affected by specific phenomena. Often the surge line can be found in a map region, where the pressure build-up characteristic of the compressor has a mathematically positive slope, i.e. pressure ratio decreases with decreasing mass flow. As a result, operating such a compressor near or at the surge line, the working point can be already slightly instable. This effect is often accompanied by rotating stall phenomena [11], affecting the determination of the exact surge line additionally. Furthermore, very slight changes in the flow field or ambient conditions can lead to self-induced surge at these points while waiting for reasonably steady conditions of the measured temperatures. Another problem of accurate surge line determination is the throttle valve adjusting the compressor operating point. Such valves can usually not be operated completely continuous due to the drive via gear wheels. Hence, the actual surge point can be located between the last stable point (valve position N) and the consecutive valve position (N+1).

In order to assess the surge line measurement determination accuracy, results of a sample test are employed. For several blade speeds, the surge line has been measured repetitively, always approaching slowly and in single throttle valve position steps from a definitively stable point. The results of an example speed line are presented in Figure 3 in comparison to a Gaussian probability distribution obtained from the respective measurement results. Mass flow is normalized with the mass flow of the expected value μ for non-disclosure reasons and the sample count of the shown example is 16. It can be seen from the image that the assumption of a Gaussian distribution is plausible, although a higher extent of the sample would probably have made this clearer. From the results of all blade speeds investigated concerning surge line measurement accuracy, a maximum standard deviation of σ=1.5 % in mass flow has been found. Hence, a surge line alteration magnitude of more than 3 % in mass flow rate measured with a different piping can be regarded as an actual surge line shift with 95 % confidence (2σ neighbourhood).

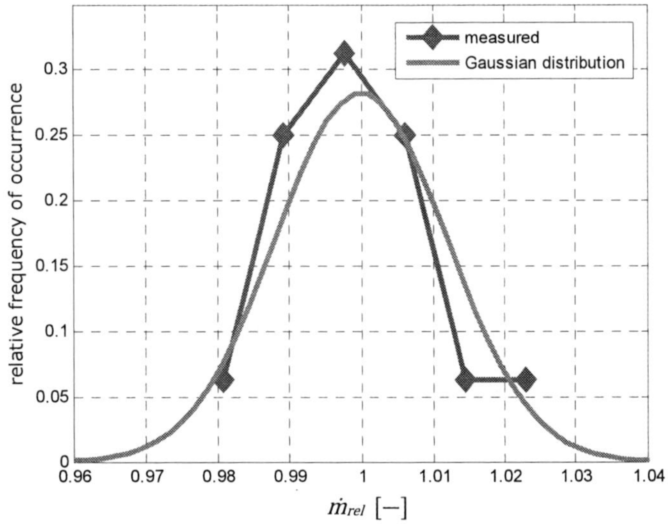

Figure 3: Frequency of occurrence of measured surge point mass flows for an example speed line

3.2 Geometry variation results

The measurement results for the different configurations explained in section 2 are presented in the form of a compressor map in Figure 4. Mass flow rates and pressure ratios are normalized employing the respective values of an arbitrary map point for non-disclosure reasons. Although the piping is different, the compressor characteristics are not altered, the difference is only the surge line and hence, the usable range of the map. This is in good agreement with Greitzer's theory [4] that changing the system and retaining the compressor may allow an operation at lower mass flows or limit it to higher ones on exactly the same characteristic.

For surge line variation assessment, a clearer representation was chosen based on the relative surge line improvement being defined as:

$$relative\ surge\ line\ improvement = \left.\frac{\dot{m}_{SL,ref} - \dot{m}_{SL}}{\dot{m}_{SL,ref}}\right|_{@actual\ blade\ speed} \quad (1)$$

with the reference case being P1s-P2L and shown in per cent. Hence, positive values indicate an improved surge line (lower mass flow) and negative ones a deteriorated surge limit. This post processing approach is illustrated in Figure 5 for a single operating point. Respective results for all measurement data are presented in Figure 5 (right) versus blade speed normalized by maximum blade speed as well. Furthermore, the three per cent confidence interval introduced in section 3.1 is illustrated. Only surge line alterations outside of this interval can clearly be regarded as an actual surge line shift. It should be noted that comparing the cases with small P2-volume directly, such a three per cent confidence interval has to be considered as well.

For a reduced P2-volume an at least slightly improved surge line can be observed from Figure 5 in any case and for every blade speed. Regarding P1 influence, for low and medium blade speeds up to U_{norm}=0.7 an unaffected or even slightly deteriorated surge limit can be found considering the three per cent confidence neighbourhood. Furthermore the values for the lowest blade speed should not be overstated because the absolute surge mass flows are very low for that speed.

Hence a small absolute mass flow change leads to a high change in relative surge line improvement (compare equation 1). Considering higher blade speeds an improved surge line can be observed. In particular for U_{norm}=0.84 an U_{norm}=0.9 a significant gain of map range can be achieved in the combination with the small P2-volume.

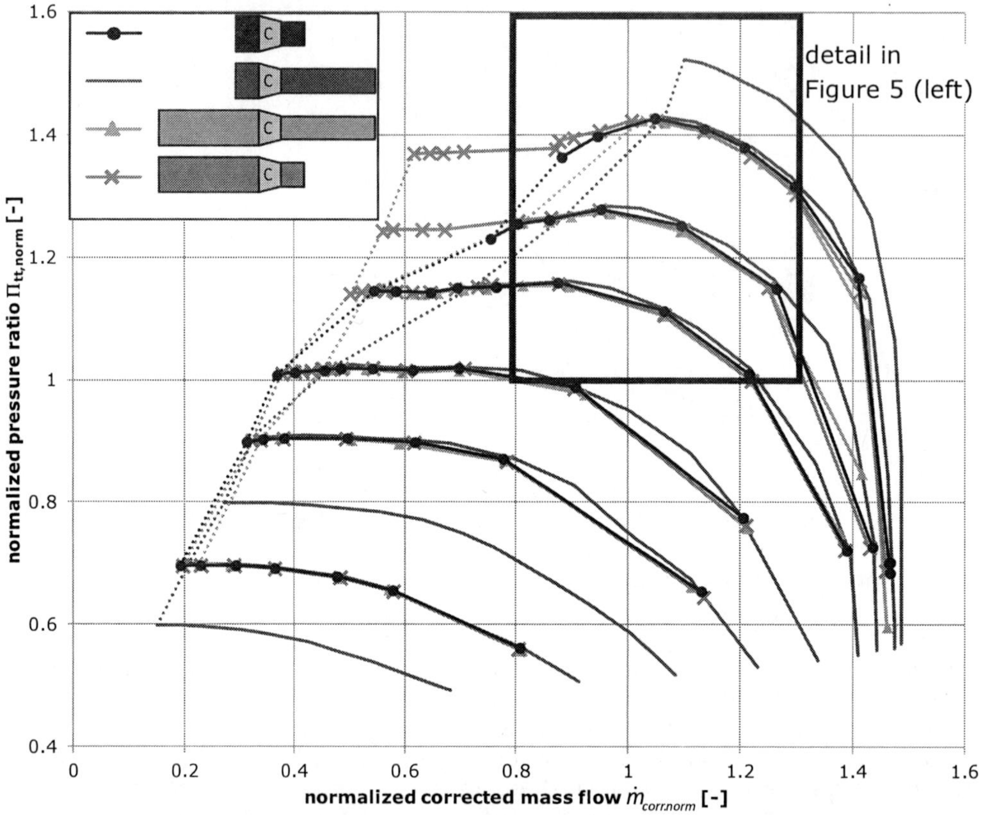

Figure 4: Piping geometry variation results – compressor map illustration

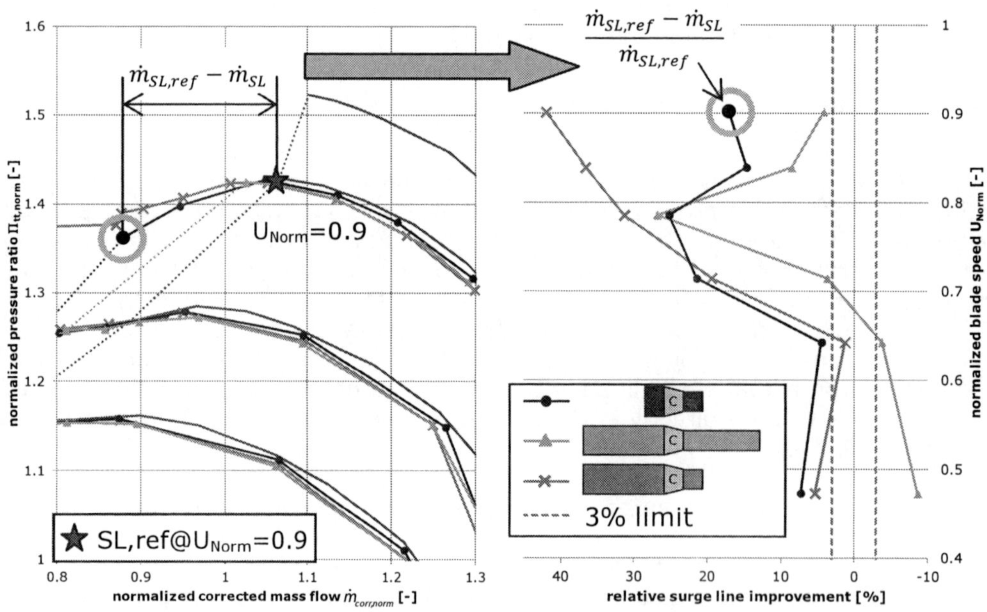

Figure 5: Surge line improvement – explanation of post processing (left and right) and measurement results (right)

A large P1-volume works as an additional energy storing element and can therefore damp possible surge or pre-surge oscillations from downstream, which may be a reason for the influence on surge line location. From the given test setup and configuration no further information on the physical background could be obtained. Possible explanations are discussed in sections 4 and 5.

3.3 Inlet elbow influence estimation

As shown by Müller [2], an inlet elbow can affect the surge line. In Figure 2 it was illustrated that such a bend had to be implemented in the P1L setup due to the reasons explained in section 2. It may be argued that the surge line improvement may result from that element. Hence, in this section its influence is estimated.

Studies found in literature dealing with the issue mostly investigated inlet elbows close to the impeller. Müller [2] observed only a slight surge line alteration, but the effect on the compressor characteristics is not negligible. A more detailed study on this topic has been conducted by Ueda et al. [12]. Here, a remarkable pressure loss was found employing the inlet elbow, but no effect on the surge line was reported at all. Because in the current study no noticeable alteration of the compressor characteristic could be observed, no effect on the surge line may be assumed as well.

In order to disprove the bend argument further, CFD results have to be considered. The full P1L piping has been simulated for the choke mass flow and was compared to a simple bend with the diameter of the measurement pipe (compare Figure 2) at the same mass flow. As a result, the pressure distribution of both cases is shown in Figure 6 in combination with tangentially projected vectors representing the secondary flow. The axial position is right downstream of the elbow for the simple bend, where the compressor inlet would be located in studies comparable to the one of Ueda et al. [12], and close to the end of the measurement pipe for the P1L case. In both images the static pressure is scaled identically, but the length of the vectors does not scale with the velocity magnitude due to visibility. For the P1L case it can clearly be seen that due to the acceleration in the pipe reduction and losses in the piping the intensity of the secondary flow is significantly reduced and the typical flow pattern is vanishing. This statement is quantified in Table 2 by results obtained from analysis planes shown in Figure 6.

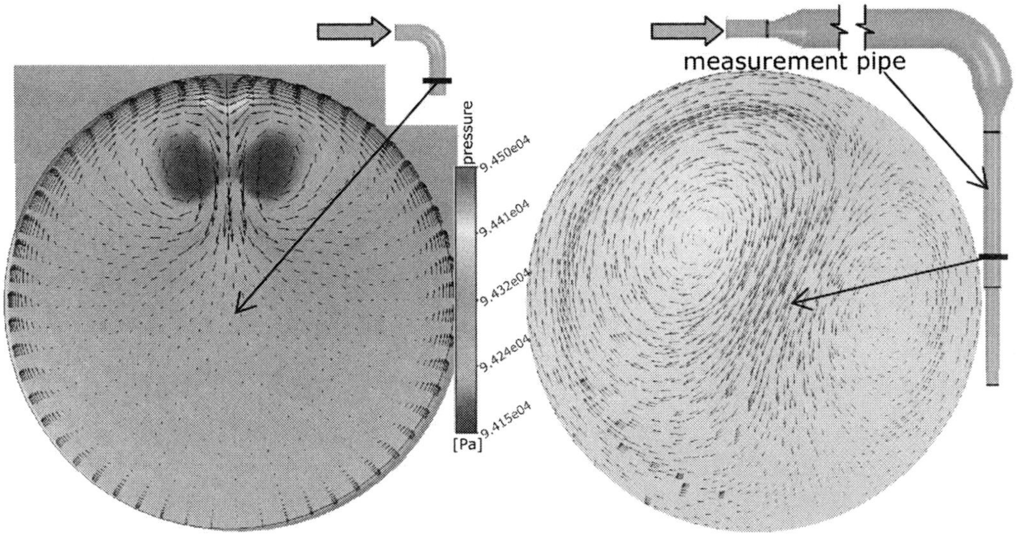

Figure 6: Flow field comparison of simple bend (left) and P1L setup (right)

The effect of secondary flow caused by typical bends (compare [12]) is that the impeller leading edge velocity triangles are changed locally and hence Euler work as well. Therefore, blade loading distribution is altered leading to a different separation behaviour and finally resulting in a possibly shifted surge line. Secondary flow intensity is reduced significantly in the P1L piping in comparison to the simple bend as discussed. Hence, a negligible effect of the inflow bent pipe on surge line can be assumed for the measurements presented in this study.

Table 2: Quantitative comparison of simple bend and P1L configuration (analysis planes identical to Figure 6)

quantity	$\frac{p_{s,min}}{p_{s,mean}}$	$\frac{p_{s,max}}{p_{s,mean}}$	$\frac{\lvert c_\varphi \rvert_{max}}{c_{ax,mean}}$	$\frac{\lvert c_R \rvert_{max}}{c_{ax,mean}}$
simple bend	0.99817	1.00189	0.25196	0.33697
P1L configuration	0.99991	1.00006	0.05509	0.05962
reduction [%]	95.1	96.8	78.1	82.3

The CFD results support the observations from the compressor characteristics and the conclusions from literature. Due to acceleration and losses, the secondary flows are reduced significantly and no influence of the bend on the compressor characteristics can be observed. Therefore, it can be stated that the bend in the P1L inflow piping should have a negligible influence on the surge limit. Hence, surge line alterations should be caused by the piping and not by the bend, supporting the conclusion that inflow piping has an effect on the surge line.

4 NUMERICAL MODEL AND SIMULATION APPROACH

As explained in the introduction, a first step for a model enabling the quantitative comparison of compressor maps measured at different test beds is the prediction of piping influence on surge line. Therefore, the investigated model will be presented in this section. In the following section 5, the numerical results will be compared to the measurements presented in section 3 and the prediction accuracy will be assessed.

The well known Greitzer model [4] was used as a baseline and results will be shown in the next section, but it does not model the intake piping and hence, it will not be discussed here in more detail. Theotokatos and Kyrtatos [7] introduced a formulation for the intake ducting, which is motivated by Greitzer's simple transient response model. Furthermore, they simulated pressures instead of pressure differences and employed the pressure ratio directly in the transient response equation. In the current study, rotational speed has been assumed constant, because according to Fink et al. [13] the influence on the surge line is negligible. Applying the non-dimensionalization rules of Greitzer [4], the resulting equation system is the following:

$$\frac{d\widetilde{\dot{m}_C}}{d\tilde{t}} = B \cdot (\widetilde{p_{ID}} \cdot \Pi - \widetilde{p_{ED}}) \qquad \textbf{(2)}$$

$$\frac{d\widetilde{\dot{m}_T}}{d\tilde{t}} = \frac{B}{G} \cdot \left(\widetilde{p_{ED}} - \left(\frac{A_C^2}{A_T^2} \cdot \widetilde{\dot{m}_T}^{\,2} + \widetilde{p_{amb}} \right) \right) \qquad \textbf{(3)}$$

$$\frac{d\Pi}{d\tilde{t}} = \frac{1}{\tilde{\tau}} \cdot (\Pi_{SS} - \Pi) \qquad \textbf{(4)}$$

$$\frac{d\widetilde{p_{ED}}}{d\tilde{t}} = \frac{1}{B} \cdot (\widetilde{\dot{m}_C} - \widetilde{\dot{m}_T}) \qquad \textbf{(5)}$$

$$\frac{d\widetilde{\dot{m}_{ID}}}{d\tilde{t}} = \frac{1}{\tilde{\tau}_{inl}} \cdot \left(\widetilde{\dot{m}_{ID,SS}} - \widetilde{\dot{m}_{ID}} \right) \qquad \textbf{(6)}$$

$$\frac{d\widetilde{p_{ID}}}{d\tilde{t}} = \frac{1}{B} \cdot \frac{V_{ED}}{V_{ID}} \cdot (\widetilde{\dot{m}_{ID}} - \widetilde{\dot{m}_C}) \qquad \textbf{(7)}$$

B is the so called Greitzer parameter and G is a geometry parameter, both being defined in equations 8 and 9 respectively. The definition of the time constant τ can be seen in dimensional form in equation 10. In the original Greitzer work [4] the included quantity of DF was interpreted as the number of revolutions it takes to fully develop instabilities leading to surge or stall. As suggested for the first time by Hansen et al. [14], this factor can be adjusted to match the experimental results better and hence, this factor was regarded as a model factor in the current study and will further be called damping factor DF. Details on the steady state mass flow in the intake duct $\dot{m}_{ID,SS}$ can be found in Theotokatos and Kyrtatos [7] and the inlet time constant τ_{inl} will be discussed below.

$$B = \frac{U}{2a} \cdot \sqrt{\frac{V_{ED}}{A_C \cdot L_C}} = \frac{U}{2 \cdot \omega \cdot L_C} \quad \textbf{(8)} \qquad G = \frac{L_T \cdot A_C}{L_C \cdot A_T} \quad \textbf{(9)} \qquad \tau = \frac{DF \cdot 2\pi \cdot r_C}{U} \quad \textbf{(10)}$$

Integration of the equation system has been conducted in Matlab R2013a employing the ODE45 solver similar to Morini et al. [15]. For the determination of the steady state pressure ratio the measured map with the best surge line (P1L-P2s) was used as baseline. This map was extrapolated by a 3rd order polynomial analog to Morini et al. [15] for positive mass flow rates and by a quadratic approach for back flow. The result is illustrated in Figure 7. Depending on the numerical behaviour it was necessary to adapt the initial conditions for each blade speed. For some rotational speeds an initial operating point of $\dot{m}_{ini} = \dot{m} + 0.08 \cdot \dot{m}_{choke}$ was advantageous and for others the initial conditions had to be set to the point of peak pressure ratio of the respective speed line. No pattern could be found to optimize this selection.

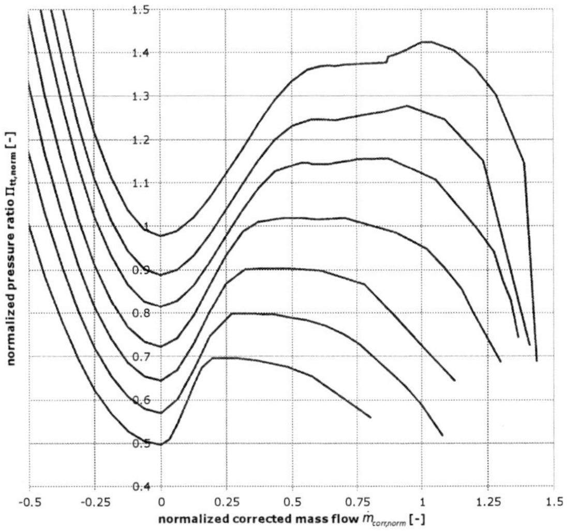

Figure 7: Extrapolated compressor map for numerical investigations

The computational results of a single point allowed the categorization whether this point was a stable one with almost no oscillation in mass flow or an unstable (surging) one showing mass flow oscillations. Hence, the surge line of a blade speed could be determined as follows. A simulation for the peak pressure ratio was conducted, which will always be stable. The respective mass flow was stored as last stable one. A very small mass flow rate was computed and stored as last surging one. Here, a mass flow of roughly eight per cent of the choke mass flow was always sufficient. Afterwards a map point determined by the algebraic mean mass flow rate of last stable and last surging operating point was simulated and the result stored as one of those "last" quantities according to its behaviour. This procedure was continued until a chosen ratio of both values was reached, which can be regarded as a computational surge line determination accuracy and was always set to be 0.1 %.

For piping geometry variations, the damping factor DF was adjusted separately for each speed line so that the numerically obtained surge point would match the measured one for a reference geometry. Afterwards the piping geometry variation was simulated retaining damping factors and the result could be compared to the respective measured surge line.

The key factor of the combined Greitzer/ Theotokatos model is the inlet time constant τ_{inl} employed in equation 6. Theotokatos and Kyrtatos derived it from the definition of axial velocity $c = L_{ID}/\tau_{inl}$ and the Bernoulli equation leading to equation 11. This formulation is based on the assumption that the velocity in the inlet pipe is responsible for the surge line shift and hence, the pipe diameter would be the crucial geometry parameter.

Following this assumption but employing the continuity equation instead of the Bernoulli one leads to the formulation in equation 12. A different approach would be to assume the inertia of the air mass in the inflow duct to be the cause of the surge line shift. Hence, the law of inertia, equation 13, would have to be adapted yielding equation 14.

$$\tau_{inl} = L_{ID} \cdot \sqrt{\frac{2}{R \cdot T_{amb} \cdot (1 - p_{ID}/p_{amb})}} \qquad p_{amb} \geq p_{ID}$$
$$\tau_{inl} = L_{ID} \cdot \sqrt{\frac{2}{R \cdot T_{ID} \cdot (1 - p_{amb}/p_{ID})}} \qquad p_{amb} < p_{ID} \qquad \textbf{(11)}$$

$$\tau_{inl} = A_{ID} \cdot L_{ID} \cdot \frac{p_{amb}}{R \cdot T_{amb} \cdot |\dot{m}_C|} \qquad \textbf{(12)} \qquad\qquad m \cdot \frac{dc}{d\tau_{inl}} = F = \Delta p \cdot A \qquad \textbf{(13)}$$

$$\tau_{inl} = \frac{L_{ID} \cdot \dot{m}_{ID}}{A_{ID} \cdot |p_{amb} - p_{ID}|} \qquad \textbf{(14)}$$

Another possibility could be the acoustic eigen-frequency of the piping. Here, the first mode of a pipe open on both sides has been considered. Thus $\tau_{inl} = 1/f_{eigen}$ would lead to equation 15. On the other hand, in the Greitzer model [4] the compressor and plenum geometries were simplified to a Helmholtz configuration with the eigen-frequency ω, consisting of actuator disk (model for compressor wheel), small diameter compressor ducting and plenum with large cross-sectional area. This approach worked good according to Greitzer [5]. Regarding the employed P1L geometry (compare Figure 2) from the impeller towards upstream direction, it consists as well of actuator disk (model for compressor wheel), small diameter ducting and large diameter piping forming a kind of P1 plenum with large cross-sectional area and hence low flow velocity. Therefore, the P1L geometry can be interpreted as a Helmholtz configuration as well, which would allow the dominant frequency to be calculated according to equation 16.

$$\tau_{inl} = \frac{2 \cdot L_{ID}}{\sqrt{\kappa \cdot R \cdot T_{amb}}} \qquad \textbf{(15)} \qquad\qquad \tau_{inl} = 2\pi \cdot \sqrt{\frac{V_{ID} \cdot L_h}{\kappa \cdot R \cdot T_{amb} \cdot A_h}} \qquad \textbf{(16)}$$

In Table 3 the nomenclature being used further is listed for clarity reasons. The computational results obtained employing all of those models will be discussed in the next section in comparison to the measurements presented in section 3.

Table 3: Summary of inlet time constant models

case name	Theotokatos	pipe velocity	inertial mass	pipe frequency	Helmholtz frequency
equation for τ_{inl}	11	12	14	15	16

5 SIMULATION RESULTS

In order to verify the validity of the presented model, a P2 geometry variation study has been conducted in comparison to the original Greitzer model. Therefore, the P1 geometry was chosen to be the P1s configuration, because Greitzer's setup did not include an extensive inlet piping. For the combined Greitzer/Theotokatos model, the pipe velocity approach (equation 12) has been chosen for inlet time constant calculation. As explained in the previous section, the damping factor DF has been calibrated for each speed line individually in order to match the measured surge line of the P1s-P2s configuration for both models. Afterwards, the damping factors have been retained and the pressurized side piping geometry has been varied.

The result is presented in Figure 8 in compressor map illustration. For clarity reasons, the simulated surge lines for the P1s-P2s configuration are not plotted, because they have been calibrated to match the measured one, as explained. It can be recognized that the simulation results do not agree with the measurements for the lowest two investigated blade speeds. On the other hand, for medium and high rotational speeds the results of the combined Greitzer/Theotokatos model fit very well to the measured data. The original Greitzer model over predicts the surge line shift. Because only for medium and high blade speeds a significant surge line alteration could be measured, the Greitzer/Theotokatos model provides good results. Hence, the model is reasonably valid for P2 geometry variation assessment. Nevertheless, the issue of bad agreement with measurement data for low rotational speeds needs further investigation.

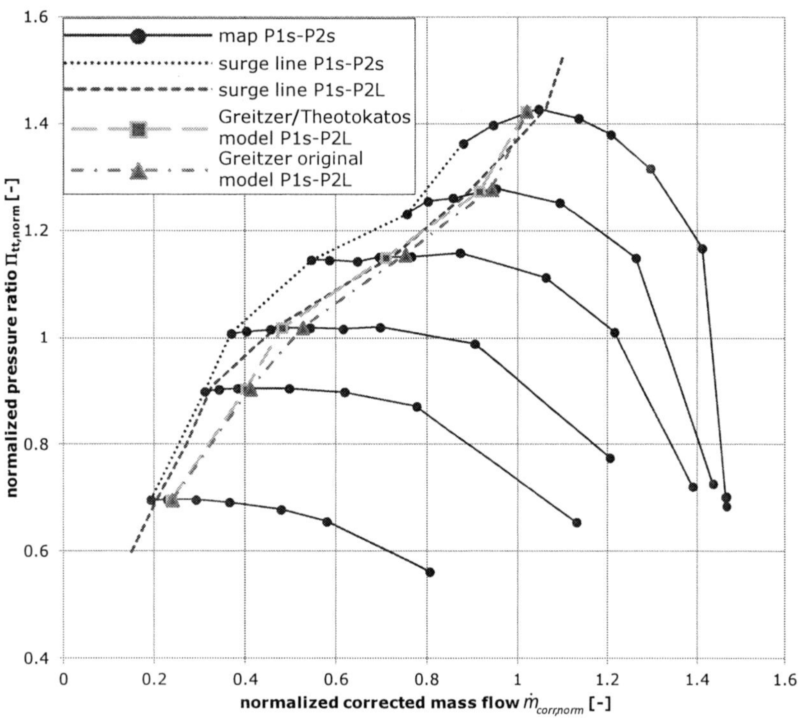

Figure 8: Results of numerical P2 geometry variation study in comparison to measurements

Furthermore, the various models for the inlet time constant listed in Table 3 have been tested in an inlet piping geometry variation study. Therefore, for each model the damping factor DF has been calibrated as before matching the measured surge line of the P1s-P2s configuration. Again, these damping factors have been retained for the simulations of the altered inflow piping setup P1L-P2s. The results of the numerical P1 variation have been post processed according to equation 1 and Figure 5, but the reference case being P1s-P2s. In Figure 9 the output is illustrated in combination with the respective measurement results. It should be noted that a single line in the graphic represents the surge line shift resulting from the geometry variation P1s to P1L for a single numerical model or the measurement results respectively. Because the Helmholtz frequency formulation could not be reasonably adapted to the P1s configuration, the respective results obtained employing the pipe frequency approach have been used as reference data. This is the most plausible option because both approaches are based on the assumption that the acoustic eigen-frequency of the P1 geometry is the cause for the surge line shift.

The illustration presented in Figure 9 shows that the Helmholtz frequency approach provides the best simulation results for the three lowest blade speeds. But for the three highest rotational speeds all models fail in a plausible prediction and they even show a wrong trend. Except for the inertial mass model at blade speed U_{Norm}=0.79 all approaches predict a deteriorated surge line although the measurements show a significantly improved one.

A problem could be the modeling of the complete inflow duct. Because due to the significantly reduced flow velocities in the large diameter pipe (compare Figure 2), which are only 16 % of the ones in the measurement pipe, this large diameter duct could already act like the ambient. Hence, the simulated P1 geometry has been adapted in a way that only the piping between the compressor and the reduction from the measurement pipe to the large diameter one has been taken into account.

The respective results are shown in Figure 10. As mentioned before, the Helmholtz approach could not reasonably be adapted to such a geometry and hence it has been omitted here. It can be recognized that except for U_{Norm}=0.71 the pipe frequency formulation does not predict a remarkable surge line shift at all. Furthermore, all other models show a wrong trend for the lowest three blade speeds. For the highest three ones the tendency is at least in part correct but always to a much too low extent.

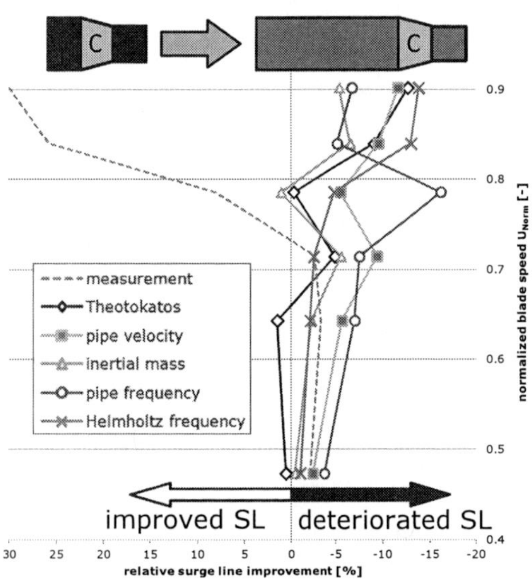

Figure 9: Simulation results of P1 variation vs. measurements (reference case respective P1s-P2s each) – original P1L geometry

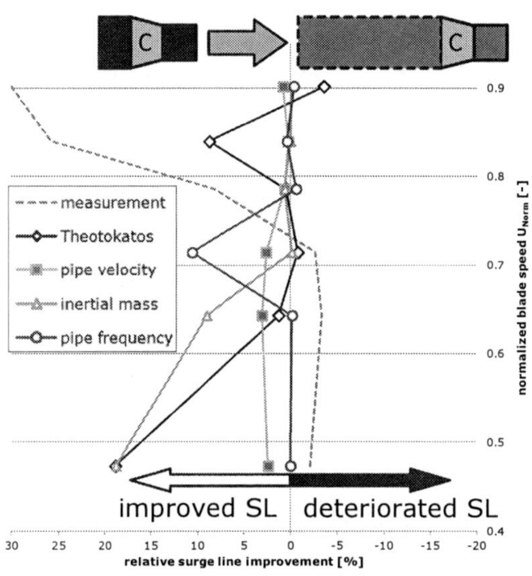

Figure 10: Simulation results of P1 variation vs. measurements (reference case respective P1s-P2s each) – adapted P1L geometry

Summarizing, as far as P2 geometry variation is concerned, the combined Greitzer/Theotokatos model provides good results for high blade speeds, where the relevant map range variations occur and the results are better than the ones obtained with the original Greitzer model. Nevertheless the prediction accuracy for low blade speeds has to be improved. Regarding inlet geometry variation, all tested approaches do not provide sufficient results and even a wrong trend can be predicted. Hence, none of the underlying theories alone can be the cause of the measured surge line shift due to P1 geometry variation. Furthermore, respective investigations based on common engine simulation tools employing the approach of Theotokatos and Kyrtatos [7] should be conducted with great care, because varying P1 geometry in such tools may lead to wrong trends as well. Further investigations aiming on the development of a valid P1 model are required.

6 CONCLUSIONS AND FUTURE WORK

The influence of the in- and outflow piping on the surge line of a small centrifugal compressor for automotive applications has been investigated experimentally and numerically. For the simulations an adapted Greitzer/Theotokatos approach employing several models for inlet time constant calculation has been utilized. The main conclusions can be summarized as follows:

- From the experiments it can be stated that an improved surge line can be observed employing a small volume for the pressurized side piping in any case.
- Furthermore, a large volume of the inlet ducting is advantageous for the surge limit in particular for high blade speeds and in combination with the small outflow piping volume.
- The computations show good agreement with measurements for high rotational speeds as far as outflow piping variation is concerned. Nevertheless, the prediction accuracy for low blade speeds has to be improved.
- All investigated numerical models fail to predict the correct surge line shift for inflow piping variation. But at least, a remarkable influence could be simulated.

Furthermore, due to the numerical results, a valid physical explanation for the surge line alteration caused by inlet ducting variation could not be presented. Hence, further investigations are required employing several carefully selected inflow geometries and extended instrumentation. In addition, work on a respective numerical model in order to obtain a physically valid explanation of the effects is of great interest.

As the main conclusion it can be stated that the inflow piping can have a significant influence on the surge line of a small centrifugal compressor for automotive applications. Hence, a model taking this effect into account correctly would sustainably help to develop a method for the quantitative comparison of surge lines measured at different test benches and for the estimation of the influence of the actual engine packaging on the surge line. Meanwhile, these operations have to be conducted with great care in particular when common engine simulation tools employing the Greitzer/Theotokatos model are involved.

REFERENCE LIST

[1] M. Ebisu, T. Shiraishi, I. Tomita and M. Furukawa, "Development of advanced centrifugal compressor for turbocharger, applying control of internal unsteady flow structure", in *Proceedings 10th International Conference on Turbochargers and Turbocharging*, London, UK, 2012.

[2] D. Müller, "Einfluss der Anordnung und des Betriebs von Abgasturboladern auf die Pumpgrenze", Dissertation, Universität Hannover, 2008.

[3] J. B. Schwarz and D. N. Andrews, "Considerations for gas stand measurement of turbocharger performance", in *Proceedings 11th International Conference on Turbochargers and Turbocharging*, London, UK, 2014.

[4] E. M. Greitzer, "Surge and Rotating Stall in Axial Flow Compressors - Part I: Theoretical Compression System Model", *ASME Journal for Engineering for Power,* pp. 190-198, April 1976.

[5] E. M. Greitzer, "Surge and Rotating Stall in Axial Flow Compressors - Part II: Experimental Results and Comparison With Theory", *ASME Journal of Engineering for Power,* pp. 199-217, April 1976.

[6] K. Kabalyk, W. Kryllowicz, G. Liskiewicz and L. Horodko, "Investigation of the Inlet Filter Influence on Unsteady Flow Phenomena in a Single-Stage Centrifugal Blower", in *Proceedings European Turbomachinery Conference*, ETC2015-055, Madrid, Spain, 2015.

[7] G. Theotokatos and N. P. Kyrtatos, "Diesel Engine Transient Operation with Turbocharger Compressor Surging", in *Proceedings SAE World Congress*, SAE2001-01-1241, Detroit, Michigan, USA, 2001.

[8] Gamma Technologies Inc., GT Suite user's manual v7.5, 2014.

[9] P. Grigoriadis, "Experimentelle Erfassung und Simulation instationärer Verdichterphänomene bei Turboladern von Fahrzeugmotoren", Dissertation, Technische Universität Berlin, 2008.

[10] H. Mai, "Parameterstudie zur Turbolader-Kennfeldvermessung", Berlin: VFI/ Technische Universität Berlin, 2009.

[11] R. A. van den Braembussche, "Centrifugal Compressors Analysis & Design", Brussels, Belgium: von Karman Institute for Fluid Dynamics Course Note 192, February 2014.

[12] K. Ueda, S. Kikuchi, Y. Inokuchi, N. Yamasaki and A. Yamagata, "The effects of upstream bent pipe on the performance of a small centrifugal compressor", in *Proceedings 12th International Symposium on Experimental Computational Aerothermodynamics of Internal Flows (ISAIF 12th)*, Lerici, Italy, 2015.

[13] D. A. Fink, N. A. Cumpsty and E. M. Greitzer, "Surge Dynamics in a Free-Spool Centrifugal Compressor System", *ASME Journal of Turbomachinery,* pp. 321-332, 114, April 1992.

[14] K. E. Hansen, P. Jorgensen and P. S. Larsen, "Experimental and theoretical study of surge in a small centrifugal compressor", *ASME Journal of Fluids Engineering,* pp. 391-395, 103, September 1981.

[15] M. Morini, M. Pinelli and M. Venturini, "Development of a one-dimensional modular dynamic model for the simulation of surge in compression systems", in *Proceedings ASME Turbo Expo2006: Power for Land, Sea and Air*, Barcelona, Spain, GT2006-90134, 2006.

SESSION 2A

SUPERCHARGING, TURBOCOMPOUND, V/G AND 2-STAGE

Ultra high efficiency two-stage turbocharging system

R Cadle, D Giebert, A Mohamed, S Pees, M Oakes
Honeywell Turbo Technologies, USA and France

ABSTRACT

Honeywell Turbo Technologies and our customer Volvo Trucks have collaborated on an ultra-high efficiency two stage serial turbocharger system under the European Union FP7 GA-2012 CORE project. The compressors of both the low pressure and high pressure turbochargers were design optimized for the highly specific operating conditions of two stage boosting. The high pressure turbine is characterized by a sector divided fixed vane nozzle, while the low pressure turbine stage features an axial turbine wheel. The two turbines were designed specifically to achieve highest possible efficiency while minimizing ducting losses between the two stages. The end result is a novel high performance boosting system that fits in a unique package.

This paper describes the approach used to set design targets for each stage of the two stage turbocharger system. The design approach used to meet these targets will be presented, and test results from gas stand will be shown measured against performance targets and performance predictions.

1 INTRODUCTION

Whether from the perspective of climate change, energy security, or economic competitiveness, fossil fuel consumption in the transportation sector has become one of the major public policy issues of our time. As light duty vehicle fuel consumption begins to decline as a result of stringent globally-enacted fuel efficiency regulations, the ever-increasing carbon emissions of heavy duty trucks is becoming a focus of concern.

With partial funding from the European Union under the FP7 GA-2012 CORE project, and in partnership with Volvo Trucks, Honeywell has developed a high-efficiency turbocharger system capable of delivering significant reduction in engine fuel consumption for heavy duty long-haul trucks.

The program goal is to achieve a 15% fuel economy improvement relative to a state-of-the-art EUV-compliant vehicle whilst meeting EUVI emissions norms. A 6-9% improvement in efficiency is expected to come from the engine platform.

Honeywell worked closely with Volvo to develop an air system that would deliver maximum efficiency in support of Volvo's combustion strategy. For the CORE program, Volvo chose to utilize a Variable Valve Actuation (VVA) system which would provide a wide range of control over valve timing events, including the ability to run Miller cycle. Due to the low effective volumetric efficiency of a Miller cycle engine, overall turbocharger pressure ratio was expected to be in excess of 4:1. In keeping with the goals for a high-efficiency engine for long-haul truck usage, the design goal for the air system was to achieve the highest possible efficiency in order to minimize engine pumping losses.

2 AIR SYSTEM ARCHITECTURE

Due to the high overall pressure ratio targets, a two stage turbocharging solution was chosen for this engine program. Compared with single stage boosting, two stage (series) compressors offer a variety of performance advantages. First, series compressors allow each compressor to be design optimized for a lower pressure ratio. This allows the aerodynamic designer to use, for example, increased backsweep on the compressor stages, which increases overall compressor efficiency and broadens peak efficiency islands. The main drawback of highly backswept compressor wheels (that is, that they do not have high pressure ratio capability) is not a concern in a two stage arrangement.

In addition to fundamentally higher component efficiency, series compressors have the potential to incorporate cooling between compression stages. As the work required to compress a gas is directly proportional to the temperature of the gas at the start of compression, a major system performance improvement can be obtained by cooling the charge air between the compressor stages. This reduces the work required for the high pressure (second stage) compressor, in turn reducing the expansion ratio across the high pressure turbine, resulting in reduced back pressure and pumping losses for the engine.

The primary drawbacks of a two stage boosting system are the packaging, and the performance of the air handling ducts between stages. These negative aspects are usually correlated, as a system that can be packaged in a modern commercial vehicle typically suffers in terms of performance, due to the tight and complex bends required to route air between turbos. The interstage duct routing gases from the high pressure turbine to the low pressure turbine tends to be the highest loss component in the system. From the beginning of the CORE Project, the highest priority was placed on ensuring that maximum overall air system performance could be achieved, with an overall size that was feasible to package into a modern on-highway truck.

After close collaboration between Volvo and Honeywell, it was decided that the best architecture for such a high performance two stage system would consist of a high pressure turbocharger with radial compressor and turbine wheels, joined with a low pressure turbocharger consisting of a radial compressor wheel and an axial turbine stage. Volvo had previously developed such a system and seen positive results. This experience proved to be an invaluable benefit to the program.

The primary advantage of the radial/axial turbine arrangement is that the problem of interstage duct losses between the turbines is tackled head-on with a completely new layout for the air handling system. In a traditional radial/radial two stage boosting system, exhaust gas must exit the high pressure turbine, and make a sharp bend into the low pressure turbine. Given the high velocity of exhaust gas post-turbine, and the degree of swirl left in the flow at off-design conditions, the total pressure drop in the interstage turbine duct results in a significant performance penalty.

By using an axial low pressure turbine stage mounted concentrically with the high pressure stage, sharp bends between the stages are avoided. Additionally, due to the design of the duct and the low pressure turbine, kinetic energy not recovered from the flow in the high pressure stage can be converted into useful work by the low pressure turbine stage. This integrated duct design allows for a compact, high performance two stage turbo system that is feasible to package in a modern vehicle chassis (see Figure 1 below).

Figure 1: Two Stage turbo system on engine (image courtesy of Volvo Trucks)

In addition to the novel radial / axial turbine and interstage duct layout, other key design features of the two stage system consist of a sector divided, fixed vane nozzle turbine for the high pressure stage, and low friction Honeywell ball bearings on both turbochargers for maximum performance (see Figure 2 below for final design architecture).

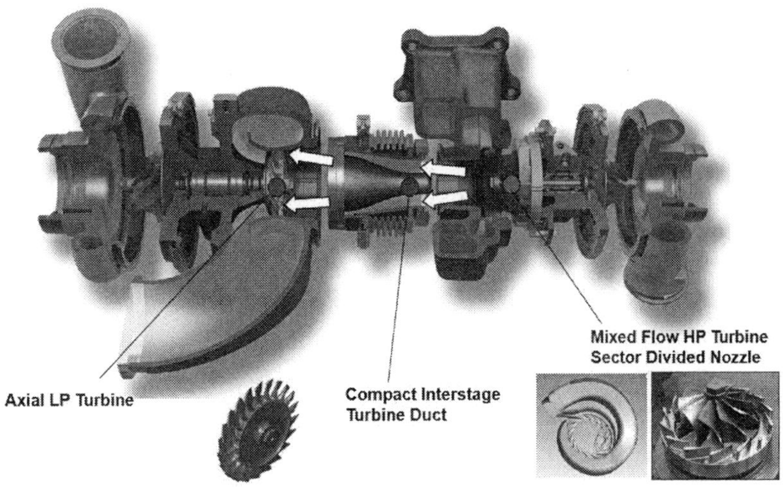

Figure 2: High Efficiency advanced Two Stage boosting system

3 DESIGN TARGETS AND APPROACH

The complete engine, including the air system, was modeled by Volvo using an in-house developed one dimensional simulation code. Optimization within this code was carried out to determine target operating points and efficiency values for the high and low pressure compressors. Target operating points with compressor range targets are shown in Figure 3 below.

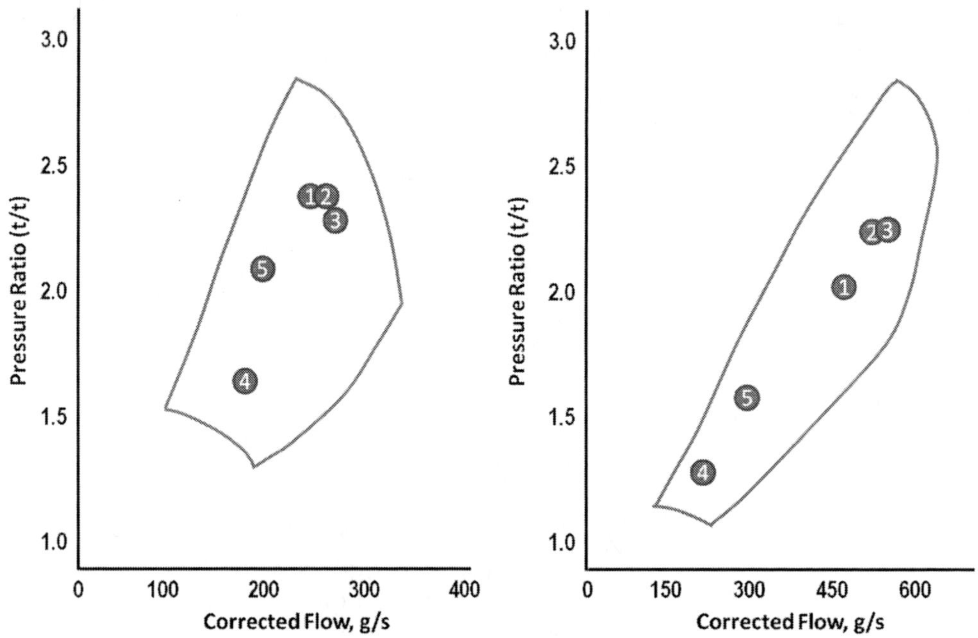

Figure 3: High Pressure (HP – left) and Low Pressure (LP – right) compressor operating targets

3.1 High pressure compressor

The High Pressure (HP) compressor performance problem statement on its own was not considered especially difficult. The targeted pressure ratio is fairly low (less than 2.5:1), with modest map width requirements. A high backward curvature design would allow high efficiency over a broad range, while still meeting the pressure ratio requirement at reasonable tip speeds. Honeywell has a wide selection of state-of-the-art compressors in its design portfolio, and the intention was to use an existing wheel that came as close as possible to meeting the performance targets, and then optimizing the design to meet the efficiency objectives. A candidate wheel was found with maximum design trim of 44, backsweep angle of 50 degrees, and an 82mm diameter. This candidate wheel was between 0.5 and 1.5 points below the Volvo efficiency target. In this new two-stage application, maximum tip speed was only expected to be 455 m/s, much lower than the qualification limits used in typical compressors designed for commercial vehicles. It was believed that by optimizing the wheel mechanically for lower maximum tip speeds, the target efficiency could be achieved.

As will be discussed in the turbine section, this design approach was not viable, because it resulted in a design point physical speed that was too high for the High Pressure (HP) turbine. In order to improve the design point efficiency of the HP turbine, the HP compressor was therefore increased from 82mm to 88mm diameter. This necessitated a reduction in design trim of the compressor from 44 to 36 in order to maintain the flow characteristic. While theoretically this reduction in trim should reduce specific speed and increase efficiency, it moves the wheel design well outside of the known automotive compressor design space and added significant technical risk to the program.

The smaller trim reduced the tip speed of the inducer, reducing the work done on the flow in the early portion of the blade. The blade length was increased to compensate for this reduction and maintain overall pressure ratio. Additionally, to optimize the performance of the new wheel for low pressure ratio operation, the ratio of outlet area to inlet area was increased relative to the baseline wheel. The resultant high trailing edge length was viable mechanically because of the lower tip speed of the design. The wheel was further optimized by adding a 4:1 elliptical trailing edge to reduce blade wake, and a "reverse clip" (shroud diameter larger than hub diameter) design was added to the exducer. This reverse clip feature increases the work factor of the wheel for a given rotational speed, further improving high pressure turbine performance. Extensive CFD optimization predicted that the new wheel should have between 1 and 1.5 points higher efficiency than the progenitor design. This was expected to allow the HP compressor to meet its performance objectives while also operating at a speed and power level well suited to the HP turbine. The final wheel design was mechanically qualified using production stress targets to a maximum tip speed of 460 m/s (see Figure 4 below).

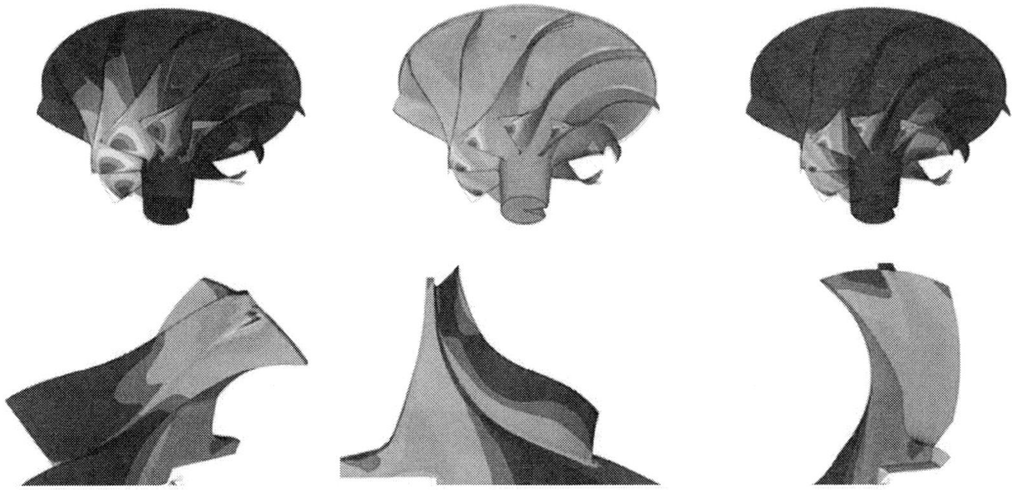

Figure 4: High pressure compressor was mechanically qualified to meet Honeywell production stress and frequency requirements

3.2 High pressure turbine

A fixed vane nozzle design was chosen for the High Pressure (HP) turbine in order to achieve maximum efficiency. It was desired to have strong pulse separation on the engine, in order to prevent cylinder "cross-talk" during the engine blow-down event. This would ensure good cylinder scavenging and reduce engine pumping losses.

Due to the complexity of designing a radially divided, nozzled turbine, a sector divided design approach was chosen. This presented challenges in the design optimization, as the flow condition under such pulsating conditions is non-uniform and time-varying. Volvo's 1-D engine simulation code provided temperature, pressure, and mass flow predictions on a crank-angle basis for each engine operating point (see Figure 5 below).

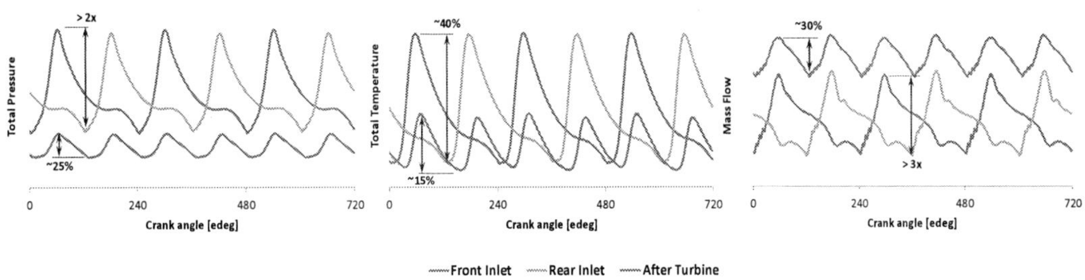

Figure 5: Crank angle resolve engine simulation data was used to set boundary conditions for HP turbine

This non-steady flow approach means that for a given point on the HP compressor map, the turbine is seeing a large variation in inlet flow conditions. Some determination had to be made in order to choose the most critical conditions for the turbine design. It was decided that at the most critical engine operating point, the turbine design would be optimized in the region around the mass flow, pressure, and temperature conditions that delivered the most work to the output shaft.

The first step in designing the high pressure turbine was selecting its diameter. Instantaneous Blade Speed Ratio (BSR) and turbine (aero) shaft power were calculated for each crank angle-resolved operating condition of the engine, assuming a range of compressor wheel to turbine wheel diameter ratios (D_c/D_t). As can be seen from Figure 6 below, the vast majority of energy from the turbine is delivered at a BSR much lower than the time averaged value. This is because while turbine physical speed (and therefore the turbine tip speed) is more or less constant during the combustion cycle, the turbine power is much higher during the blowdown portion of the engine's combustion cycle.

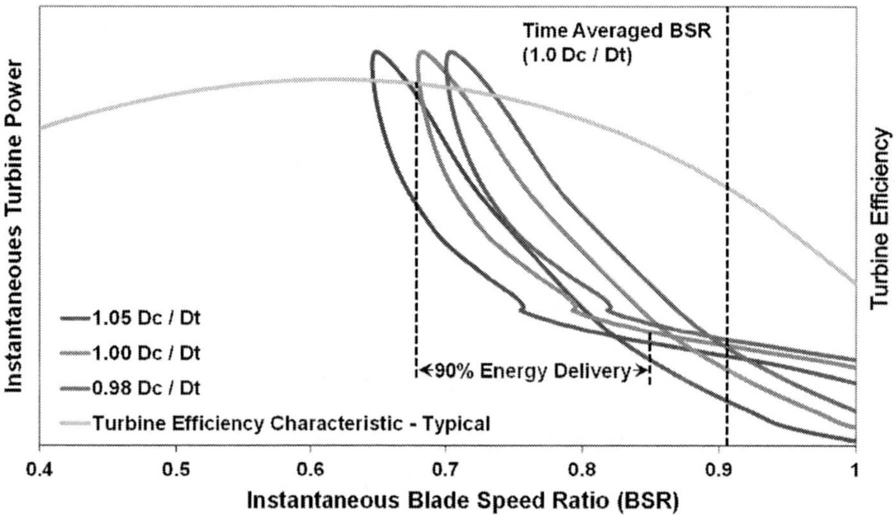

Figure 6: Instantaneous Power versus BSR for HP turbine

The efficiency characteristic of a typical radial turbine is also shown in Figure 6. As can be seen in the Figure, for a 1.0 wheel diameter ratio, 90% of the turbine energy is delivered at a blade speed ratio between 0.68 and 0.84. As the turbine wheel diameter is reduced relative to the diameter of the compressor, the BSR is reduced, and the majority of turbine work is done in a region better aligned with the turbine's efficiency characteristic. However, as the turbine wheel diameter is decreased in absolute terms, the specific speed of the stage increases. Based on Honeywell design experience, it was believed that this increase in specific speed would lead to poor aerodynamic efficiency.

An obvious solution would be to grow both the compressor and turbine wheels in diameter proportionately, allowing a reduction in turbine specific speed without altering the blade speed ratio. However, this would result in a significant reduction in the specific speed of the HP compressor wheel, requiring an even smaller trim design.

Given these considerations, a mixed flow turbine wheel was selected to allow efficient operation at high specific speed while keeping the blade speed ratio close to the optimum. Full stage CFD optimization was carried out to refine and optimize the turbine stage. The design was optimized on the basis of total-to-total efficiency, instead of total-to-static, as is common in automotive turbines. This is because the packaging arrangement allows the LP turbine to efficiently utilize any dynamic pressure left from the HP stage. Final performance predictions are shown in the plot below. Of special note is that the design was optimized to have a gradual efficiency roll-off towards high blade speed ratios, in line with the expected on-engine operation points. This was achieved by adding forward curvature to the turbine wheel in order to offset the impact of incidence changes at higher blade speed ratios.

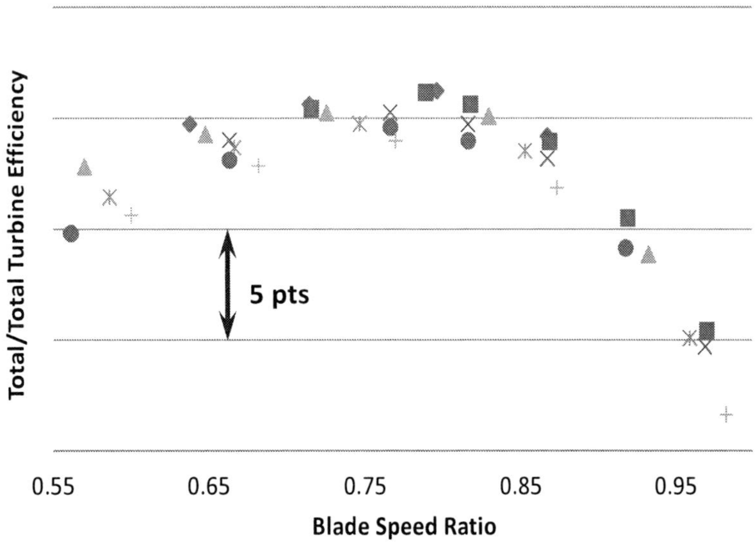

Figure 7: Final predicted efficiency for high pressure turbine stage

As the stated goal of the CORE program was to develop technology feasible for full-scale production by 2020, Honeywell validated that the stage meets all current design criteria for centrifugal and thermal stresses. Additionally, balanceability and pullability studies were completed, ensuring that series production would be possible with current manufacturing techniques. Finally, a forced response analysis was carried out using the same High Cycle Fatigue (HCF) methodology that Honeywell applies to Variable Nozzle Turbine (VNT) designs.

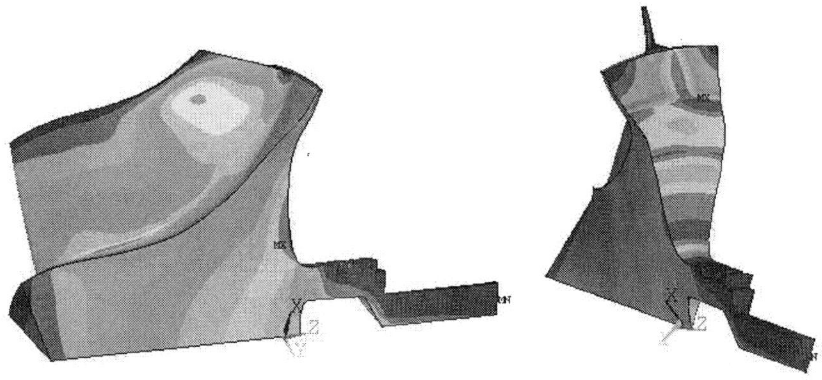

Figure 8: Centrifugal + thermal stress analysis, including impact of backdisk balance cuts

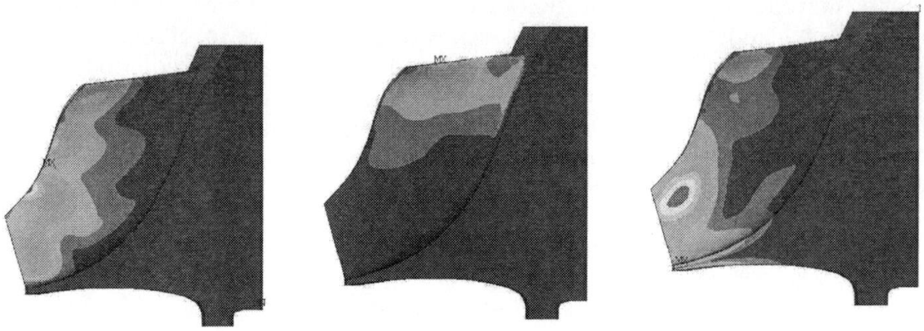

Figure 9: Forced response analysis is shown for modes 4, 5, and 6 (left to right) for the final design

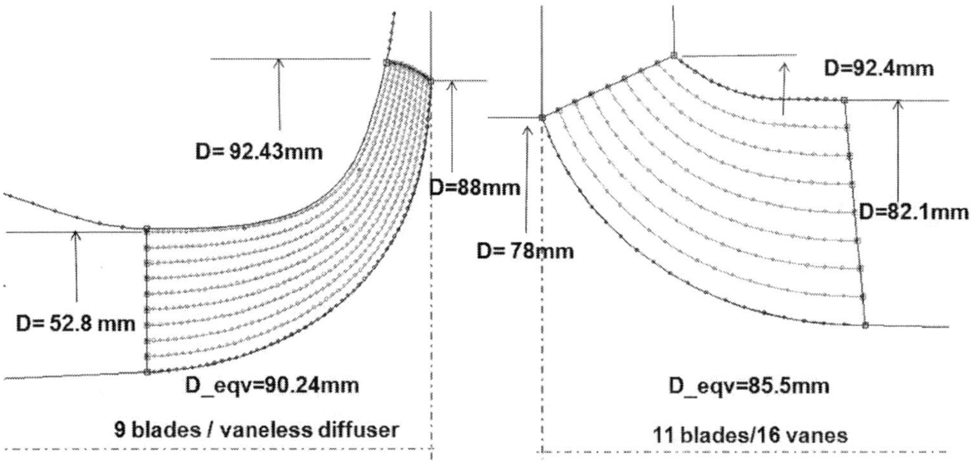

Figure 10: HP turbocharger aerodynamic configuration

3.3 Low pressure compressor

The Low Pressure (LP) compressor design targets are similar to those of the HP stage. The initial design intent was to therefore to use a 108mm version of the compressor stage planned for the HP. Performance projections showed that such a scaled wheel would be less than 1 point below the efficiency targets at higher pressure ratios, and 1.5 to 2 points lower at the low pressure ratio points. Even with the high backsweep of the initial design, maximum tip speed at altitude was projected to be less than 465m/s. The design team believed the performance objectives could be met by stress qualifying the existing wheel for a lower tip speed, thinning the blades and reducing the blade root fillet diameter. The low pressure ratio performance could be improved by using a larger A/R volute and increasing the compressor wheel exit tip width (which again would be facilitated by the low maximum tip speed). At the critical design point of the engine, the low pressure compressor operating speed was projected to be approximately 66,000 RPM.

As will be discussed in greater detail below, the low pressure turbine design optimization did not allow a compressor that would run that fast. The turbine optimization study recommended a design point speed of 58,000 RPM. This could be achieved by scaling the existing 82mm 44 trim compressor to 123mm. However, to maintain flow range, the wheel would need to be designed with only a 34 trim. A different design approach was taken instead, focusing on a 118mm wheel with a less backward curvature (nominally 42 degrees versus 50 for the original wheel), and a larger (40) design trim. This reduced backward curvature was still high enough to provide good compressor map width and efficiency at the targeted pressure ratios, but would allow the wheel design speed to be reduced to better suit the turbine. The wheel features a "reverse clip" exducer design, where the shroud side diameter is larger (124mm) than the hub side diameter (118mm). This increases the work done by the wheel, allowing the stage to achieve the required pressure ratio at the targeted design point speed of 58,000 RPM.

As with the high pressure compressor stage, the low pressure compressor was designed with a large blade exit tip width, to optimize performance for low pressure ratio. It was also stress qualified as per Honeywell's production stress acceptance criteria to a maximum tip speed of 450 m/s. This is significantly lower than normal for Commercial Vehicle compressor wheels, but was deemed sufficient given the lower speed requirements of a two stage system. The final wheel design is shown in Figure 11 below.

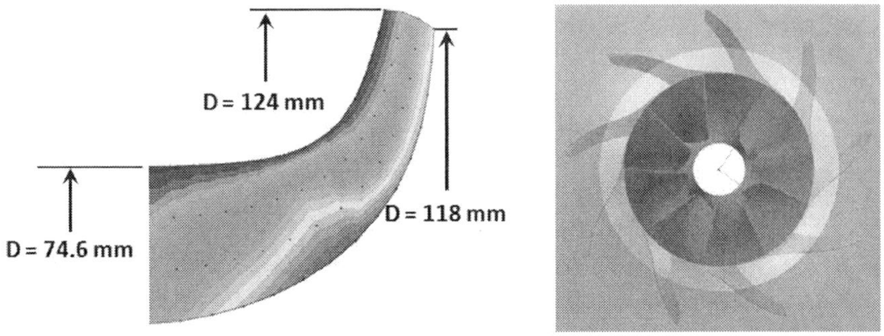

Figure 11: New low pressure compressor wheel design, with normal thickness distribution shown (left)

Figure 12: Low pressure compressor was mechanically qualified to meet Honeywell production stress and frequency requirements

3.4 Low pressure turbine

The low pressure axial turbine stage design was particularly challenging for Honeywell Turbo Technologies. Although HTT has designed low inertia, high flow axial turbine wheels for the passenger vehicle space in the past[1], a more traditional, high efficiency axial turbine stage was an entirely new undertaking. To address this knowledge gap, the turbocharger team tapped into the extensive turbomachinery design expertise of its Honeywell's aerospace division.

Honeywell Aerospace has developed a proprietary integrated conceptual gas turbine design software. "Fast Accurate Simulation of Turbine Engine Systems 1D", or FAST1D[2], integrates thermodynamic, aerodynamic, and mechanical design into a single software package. The software improves and accelerates the conceptual design process and is applicable to a wide variety of configurations. However, this new axial turbine design presented a unique challenge to the FAST1D program, as the power and mass flow requirements made it necessary to use the tool in a region of the design space that fell outside the experience of Honeywell Aerospace and therefore outside of the range of data that were used for the code's calibration.

A primary variable in the design of the axial turbine wheel was the targeted speed of the wheel at the chosen aerodynamic design point. To evaluate the impact of the turbine speed on turbine performance and mechanical risk, an optimization study was performed by coupling FAST1D with a commercially available optimization tool to arrive at an optimized design for a variety of operating speed targets.

Boundary conditions were applied to the design based on the engine system simulation provided by Volvo. These boundary conditions were turbine shaft power, inlet total temperature and pressure (based on expected outlet conditions from the HP turbine), mass flow, and turbine physical speed. For the study, design point speed was varied from 58,000 to 65,000 RPM. The result of the study was a unique, optimized wheel design for each of the targeted design speeds. Design parameters for the stage that were allowed to vary within a given range during the optimization are shown in Figure 13 below. The objective of the optimization was to maximize the design point total-to-static stage efficiency.

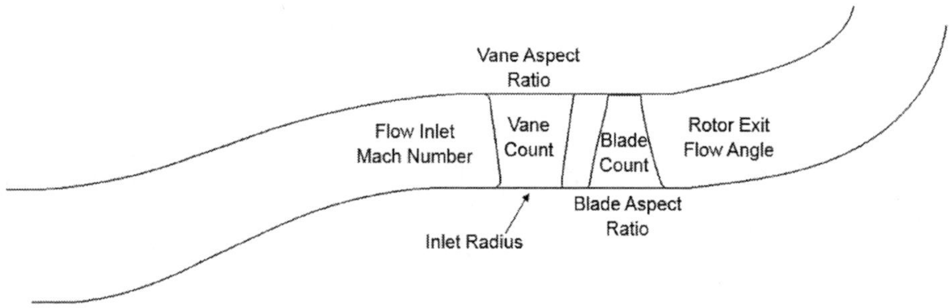

Figure 13: Turbine stage layout with design variables shown

A number of design constraints were applied to the optimization in order to minimize the chances of converging on an infeasible design. Honeywell Aerospace experience was used as a guide in setting the majority of these constraints. Upper and lower constraint values were set based on the perceived risk (low, medium, or high) that these values would bring to the design. While the program goal was to develop a high efficiency axial turbine, the fact that the design was falling outside of Honeywell's experience space mandated that performance objectives be balanced against program risk.

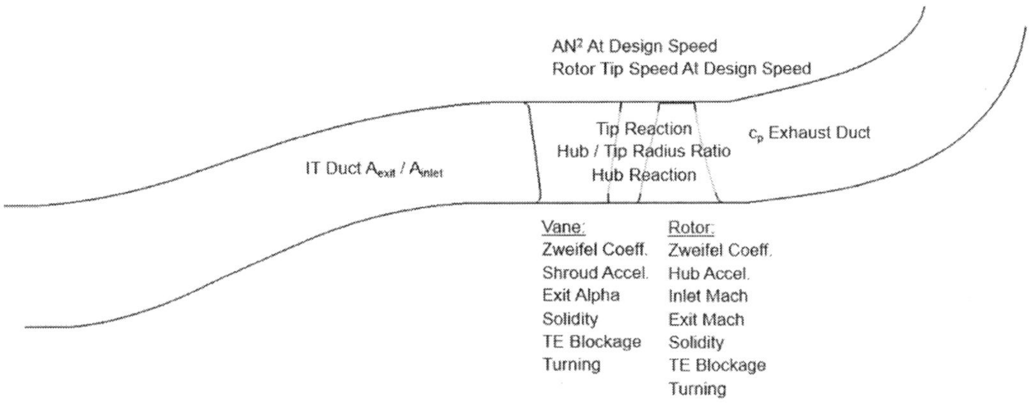

Figure 14: Design constraints applied during turbine stage optimization

The three turbine stages designed in the 1-D optimization study were down selected to determine which would carry forward to aero/mechanical 3-D design optimization. The primary down selection criteria were around expected turbine performance (total-to-static efficiency) and perceived mechanical design risk (rotor tip speed and $A*N^2$). $A*N^2$ is a parameter commonly used in axial turbine design to gauge rotor airfoil centrifugal stresses and therefore mechanical risk; with A being the rotor flow path annulus area and N a parameter that is proportional to the angular speed. Other important constraints in the design were the interstage turbine duct area ratio, static pressure recovery coefficient of the exhaust duct, and the trailing edge blockage, solidity, and degree of turning in both the vane and the rotor.

Figure 15 below shows each of the three designs compared on the basis of efficiency and mechanical risk. As can be seen in the left most plot, all three turbine designs achieve similar levels of efficiency, with a slight improvement gained by moving to the lower speed design. However, the difference in predicted efficiency (only about 0.2%) is not sufficient to select the design.

As the wheel speed increases, the mean diameter must be reduced accordingly in order to maintain aerodynamic power. For a set mass flow this is most optimally achieved by reducing both the tip and hub diameter, as this approach allows for adjusting the flowpath height in order to reduce flow velocities and aerodynamic losses like rotor tip gap leakage. As a consequence, the higher speed designs are characterized by their turbine flowpaths being located on a smaller mean diameter with increased flowpath height which gets more pronounced as the rotational speed goes up. This creates a longer blade design with a larger annular flow area and therefore a higher $A*N^2$ value as well as a higher tip speed as shown in the center and far right plots in Figure 15 below. These factors greatly increase the mechanical design complexity of the wheel and increase the risk that subsequent stress and forced response validation will result in missed performance objectives. These considerations led to the selection of a low speed turbine, with a design point speed targeted at 58,000 RPM. It is possible that the wheel would continue to improve both mechanically and aerodynamically as speed was further reduced, but the 58,000 RPM target created significant challenges for the compressor design, so exploration of a lower speed design space was not pursued.

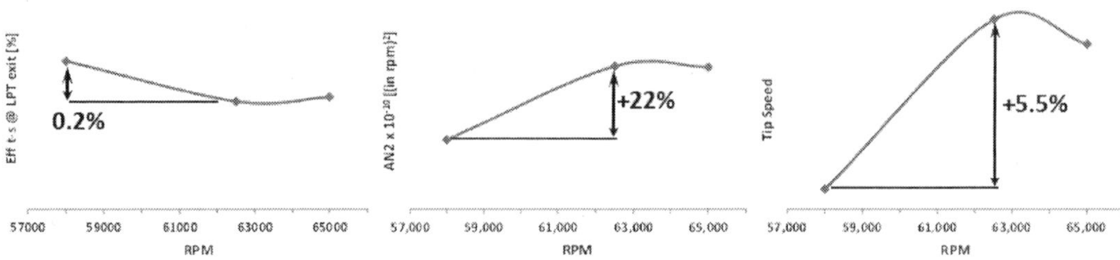

Figure 15: Low speed turbine design leads to lower mechanical risk with minimal performance impact

Significant optimization of the stage selected during the 1-D design phase was conducted using 3-D full stage CFD analysis. 1.2% additional improvement in efficiency was obtained by optimizing for stator lean, turbine wheel hub slope, and exhaust collector geometry and diffusion refinement.

While the high pressure turbine was designed for single point operation, in reality, it is subjected to a highly variable flow inlet conditions because of the unsteady flow through the engine cylinders. While the stage is optimized for performance under the conditions that correspond to maximum aero energy delivery, in reality, the HP turbine stage exit swirl angle varies between -10 degrees and +30 degrees over the duration of the pulse. On a time- (not power-) averaged basis the HP turbine outlet flow angle is actually +18 degrees. To understand the impact that this variation in flow angle can have on the downstream component, the low pressure turbine performance was analyzed over a variety of inlet flow angle conditions. It was found that the LP turbine was seeing a significant change in performance due to the variation in swirl coming from the HP stage under the initial assumption of direction of rotation. By reversing the direction of rotation of the low pressure turbine it was found that the time-averaged performance could be improved and the turbine would be more robust to variation.

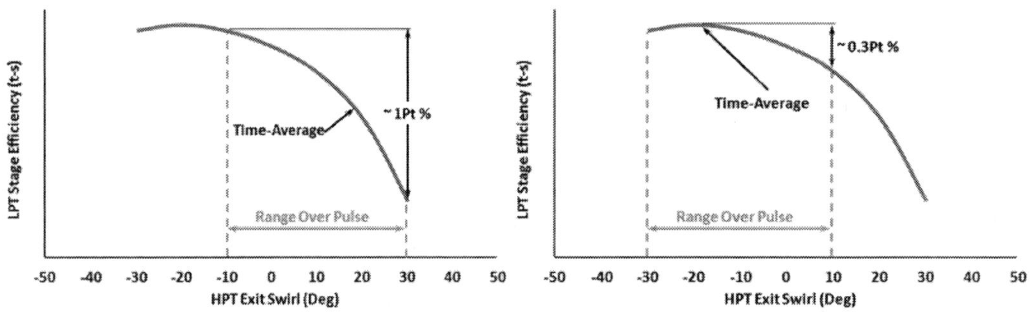

Figure 16: LPT sensitivity to HPT exit swirl for conventional (left) and reverse (right) rotation cases

After full-stage CFD design optimization on the LP turbine was complete, the wheel was mechanically qualified for low- and high-cycle fatigue according to Honeywell's standard production validation criteria. Centrifugal and thermal stress analyses were carried out including the impact of balance cuts on the rotor disk stresses. In addition, a forced response analysis was conducted to ensure that blade strains were reasonable under excitation conditions from the stator vanes. Results are shown in Figures 17 and 18 below.

Figure 17: Centrifugal + thermal stress results for pressure and suction sides of LP turbine

Figure 18: Mode 1 forced response analysis for pressure and suction sides of blade

In addition to the analytical results, light problem testing was conducted in Honeywell's gas stand to ensure that the blade strains and mode shapes of the low pressure turbine were in line with expectations. Response from all blades was measured using three different wheels. The measured blade strains were within acceptable limits. A very high degree of correlation was observed between predicted and actual mode shapes.

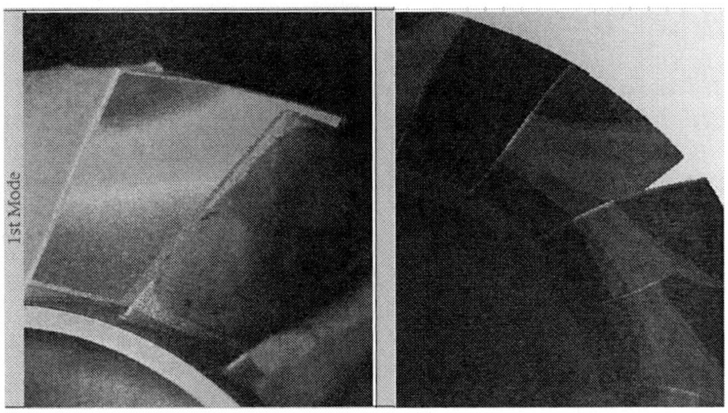

Figure 19: Measured first mode shape from Laser Scanning Vibrometer (LSV) (left) well correlated to predictions (right)

Bearing System

Because efficiency was the primary objective of this development program, low-loss, cartridge-type ball bearings were chosen for both high- and low-pressure turbochargers. A 16mm shaft diameter cartridge was chosen for the low pressure stage, while a 12mm shaft diameter cartridge was used for the high pressure stage. Both bearings were taken "off the shelf" with no bearing system development required as part of the CORE program. Ball bearings provide the greatest increase in efficiency at low turbine expansion ratios, where bearing power loss is as a percentage of turbine aerodynamic power is highest[3]. The relatively low speed operation of series turbos therefore favors the use of ball bearings.

4 TEST RESULTS AND COMPARISON TO PREDICTED PERFORMANCE

Performance test results for the compressor stages are shown in Figure 20 below. Tests were carried out in a conventional steady flow gas stand using standard instrumentation and test procedures. Overall, both compressors exhibited excellent performance on gas stand, with predictable and stable operation around the desired operating points. Adiabatic efficiency is in excess of 80% at all engine operating points, with the exception of Point 4 on the low pressure compressor map. This point, however, is at a very low pressure ratio, where accurate efficiency measurements are difficult to make due to heat transfer and other effects.

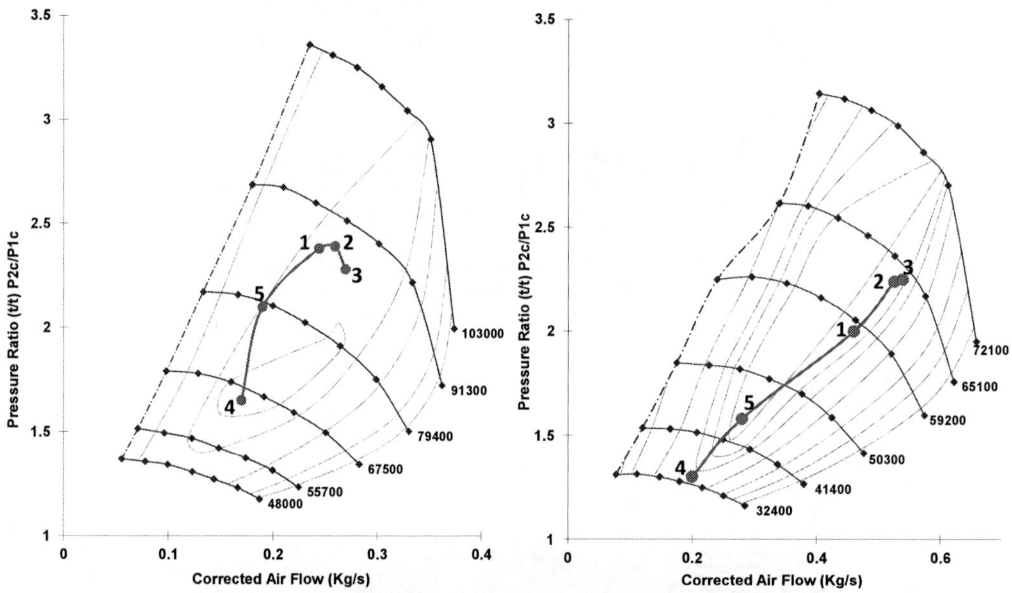

Figure 20: Performance gas stand test results for high (left) and low (right) pressure compressors, with engine operating points shown

Compressor efficiency is compared against performance targets in Table 1 below. As can be seen from the Table, the HP compressor was able to meet or exceed its performance targets at most engine operating conditions. The LP compressor, however missed its performance targets by between 1 and 2 percent. Both compressors achieved similar, absolute performance levels. However, the efficiency target for the LP compressor was set higher, in expectation that the larger diameter would lead to higher performance.

Table 1: Compressor test results compared with target efficiency levels

Operating Point	High Pressure Compressor: Normalized Efficiency Target	High Pressure Compressor: Normalized Efficiency Achieved	Low Pressure Compressor: Normalized Efficiency Target	Low Pressure Compressor: Normalized Efficiency Achieved
4	100.0%	100.1%	98.7%	93.8%
5	99.8%	97.3%	100.0%	98.1%
1	97.8%	96.9%	98.9%	98.0%
2	97.6%	97.6%	98.1%	96.8%
3	97.5%	98.2%	96.8%	96.8%

HP turbine gas stand performance results are shown compared with CFD predictions in Figure 21 below. The CFD and actual performance test data had very good agreement for turbine flow. Aerodynamic efficiency predicted from the CFD analysis was modified to reflect the impact of bearing system power losses, based on HTT bearing power loss rig data. As can be seen from the Figure, the measured turbine efficiency is approximately 5.5 points below the predicted value. It is quite difficult to make a valid comparison between the gas stand and CFD efficiency predictions, as the HP turbine has been design optimized to run under high inlet pulse pressure, and high turbine outlet pressure conditions, whereas the gas stand mapping is run under much different boundary conditions, which led to substantially different blade speed ratios between CFD and test. While in Honeywell experience it is typical for gas stand tests to show 3-4 points lower in efficiency than CFD predictions, the HP turbine stage performance deviation is higher than expected, indicating that the stage may be 1 to 1.5 points lower in efficiency than targeted during the design.

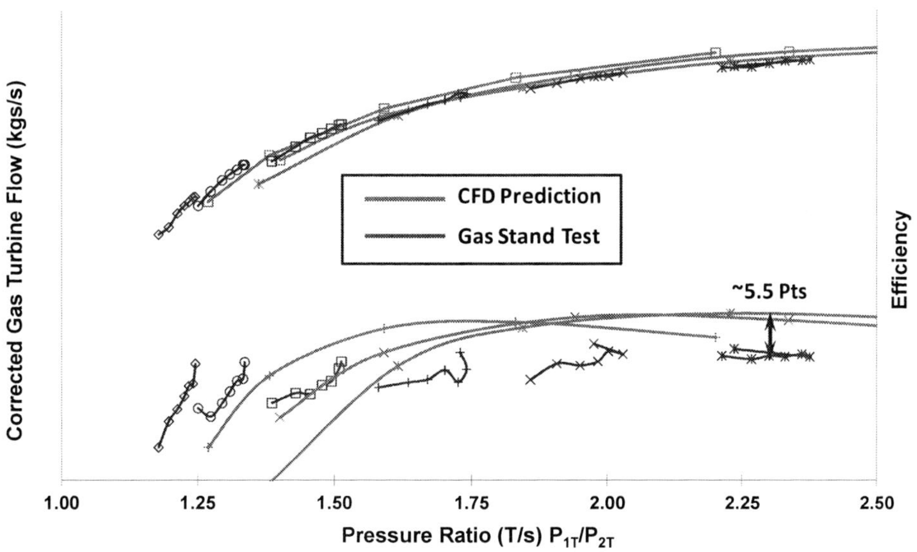

Figure 21: HP turbine measured performance against CFD predictions

Low pressure turbine gas stand performance results are compared with CFD predictions in Figure 22 below. A maximum 2.5% difference in flow between analysis and test result can be observed at higher pressure ratio conditions. As can be seen from the Figure, the measured turbine efficiency is approximately 4 points below the predicted value. This is in line with Honeywell's experience and suggests that the LP turbine is meeting performance expectations.

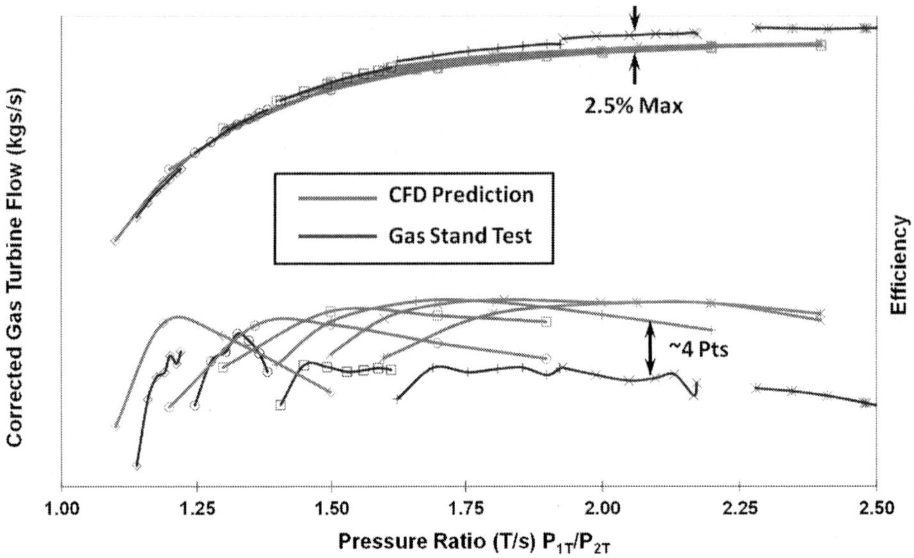

Figure 22: Low pressure turbine performance predicted from CFD (red) and measured on gas stand (blue)

5 SUMMARY AND CONCLUSIONS

An advanced two stage turbocharger system was designed and optimized to deliver reduced fuel consumption to a heavy duty on-highway truck. The novel approach in this project is the complete end-to-end system approach taken to the design optimization. Close collaboration between Volvo and Honeywell enabled the air system to be specifically designed for the unsteady operating conditions of the engine. Additionally, an iterative design approach between HP and LP turbochargers, and turbine and compressor stages, allowed individual component design targets to be compromised in order to achieve the maximum possible overall air system efficiency. The high pressure compressor stand and low pressure turbine stage fully met design targets, whilst the low pressure compressor and high pressure turbine were approximately 1.5 points below target. Honeywell would like to extend special appreciation to Martin Bauer and Johan Engström from Volvo Trucks, without whose support this project would not have been possible.

REFERENCE LIST

(1) Lotterman, J, Schorn N., Jeckel D., Brinkmann F. and Bauer, K.-H., "New Turbocharger Concept for Boosted Gasoline Engines", 16th Supercharging Conference, Dresden, 2011

(2) Cunningham, Jr, T. G., Sandefur, M. H., Brockett, T. S. and McKeever, C. W., "Integrated Conceptual Design Software for Gas Turbine Engines" Proceedings of the XIX International Symposium on Air Breathing Engines 2009 (ISABE 2009) September 7-11, Montreal, Canada. ISBN: 978-1-61567-606-4

(3) Griffith, R. C., Slaughter, S. E. and Mavrosakis, P. E., "Applying Ball Bearings to the Series Turbochargers for the Caterpillar Heavy-Duty On-Highway Truck Engines", SAE Paper Number 2007-01-4235, 2007

VTG turbocharger evolution of BorgWarner Turbo Systems

R Christmann, R Kopietz, N Buhl, H Breitbach
BorgWarner Turbo Systems, Germany

ABSTRACT

The trend of downsizing and downspeeding diesel engines continues, with specific power levels increasing to 80kW/l for engines with a single variable turbine geometry (VTG) turbocharger. This requires the development of VTG turbochargers with improved turbine efficiencies, in order to meet boost pressure demands while improving low-end torque capability and transient response.

This paper describes the development of the sixth-generation VTG turbocharger (Gen6 VTG) from BorgWarner Turbo Systems, with key objectives being cost reduction, stable minimum flow rates, and reduced vane-side clearance. Further thermodynamic gains were realised with two update packages involving optimisation of the VTG cartridge and turbine wheel. The increased turbine efficiency allows smaller turbine wheels to be used, reducing the overall turbocharger size and rotor inertia. The paper concludes with an overview of the potential for the VTG Gen6 cartridge to fulfil both gasoline VTG and high-temperature diesel applications through the use of high-temperature materials.

NOMENCLATURE

CFD	Computational Fluid Dynamics
CHT	Conjugate Heat Transfer
DFMEA	Design Failure Mode and Effects Analysis
DVP&R	Design Verification Plan & Report
EGR	Exhaust Gas Recirculation
Evo	Evolution
FEM	Finite Element Method
FGT	Fixed Geometry Turbocharger
Gen	Generation
HP	High Pressure
LET	Low-end torque
LP	Low Pressure
NOx	Nitrogen Oxide
PGVR	Pre-guide Vane Ring
PFMEA	Process Failure Mode and Effects Analysis
R2S	Regulated Two-stage Turbocharger
RDE	Real Driving Emission
TMF	Thermo-Mechanical-Fatigue
VSC	Vane-Side Clearance
VTG	Variable Turbine Geometry

1 INTRODUCTION

The trend of downsizing and downspeeding diesel engines continues, placing increased demands on the boosting system to enable specific power to be increased up to 80kW/l for diesel engines with a single turbocharger, and up to 110kW/l for multi-stage systems. Meanwhile, the low-end torque capability and transient performance of the turbocharger is to be maintained, necessitating the use of variable turbine geometry (VTG). As such, VTG turbochargers with improved turbine efficiency need to be developed [1]. Improvements in turbine stage efficiency can enable higher boost pressures to be achieved, or alternatively, a smaller turbine wheel can be selected to obtain the same boost pressure, whilst reducing inertia and improving transient performance.

The sixth-generation VTG (Gen6 VTG) of BorgWarner Turbo Systems has been recently introduced for turbocharged diesel combustion engines. The Gen6 VTG was designed as a new product line rather than simply an evolution of the previous model (Gen4), with major changes to the cartridge assembly and mechanism kinematics. These changes have resulted in improved thermodynamic performance, improved mechanism reliability over the lifetime of the vehicle (even under higher exhaust temperatures and thermal gradients), reduced actuation forces, reduced drift in minimum flow rate over time, and reduced boost pressure hysteresis.

This paper details key design features of the Gen6 VTG cartridge, highlighting improvements over previous VTG cartridges. Further modifications to the turbine wheel and volute, and the resulting improvements in thermodynamic performance, are then described. These changes were introduced with the Gen6 Evo and Gen6 Evo+ turbochargers, with the aim of increasing turbine efficiency at high mass flow rates, as well as increasing the range of flow rates through the turbine.

Due to the anticipated introduction of Real Driving Emissions (RDE) standards, the market trend of engine downsizing is expected to slow, as smaller engines emit higher levels of nitrogen oxides (NOx) [2, 3], and as such require more sophisticated and expensive aftertreatment. Instead, a focus on "right-sizing" engines is discussed as a future trend - that is, engines with lower raw NOx emissions and reduced specific power

As such, two different air handling options are feasible. For engines with medium power density (up to 80kW/l), a single-stage VTG turbocharger combined with both high-pressure and low-pressure EGR is preferred. Alternatively, raw NOx emissions could be reduced through a decrease in specific power to 55-70 kW/l, allowing the low-pressure EGR loop to be eliminated. This requires the use of a regulated two-stage (R2S) boosting system for high levels of boost pressure and for control purposes. However, this two-stage system still requires at least one VTG stage, as either the high-pressure or low-pressure turbine stage [3]. Therefore both development paths require high-efficiency VTG turbochargers to meet requirements for specific power, while maintaining good low-end torque and transient response.

2 VTG DEVELOPMENT

Inlet guide vanes have been used in turbines since the early days of turbine technology, but the application of variable turbine geometry using inlet guide vanes in diesel engine turbochargers only gained commercial adoption in the late 1990's. Today, VTG is an almost standard technology for diesel engine turbochargers, such that the majority of diesel passenger cars sold today make use of the technology. As well as single-stage VTG turbochargers, there is also a trend to use a VTG unit as the low-pressure stage of a regulated two-stage boosting system [4]. As such, development of VTG technology needs to account for its application in both single-stage and multi-stage applications.

Figure 1 shows the development of BorgWarner's VTG mechanism since 2002. The main area of development between the third and fourth generations was to the inlet guide vane profile. The introduction of so-called "S-vanes" resulted in improved thermodynamic performance and also benefitted controllability. The load profile over the S-vane is such that in the event of actuator failure, the vanes will default to the open position due to the pressure distribution over the vane surface and the location of the pivot point. This provides a fail-safe design which prevents sudden increases in exhaust pressure if the actuator or linkage fails. Other changes include the method of fixing the VTG cartridge within the turbine housing. This was changed from a contour sleeve to a cup spring, which presses the VTG cartridge against a machined surface in the turbine housing. Maintaining the correct level of spring force is essential to prevent vane sticking, and as such this design requires precise manufacturing tolerances and consideration of relaxation of the cup spring under high exhaust gas temperatures to ensure proper VTG function over the lifetime of the engine. These refinements made to the Gen4 design were carried forward to the Gen6.

Other changes introduced with the Gen6 VTG, such as the introduction of a pre-guide vane ring (PGVR) and the elimination of rollers from the mechanism, are discussed later in this paper. These are also presented in Fig. 1 for reference.

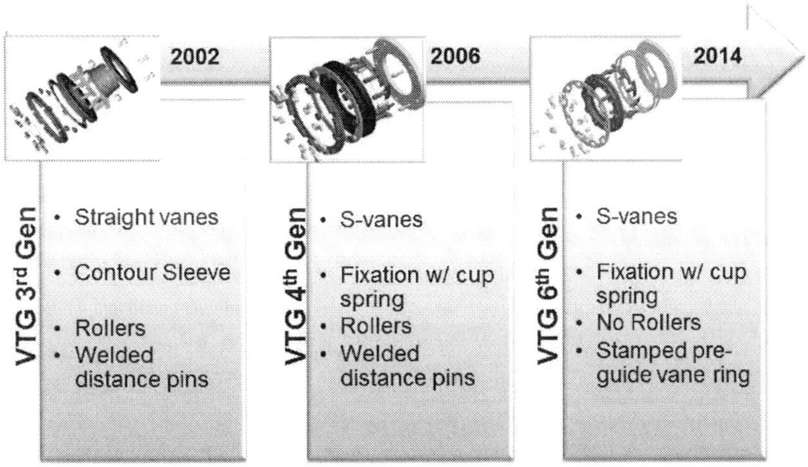

Figure 1: Evolution of BorgWarner Turbo Systems VTG cartridges

2.1 Performance objectives for new VTG generation

The main reasons for the introduction of the VTG Gen6 were an improvement in thermodynamic performance, especially for low-end torque conditions, and significant cost reduction compared to the previous Gen4 cartridge. In all other aspects, performance had to maintained or improved compared to the Gen4 design. The performance criteria for a VTG cartridge are as follows.

- Good turbine efficiencies under engine operation conditions
- No reduction in maximum mass flow rate
- Excellent robustness and wear resistance
- No sticking of vanes over vehicle lifetime
- Low Min-Flow drift over lifetime
- Benchmark controllability over the whole operating range of the engine
- Low actuation forces
- Low boost pressure hysteresis

Each of these criteria can result in different design requirements, and as such, compromises need to be made between the different criteria. For example, the profile and height of the inlet guide vane needs to ensure both aerodynamic efficiency and the low actuation forces (from both friction and aerodynamic loading) required for controllability.

2.2 Development process

As the VTG Gen6 was a completely new design, starting from the concept development stage, the requirement to match or exceed Gen4 performance in all areas is not trivial. For example, the benchmark controllability of the Gen4 VTG cartridge was identified as essential to maintain, even though the Gen6 introduced a new kinematic design and significant cost reduction, resulting in a reduction in the number of parts, fewer precision machining operations and simplified assembly processes. As such, significant emphasis was placed on intensive validation procedures, manufacturing techniques and assembly processes even at the concept development stage. The development process is as follows; however, note that rather than a linear process, many steps are carried out concurrently.

- Identification of promising development paths, concept development and selection. These are informed by the performance achieved and lessons learned from the development of the previous VTG generation.
- Evaluation of several concepts by simulation (CFD and TMF analysis)
- Construction of prototypes and performance evaluation by hot gas-stand tests
- Modification of concepts based on test results, and repeat of simulation and performance tests in an iterative process. Combined with an intensive DFMEA to identify and resolve weaknesses in the design, an optimum final design is developed after several cycles.
- Validation testing and compliance with thorough DVP&R program
- Production concerns: Process capability studies, tolerance stack-up investigations, PFMEA and determination of manufacturing processes, assembly plants and supply chains. The ability to assemble the turbocharger at multiple plants worldwide is a basic requirement of the design.
- Production approval.

Besides the design and validation of the new VTG cartridge, the design and validation of the complete turbocharger in terms of efficiency and mass flow capability is key to meeting the thermodynamic requirements of turbocharged engines. Therefore, a holistic approach needs to be adopted, with consideration of all performance-influencing factors for the whole turbocharger and even at an engine system level, including the turbine wheel, the turbine housing and resulting tip clearances, the bearing system and the compressor stage. The ratio of turbine wheel diameter to impeller diameter also has a significant effect on overall performance. Adaptation of the turbine wheel to the new VTG cartridge designs is presented here; the compressor and bearing system are not covered in this paper.

3 VTG GENERATION 6 CARTRIDGE DESIGN

The design process detailed above produced the specific cartridge design shown in Fig 2. The most critical requirement to avoid vane sticking is to maintain stability of the channel height – that is, the distance between the vane ring and disc - under high temperatures, high temperature gradients, external forces and combustion deposits. The channel height is maintained using welded distance pins in the VTG Gen4. Depending on cartridge size, a different number of pins were required to maintain the channel height over the lifetime of the product. Consideration of both elastic deformation and plastic deformation under applied loading is required, in order to properly specify the correct number of pins for each cartridge size.

Substantial validation through both structural analysis and endurance testing is required to ensure there is minimal creep and minimal fatigue risk over product lifetime, under high temperatures, aerodynamic loading and thermal gradients.

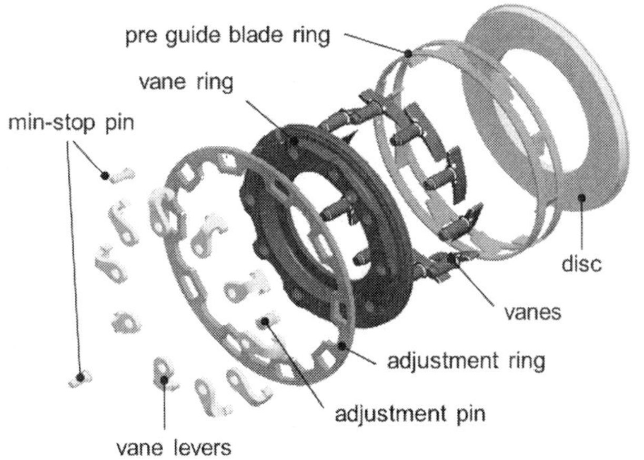

Figure 2: VTG Gen6 cartridge

The VTG Gen6 replaces the welded distance pins with a single generic component, known as the pre-guide vane ring (PGVR). The PGVR is primarily a structural component, manufactured from a nickel superalloy with high creep strength at the temperatures encountered in the turbine stage. As a result, it demonstrates minimal plastic deformation and maintains a stable channel height. Figure 3 shows the simulated plastic deformations of the vane ring. The elastic deformation is likewise simulated in TMF calculations with severe temperature and mass flow gradients.

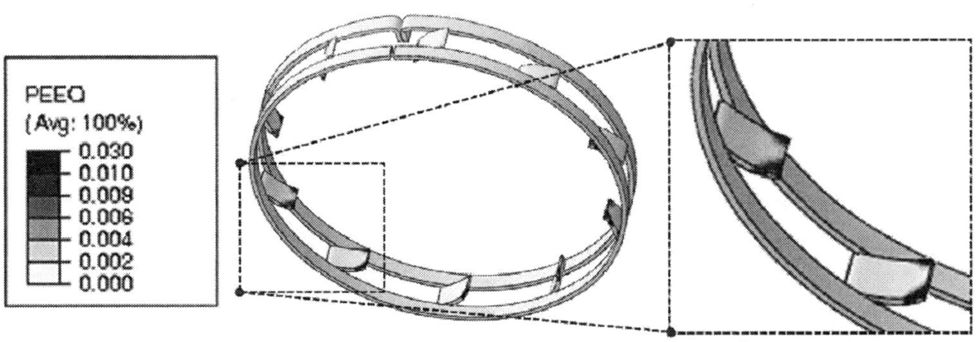

Figure 3: Deformation of the PGVR

The introduction of the PGVR significantly increased the axial load capability of the VTG cartridge. However, the vanes of the PGVR restrict the inlet area and limit the maximum possible mass flow. The detriment to high mass flow performance is not significant compared to the benefits of the PGVR, which improves the performance of the VTG cartridge in two aspects. Firstly, there is reduced variation in flow rate over the lifetime of the product. Secondly, it allows the nominal vane-side clearance (VSC) to be reduced. The VSC is the gap between the end face of the vane and the disc, and has a significant effect on LET performance. There is a trade-off to be made between minimising the VSC to ensure peak efficiency, and maintaining sufficient clearance to avoid sticking at high temperatures. As the PGVR is a single component and permits a simplified assembly procedure over the welded pins, it allows the nominal VSC to be reduced without the risk of vane sticking with parts at the limits of manufacturing tolerances.

Besides the introduction of the PGVR, the VTG linkage was also redesigned, simplifying the mechanism and reducing the forces acting on the vane shafts. The reduction in part count is achieved by designing the VTG mechanism so that multiple kinematic functions are combined in the one interface. The mechanism design is shown in Fig 4. The adjustment levers (labelled "a") slide along a curved profile formed by stamped windows in the adjustment ring (labelled "b"). These curved profiles allow the lever to rotate around 30°. The end points of the curves the lever contacts against the edge of the stamped window in the adjustment ring, which provides a mechanical stop to the motion of the lever. Also, in the previous Gen4 VTG, the coaxiallity of components in the VTG cartridge was controlled by a set of rollers which prevented radial motion of the adjustment ring. In the Gen6 VTG, the rollers have been eliminated, and their function is transferred to the levers. Contact between the outer profile of the lever and the stamped window of the adjustment ring prevents the adjustment ring from moving off-centre. As such, the contour shape of the levers and the windows of the adjustment ring allow three functions to be combined in the interface between two components

a) Rotational motion is enabled
b) Radial motion is restricted
c) A maximum stopper position is provided.

The support of radial forces over the surface of the adjustment lever leads to a reduction in contact stresses compared to the rolling contact condition provided by the rollers.

The bearing forces on the vane shaft were also reduced. This was achieved by moving the adjustment ring axially towards the midpoint in the shaft, requiring a 90° bend in the adjustment lever (Fig 4). The additional leverage provided by the 90° bend in the lever results in a more even loading distribution on the vane shaft. This has the benefit of reducing the required actuation force and reducing wear on the vane shafts over their lifetime. The ratio between the diameter and length of the guide vane shaft was also optimised to reduce friction and wear. The high precision blanking and bending processes used to manufacture these components also offers very low levels of surface roughness and low friction coefficients. The combined effect of all these changes allows a great reduction in frictional forces to be realised with this design.

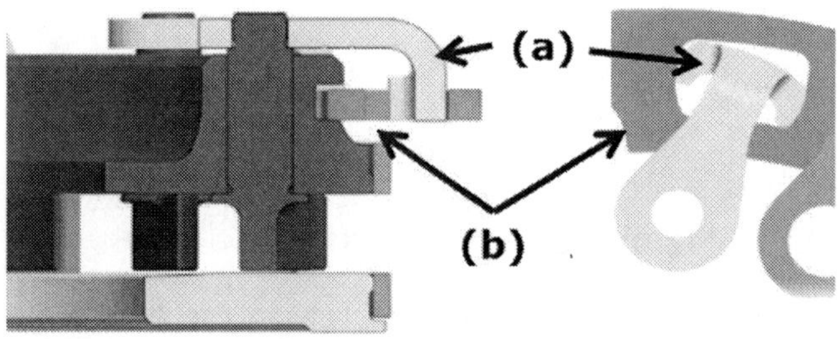

Figure 4: Mechanical design of the VTG cartridge

As frictional coefficients have a significant impact on both the mechanical design of the cartridge and wear over product life, the selection of suitable friction pairings with high temperature capability is essential, and necessitates the use of high-temperature-grade materials to withstand current diesel exhaust temperatures of 860°C and above. A further increase in temperature capability to permit use in gasoline applications can be achieved by replacing the material used in particular components with a higher temperature grade without requiring design changes to the VTG cartridge, and is further discussed in Section 4.3.

4 VTG GEN6 EVOLUTION

After the successful launch and application of the Gen6 VTG cartridge in 2014, the focus of development shifted towards meeting future industry trends: in particular, meeting anticipated demands for specific power of up to 80kW/l for diesel engines. This required an enhancement of the base Gen6 VTG cartridge to provide improved turbine stage thermodynamic performance and higher exhaust temperature capability. The good performance of the base Gen6 VTG in terms of controllability and reliability were to be maintained, and a further constraint was that the new "Gen6 Evo" cartridge was to be fully interchangeable with the basic Gen6 hardware, meaning that the external dimensions of the cartridge could not be changed.

The development process is similar to before. Concept development and selection, followed by simulation (aerodynamic, conjugate heat transfer and structural analyses). The simulation results inform modifications to the design, various prototypes are manufactured, and performance is evaluated through special component tests, hot gas stand tests and on-engine dynamometer testing. Analysis of these results, combined with past experience, allows final optimisation of the design. Analysis of the aerodynamic performance of the previous generation and lessons learned from its development permit the identification of the most promising areas for development to improve thermodynamic performance.

4.1 VTG Gen 6 Evo: Performance improvements

The targeted improvement in thermodynamic performance of the VTG Gen6 Evo is shown in Fig 5. The turbine efficiency of the previous Gen4 and Gen6 VTG units is also presented for comparison. As can be seen, the Gen6 VTG already improved the LET capability through increased efficiency at low mass flow rates. An improvement in peak efficiency was also realised. However, the introduction of the PGVR led to a decrease in efficiency at high mass flow rates.

The development target for the Gen6 Evo, as presented in Fig 5, was to improve the turbine efficiency at off design conditions while maintaining the same peak efficiency. As such, both the LET capability (at low mass flow rates) and the performance at the engine's rated power condition (high mass flow rates) would be improved.

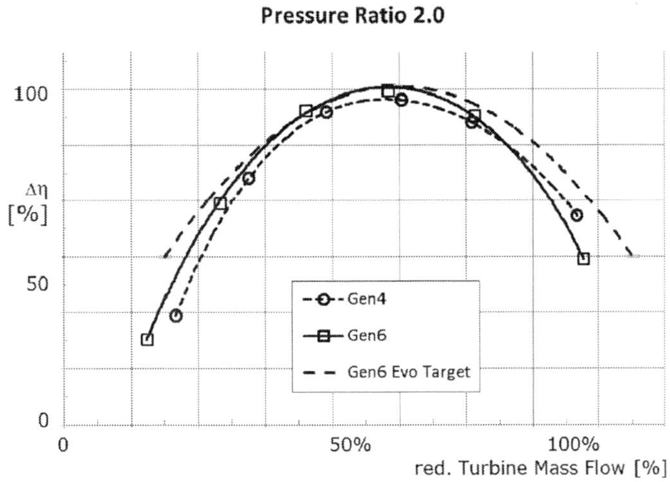

Figure 5: Turbine efficiency $\eta_T\eta_M$ of Gen4 and Gen6 VTG stages

The most promising options for improvement of thermodynamic performance were identified as

- Further reductions in VSC without risking sticking issues.

- Refinements of the profile shape of the PGVR.
- Improved manufacturing tolerances for components, leading to tighter clearances due to a reduction in the tolerance stack-up in assembly of the cartridge.
- Inlet guide vane profile optimisation.
- Aerodynamic improvement of the turbine wheel.

Three of these possible development routes are discussed further: namely, the optimisation of the PGVR profile, the reduction of tolerance stack-up in the VTG cartridge assembly, and optimisation of the turbine wheel.

The primary function of the PGVR is to maintain the height of the flow channel over the vehicle lifespan. However, the presence of the pre-guide vanes lead to losses in the air flow. Figure 6 shows the refinements made to the shape of the PGVR. The shape of the pre-guide vane was changed in three aspects.

- Reduced chord length, resulting in lower frictional losses.
- Reduced stagger angle.
- Improved trailing edge shape, reducing the wake loss from the trailing edge.

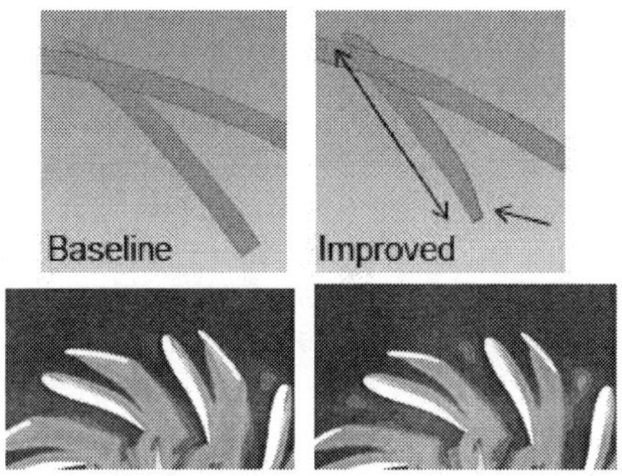

Figure 6: PGVR modifications (left: VTG Gen6; right VTG Gen6 Evo)

Figure 6 also shows the velocity distribution around both the Gen 6 and Gen 6 Evo pre-guide vanes at high mass flow rates. It can be seen that the improved pre-guide vane profile results in reduced losses of dynamic pressure. As such, the optimised design provides improved high mass flow characteristics while maintaining the stability of the channel height.

The reduction in plastic deformation also permits an improvement in LET capability through reduction of the VSC by approximately 10%. As the VSC is the clearance between the moving inlet guide vane and the disc, it can be further reduced by more precise component manufacturing tolerances, as well as improved fixturing and welding tolerances in the assembly of the cartridge itself. The more precise control over tolerances also reduces product-to-product variation in thermodynamic performance. Overall, improved control over manufacturing and assembly tolerances allowed the VSC to be reduced by a further 5%.

The changes made to the VTG cartridge also required the turbine wheel to be adapted with respect to the number of blades, the blade shape and the blade inlet height. The interaction between the moving inlet guide vanes and the turbine wheel needs to be optimised in order to achieve the best possible turbine efficiency. The newly developed turbine wheel and optimised PGVR resulted in improved thermodynamic performance which even exceeded that predicted by CFD simulations during the development process. Trends seen in the CFD results were reflected in physical testing, but absolute efficiency values were slightly better than expected. The performance of the Gen6 Evo turbine stage is compared to the standard Gen6 in Fig 7.

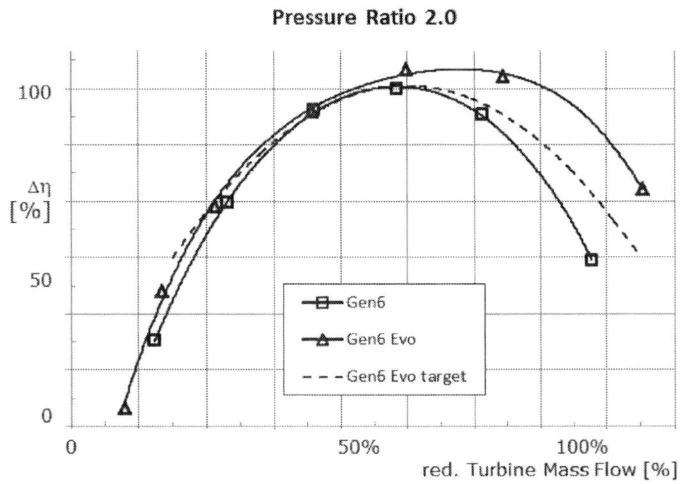

Figure 7: Turbine efficiency $\eta_T\eta_M$ of Gen6 Evo compared to previous design

4.2 Development of the VTG Gen6 Evo+ turbine stage

As one of the requirements for the Gen6 Evo cartridge was to maintain exchangeability with the basic Gen6 cartridge, further improvements in thermodynamic performance were limited by existing design constraints. However, measures had already been identified which would improve thermodynamic performance without changing the basic architecture of the Gen6 cartridge. Namely, an increase in the vane height, and consequently increased channel height, showed significant thermodynamic improvements on CFD simulations. This violated the principle of 100% part exchangeability, but the only necessary modifications to the existing turbocharger assembly were a machining operation on the turbine housing. The taller vanes also required a change of turbine wheel to one with a larger blade height at inlet, and a detailed CFD parameter study was conducted to optimise the new wheel to the taller inlet guide vane design. The definition of other turbocharger components remained unchanged. The performance of the optimised Gen6 Evo$^+$ turbine stage is compared to the basic Gen6 and Gen6 Evo in Fig 8.

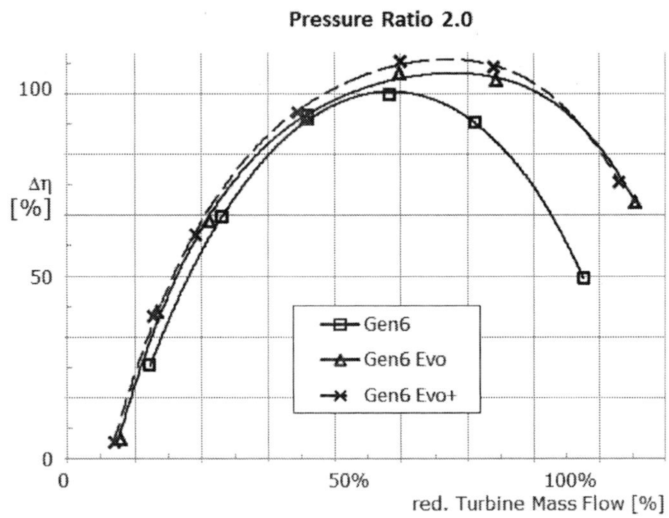

Figure 8: Turbine efficiency $\eta_T\eta_M$ of Gen 6, Gen6 Evo and Gen6 Evo$^+$ designs

However, a side effect of the taller vanes is an increase in the actuation forces for the Gen6 Evo cartridge due to increased aerodynamic loading. The trade-off between improved thermodynamic performance and higher actuation forces needs to be considered for each new vehicle application.

The performance increase showed that a step down to one smaller VTG size should be possible, which decreases the turbine wheel moment of inertia and therefore improves the time-to-torque capability. This is only possible because the overall efficiency does not decrease compared to a bigger VTG turbocharger size. In particular, efficiency is increased at low mass flow rates, which is important to achieve good LET performance. However, it is only because there is also a parallel increase in efficiency at high mass flow rates that a smaller turbine size can be considered while still fulfilling the rated power operation point for the engine. With the opportunity to decrease the turbine wheel diameter usually the compressor wheel is also reduced in diameter, but then the compressor wheel suffers from reduced performance at the choke limits on the compressor map. This bottleneck can be resolved by using compressor wheels with a relatively small trim and a wide compressor map, or to consider higher diameter ratios of compressor wheel vs. turbine wheel.

4.3 VTG Gen6 for Gasoline applications

Due to the introduction of tight CO_2 emission legislation targets such as EURO 7 legislation, it is necessary to improve the efficiency of gasoline engines. Downsizing, especially if cylinder numbers can be reduced, and down-speeding are seen as major contributors to meet emissions targets. Fuel economy can be further improved by increasing the base engine compression ratio to enable better engine low- and part load fuel efficiency [5]. To avoid penalties in high engine load fuel efficiency, variable compression ratio or adaptation of real compression ratio vs. geometrical fixed compression ratio is currently under consideration. As fully variable compression ratio introduction is still pending, alternative combustion processes like Miller or Atkinson cycle have to be utilized [6, 7]. Especially for these new combustion processes a VTG TC has remarkable advantages. On the basis of the wider operating range of a VTG compared to a conventional wastegate TC, high boost pressure for good LET behaviour can be achieved and simultaneously a high rated power output is possible [5]. The major advantage of gasoline VTG compared to a fixed geometry TC (FGT) is given by the boost pressure control via controlled variable guide vanes at full exhaust mass flow where the FGT uses the wastegate for turbine power control, resulting loss in mass flow and thermal energy.

The first introduction of VTG in 2005 in a Porsche application was also derived in the beginning from a previous diesel VTG design [8, 9]. However, the modifications made to the then-contemporary Gen3 VTG architecture were much more difficult than they would be today with the Gen6 design. The modifications required to adapt a VTG turbocharger for gasoline application are firstly, adaptation of the VTG cartridge and turbine housing to withstand the higher exhaust gas temperatures, and secondly to take the different boosting requirements of gasoline engines into account.

First of all, as gasoline engines generally require a higher mass flow capability, an adapted high flow turbine wheel is needed. In addition, low rotor inertia is vital for transient performance. Therefore the turbine wheel for the turbocharged gasoline engine needs to be as small as possible in diameter, and the turbine wheel design itself should lead to a lower moment of inertia, while maintain or improving upon the efficiency of the previous larger wheel design. In comparison to the diesel turbine wheel, the high mass flow turbine wheel used for gasoline applications has reduced performance at low mass flow rates and a slightly lower peak efficiency. However, the gasoline VTG is not usually operated in very closed VTG positions, and so the loss of low mass-flow efficiency is not critical. In contrast, high performance at high flow rates and wide-open VTG positions is essential. This is because the wide open position of the VTG defines the control margin required to maintain performance at high altitudes.

An alternative to using a high-flow turbine wheel for the gasoline VTG is to use a wastegate in combination with the VTG cartridge, allowing flow to bypass the wheel at wide-open throttle and providing an additional means of controlling boost pressure. However, the gain in performance needs to be offset by the cost and difficulty of installing an extra wastegate flap and actuator in addition to the VTG actuator. As such, the development of high-flow turbine wheels for use in gasoline VTG applications is essential. The desired performance characteristic for the high-flow wheel is shown in Fig 9, and is compared to both an equivalent wheel for a diesel VTG application, and the initial development targets for a high-flow wheel for a specific application.

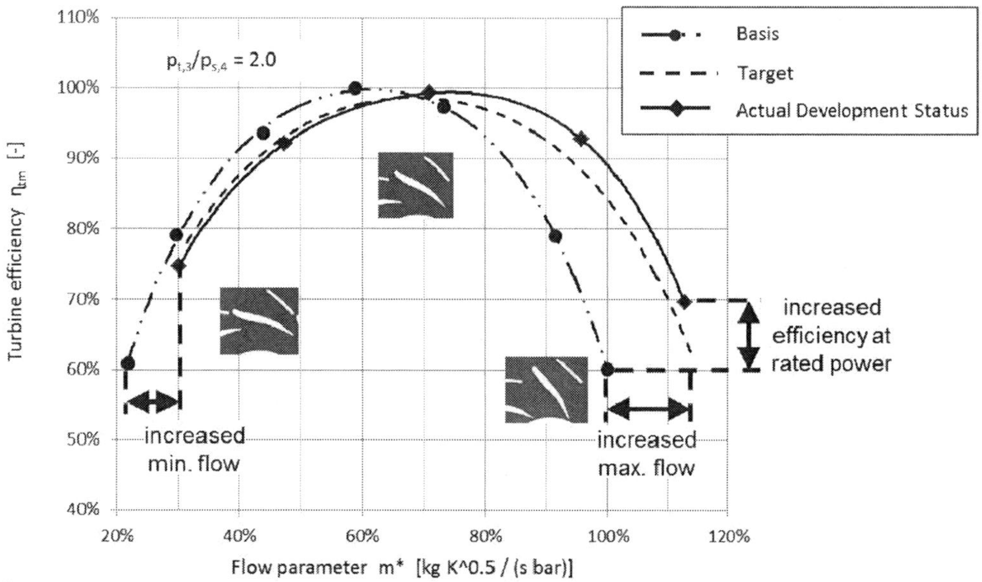

Figure 9: Comparison of Turbine efficiency $\eta_T\eta_M$ for standard and high-flow turbine wheels

As can be seen, the peak efficiency point of the turbine has been shifted to higher mass flow rates, with a slight decrease in peak efficiency. Most importantly, the swallowing capacity has increased by more than 10%, permitting a wide operating range, while, maintaining good transient response and sufficient boost pressure at the rated power point.

In comparison to previous VTG architectures adapted for gasoline application, the modifications required for the VTG Gen6 are mainly material upgrades, and depending on the exhaust temperature range, may be relatively minor. In fact, these material upgrades are also expected to be required in diesel applications due to an increase in diesel exhaust temperatures to 900°C in the next five years. In particular, regulated two-stage (R2S) applications with high specific power output at least one VTG stage are expected to operate at 900°C [4].

The specific components which require material upgrades are the PGVR and, depending on the application, the vane ring, vanes and disc. The adoption of the Miller cycle in gasoline engines leads to a decrease in exhaust temperatures compared to conventional gasoline engines. Depending on the specific power output of the engine, exhaust temperatures can typically vary between 900°C and 950°C, and even up to 980°C, the change in materials is quite straightforward.

As an example, the PGVR is required to maintain the channel height, with minimal plastic deformation, acceptable stress levels, and low risk of thermomechanical fatigue under high temperature gradients. In this case existing structural models can be utilised, and the material grade that is required to meet the higher temperature requirement has already been identified. This result has been verified through prototype tests where the PGVR was subjected to a static axial load at high temperatures, and the change in height monitored over time.

Figure 10 shows the comparison between the levels of deformation exhibited over a number of hours by the existing, diesel application PGVR and the new high-temperature grade component. The applied load was equivalent to the force exerted by the cup spring and the temperature was 900°C. As shown, under the maximum load applied in application at a temperature of 900°C, the new material exhibits 45% lower plastic deformation compared to the existing PGVR.

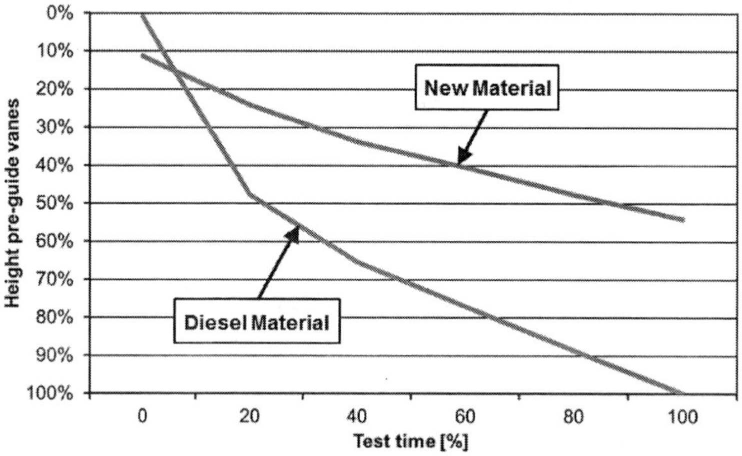

Figure 10: Stability of PGVR under static load tests

For cyclic mechanical loads and temperature cycles, TMF analysis has already identified areas where the design can be optimised and weaknesses of the material chosen. Validation tests, both on hot gas stands and on the engine dynamometer support the results of this simulation work so far. Further tests are planned and validation is ongoing.

High-temperature yield strength and creep strength are not the only factors affecting material selection for gasoline application. Friction coefficients at high temperatures, high-temperature oxidation resistance, and the effect of different thermal expansion coefficients needs to be considered as well. The turbine housing material is usually changed to stainless steel, and the bearing housing requires water cooling, resulting in significantly altered temperature distributions and gradients. Differential thermal expansion results in a reduction of the VCS at high temperature. Combined with an expected increase in both elastic and plastic deformation over the lifetime of a gasoline engine, this means a larger nominal VSC needs to be adopted in order to ensure a similar level of VSC to that used in diesel applications, having said this, a tight VCS is not as critical in gasoline applications as the VTG is rarely operated in very closed positions. The different composition and level of deposit build-up during the lifetime of a gasoline engine compared to a diesel VTG also needs to be considered when sizing the VSC in a gasoline turbocharger.

This summarises the main changes required to meet gasoline application requirements with a peak exhaust temperature of 980°C. For units operating above 980°C, the changes are much more complex. The final design of the gasoline VTG Gen6 will be presented as soon as all validation testing is finished.

5 CONCLUSION AND OUTLOOK

The history of VTG development by BorgWarner Turbo Systems since 2002 has been presented, up to the introduction of the newest generation of VTG turbochargers. In addition to the robust new VTG cartridge design, the impact of thermodynamic updates has been presented, with the resulting performance improvement capable of meeting the latest diesel application requirements. The potential for adapting the Gen6 VTG cartridge to both gasoline and high-temperature diesel applications has been described. Only moderate technical modifications and on-cost are required to achieve an increase in exhaust temperature capability suitable for gasoline engines using the Miller cycle, as well as for diesel applications with exhaust temperatures up to 960°C, which are anticipated within the next five years. With further material upgrades and design modifications, temperature capability can be increased. These materials are capable of permitting the VTG Gen6 to replace the first production gasoline VTG, which is currently still based on the VTG Gen4, in the near future. As such, the current VTG Gen6 has potential to cover the requirements of both highly boosted diesel engines and also gasoline turbocharged engines.

Aerodynamic requirements for gasoline VTG applications - in particular - turbine wheel design - have also been discussed. The demand for improvements in transient performance results in a trend towards using smaller turbochargers in order to reduce rotor inertia, while maintaining peak efficiency and good LET capability. The requirements for improved transient response, high boost pressures and high efficiency at LET conditions drives the introduction of VTG in gasoline applications, especially for engines using the Miller cycle.

Work on the next generation of VTG is ongoing, and the new generation is expected to lead to further reductions in VSC and both axial and radial clearances in the turbine and compressor. In addition to further refinement of the VTG cartridge and turbine wheel, further improvements in efficiency, transient response and LET capability are expected through advances in bearing technology, with significant improvements in performance at LET and part-load conditions expected through the development of both ball bearings and low-friction journal bearings.

REFERENCE LIST

[1] Christmann, R.: New VTG Generation Turbocharger of BorgWarner Turbo Systems, The 14th Hyundai-Kia Powertrain Conference, Namyang, 2014.

[2] Holderbaum, B.: Emission reduction strategies for passenger car Diesel engines to meet future EU and US emission regulations, The 14th Hyundai-Kia Powertrain Conference, Namyang, 2014.

[3] Stütz, W., Steinparzer, F., Brüne, H.-J., Raschl, P.: Diesel engines for the Future of Mobility. New Technology Developments at BMW, 26th International AVL conference "Engine & Environment", Graz, 2014.

[4] Nefischer, P. , Hall, W., Honeder, J., Steinmayer, T.: The 1st Diesel Engine with 2-stage Turbocharging and Variable Turbine Geometry in Passenger Cars, 18. Aachener Kolloquium Fahrzeug- und Motorentechnik, Aachen, 2009.

[5] Pucher, H., Zinner, K.: Aufladung von Verbrennungsmotoren, Springer Verlag, 4. Auflage, 2012.

[6] Schutting, E., Neureiter, Fuchs, C., Schatzberger, T., et al: Miller- und Atkinson-Zyklus am aufgeladenen Dieselmotor, MTZ, Volume 68, 06/2007.

[7] Scheidt, M., Brands, C., Lang, M., et al.: Kombinierte Miller-/Atkinson-Strategie für zukünftige Downsizing-Konzepte, Internationaler Motorenkongress, 2014.

[8] Gabriel H., Lingenauber, R., Ramb, T.: Der Turbolader mit variabler Turbinengeometrie (VTG) für den neuen Porsche 911 Turbo - Ein Meilenstein in der Ottomotorenaufladung, 11. Aufladetechnische Konferenz, 2006.

[9] Sterner, A., Hofstetter, M., Kerkau, M., Beer, M., Ronneburger, R., Knirsch, S.: Die variable Turbinengeometrie für die ottomotorische Anwendung beim neuen 3,6 l Biturbo - Motor des Porsche 911 Turbo, 11. Auflade-technische Konferenz, 2006.

SESSION 2B

GAS STAND AND NOVEL TESTING METHODS

The E-drive: An isentropic compressor performance test rig

M Bergmann, P Parma
BorgWarner Turbo Systems Engineering GmbH, Global Strategies & Methods, Germany

ABSTRACT

The majority of the operating points of current Passenger Car drive cycles are beyond the boundaries of the conventional compressor map as measured on a standard gas stand (Figure 1). With such measured compressor maps, it is not possible to describe these points without interpolation tools for the pressure ratio and the efficiency of lower speed lines. This latter data, however, is needed for engine simulation by the OEM to simulate emissions, fuel economy and performance over the entire operating range of the engine. The E-drive compressor performance test rig is capable of measuring compressor maps at low mass flows and pressure ratios not previously available with conventional gas stand equipment. This paper shows a comparison between compressor map measurements on the E-drive test rig and a gas stand as well as the possibility to calculate the heat flux influence from turbine to compressor side out of the measured data. A combination of compressor maps measured on the two different test rigs allows characterisation of all operating points of a current drive cycle.

NOMENCLATURE

c	flow velocity
d	diameter
g	gravitational acceleration
h	specific enthalpy
I	current
$\dot{m}$	mass flow
n	rotational speed
p	pressure
P	power
q	specific heat
T	temperature
u	compressor wheel tip speed
U	voltage
w	specific mechanical work
z	height
η	efficiency
λ	powerfactor
φ	flow coefficient
ρ	density

Subscripts

1	compressor stage inlet
2	compressor stage outlet
12	difference state between 1 and 2
3	turbine stage inlet
4	turbine stage outlet
C	compressor
CW	compressor wheel
E	electrical
ED	E-drive compressor performance test rig
GS	gas stand
is	isentropic
L	losses
M	electric motor drive
t	total
u	circumferential

Abbreviations

CFD computational fluid dynamics
EGR exhaust gas recirculation
FTP federal test procedure
GS gas stand

1 INTRODUCTION

At the beginning of each turbocharger development process between the OEM and the manufacturer, input data is needed for simulation work in order to minimise the number of development loops.

Figure 1 gives an overview of the current output post-processing of simulation of the FTP 75 test cycle for a diesel engine with high and low pressure EGR, based on measurements on the engine test stand, chassis dynamometer and gas stand. The compressor and the engine maps show the operating points of the drive cycle and the full load curve. The applied test cycle is shown in the lower plot.

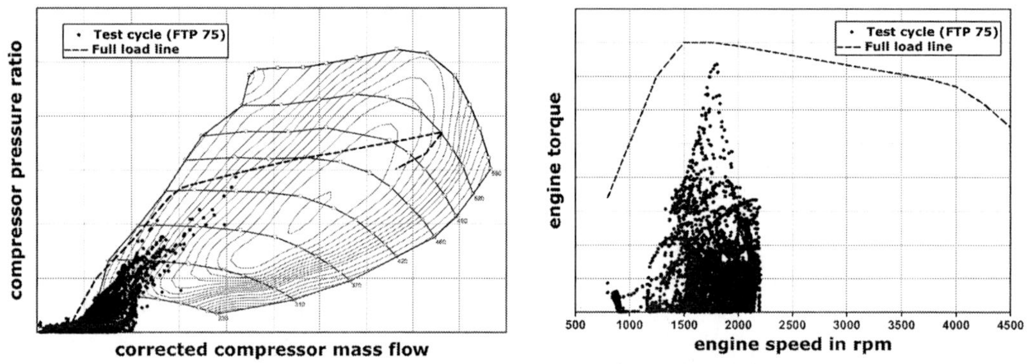

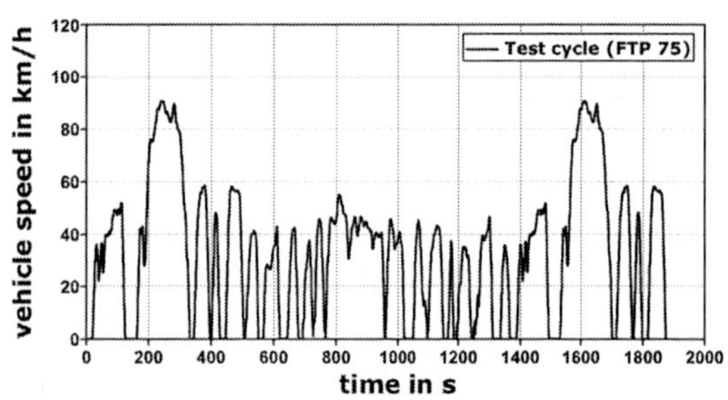

FTP 75 (Federal Test Procedure) was developed from FTP 72 and has been the official American driving cycle since 1975. The cycle uses a warmed-up engine and a maximum test vehicle speed of about 90 km/h.

Figure 1: Operating points of current American drive cycles compared to a compressor and engine maps

It is obvious, that the majority of the operating points are beneath the compressor map, at low mass flows and pressure ratios.

Simulation work or a different measurement method is needed to characterise the compressor at these points, because the accuracy of "low end" compressor mapping on a gas stand is severely limited by the impracticality of accurate temperature control at low mass flows, as well as by a negative heat influence of the turbine temperature on the compressor efficiency-calculation under these conditions. A further reason to use an alternate measurement method is that natural gas burners are inefficient below a power output of 5 kW.

One way to generate the needed low power and to significantly reduce the heat influence is to use an electrical motor drive. In this manner, no turbine wheel is needed to accelerate the compressor. The compressor is directly driven by the electrical motor.

2 COMPARISON OF THE TEST RIGS

The essential element of the E-drive test rig is the electrical motor drive used to drive the compressor wheel.

The key properties of the motor are:

- Power: up to 18 kW (nominal)/25 kW (peak)
- Speed: 24.000 – 125.000 rpm
- Water cooling
- Foil-air bearings

Figure 2: Build-up and key properties of E-drive the test rig

Different compressor wheels can be adapted to the rotor and the proper compressor housing is attached directly to the electrical motor drive. The rotor has to rotate with at least 24,000 rpm, because of the load capacity of the foil-air bearing system. Below this minimum speed, mixed friction occurs. On the inlet and outlet sides of the compressor, measurement instrumentation for compressor mass flow, temperature, and pressure sensors are installed. A back pressure unit allows a variable adjustment of the compressor outlet pressure.

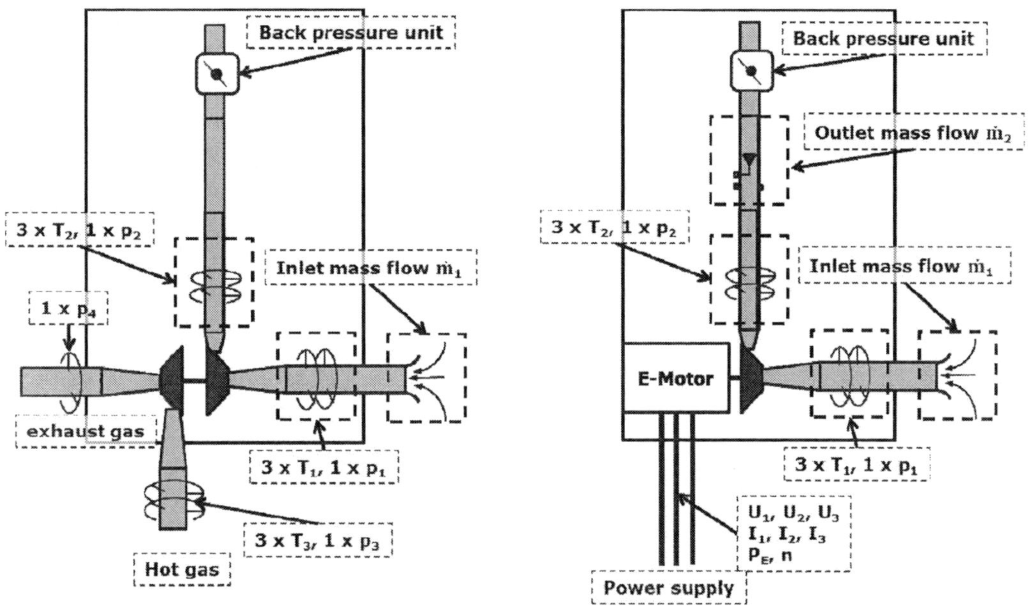

Figure 3: Schematic overview of measuring equipment on a gas stand (left) and E-drive compressor performance test rig (right)

The electrical motor drive is supplied with 3-phase alternating current. To measure the actual current and voltage and the corresponding input power P_E, additional instrumentation is installed. Figure 3 shows the differences in the instrumentation of a gas stand and the E-drive compressor performance test rig. For the best possible comparison of the two test rigs, the same design of measurement stacks at intake and outlet are used.

The water-cooled electrical motor drive allows a compressor map measurement with minimal heat influence on the compressor flow. The minimum speed of 24.000 rpm also allows measurements at low tip speeds (e.g. down to 75 m/s with a compressor wheel diameter of 60 mm).

The well-known definition of the isentropic compressor efficiency on a gas stand is

$$\eta_{is,C,GS} = \frac{h_{2t,is} - h_{1t}}{h_{2t} - h_{1t}} = \frac{\Delta h_{12t,is}}{\Delta h_{12t}} \quad (1)$$

but for the E-drive test rig, a new definition is needed.

The electrical power of the electrical motor drive is given by the power of the electrical supply minus the losses:

$$P_M = P_E - P_L \quad (2)$$

P_L is the sum of all the losses in the electrical motor drive and the bearing system, such as copper losses, core losses, friction losses and others. The power of the electrical motor drive equals the power of the compressor wheel:

$$P_{CW} = P_M \quad (3)$$

The power of the compressor wheel P_{CW} can be calculated by solving the Euler equation, which results in

$$P_{CW} = \dot{m}\, u_2\, c_{u2} = \dot{m}\left[\Delta h_{12} + \frac{1}{2}(c_2^2 - c_1^2)\right] = \dot{m}\, \Delta h_{12t} = P_E - P_L \quad (4)$$

Transforming equation (4) leads to an equation for the total enthalpy:

$$\Delta h_{12t} = \frac{P_E - P_L}{\dot{m}} \quad (5)$$

At this point it is possible to replace the total enthalpy Δh_{12t} in equation (1) with equation (5) and the isentropic efficiency for the E-drive can be defined as:

$$\eta_{is,C,ED} = \frac{\Delta h_{12t,is}}{\Delta h_{12t}} = \frac{\Delta h_{12t,is}}{\frac{P_E - P_L}{\dot{m}}} \quad (6)$$

Equations (1 - 6) show, that the efficiency of the E-drive test rig is calculated by the compressor inlet temperature, the inlet and outlet pressures, the mass flow and the electrical power of the motor and its losses.

On a gas stand, the compressor efficiency is completely calculated with both enthalpies, as in equation (1). A comparison of the two definitions of the efficiency shows, that the difference in the calculation is the determination of the total enthalpy difference Δh_{12t} (comparison of equation (6) and (1)).

$$\Delta h_{t12_{ED}} = \frac{P_E - P_L}{\dot{m}_C} \neq \Delta h_{t12_{GS}} \tag{7}$$

Figure 4 shows the result of a study of the efficiency behaviour of the E-drive test rig immediately after a load change, where the compressor efficiency is calculated with equations (1) and (6). The efficiency calculated with the electrical power P_E immediately takes a constant value. The reason is, that the current and voltage do not need any observable stabilising time. The efficiency calculated with the enthalpies (e.g., on a gas stand), rises to a high value and then slowly decreases until it converges with the line of efficiency calculated with the electrical power. The explanation is the stabilising time of the temperature at the compressor outlet.

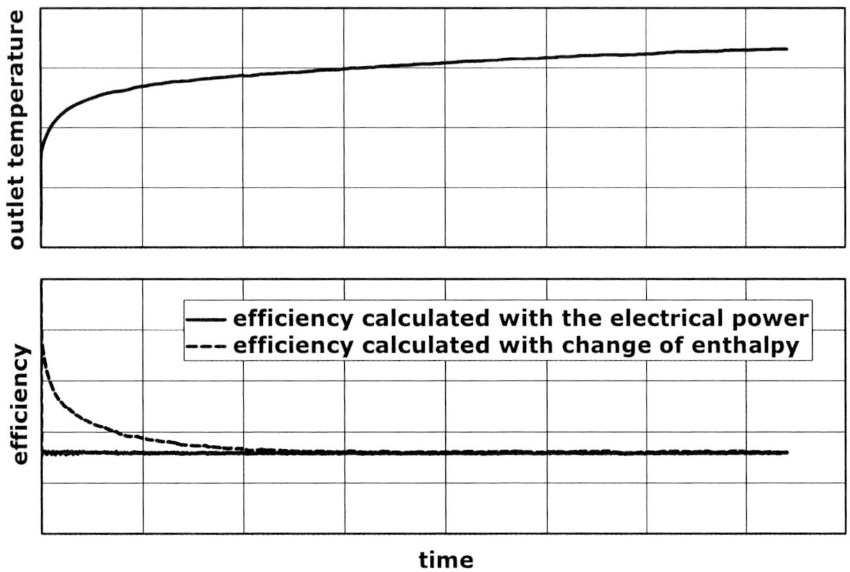

Figure 4: Comparison of the efficiency calculations

It is an important finding for the comparison of the two test rigs, that the two definitions of the efficiency deliver the same value, if the system is under thermodynamic steady-state conditions.

The energy balance of the two systems (E-drive and gas stand) is given by the first law of thermodynamics:

$$q_{12} + w_{12} = h_2 - h_1 + \frac{1}{2}(c_2^2 - c_1^2) + g(z_2 - z_1) \tag{8}$$

Assuming that the mechanical work of the compressor is the same on both test rigs as well as the friction losses, it is possible to determine the heat influence on the gas stand by calculating the difference of total enthalpy between the two test rigs. With the definition of the total enthalpy

$$\Delta h_{t12} = h_2 - h_1 + \frac{1}{2}(c_2^2 - c_1^2) \tag{9}$$

the specific heat flux can be written as:

$$q_{heat\ flux} = \frac{P_E - P_L}{\dot{m}_C} - \Delta h_{t_{GS}} \tag{10}$$

3 INFLUENCE OF HEAT FLUX ON A COMPRESSOR MAP

Conventional measured compressor maps are based on the measurement of temperature, pressure and mass flow before and after the compressor stage. With those values, it is possible to calculate the compressor power consumption and the efficiency of the compressor stage.

At the gas stand the hot exhaust gas heats the turbine wheel and the turbine housing. The turbine housing will lose a proportion of this heat to the environment due to natural convection. Another part of the heat transfers to the compressor side. Additional losses are due to heat exchange between the bearing system and the oil, as well as between bearing housing and environment. The dominant heat transfer effects are conduction and convection, radiation plays a minor but not negligible role (1). Figure 5 shows the different heat conduction processes in a turbocharger (2).

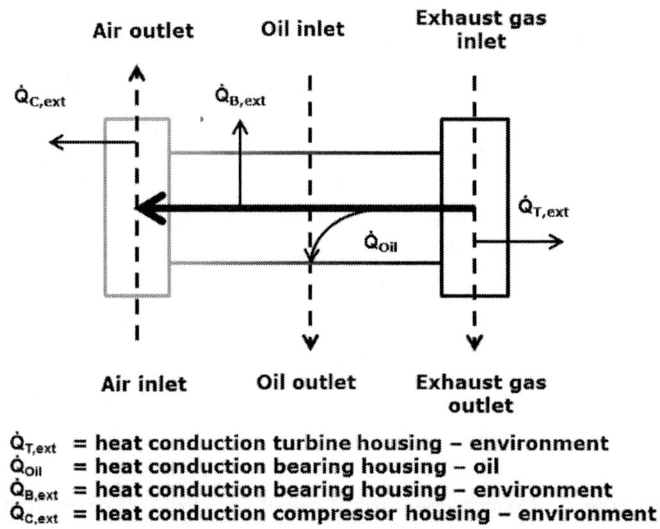

Figure 5: Heat conduction processes in a turbocharger

This leads to a rise of temperature of the gas, which flows through the compressor wheel, in addition to the heating by the compression process. As a result the temperature measurement behind the compressor wheel is not accurate and leads to an inaccurate calculation of the efficiency (1). This apparent efficiency is not identical to the real efficiency determined under adiabatic conditions ((3),(4),(5)).

The non-adiabatic change of state leads to a higher change of enthalpy, which results in a lower efficiency of the compression process. With increasing rotational speed, the heat flux decreases because the temperature at the compressor outlet rises with higher compression rates reached at higher speeds. Another reason is the expansion in the turbine, which results in less heat conduction into the turbine housing and the compressor side. Therefore, the temperature difference between turbine and compressor sides decreases and therefore the heat transfer reduces also ((6),(7)).

Various work has been done to simulate the heat transfer in a turbocharger or to calculate the real efficiency from the measured apparent efficiency. One approach is to consider the efficiencies for different tip speeds plotted versus the outlet flow coefficient. The flow coefficient (equation (11)) describes the flow rate through the turbocharger and is similar to the mass flow and the volumetric flow. This coefficient is established as a design parameter for turbo machines and allows comparison over all frame sizes.

$$\varphi_2 = \frac{4}{\pi} \frac{\dot{m}}{\rho_2 \, u_2 \, d_2^{\,2}} \tag{11}$$

One can see, that for high speeds, where the heat flow is negligible (i.e. u_2, u_3 and u_4), the peak efficiencies are all lined up. By vertically shifting the peak efficiencies for low speeds (i.e. u_1), which differ from this line, until they meet this line, an approximation for the real efficiency can be made (Figure 6). Another way is to determine the heat flow directly, for example, by measuring the temperature profile and applying the Fourier's heat flow law. Further information can be found in (8).

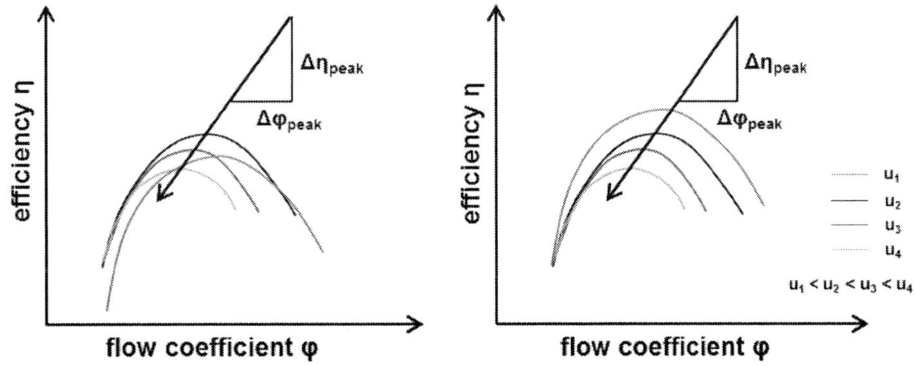

Figure 6: Apparent efficiency (left) and efficiency corrected by shifting the tip speed line (right) (8)

The most promising approach is to eliminate the heat conduction, so that the real efficiency can be measured directly without any further recalculation or simulation. That is the motivation for development of this E-drive compressor test rig, which uses an electrical motor to drive the compressor wheel.

This electrical motor drive replaces the turbine wheel and eliminates the heat influence of the hot exhaust gas. As a result, the temperature at the compressor stage outlet is affected only by the compression process itself. Using this method, it is possible to directly calculate the real efficiency from the measured values.

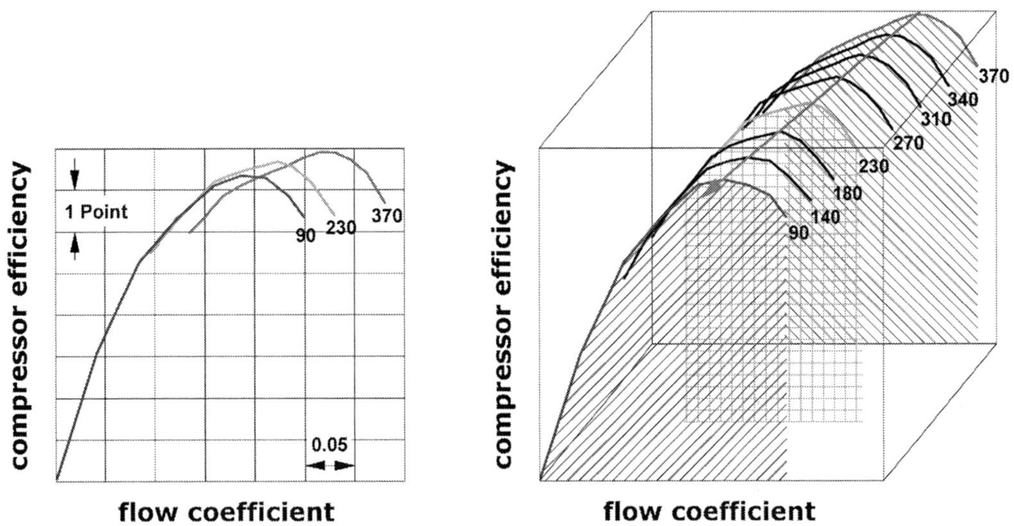

Figure 7: Compressor efficiency for different tip speed measured on the E-drive test rig

Figure 7 shows the compressor efficiency plotted versus the flow coefficient for a measurement on the E-drive test rig. For better visibility, only three lines are plotted in the left diagram. In the diagram on the right, all measured speed lines are staggered. One can easily recognise, that the peak efficiencies for different speeds correspond to the same line. According to (8), this proves that the measurements on the E-drive test rig are free from heat influence.

To analyse the differences between the E-drive compressor test rig and the gas stand, one can look at the overlapping area of the two compressor maps. Figure 8a shows the two measured maps, where the red map is from the E-drive test rig and the blue map is from the gas stand. In Figure 8b, the overlapping area in green shows where it is possible to make a comparison of the efficiency. The maximum tip speed on the E-drive compressor test rig is limited by the maximum rotational speed of the electrical motor driver. Therefore, the overlapping area lies between 230 m/s and 370 m/s.

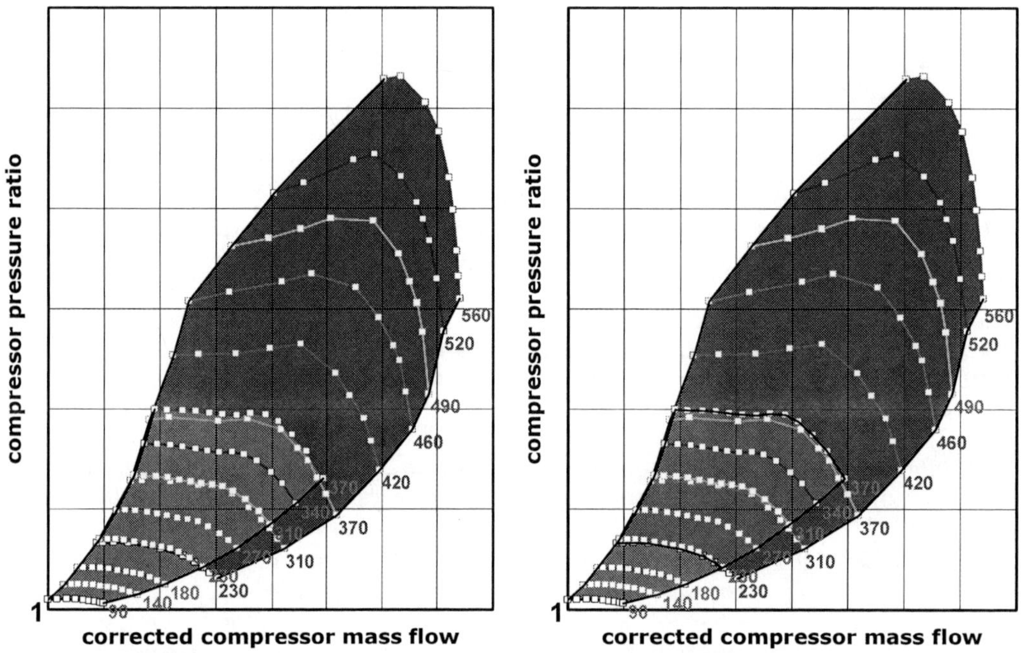

a) Compressor maps (blue GS and red E-drive)

b) Overlapping area (green) of the compressor maps

Figure 8: Compressor maps on E-drive test rig and gas stand

On the chart below (Figure 9), the differential map is presented. To create this map the following equation is needed:

$$\Delta\eta_{is,C} = \eta_E - \eta_{GS} \tag{12}$$

For low tip speeds and mass flows, the efficiency on the E-drive test rig is up to 4 points higher than on the gas stand. The efficiency difference decreases with increasing tip speed and mass flow, as expected. In other words, the efficiency values on the gas stand are reliable for high tip speeds and mass flows. The compressor map of the E-drive test rig can be used as a correction and extension for low tip speeds and mass flows.

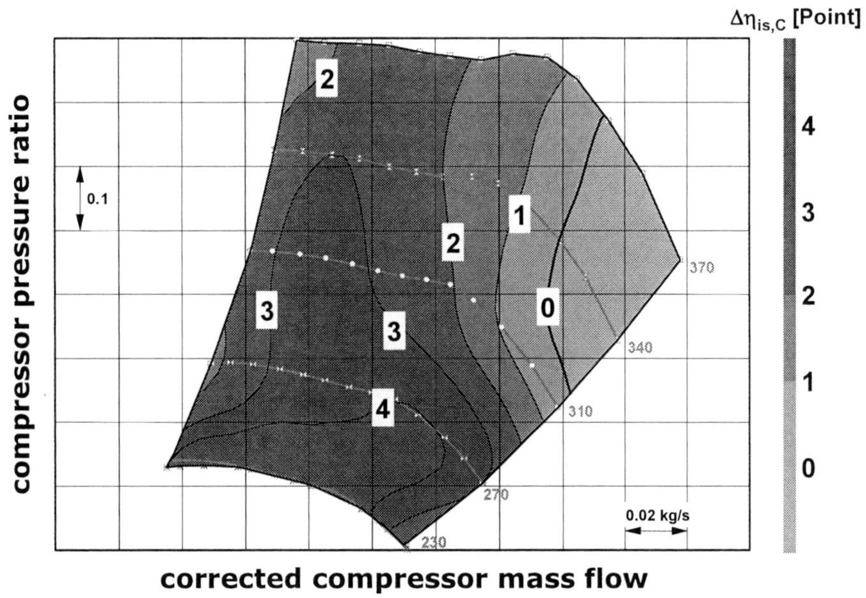

Figure 9: Differential map of the compressor efficiency (E-drive – GS)

With the results from the E-drive test rig, equation (11) can now be used to calculate the heat input from the turbine side into the air flowing through the compressor housing on the gas stand. Figure 10 shows the calculated specific heat flux on the vertical and the mass flow on the horizontal axis. This sample calculation is done for one compressor speed line, with a circumferential speed of 230 m/s. The heat conduction into the compressed air decreases steadily with increasing mass flow. That means, that the influence of the heat from the turbine side on the compressor outlet temperature fades away with rising mass flow, and the measured efficiencies from GS and E-drive converge.

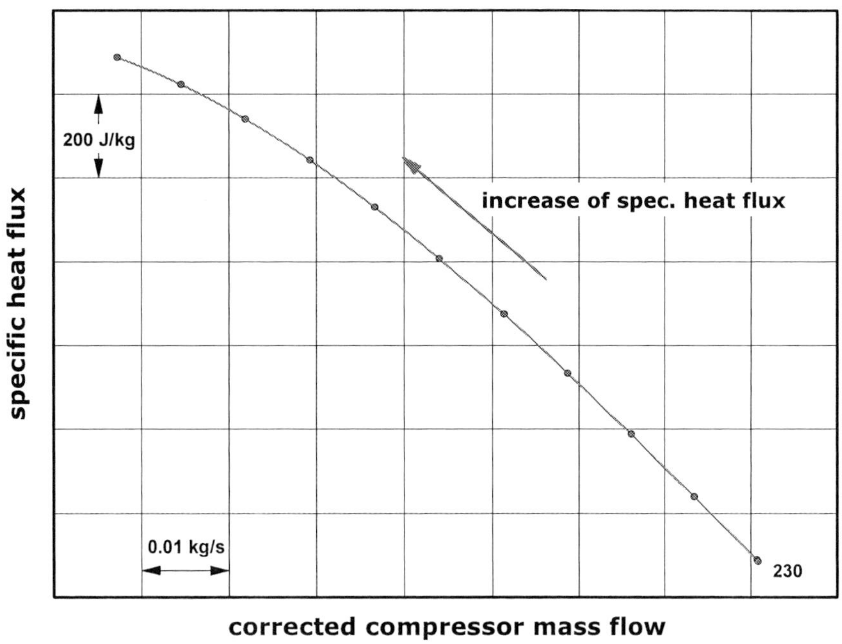

Figure 10: Calculated specific heat flux on a compressor speed line

4 EXTENSION OF THE COMPRESSOR MAP IN THE LOWER LEFT SIDE

Figures 11 and 12 show the measured results from the gas stand (blue) and the E-drive (red). The dashed line in the diagrams on the left hand side in Figure 11 and 12 describes the surge line, and in the diagrams on the right hand side it describes the envelope of the efficiencies.

At the E-drive test rig, the compressor map is measured from tip speeds of 90 m/s to 370 m/s, and the measured speed lines on the gas stand are in the range of 230 m/s to 560 m/s. This gives three lines of equal speed, which is shown in Figure 8b. One can see, that the increase of the peak efficiency between the lines 230 m/s and 370 m/s on the gas stand is much higher than on the E-drive test rig. As mentioned above, this is a result of the higher heat influence at lower mass flows.

For the combination of the two compressor maps, the values from the E-drive test rig are used for the speed lines up to 370 m/s. For higher speeds, it is acceptable to use the measured values of the gas stand because of the high mass flows and compressor outlet temperatures and therefore the lower heat influence. The combination of the two efficiency envelopes in Figure 13 validates this approach.

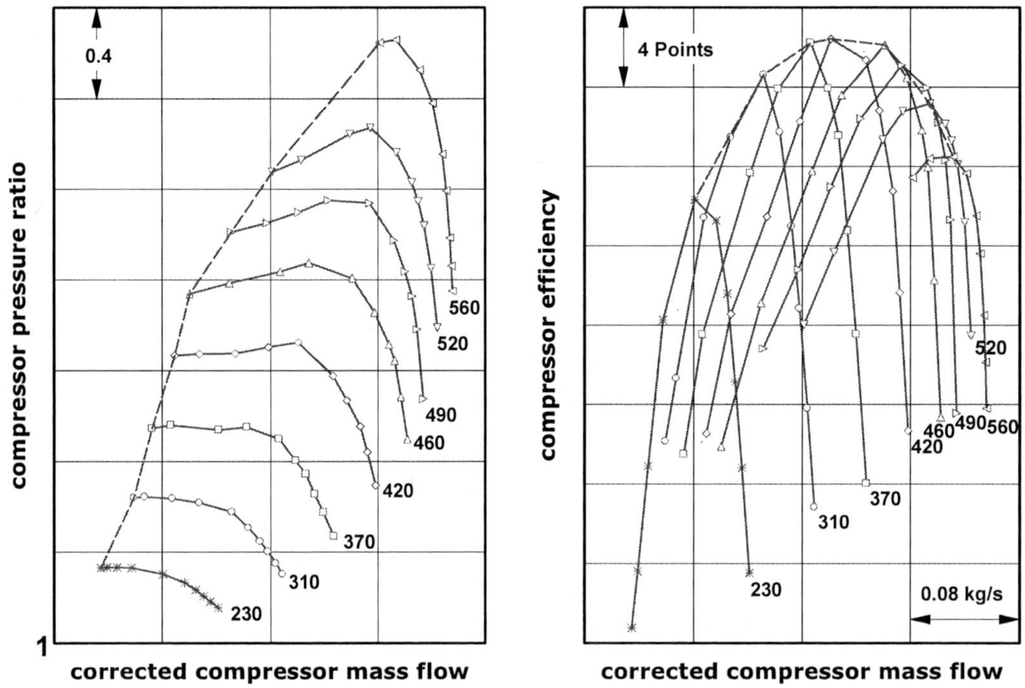

Figure 11: Measurement results from gas stand

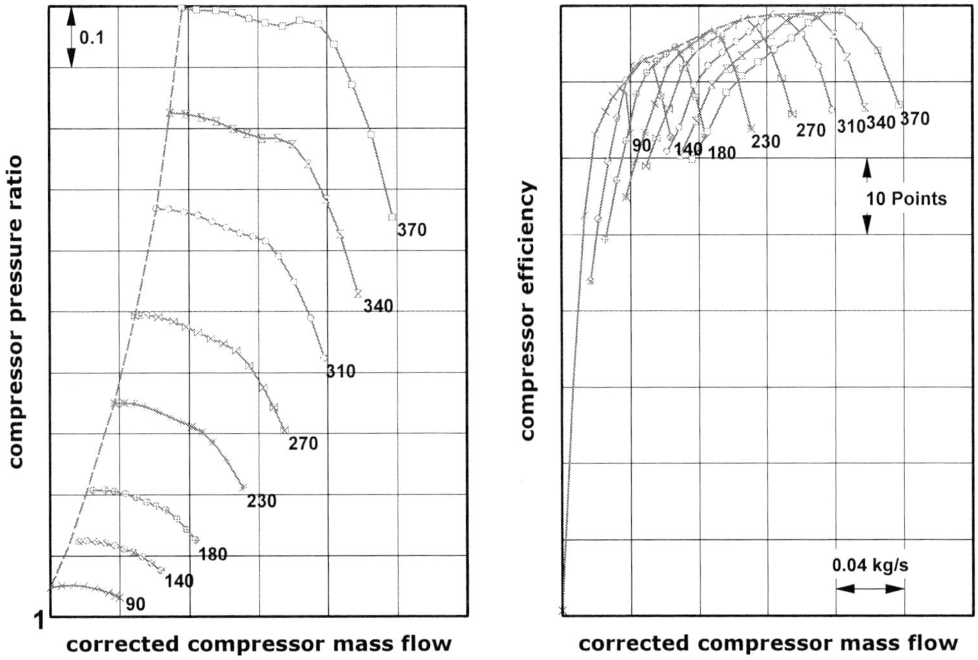

Figure 12: Measurement results from E-drive

The blue dashed line in Figure 13 is the envelope of the efficiencies on the gas stand and the red dashed line is from the E-drive test rig. After combining the two compressor maps, the green solid line describes the new envelope of the combined compressor map.

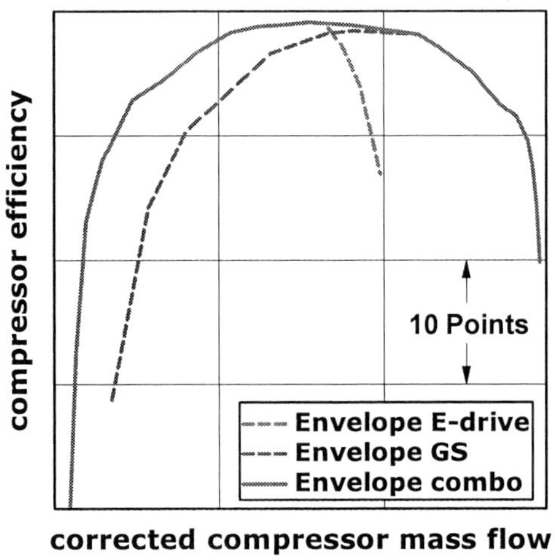

Figure 13: Envelope of the efficiencies

Figure 14 shows the combination of the two compressor maps. There is a step in the choke line between the E-drive test rig and the gas stand. This step is a result of flow restrictions in the compressor discharge line on the E-drive test rig. On the other hand, it is possible to reach higher mass flows with a given tip speed at the gas stand because of less throttling in the piping.

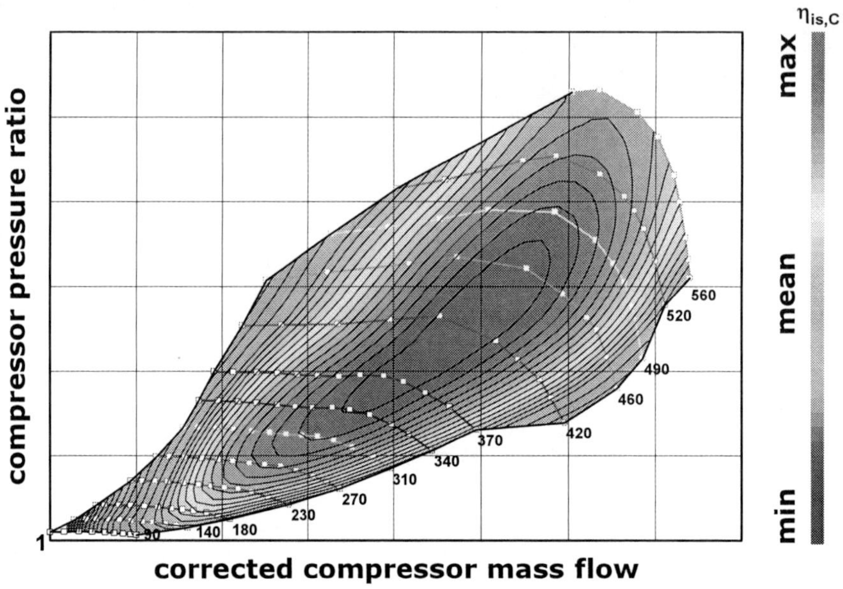

Figure 14: Combined compressor map (E-drive and GS)

This combined compressor map allows characterisation of all operating points of a current drive cycle.

5 REAL-TIME SURGE DETECTION

On the E-drive test rig, the speed of the compressor wheel is set by the frequency of current and voltage. Depending on the power needed to drive the compressor wheel at a certain operation point in the map, the control unit creates the necessary torque by adjusting the amplitude of current and voltage. The output power of the electrical motor drive is calculated as the sum of the products of voltage and current for each phase:

$$P_E = \sum_{i=1}^{3} U_i I_i \lambda_i \tag{13}$$

where λ is the power factor, which is a measure for the phase difference between current and voltage. If the circumferential speed is kept constant and the operating point of the compressor changes by opening or closing the throttle valves, the power electronics regulates the output current and, as a result, the output power.

If, at constant circumferential speed, the mass flow through a compressor wheel is reduced, the flow velocity decreases. This leads to a change in the angle of incidence. If this change exceeds a certain limit, the flow stream will detach from the blades, which results in stall and a blocking of blade passages. Further reduction of mass flow intensifies the stall, more blade passages will be blocked until the mass flow equals zero. Because of the still existing pressure ratio a backflow of air through the compressor begins. The backflow continues until pressure is reduced enough, so that the flow attaches the blades again. If the boundary conditions do not change, the pressure will start to build up again, the flow velocity decreases and the cycle starts once more. This phenomena is called surge and can be indicated by a rise of temperature in front of the compressor wheel or a high oscillation in the pressure signal (10). More detailed information about surge and stall can be found in (11) and (12).

In the measurements in Figure 16 the throttle valve of the back pressure unit is closed continuously to reduce the mass flow. To determine if there is any backflow of air through the compressor, a thermocouple is placed directly in front of the compressor wheel, as shown in Figure 15.

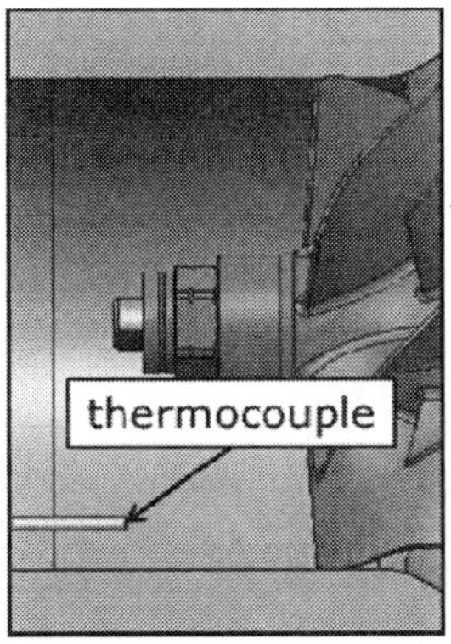

Figure 15: Position of thermocouple in front of compressor wheel

Figure 16 shows, that the inlet temperature rises significantly at a certain level of mass flow, indicated by the left green line in the diagram. At the same time, the fluctuation in the pressure ratio and mass flow increases significantly. Both are indicators of backflow of air and the beginning of stall. There is also a sharp increase in the electrical power, while the gradient stays the same. After the first step, the gradient of the inlet temperature decreases and stays on a constant, positive level until the closing of the throttling valve stops. When the valve is opened again, electrical power and mass flow increase and the inlet temperature decreases, which means, that the backflow of air is reduced.

When the mass flow again reaches the level indicated by the second green line in the diagram (which equals the level of the first green line), there is a sharp decrease in the electrical power and the inlet temperature, while the strong oscillation in the pressure ratio and mass flow stops. The electrical power then increases, because of the rise of the mass flow, but the inlet temperature stays constant. This means, that the compressor runs in stable operation, and there is no more backflow. The speed signal shows, that the number of revolutions does not change during the whole run, because the power electronics is reactive enough to keep it on a stable level.

The measurement shows, that a rise of the electrical power of the motor drive can be used as a surge indicator on the E-drive test rig. The sharp rise of the electrical power is a result of the sudden high pressure oscillations. The control electronics need more power to keep the compressor on a constant speed, because of the high accelerations due to the pressure oscillation.

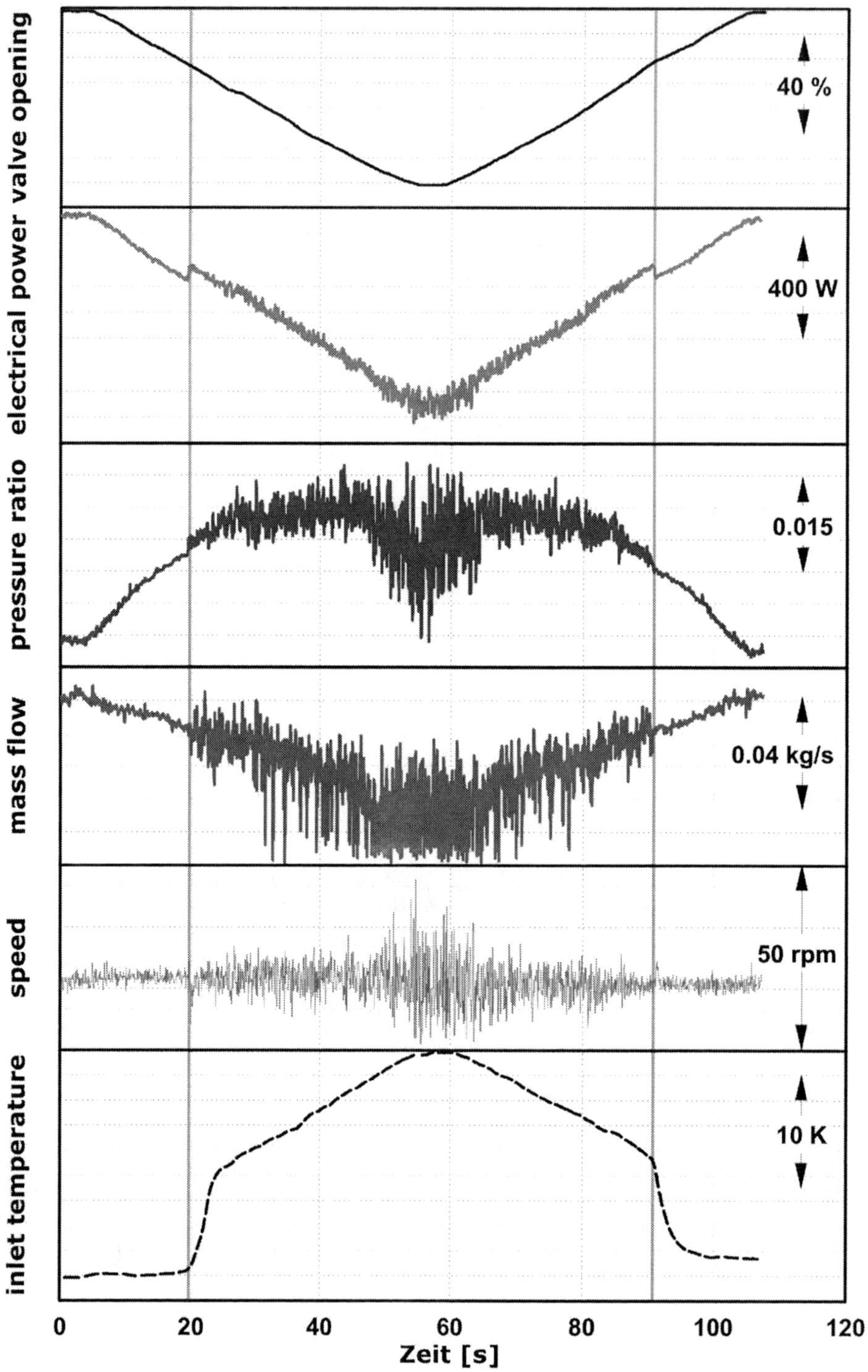

Figure 16: Surge detection on the E-drive test rig

6 CONCLUDING REMARKS

This paper has showcased significant differences between a standard gas stand and the new E-drive compressor test rig in evaluation and instrumentation. In this context, a new definition of the compressor efficiency for the E-drive compressor test rig is suggested and validated.

With the new test rig, it is possible to measure compressor maps under isentropic conditions and at lower speed lines than on a gas stand. These maps can now be used for improved engine power simulations. In the past those simulations had to be done using extrapolated compressor maps and with an approach to correct the maps from the heat influence.

Another use of these isentropic compressor maps is to quantify the important parameters for 1D heat flux simulation tools, which describe the heat flux from the turbine through the bearing housing to the compressor. This new test rig also benefits the compressor stage development process through improved validation of CFD simulations. The maps are also helpful for the matching of turbochargers, because with the more accurate measured data of the compressor efficiency, it is possible to recalculate the turbine efficiency and improve matching for each engine.

The control electronics of the used electrical motor drive are able to keep the compressor speed constant during instabilities. Therefore additional electrical power is needed, which can easily be measured. The rise of power can be used as an indicator for compressor instabilities.

REFERENCE LIST

(1) **Lüddecke, B. Filsinger, D., Bargende, M.:** *On wide mapping of a mixed flow turbine with regard to compressor heat flows during turbocharger testing.* Proc. 10th International Conference on Turbochargers and Turbocharging of the IMechE, 2012

(2) **Nakhjiri, M., Pelz, P. F., Matyschok, B., Däubler, L., Horn, A.:** *vATL – ein virtueller Abgasturbolader auf Basis physikalischer Modelle*. Unpublished research report, TU Darmstadt, 2012

(3) **Sirakov, B., Casey, M.:** *Evaluation of Heat Transfer Effects on Turbocharger Performance*. ASME GT2011-45887, ASME Turbo Expo, Vancouver, Canada, June 06-10, 2011

(4) **Casey, M. V., Fesich, M.:** *On the efficiency of compressors with diabatic flows*. ASME GT2009-59015, ASME Turbo Expo, Orlando, USA, June 8-12, 2009

(5) **Serrano, J.R.; Olmeda, P.; Arnau, F.J., Dombrovsky, A., Smith, L.:** *Analysis and methodology to characterize heat transfer phenomena in automotive turbochargers*. Journal of Engineering for Gas Turbines and Power Volume: 137 Issue: 2 Published: 2015

(6) **Bohn, D., Heuer, T., Kusterer K.:** *Conjugate flow and heat transfer investigation of a turbo charger: Part I: Numerical results*. Proceedings of ASME Turbo Expo, 2003

(7) **Cheese, P., Chalet, D., Tauzia, X.:** *Impact of the heat transfer on the performance calculations of automative turbocharger Compressor*. IFP Energies nouvelles, Vol. 66 (2011), No. 5, pp. 791-800

(8) **Nakhjiri, M., Pelz, P. F., Matyschok, B., Däubler, L., Horn, A.:** *Apparent and real efficiency of turbochargers under influence of heat flow*. ISROMAC-14, 2012

(9) **Greitzer, E.:** *Surge and rotating stall in axial compressor: Part I –Theory, Part II – Experiments.* ASME 75-GT-9;75-GT-10;ASME J. Eng Gas Turbines Power, 98(2), 1976

(10) **Grigoriadis, P.:** *Experimentelle Erfassung und Simulation instationärer Verdichterphänomene bei Turboladern von Fahrzeugmotoren.* Dissertation, TU Berlin, 2008

(11) **Day, I.J.:** *Stall, surge and 75 years of research.* Proceedings of AMSE Turbo Expo, 2015

(12) **De Jager, B.:** *Rotating stall and surge control: A survey.* Proceedings of the 34th Conference on Decision & Control, 1995

An approach to turbine housing validation through the measurement of residual strains using neutron diffraction and operationally induced strains using high temperature strain gauges

K Gannon [1,2], K Shoghi [1], S H Kilcoyne [2], T Pirling [3], S Zhang [4]

[1] BorgWarner Turbo Systems, UK
[2] The University of Huddersfield, UK
[3] Institut Laue-Langevin, France
[4] Rutherford Appleton Laboratory, UK

ABSTRACT

In order to fully understand turbine housing behaviour it is necessary to appreciate the complex behaviour of the strains within the turbine housing both prior to and during operation. In this paper neutron diffraction techniques have been used to measure surface and internal strains in the turbine housing prior to operation (residual strains) in an area prone to fatigue crack initiation, while high temperature strain gauges have been used to measure surface induced strains due to thermal cycling on gas stand test. Service strain measurements have been compared with simulation results. The direction and magnitude of the measured residual strain is explored in relation to the effect on fatigue crack initiation/propagation.

NOMENCLATURE

d_0: Stressed/unstressed interplanar spacing parameter (m)
d_{hkl}: Interplanar spacing between crystallographic planes (m)
E: Elastic Modulus (N/m^2)
hkl: Miller Indices defining a family of crystallographic planes
K: Gauge Factor
R_i: Gauge resistance(s) (Ohms)
ν: Poissons Ratio
V_{ex}: Excitation voltage (V)
V_{out}: Output voltage (V)
$\sigma_x\sigma_y\sigma_z$: Normal Stress in X, Y, Z directions (N/m^2)
ε : Strain
2θ : Diffraction angle (Deg)
λ: Wavelength (m)

1 INTRODUCTION

In the EU, heavy duty commercial vehicles such as trucks, buses and coaches produce around 25% of the CO2 emissions and around 5% of the total greenhouse gas emissions [1]. As a result, EU emissions legislation is becoming more stringent and the operating environment of the turbine housing component of the turbocharger is set to become more severe, with exhaust gas temperatures expected to increase as the demands for cleaner more efficient engines becomes paramount, it is therefore important to actively explore new ways to optimise turbine housing durability.

The turbine housing is an expensive part of the turbocharger due to its size, intricate geometry and material. With design complexity increasing as wall thicknesses reduce, the durability of the turbine housing is of significant concern. During service, the turbine housing is subjected to harsh thermo-mechanical fatigue as a result of the combination of both transient thermal and mechanical loading. To ensure high levels of reliability and durability of the turbine housing, it is imperative that the behaviour of the internal stresses and strains induced both during service and before service (residual stresses) are fully characterised and understood. Manufacturing process variables such as heat treatment that can affect residual strain/stress distribution should also be investigated.

In the turbine housing fatigue strength is of particular importance. Fatigue cracking normally initiates at a surface, and, in the case of the turbine housing this surface is hidden within the housing making it difficult to access to apply standard strain measurement techniques such as strain gauges. In this paper Neutron diffraction based techniques have been used to obtain non-destructive strain measurements from internal locations within the turbine housing, which allows the subsequent calculation of stress. The technique presented is an attempt at gaining baseline measurements of the residual stress distribution which can later be compared with simulations. Neutron diffraction had not previously been used on such a complex geometry and, as a result of this work a measurement methodology now exists.

It is known that compressive residual stresses can be beneficial in improving fatigue life, while tensile residual stresses can combine with service generated stresses and cause unexpected failures during operation [2]. Although Finite Element Analysis (FEA) methods exist for the determination of service induced stresses due to thermo-mechanical loading of the turbine housing, the pre-loading of residual stress is not currently considered and work to accurately validate such simulation models is difficult, as internal stresses within the housing are challenging to measure accurately. Ultimately the aim is to use neutron measurements to validate FEA simulations.

This work is a continuation of the work by K. Shoghi [3] in which gas stand test thermocouple measurements were used to validate FEA temperature simulations. High temperature strain gauges have since been utilised to enable a more complete study of turbine housing surface strains during gas stand thermal cycle testing to be carried out, and a comparison of these measurements with FEA simulations has been completed. It is hoped that this work combined with the study of residual stress in the turbine housing at both surface and internal locations, will provide a detailed picture of residual stresses within a housing and hence a means to both validate and improve current simulation methodologies.

2 RESIDUAL STRESS

The turbine housings considered in the study are made from ductile cast iron and designed for heavy duty commercial usage with exhaust gas temperatures up to 760°C. Turbine housing manufacturing methods generally involve the casting of the component in sand moulds. The casting is then either slowly cooled in the mould (self-annealed) to achieve the desired mechanical properties without heat treatment or heat treated through an annealing process. During solidification and cooling residual stresses can form due to non-uniform phase transformations and contraction rates as a result of different cooling rates in sections of varying thickness, as can be found in the interior of the turbine housing. The design and casting process have the biggest impact on residual stresses.

Residual stresses can normally be controlled either by ensuring that the component cools uniformly and by having a substantial holding time in the mould sufficiently long enough to reduce the temperature differential in each region of the casting [4] or by heat treating and slowly cooling the housing uniformly. Techniques such as shot peening can affect residual stresses and have been shown to induce compressive residual stresses as a result of plastic deformations on the applied surface which have been shown to improve fatigue life [5].

In order to increase the durability of the turbine housing the fatigue strength is of principal concern. Fatigue cracks nearly always initiate and propagate normal to applied tensile stresses, and in order to improve fatigue life residual stresses that are in a direction such that they reduce the principal applied stress will provide the biggest advantage assisting with inhibiting crack growth and propagation [6], the extent to which residual stresses can be of benefit is highly dependent on the magnitude and direction of the residual stress. It should be noted that residual stresses are likely to relax to some extent if not completely during thermal cycle operation, this in itself can cause issues with dimensional stability and hence further importance is placed on understanding residual stress behaviour. Any cyclic residual stress relaxation during operation will reduce the benefits of any compressive residual stresses, however in some cases it has been shown that despite partial relaxation of compressive residual stress, there was still found to be a beneficial effect on fatigue life [7] this is very dependent on the yield strength of the material and the stress amplitudes experienced during cyclic operation [8]. Evaluation of the effect of residual stresses and their relaxation behaviour could lead to a means to optimise and improve the durability of the housing.

3 NEUTRON DIFFRACTION

There are numerous techniques for the measurement of strain each with their own advantages and disadvantages. Most methods involve cutting or drilling however, a method exists that can measure internal strains non-destructively known as neutron diffraction.

Neutron diffraction can be used to measure elastic strain within a defined volume in a crystalline solid. If the Young's modulus and Poisson's ratio for the material are known it is then possible to calculate the mean stress in the studied volume. Neutrons are able to penetrate very deeply into materials (about 1000 times deeper than lab x-rays in most materials) [9], this allows sampling within the bulk of the material. This is advantageous as the stress field near the surface will often differ from that in the bulk of the material. Full analysis of the strain tensor requires measurements of the elastic strain in at least six independent directions however, if the principal strain directions within the component are known, measurements along three principal directions will suffice. Due to the relatively low intensity of neutron flux the volume of the component studied is required to be relatively large; this makes it possible to study a large number of grains within the material providing an average value of strain that is statistically representative, the resulting stress measurements are therefore considered macro-stresses.

Constructive interference occurs when neutrons of wavelength λ are scattered from adjacent planes of atoms of interplanar lattice spacing d_{hkl} in a crystalline material and remain in phase. This will occur when the paths travelled by the scattered waves differ by integral multiples of λ. From Figure 1 it can be seen that the extra path length travelled by the wave scattered from the lower scattering plane is 2dsinθ. If this distance is equal to $n\lambda$ (where n is an integer), the result is Bragg's Law [10], given by equation (1).

$$n\lambda = 2d_{hkl}\sin\theta \tag{1}$$

Higher order (n=2,3,4 etc.) scattering from crystal planes d_{hkl} is indistinguishable from first order (n=1) scattering from planes $d_{hkl}/2$, $d_{hkl}/3$, $d_{hkl}/4$ and so forth, and therefore for neutron scattering the Bragg equation reduces to:

$$\lambda = 2d_{hkl}\sin\theta \tag{2}$$

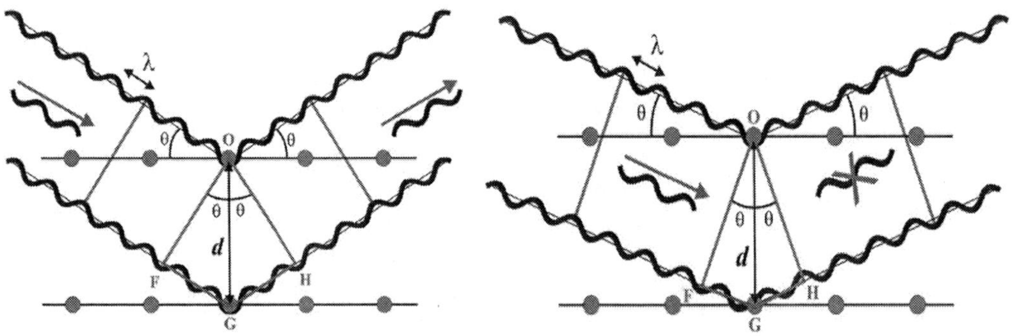

Figure 1: Diagram demonstrating the Bragg equation and the conditions for a) constructive interference and b) destructive interference.

If the conditions for constructive interference are satisfied a diffraction pattern with sharp peaks of intensity will be produced. Each peak of intensity will correspond to diffraction from atomic planes with interplanar spacing d_{hkl} .

The lattice spacing d_{hkl} can be used as a strain gauge and measurements of the lattice spacing can be achieved through the use of constant neutron wavelength sources in which the scattering angle is measured or a pulsed neutron spallation sources in which the scattering angle is fixed and the time of flight of each neutron is measured, and used to determine its wavelength. The d-spacing will vary depending on the stress state of the material, for example it will increase when the material is under tension and reduce when under compression, in this way the lattice planes act as internal strain gauges and a shift in the diffraction peaks is seen. The elastic lattice strain is therefore derived from the change of the stressed lattice spacing with respect to a stress-free reference lattice spacing d_0 as shown in equation 3.

Strain can then be determined from an analysis of individual peaks in the diffraction pattern or from a Rietveld refinement of the entire pattern. For a single peak analysis, the values appropriate to the specific (*hkl*) crystallographic plane on which the strain is being measured are used; for a multi peak Rietveld refinement analysis bulk values are used. To determine elastic strain ε a change in the lattice spacing Δd_{hkl} will result in a shift $\Delta\theta$ in the angular position of the Bragg reflection [11] so that the lattice strain in the direction of the scattering vector is given by equation (3).

$$\varepsilon = \left(\frac{d-d_0}{d_0}\right)_{hkl} = \left(\frac{\Delta d}{d_0}\right)_{hkl} = -\cot\theta\Delta\theta \quad (3)$$

Strain can then be converted to stress for each of the principal directions using:

$$\sigma_x = \frac{E}{(1+v)(1-2v)}\left[(1-v)\varepsilon_x + v\left(\varepsilon_y + \varepsilon_z\right)\right] \quad (4)$$

$$\sigma_y = \frac{E}{(1+v)(1-2v)}\left[(1-v)\varepsilon_y + v(\varepsilon_x + \varepsilon_z)\right] \quad (5)$$

$$\sigma_z = \frac{E}{(1+v)(1-2v)}\left[(1-v)\varepsilon_z + v\left(\varepsilon_x + \varepsilon_y\right)\right] \quad (6)$$

4 EXPERIMENTAL PROCEDURE

4.1 Residual strain measurements

Residual strain measurements were obtained using: 1) the pulsed neutron source at the ISIS spallation facility using the ENGIN-X instrument at the Rutherford Appleton Laboratory, UK (Figure 2) and 2) the reactor neutron source at the Institut Laue-Langevin (ILL) on the SALSA strain scanner [12].

The experiment on ENGIN-X focussed on axial surface strain measurements on a finished (machined and heat treated) twin volute housing; the measurement points were located in an area on the outer wall of the volute.

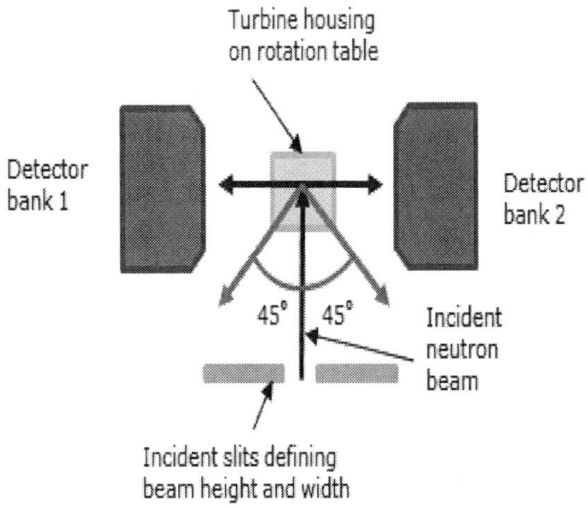

Figure 2: Diagram depicting layout of instrumentation on the ENGIN-X strain scanner at ISIS.

ENGIN-X benefits from the use of the simulation and control software SScanSS [13,14] which, when used in conjunction with a triple axis 'cybaman' robotic positioner, allowed for the automatic positioning and orienting of the housing. A coordinate measurement machine was used to probe points of the surface of the housings; the resulting points, in conjunction with CAD models of the interior geometry of the housing, were then converted to a surface mesh and imported into the SScanSS virtual laboratory. As the neutron path-length is limited to 60mm in iron alloys, the software was used to calculate the optimum orientation, in terms of rotation about the instrument q-vector that minimised attenuation of the neutron beam through the turbine housing. A gauge volume of 4x4x4 mm^3 was employed as defined by height and width of the incident neutron beam slit and the choice of 4mm radial collimators.

A full diffraction pattern was obtained from each point of interest, this was then analysed through the Rietveld method with the General Structure Analysis system (GSAS). The strain free d-spacing, d_0, was obtained from a location on the inlet flange of the housing, in an area considered to be relatively free of strain. The d_0 was measured in all three principal directions (X, Y, Z) multiple times for an average value to be obtained.

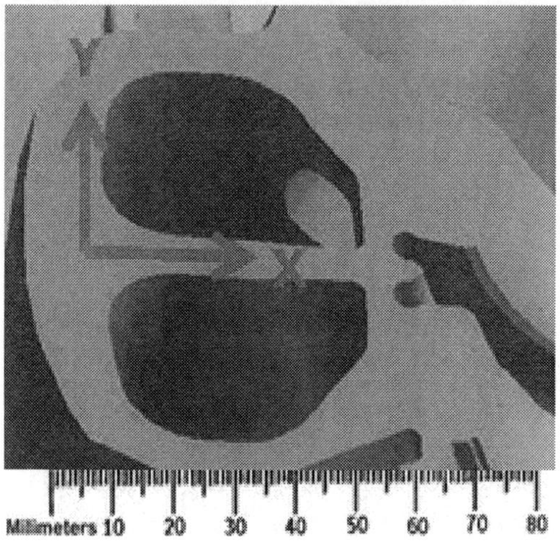

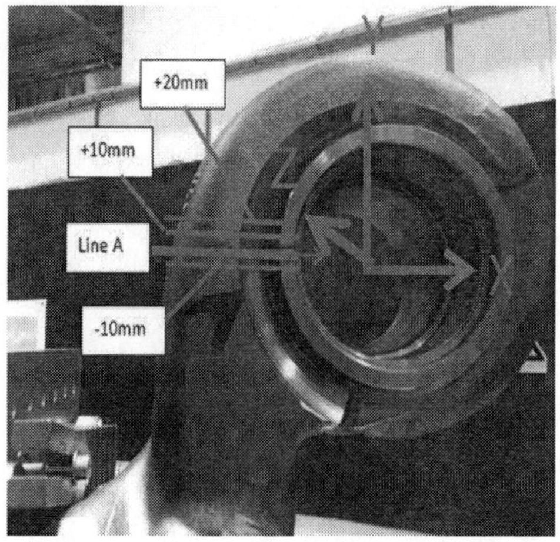

Figure 3: (left) showing line (X) progressing from the outer wall into the divider wall. A change in section thickness is evident at ~ 15mm. (right) Repeat measurements of the same line at various intervals (Y).

The residual strain measurements on SALSA focused on the same twin volute housing, but, in this experiment, measurements were made in the interior of the turbine housing with data collected in a line from the outer wall of the housing into the divider wall close to the tongue as shown in Figure 3. This line was repeated at 10mm intervals above and below A as shown in the figure. Measurements were made in 3mm steps from the outer wall inwards with a gauge volume of 3x3x3 mm^3. On this occasion, a sectioned housing was examined to determine d_0, 20 measurement points were recorded to obtain an average d_0 value with measurements in all three principal directions (X, Y, Z). Residual stresses were measured in two machined housings, one heat-treated and the other self-annealed.

In order to rotate the sample to measure in all strain directions SALSA uses a unique hexapod design which has a significant advantage over traditional X,Y,Z translators. The hexapod has six independently driven hydraulic legs which can move large engineering components with a precision of the order of a few microns [15].

Measurements of the d-spacing shift were converted to residual strains using equation (3) and subsequently stress values (for the ENGIN-X data only) were obtained using equations (4), (5) and (6). Figure 4 shows the principal strain directions considered in the analysis.

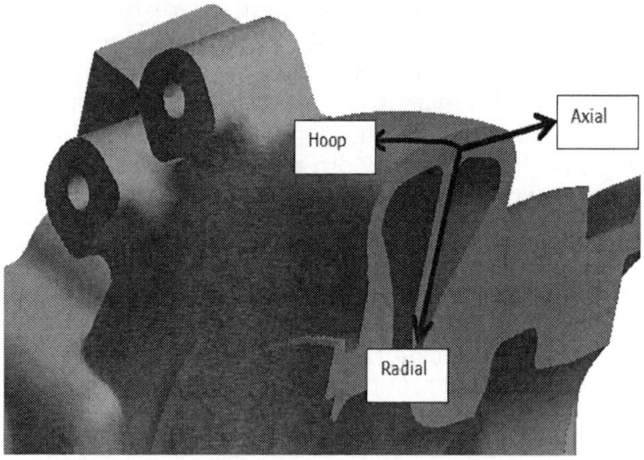

Figure 4: The three principal strain directions.

4.2 Measurement of strain during gas stand testing

We also wished to investigate the strain in the turbine housing during thermal cycling. High temperature strain gauges were added to a turbine housing to measure surface strains during thermal cycling in order to validate simulation methods following the methods developed by K. Shoghi [3].

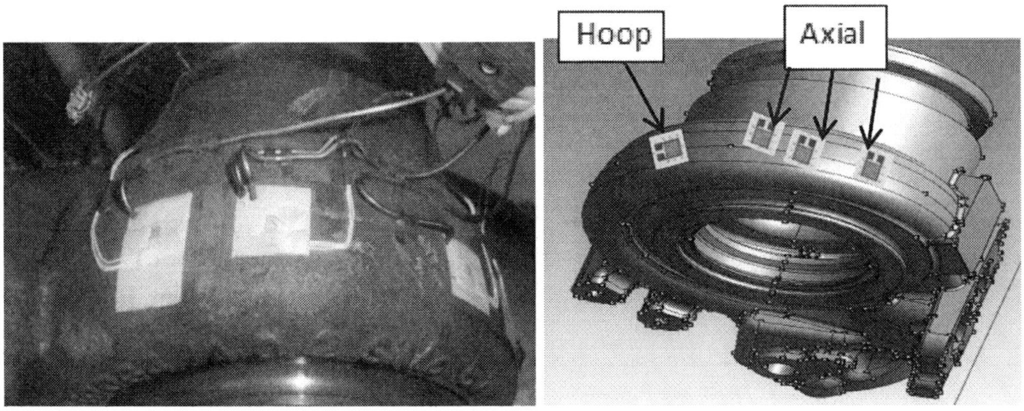

Figure 5: Strain gauge locations.

In this experiment a smaller twin volute housing was fitted with strain gauges. The turbine housing was fitted with three strain gauges positioned in the axial direction and one in the hoop direction as detailed in Figures 5 and 6, to accompany the eight thermocouples located around the housing. Half bridges with two active gauges were used, with each half bridge wired to a bridge completion unit. The half bridge set up consists of two active gauges, one gauge aligned in the direction of the strain (i.e. axial, hoop) to be measured, and the other aligned in the transverse direction and subject to Poisson's strain v . The presence of the second gauge corrects for the change in gauge resistance due to temperature.

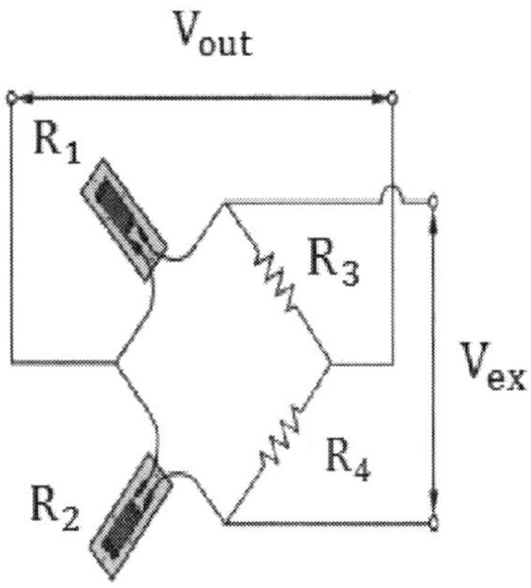

	Description
TC 1	gas in passage
TC 2	wall in passage
TC 3	divider wall in passage
TC 4	inside near end divider wall, near SG4
TC 5	outside skin, near SG5
TC 6	inside near divider wall, near SG5
TC 7	inside skin, near SG6
TC 8	outer skin
TC9	Additional TC near SG 7
SG4	Axial Strain near TC4
SG 5	Axial Strain near TC 5
SG6	Axial Strain near TC 7
SG7	Hoop Strain near TC 9

Figure 6: Strain gauge set up and description of thermocouple (TC) and strain gauge (SG) locations.

The voltage output from a Wheatstone bridge is given in [16] and shown in equation (7):

$$V_{out} = \left(\frac{R_2}{R_1+R_2} - \frac{R_4}{R_3+R_4}\right) V_{ex} \qquad (7)$$

R_1 and R_2 are the active gauges; R_3 and R_4 are the fixed resistances (figure 6). V_{ex} is the supply (excitation) voltage to the bridge.

The bridge is balanced when:

$$\frac{V_{out}}{V_{ex}} = 0 \tag{8}$$

Which occurs when:

$$R_1 = R_2 = R_3 = R_4 \text{ or } R_1 : R_2 = R_3 : R_4 \tag{9}$$

If the resistors vary, the bridge is unbalanced and an output voltage V_{out} will appear. It can be assumed that the variation in resistance ΔR_i is much smaller than the resistance R_i therefore:

$$\frac{V_{out}}{V_{ex}} = \frac{1}{4}\left(\frac{\Delta R_1}{R} - \frac{\Delta R_2}{R} + \frac{\Delta R_3}{R} - \frac{\Delta R_4}{R}\right) \tag{10}$$

With R_1 and R_2 being the only active gauges:

$$\frac{V_{out}}{V_{ex}} = \frac{1}{4}\left(\frac{\Delta R_1}{R} - \frac{\Delta R_2}{R}\right) \tag{11}$$

The relative change of the resistance of a strain gauge is described by equation (12).

$$\frac{\Delta R_0}{R_0} = K\varepsilon \tag{12}$$

Combining equation 11 and 12:

$$\frac{V_{out}}{V_{ex}} = \frac{K}{4}(\varepsilon_1 - \varepsilon_2) \tag{13}$$

In this case ε_1 is the strain applied to strain gauge 1, normal to this a contraction will occur in ε_2 which is subjected to Poisson's ratio v hence:

$$\varepsilon_1 = \varepsilon$$

$$\varepsilon_2 = -v\varepsilon \tag{14}$$

Therefore the output voltage can be defined as:

$$V_{out} = \frac{V_{ex}}{4}(1 + \nu)K\varepsilon \tag{15}$$

Prior to the test when the turbocharger was at room temperature, the strain gauge readings were zeroed and a calibration thermal cycle was performed after which the housing was allowed to cool again to almost room temperature. There was found to be some shift in the strain gauge readings, data readouts were corrected to account for this.

The thermal cycle performed during the test is shown in Figure 7. It was ensured that there was enough soak time at maximum temperature to allow enough time for the heat to travel to all parts of the turbine housing.

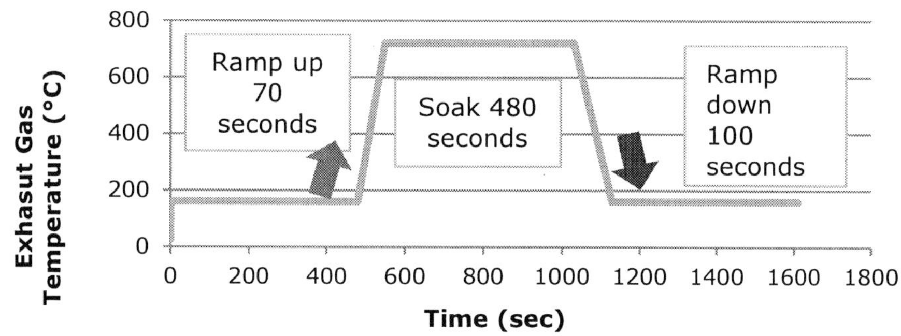

Figure 7: Thermal cycle used for gas stand testing and simulation.

4.3 FEA simulation methodology

The FEA simulation methodology consisted of a transient thermal analysis coupled with a static structural analysis. Using both CFD and 1-D approach [17], all values of temperatures and heat transfer coefficients were calculated and mapped to the transient heat transfer model. The results of temperature distribution from the transient thermal analysis were then used as an input for the structural part of the model. In order to prevent rigid body motion, the dummy cylinder head was used to constrain the model in all degrees of freedom. Kinematic hardening material model with temperature dependent stress-strain values were used for the turbine housing, dummy manifold and dummy cylinder head. The Bearing housing material was defined separately due to its material property being different to the rest of the assembled components.

For ease of analysing, a cylindrical coordinate system was used. Areas corresponding to the strain gauge locations were probed on the FEA model and compared with the gas stand test measurements with uni-axial strains in the relevant directions recorded. Solution was also obtained for accumulated equivalent plastic strain with areas of accumulated plasticity compared against cracked locations in the interior of the housing after gas stand testing. From the magnitude of accumulated plastic strain it can be established based on previous fatigue data and experimental results if cracking will occur.

5 RESULTS AND DISCUSSION

5.1 Residual strain measurements prior to service

Residual strain measurements taken on the ENGIN-X instrument at the surface of the turbine housing as shown in Figure 8 were compressive in direction and although it was not possible to obtain a full strain tensor, as only the axial direction was measured, it has served as a comparison to the results obtained on the SALSA instrument. The surface strains were on average slightly higher in compressive magnitude than those measured just below the surface.

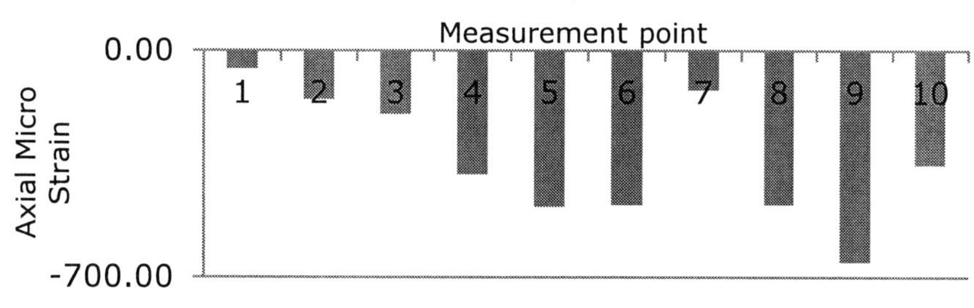

Figure 8: Showing axial residual strain surface/near surface measurements on ENGIN-X at ISIS.

Residual strain measurements taken in the interior of the housing in the three principal directions were collected on both heat treated and self-annealed housings on the SALSA instrument. Figure 9 shows the strain values from one of the measurement lines (line A) as shown in Figure 3. It could be seen that most of the points exhibited compressive strains. It was evident that the magnitude of the strain was reduced by heat treatment from its compressive state toward zero when compared with the self-annealed parts. In some cases the heat treatment appeared to induce tension in previously compressive areas. However it should be noted that the heat treated and self-annealed housings were different components and it is possible that a slight variation in material composition can affect the values obtained.

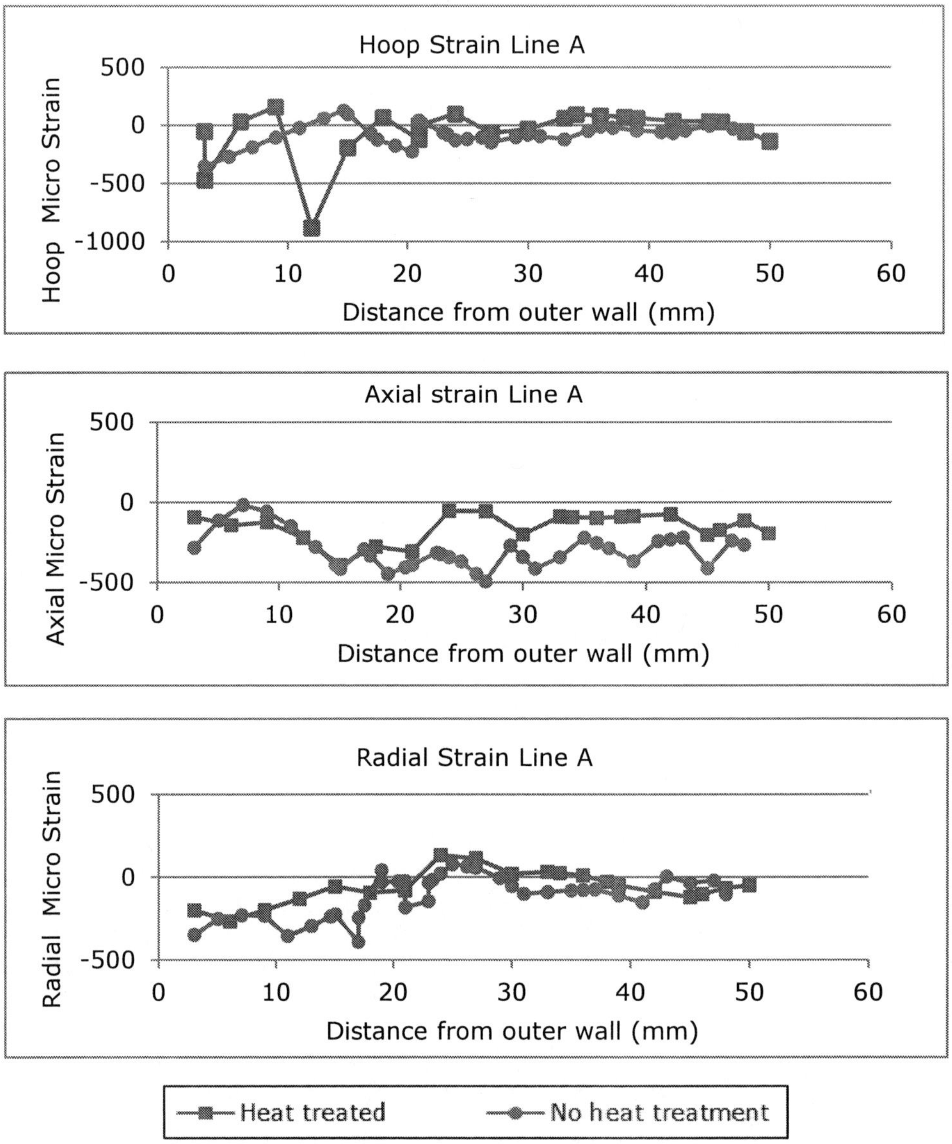

Figure 9: Interior residual strain measurements of a heat treated and self-annealed turbine housings obtained on SALSA. The experimental uncertainy in the measrements in indicated by the size of the measurement point.

It was also found that the axial and hoop residual strains were, on average, the largest in magnitude. Only the axial strain values for both the heat treated and self-annealed were completely compressive in nature over all measured lines. Measurements taken on the heat treated part showed a region of larger compressive strains at a point approximately 15mm into the housing; this was evident in all directions but most prominent in the hoop direction. This could be due to a change in the section thickness which occurs at this point as shown in Figure 3 and will be the subject of further investigation.

As a full strain tensor was obtained a calculation of the internal residual stress within the divider wall was possible. Contour plots of the percentage of maximum residual stresses are shown in Figure 10. Tensile residual stresses were found to be less than 15% of the yield strength of the material. Compressive stresses were larger in magnitude, with the maximum values close to 40% of the yield strength of the material.

It was evident that close to the surface residual compressive stresses dominate but further into the divider wall stresses reduced in compressive magnitude. Heat treatment appears to have reduced compressive stresses at the majority of points becoming slightly tensile in a few areas. It can be seen in the heat treated contour maps the areas of higher compressive stress corresponding with a reduction in section thickness, however this was only seen on the heat treated housing.

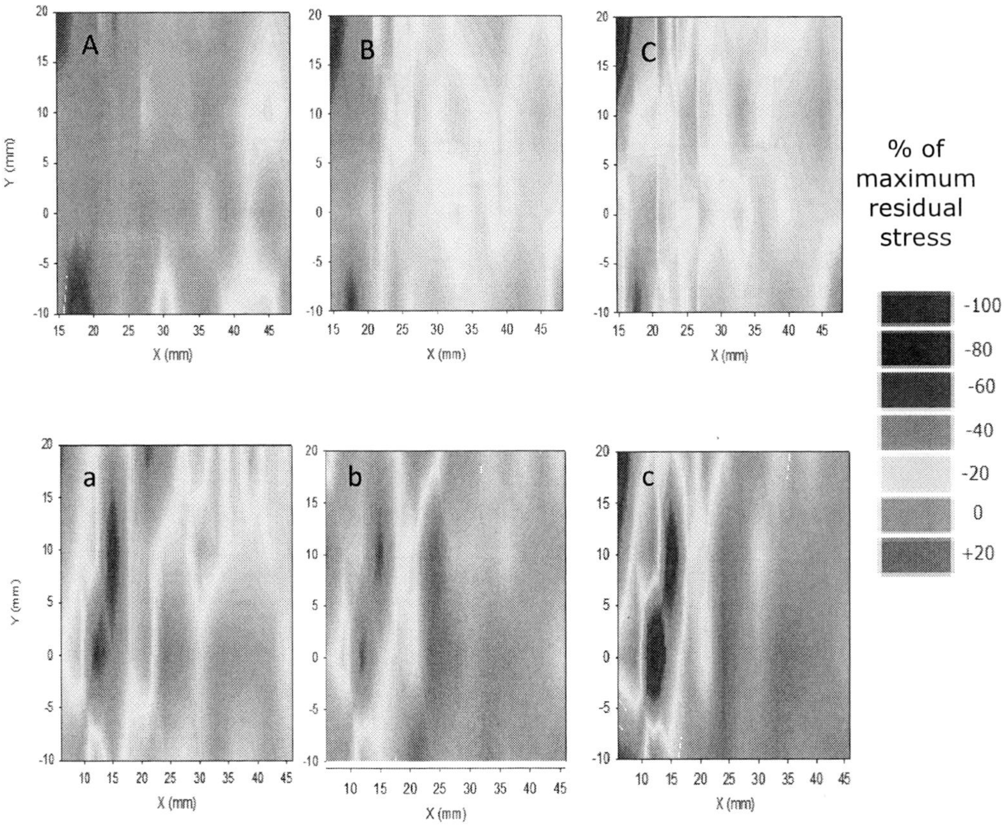

Figure 10: Contour maps of residual stress taken at various slices through the housing before heat treatment A) Axial B) Radial and C) Hoop directions and residual stresses after heat treatment in a) Axial b) Radial and c) Hoop directions.

5.2 Strain gauge measurements during operation

The correlation between the FEA model and the thermocouple readings an example of which is shown in Figure 11 and presented by K. Shoghi [3] was found to be good. After gas stand temperatures had stabilised measurements were found to be within an average of 3% of FEA values, in the thermocouples close to the strain gauge locations with CFD and 1D values showing good agreement.

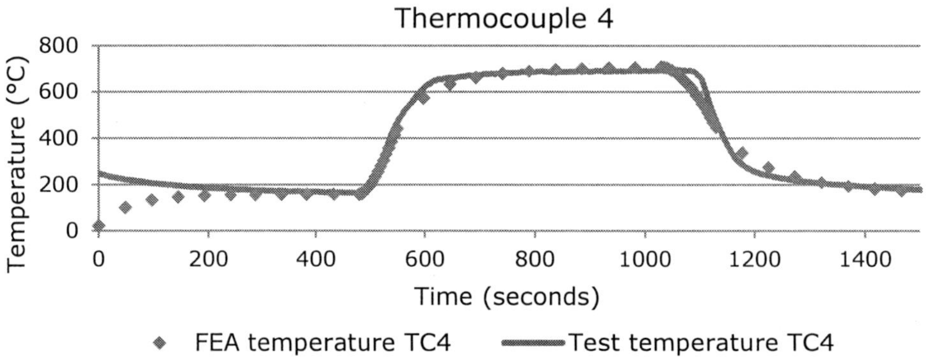

Figure 11: Showing correlation between simulated temperatures and thermocouple measurements.

Figure 12 shows the strain gauge profiles measured during thermal cycle gas stand testing, alongside the exhaust gas inlet temperature, it can be seen that all strain gauge reading are in phase with the applied temperature.

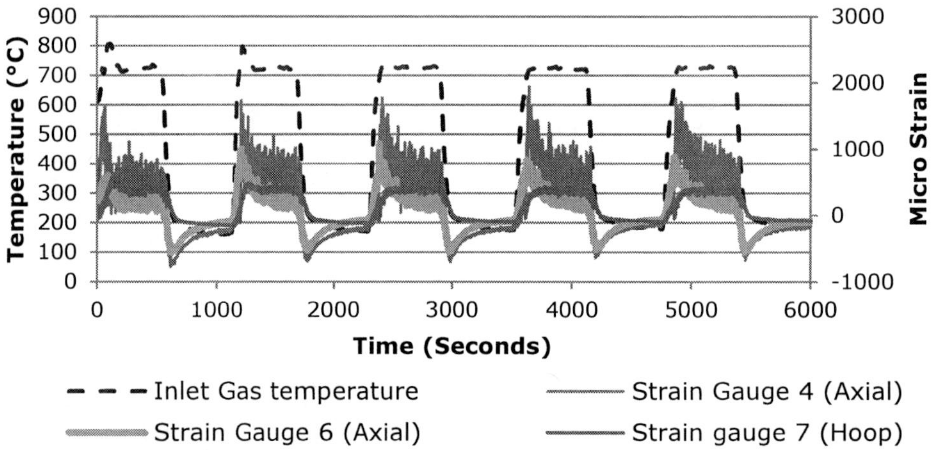

Figure 12: Showing gas stand strain gauge measurements and inlet gas temperature.

Figure 13 shows the correlation between strain gauge measurements and simulation predictions of uni-axial strain in the relevant direction. The results from strain gauge 4 correlated well with the test values; the test signal was however very noisy with electrical noise masking any clear transition points. Strain gauge 6 FEA results were found to over predict strain magnitudes, whereas FEA under predicted the strain at strain gauge 7 location although to a smaller degree. Unfortunately strain gauge 5 developed a fault and no valid signal was obtained.

Figure 14 shows further that the correlation between the actual test crack locations and simulation accumulated plastic strains was reasonable, with the FEA model successfully locating the majority of critical locations where the housing has cracked.

Numerous possibilities could explain the slight discrepancy in results. For example the geometry of the model will always be slightly different to the gas stand tested part due to design wall thickness tolerances. Also, the FEA material properties used in the FEA model could be slightly different to that of the tested housing. It was also evident that the gradient of strain in the FEA model was high over a very small region and although every care was taken to make sure the same areas were selected in the model as were fitted with strain gauges during testing the location may be slightly different. FEA temperatures in the areas investigated did not match the simulation exactly and as a consequence the strain values would be slightly different. Finally, these simulations only considered one thermal cycle, future simulations will run for multiple cycles to account for the lower compressive strain preceding the start of each cycle.

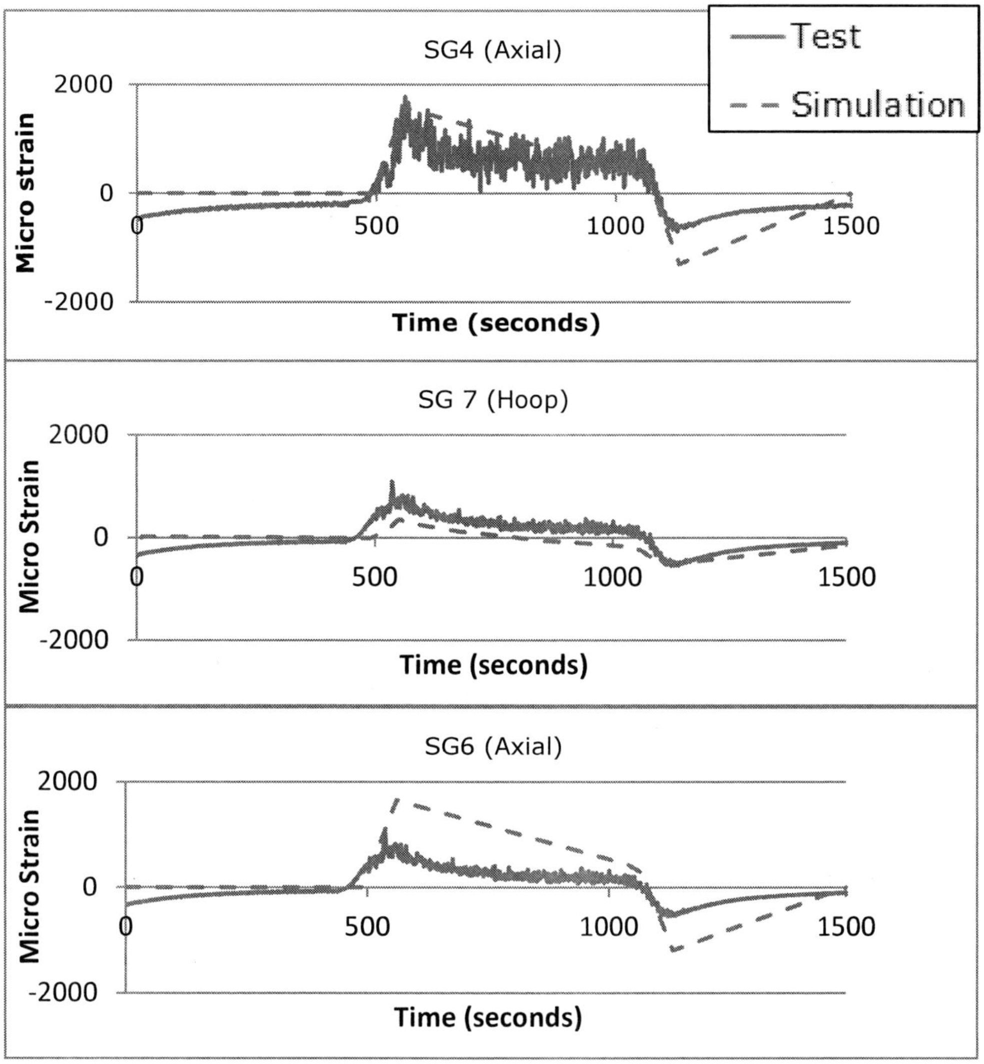

Figure 13: Showing high temperature strain gauge (SG) test measurement correlation with simulation results.

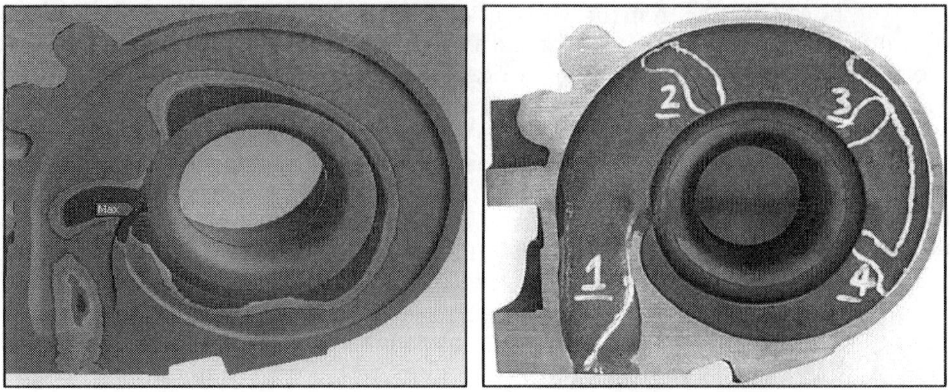

Figure 14: Correlation between FEA simulations (left) and actual crack locations (right) from gas stand tested housing.

6 CONCLUSION

In this paper, two methods of measuring strains in turbine housings have been presented: (1) on the surface of the housing during thermal cycle testing using strain gauges and (2) on the surface and within the turbine housing using the non-destructive neutron diffraction technique prior to operation. The use of neutron diffraction to measure internal strains on such a complex geometry had not previously been performed and this has enabled the development of a methodology to facilitate future neutron diffraction measurements on turbine housings.

The results of the strain gauge measurements during operation correlated reasonably well with FEA results. It was evident that during service, in the locations investigated the outer wall of the housing is in tension at maximum temperature loading. It was found that residual stress surface measurements although small in magnitude opposed strains induced at maximum loading during thermal cycling in the areas investigated. The maximum compressive residual stress was close to 40% of the yield strength of the material. This could be beneficial because compressive residual stresses on the surface of the housing could work to a degree against the tensile stress induced in the surface at maximum temperature loading during thermal cycling. However in the bulk of the material in the divider wall, residual stresses were found to be smaller in compressive magnitude and on occasion tensile in nature and could work to further accelerate any crack propagation. In reality residual stresses will likely relax substantially during repeated thermal load and this will need to be further explored. In all three measurements directions as suspected, heat treatment was found primarily to reduce the magnitude of residual stress toward zero.

7 FUTURE WORK

Future work will measure residual stress on the same housing as was subjected to gas stand testing. Other residual stress measurement techniques will be employed, allowing mapping of larger areas within the housing to generate a complete picture of how residual stresses change throughout the internal geometry of the housing. Simulation of residual stress will then be performed and compared with measured values. FEA simulation will be performed using measured values to simulate service conditions with the pre-loading of residual stress, to investigate the effect on accumulated plastic strains and further investigations into stress relaxation during service will be performed. In order to improve the accuracy between FEA and measured strain values from the gas stand, the exact locations of the strain gauges could be measured with a coordinate measuring machine.

8 ACKNOWLEDGMENTS

The authors would like to thank John James from The Open University, for his assistance with the measurements obtained on the ENGIN-X instrument with the SScanSS software at the ISIS facility. Financial support was provided for this research by Innovate UK as a result of a Knowledge Transfer Project.

9 REFERENCES

[1] European Commission, "Reducing CO2 emissions from Heavy-Duty Vehicles," 2015. [Online]. Available: http://ec.europa.eu /clima/policies/transport/ vehicles/heavy/index_en.htm. [Accessed: 04-Jan-2016].

[2] D. Lohe and O. Vohringe, "Stability of residual stresses," in *ASM International, Member/Customer Service Center, ...*, 2002, pp. 54–68.

[3] S. Kiumars, "Shoghi, K. (2014). Influence of holding time in thermal cycle proven by state-of-the-art thermomechanical fatigue calculation in the turbine housing of a turbocharger," *11th Int. Conf. Turbochargers Turbocharging*, pp. 461–469, 2014.

[4] T. Siyi and L. Lin, "Study and practise of decreasing residual stress with residual heat of casting," vol. 5, no. 3, pp. 4–7, 2008.

[5] J.-C. Kim, S.-K. Cheong, and H. Noguchi, "Residual stress relaxation and low- and high-cycle fatigue behavior of shot-peened medium-carbon steel," *Int. J. Fatigue*, vol. 56, pp. 114–122, 2013.

[6] M. Ekstrom, *DEVELOPMENT OF A FERRITIC DUCTILE CAST IRON FOR INCREASED LIFE IN EXHAUST APPLICATIONS Madeleine Ekström Licentiate Thesis Stockholm , Sweden*. Sweden: Royal Institue of Technology, 2013.

[7] W. Z. Zhuang and G. R. Halford, "Investigation of residual stress relaxation under cyclic load," *Int. J. Fatigue*, vol. 23, pp. 31–37, 2001.

[8] K. Dalaei, B. Karlsson, and L.-E. Svensson, "Stability of shot peening induced residual stresses and their influence on fatigue lifetime," *Mater. Sci. Eng. A*, vol. 528, no. 3, pp. 1008–1015, 2011.

[9] R. Dakhlaoui, V. Klosek, M. H. Mathon, and B. Marini, "Orientation stress field analysis in polycrystalline bcc steel using neutron diffraction," *Acta Mater.*, vol. 58, no. 2, pp. 499–509, 2010.

[10] W. H. Bragg, "The Reflection of X-rays by Crystals. (II.)," *Proc. R. Soc. A Math. Phys. Eng. Sci.*, vol. 89, no. 610, pp. 246–248, 1913.

[11] P. J. J. Withers and H. K. D. H. K. D. H. Bhadeshia, "Residual stress Part 1 – Measurement techniques," *Mater. Sci. Technol.*, vol. 17, no. 4, pp. 355–365, 2001.

[12] S. Kilcoyne, K. Gannon, K. Shoghi, and T. Pirling, "Measurement of resiudal stress in the turbine housing of turbochargers (experimental report).," *Inst. Laue-Langevin*, 2013.

[13] J. A. James, "Application of robot kinematics methods to the simulation and control of neutron beam line positioning systems," *Nucl. Instruments Methods Phys. Res. A*, vol. 571, pp. 709–718, 2007.

[14] J. A. James, J. R. Santisteban, L. Edwards, and M. R. Daymond, "A Virtual Laboratory for Neutron and Synchrotron Strain Scanning.," *Phys. B Condens. Matter*, vol. 350, no. 1–3, pp. 743–746, 2004.

[15] T. Pirling and J. S. Robinson, "Precise Determination Of Residual Stresses In Large Specimens By Neutron Diffraction," no. NOVEMBER, 2009.

[16] K. Hoffmann, "Applying the wheatstone bridge circuit," *HBM S1569-1.1 en, HBM, Darmstadt, Ger.*, pp. 1–28, 2001.

[17] C. Oberste-Brandenburg, S. Kiumars, M. Gugau, and F. Kruse, "On the influence of thermal boundary conditions on the Thermo Mechanical Analysis of turbine housing of a turbocharger," *10th Int. Conf. Turbochargers Turbocharging*, pp. 83–95, 2012.

10 IMAGES

Figure 1: Crystallography CSIC (2016) obtained from http://www.xtal.iqfr.csic.es/Cristalografia/parte_05_5-en.html, licensed under http://creativecommons.org/licenses/by-nc/4.0/

SESSION 3

ADVANCED SIMULATION

Obtaining bulk flow based heat transfer coefficients for thermal evaluation of turbochargers

A Karamavruc
BorgWarner Inc., USA

ABSTRACT

An accurate transient thermal evaluation of a turbine housing, which is a component of a turbocharger, needs transient conjugate heat transfer analysis (CHT). However, a CHT calculation of a turbocharger is very time consuming, therefore not feasible to use on design iterations. Currently, the most efficient way to calculate transient thermal analysis is to use transient thermal calculations in a finite element analysis (FEA) environment. However, thermal analysis in FEA requires accurate local heat transfer coefficients (HTC) and reference fluid temperature (bulk temperature).

A local HTC and bulk temperature can be obtained via an adiabatic CFD analysis. Most commercial codes evaluate HTC using dimensionless temperature T^+ which is function of y^+. The requirement for the y^+ values may vary depending on the turbulence model used. A wall function turbulence model requires larger y^+ values compared to the low-Reynolds number approach. This makes near-wall based HTC calculations mesh dependent.

A bulk flow based HTC calculation was proposed using an adiabatic CFD calculation. HTC was calculated using y^+ value in a fully turbulent region. The resultant HTC was mesh independent. The proposed HTC, along with bulk temperature, resulted in accurate transient thermal simulations for turbine housings. In this present work, HTC obtained using the CFD code CFX was compared with the proposed HTC for wall function and low-Re number approaches. The transient thermal calculations of turbine housings using proposed HTC were correlated with full transient CHT calculations and test data. The test data were measured using several thermocouples around a turbine housing. Bulk flow based transient thermal predictions resulted in good agreement with CHT calculations and test data. The proposed method also provided consistent results when using different turbulence models, such as using wall function or low-Re number approaches.

NOTATIONS

y^+= Dimensionless wall distance coordinate
u^+= Dimensionless velocity
τ_ω= Wall shear stress [kg/ms^2]
ρ= Density [kg/m^3]
n= Coefficient
β= Coefficient
Γ= Coefficient

κ= Von Karman constant
Pr= Prandtl number
Cp= Specific heat capacity [J/kgK]
T_w= Temperature at wall [K]
T_{nw}= Near-wall temperature [K]
T^+= Dimensionless temperature
T^{+spec}= Dimensionless temperature at specific y^+
u*= Friction velocity [m/s]
Δn= Vertical distance from the wall to the next node [m]
q_w= Wall heat flux [W/m^2]
h_{nw}= Near-wall heat transfer coefficient [W/m^{2}K]
h_B= Bulk flow based heat transfer coefficient [W/m^{2}K]

1. INTRODUCTION

A turbocharger is one of the most significant add-on devices for engine components. It can significantly increase the engine horse power without adding significant extra weight. The basic principle of turbocharging is to squeeze additional air into the cylinder so that more fuel can be added to make powerful explosions during the combustion process. The result is not only an increase of the engine power but also less environmental pollution through a reduction of NOx and CO_2. Recently, the Environmental Protection Agency (EPA) has pushed for reduced emissions through stricter regulation of the chemicals released from vehicles. Consequently, many car manufacturers recognize the turbocharger as an immediate remedy for pollution issues.

With the increased interest and attention on turbochargers, extensive research has resulted in, not only understanding the aerodynamic behavior, but also optimization of the cycle life. Due to the harsh thermal environment of turbo chargers, specifically the turbine housing (up to 1050 C), the durability and cycle life calculations of turbine housings became one of the core development areas. Many researchers developed Thermal Mechanical Fatigue (TMF) methodologies to estimate a cumulative damage and predict the cycle life to crack initiation under thermal load conditions [1,2]. These methodologies are successfully applied to estimate the life of turbine housings under complex transient thermal loadings [3,4].

All of the cycle life calculation methodologies of a turbine housing first require a reliable thermal evaluation. Currently, the thermal evaluation approaches are based on three main methodologies. The first method is to use transient Conjugate Heat Transfer analysis (CHT) [4, 5]. CHT is the most accurate method due to inclusion of both the solid and the exhaust gas in the thermal calculations. Some designs require several iterations within a short period of time before ordering a prototype. If CHT is used in the design iteration process, it will be very cumbersome and time consuming. Additionally writing enough transient files to capture cycle calculations might consume significant disc space.

The second method uses an adiabatic CFD approach without inclusion of the turbine housing. The heat transfer coefficients (HTC) and the bulk temperatures are extracted at power rated and idle conditions and used as boundary conditions in Finite Element Analysis (FEA) software to evaluate thermal cycle calculations. As will be shown later, the main commercial CFD codes (CFX, FLUENT, STAR CCM+) output the HTC as a variable. The HTC is calculated using a non-dimensional temperature distribution which is a function of a dimensionless number called y^+. Depending on which turbulence method is used, wall function or low-Reynolds number method, the y^+ requirements can vary. The Shear Stress Transport (SST) method, which is a low-Reynolds number turbulence modeling methodology, requires much finer mesh near the wall (about $y^+=1$) compared with K-ε, which is a wall function methodology ($y^+\geq30$). Due to difference in y^+ in both approaches, HTC calculations are mesh dependent.

The third method to evaluate thermal calculations is to use analytical approaches [6]. The turbine housing can be partitioned into several different sections. Based on basic pipe flow assumptions, the Nusselt number can be calculated using Gnielinski [7], Petukhov [8], or Dittus-Boelter [9] equations. In a similar way, bulk temperature can be estimated for each region using the first law of thermodynamics. The analytically calculated HTC and bulk temperature can be applied as boundary conditions in FEM to simulate thermal response. The drawback of this type of analytical approach is that it cannot accurately capture the local temperature.

The objective of this study is to investigate a variation of the second approach, using a bulk flow based HTC rather than a near-wall based HTC. This is accomplished by modifying the output of the HTC calculation by the CFD code without violating the heat flux calculations. Once an accurate HTC is obtained from an adiabatic CFD approach, thermal evaluation of a turbine housing can be processed in FEA environment. The corrected HTC will serve as a thermal boundary condition in FEA. The proposed method not only enables an efficient turbine housing design iteration process but also accurately predicts the local temperature distributions. All of the investigations in the present work were accomplished using Ansys CFX code.

2. NEAR-WALL MODELING METHODOLOGY and NUMERICAL WALL HEAT TRANSFER CALCULATIONS

Most CFD codes have the capability to output a HTC that can be further used in other analytical or numerical thermal evaluations. It is of vital importance to understand whether the HTC provided by the CFD simulations truly matches experimental measurements. Despite having slight differences in their heat flux calculations, CFD codes use a consistent relationship at the boundary layer flow between the turbulent momentum and heat transfer [10,11]. This relationship, called Reynolds analogy, enables heat transfer coefficients to be predicted by calculating the pressure losses due to friction in the flow. According to this analogy, the non-dimensional turbulence velocity profile at the boundary layer is very similar to the non-dimensional temperature profile. Both will follow the universal law of the wall in the fully turbulent region. A brief review of a turbulence boundary layer simulation methodology below will help the reader understand and follow the main objective of this present work.

Due to a no slip boundary condition at the wall, the variation of the velocity and heat transfer is largest in the near-wall regions. Therefore, capturing of the velocity and thermal gradients at the boundary layer is vital for the accuracy of CFD simulations. The flow near the wall has been analyzed in terms of three layers as described by Tennekes and Lumley [12], namely the inner layer, logarithmic layer, and outer layer [10]. The inner layer is the smallest layer which is dominated by viscous shear.

In the inner layer, the dimensionless velocity profile is described as;

$$u^+ = y^+ \qquad 0<y^+<5 \qquad (1)$$

where y^+ is the non-dimensional number similar to the local Reynolds number and is given as;

$$y^+ = \frac{u^* y}{\nu} \tag{2}$$

where v is the kinematic viscosity, y is the vertical distance from the wall, u* is the friction velocity and described as;

$$u^* = \sqrt{\frac{\tau_\omega}{\rho}} \tag{3}$$

where τ_ω is the wall shear and ρ is the density.

The second important layer is the logarithmic layer, where the turbulence shear is dominant. This layer is sometimes called the overlap layer because it connects the inner layer and outer layer smoothly. The dimensionless velocity profile in the logarithmic layer is given as [13];

$$u^+ = \frac{1}{\kappa}\ln(y^+) - B \qquad 30<y^+<350 \tag{4}$$

where κ (Von Karman constant) and B are universal constants. Coles and Hirst [14] use the values of 0.41 for κ and 5.0 for B.

Finally, the last layer into the fluid domain is the outer layer, or defect layer, where large scale turbulent eddy and shear effects are dominant (see White F. M. [13]). The outer layer is described as;

$$\frac{U-u}{u^*} = 2.5\ln\left(\frac{y}{\delta}\right) \tag{5}$$

where U is the centerline velocity, u* is the friction velocity, y is the local distance from the wall, u is the local velocity, and δ is the thickness of the boundary layer.

There are two main modeling methodologies to resolve turbulent boundary layers in CFD calculations, namely Low- Reynolds number (low-Re) modeling and wall function theory. Wall function theory [9] allows CFD models to interpret behavior near a wall without a need for a very fine mesh. Therefore, first mesh nodes will be in the logarithmic turbulence region. The result of this approach is significant reduction in calculation time. However, this modeling is not recommended for problems with adverse pressure gradient, strong curvature to the flow, or jet flows.

In the low-Re approach, the viscous sub layer is finely meshed so that the first calculated mesh nodes are embedded within the laminar viscous sub layer. The height of the first cell is generally recommended around $y^+=1$. This will lead to a high number of elements that can cause longer convergence time. However, the low-Re approach predicts complex turbulent flows much better than the wall function approach. In this present work, the SST method of Menter [15] is primarily used. An SST model works by solving a turbulence/frequency-based model (k-ω) by Wilcox [16] at the wall and k-ε in the bulk flow. A blending function in SST ensures a smooth transition between the two models.

Because the focus of this present work is the HTC calculation, we will review briefly how temperature profiles are modeled in both turbulence modeling methodologies in CFX. A non-dimensional temperature profile for wall functions in CFX follows the logarithmic law of the wall [17] as:

$$T^{+}(y^{+},\mathrm{Pr}) = 2.12\ln(y^{+}) + \beta \quad (6)$$

Similarly, non-dimensional temperature for the Low-Re approach in CFX is modeled by blending the viscous sub-layer and the log-law as described by Kader [18];

$$T^{+}(y^{+},\mathrm{Pr}) = \mathrm{Pr}\, y^{+} e^{(-\Gamma)} + \left[2.12\ln(y^{+}) + \beta\right] e^{(-1/\Gamma)} \quad (7)$$

$$\beta = (3.85\,\mathrm{Pr}^{1/3} - 1.3)^{2} + 2.12\ln(\mathrm{Pr}) \quad (8)$$

$$\Gamma = \frac{0.01(\mathrm{Pr}\, y^{+})^{4}}{1 + 5\,\mathrm{Pr}^{3}\, y^{+}} \quad (9)$$

$$\mathrm{Pr} = \frac{\mu c_{p}}{\gamma} \quad (10)$$

where Pr is the Prandtl number, μ is the dynamic viscosity, c_p is the heat capacity, γ is the fluid thermal conductivity, and both β and Γ are the coefficients.

Heat flux at the wall for both turbulent modeling of the low-Re approach and the wall function can be modeled in CFX as [7];

$$q_{w} = \frac{\rho c_{p} u^{*}}{T^{+}(y^{+},\mathrm{Pr})}(T_{w} - T_{nw}) \quad (11)$$

where T_w is the wall temperature, T_{nw} is the near-wall temperature, c_p is the heat capacity and u* is the friction velocity. From equation (11), it is easy to see that the HTC can be described as;

$$h_{nw} = \frac{\rho c_{p} u^{*}}{T^{+}(y^{+},\mathrm{Pr})} \quad (12)$$

h_{nw} in equation (12) will be called, in this present work, the near-wall heat transfer coefficients because T^+ in the equation (6) and (7) uses y^+ of the first element nodes next to the wall as reference points. y+ can change throughout the calculation domain depending on the friction velocity and the size of the element next to the wall, as shown in equation (2).

A plot showing the relationship of the non-dimensional temperature T^+ and non-dimensional velocity u^+ along with near-wall heat transfer coefficient h_{nw} is plotted in Figure 1 as a function of y^+. The input data and the details for this plot are given in Section 4.2 Case 2. Both T^+ and u^+ confirm their logarithmic behavior at the boundary layer. The similar distribution between T^+ and u^+ at the boundary layer also confirms the analogy that exists between the momentum and heat transfer. Notice also the behavior of the heat transfer coefficient. As y^+ gets smaller toward the wall, the heat transfer coefficient increases. Maximum heat transfer coefficients will be the ones calculated using first nodal distances from the wall. One can also clearly say that in the turbulence overlap and outer regions ($y^+>30$) the rate of change of the HTC is very small. Depending on the turbulence model used, such as low-Re turbulence modeling or wall function approach, one will obtain different near-wall heat transfer coefficients h_{nw} in CFX due to the distance from the first element nodes next to the wall. As mentioned before, SST requires $y^+=1$ for an accurate simulation of a boundary layer whereas the K-ε requires $y^+>30$. In order to obtain an accurate HTC that can be used in FEA environment, the calculation of HTC should be based on bulk flow characteristics as will be explained in next section.

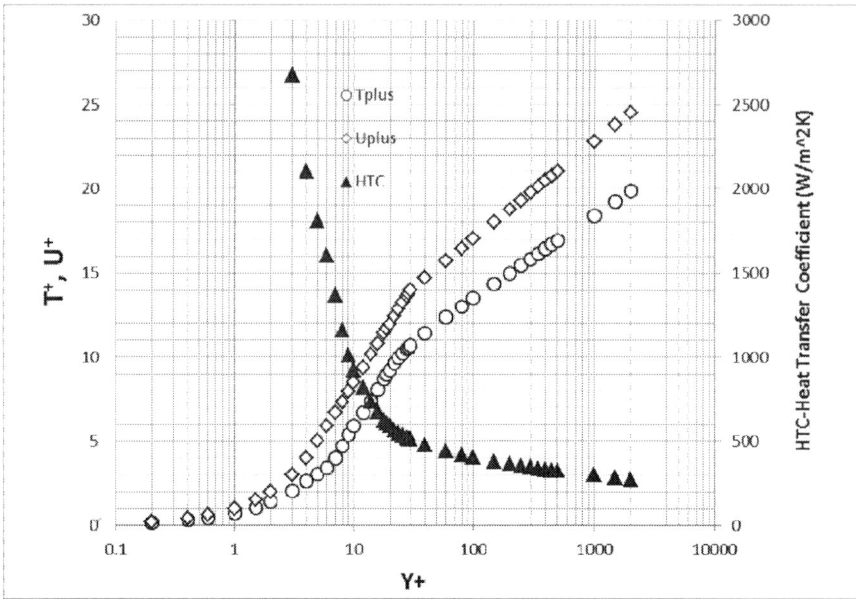

Figure 1 - Dimensionless temperature and velocity along with heat transfer coefficient as a function of dimensionless wall distance

3. PROPOSED WALL HEAT TRANSFER COEFFICIENT AND BULK TEMPERATURE

Whenever one measures heat flux or heat transfer coefficients of a fluid in motion, reference temperature will be based on the free stream or bulk flow temperature [19]. Equation (12) shows how CFX calculates the heat transfer coefficients using the first nodal distance from the first layer element off the wall. In the low-Re approach, that element will be completely embedded in the viscous sub-layer region. In this present work, it is proposed that heat transfer coefficients be calculated using a non-dimensional temperature $T^{+spec}(y^+=250,\Pr)$ evaluated further away from the viscous sub-layer region. The equation for $T^{+spec}(y^+=250,\Pr)$ is the same as equation (7). The only difference in $T^{+spec}(y^+=250,\Pr)$ is setting the y^+ value as 250 in equation (7). As shown in Figure 1 and pointed out in the previous section, the changes in HTC in the turbulence regions (overlap and outer layers), is not significant. Referencing the heat transfer calculations to the transition region from turbulent overlap layer to the outer layer should result bulk flow representative HTC. Without violating CFX heat flux calculations in equation (11), bulk flow temperature based heat flux can be written as;

$$q_w = \frac{\rho c_p u^*}{T^+(y^+,\Pr)}(T_w - T_{nw}) = \frac{\rho c_p u^*}{T^{+spec}(y^+ = 250,\Pr)}(T_w - T_B) \qquad (13)$$

Or one may express equation (13) as;

$$q_w = h_{nw}(T_w - T_{nw}) = h_B(T_w - T_B) \qquad (14)$$

where T_B is the bulk flow temperature, T^{+spec} is the dimensionless temperature evaluated in the overlap layer, and h_B is the bulk flow based heat transfer coefficient. It is assumed that the y^+ value chosen in the transition from overlap to outer layer for HTC calculations should provide mesh independent HTC whether one uses low-Re number or wall function turbulence modeling techniques. After a mathematical manipulation of equation (13) and (14) the new bulk flow based HTC can be obtained as;

$$h_B = h_{nw} \frac{T^{+}(y^{+}, \mathrm{Pr})}{T^{+spec}(y^{+} = 250, \mathrm{Pr})} \qquad (15)$$

Or equation (15) can also be expressed as;

$$h_B = \frac{\rho c_p u^{*}}{T^{+spec}(y^{+} = 250, \mathrm{Pr})} \qquad (16)$$

where h_{nw} is the near-wall HTC as calculated by the CFX code. Equation (15) modifies CFX calculated heat transfer coefficients h_{nw} whereas equation (16) recalculates bulk flow based HTC. Both of the equations provide the same answers. The contour plots provided by equation (15) might be more smooth due to smoothing algorithms done in CFX.

The proposed method is first tested for its consistency between SST and K-ε turbulence modeling. Figure 2 shows the contour plots of y+, HTC, and proposed HTC calculations using SST turbulence modeling of a turbine housing. The conditions and the details of the input parameters for these comparisons are given in section 4.2 Case 2. Figure 2-A shows the contour plot of y^+ distribution. Other than the small gaps between the wheel tip and shroud, y^+ values are below 5. Figure 2-B shows the HTC distributions calculated by CFX. Clearly the HTCs are unrealistically high due to small y^+ values required by the SST calculations. When one corrects the HTC in Figure 2-B using the method proposed in equation (15), then one will obtain bulk flow based HTC as shown in Figure 2-C. The same analysis was done using wall function based approach, namely the K-ε method, by re-meshing the same fluid volume with coarser grids and rerunning it with exactly the same input parameters as were used in the SST method. Figure 3-B shows the CFX calculated HTC using the K-ε method. Clearly the HTC in Figure 3-B is much lower than the HTC in Figure 2-B. By correcting the HTC in Figure 3-B, using equation (15), one can obtain very similar HTCs, as shown in Figure 3-C, compared with HTC in Figure 2-C which is obtained using SST method. The red circles in Figure 2-C show minor differences between the two methods. The basic reason for these differences is the predictions of the wall shear between the SST and K-ε methods. In the red circled regions, SST predicted higher wall shear compared with K-e. Therefore, the estimations of HTC using the K-ε method at those regions are slightly lower than the SST method. This test confirms that the method proposed in equation (15) or (16) will provide mesh independent HTC calculations.

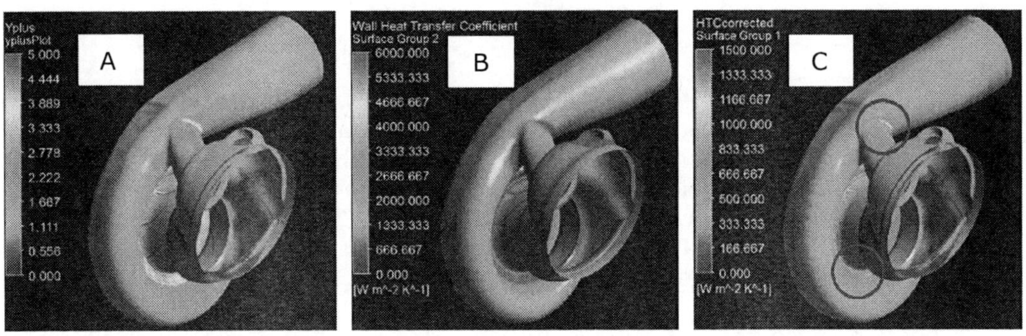

Figure 2 - Low-Re turbulence modeling (SST) A) y^+ distribution, B) Wall based HTC calculated by CFX code, C) Bulk flow based HTC calculation-see equation (15)

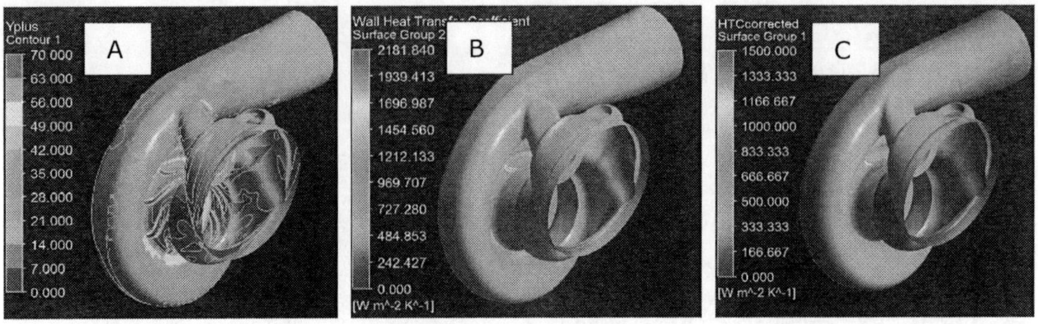

Figure 3 - Wall function based turbulence modeling (K-ε) A) y^+ distribution, B) Near-wall based HTC calculation by CFX code, C) Bulk flow based HTC calculation-see equation (15)

The next step is to confirm the accuracy of this proposed method when used in transient thermal simulations of turbine housings. Besides corrected HTC, bulk flow temperature is also needed as a reference temperature to proceed with thermal calculations of a turbine housing in FEA. As will be explained in the next section, steady state adiabatic CFD analyses are needed at idle and rated power conditions. The HTC and the bulk temperature will be extracted at these steady state operating points. At these conditions, the heat loss of turbine housing is minimal due to negligible temperature gradient through the wall. Besides heat conduction to the shaft and to the bearing housing, the main factor effecting the exhaust temperature at these conditions is the conversion between the enthalpy and the kinetic energy. As the exhaust flow speeds up temperature will drop and vice versa. Due to assumption of adiabatic flow at idle and rated power conditions, there is no significant differences in temperature between the core and near wall boundary flow. Therefore near-wall temperature in an adiabatic CFD calculation of a turbine housing is a good estimation of bulk temperature.

4. FEA AND CFD PROCESSES ALONG WITH EXPERIMENTAL SETUP FOR THE CASES INVESTIGATED

The proposed approach discussed in section 3 was tested with several different cases for its accuracy and applicability of the calculations to turbine housings. In this present work, 2 cases are presented for correlation of the new approach. The first case was a correlation to measured temperature data, whereas the second case was correlation to a full transient CHT calculation.

4.1 Case 1

For the first case, the selected turbine housing was for a commercial diesel application which is much bigger in size when compared with a passenger car application. This turbine housing was instrumented with several thermocouples and strain gages as shown in Figure 4. A test was performed on a gas stand for about 1200 cycles. Each cycle was about 1240 sec. Each cycle started with a 70 sec ramp-up from 160 C to 720 C. The exhaust temperature was held constant at 720 C for about 535 sec then cooled to a temperature of 160 C over 100 sec. It was held at that temperature another 535 sec. The mass of inlet flow was 0.21 kg/sec in the heating phase and 0.27 kg/sec in the cooling phase. The readers are referred to the publication by K. Shoghi [20] for further details about the test setup and thermocouple specifications.

Figure 4 - Instrumented turbine housing with thermocouples and strain gages (see the details by Shoghi [20])

Temperature measurements of the case 1 were compared with full transient CHT and proposed CFD approaches. CHT simulation began with the creation of fluid and solid volumes as shown in Figure 5 from the turbine housing shown in Figure 4. Both the CHT and the CFD model included additional outlet fluid extension to help with the convergence and stabilization of the outlet flow. The SST turbulence model was applied to all CFD and CHT cases due to its accuracy with respect to prediction of boundary layer flow [15]. A frozen rotor interface model [17] was chosen as coupling between the rotational and stationary fluid domains. For proposed CFD method (without solid turbine housing and bearing housing volume), two steady state CFD cases were run; one at the idle condition (160C) and the other at the rated power (775 C) condition. The inlet and outlet boundary conditions were specified as mass flow (see Table 1) and relative pressure respectively. In proposed CFD cases, the outside walls were specified as adiabatic. In CHT case, a constant convective boundary condition (HTC=20 W/m^{2}K, T_{air}=32 C) was applied to all outside wall boundaries. Once the CFD run was completed, the proposed HTC shown in equation (15) or (16) was calculated in CFX post and exported to a file, along with wall adjacent temperatures which were used as an input for FEA.

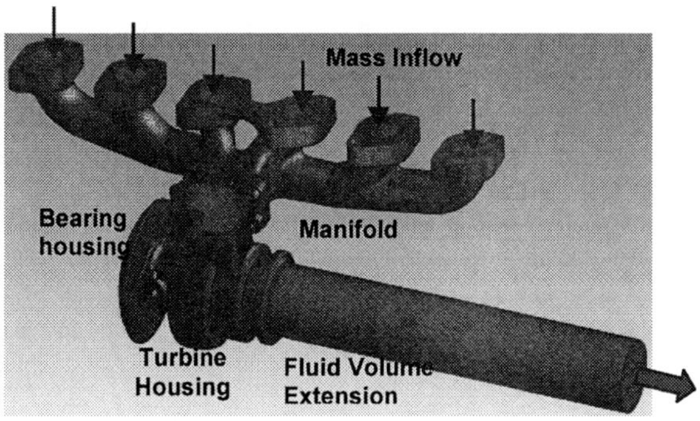

TABLE.1 CFD Inputs data-Case1	
Mass flow (hot) (kg/sec)	0.21 @ 775 C
Mass flow (cold) (kg/sec)	0.27 @ 160 C
Rump up (s)	70
Rump down (s)	100
Wheel Rotation (hot) (rpm)	126474 @775 C
Wheel Rotation (cold)	12500 @160 C

Figure 5 - CHT Mesh and inputs table for Case 1

The model was meshed in Ansys Workbench using tetrahedral elements. Special care was given to the mesh size and mesh quality around the turbine blade due to high velocity gradient. Specifically the gap between the turbine wheel and the shroud (0.5mm) needs to be meshed using fine meshes (17 layers of elements) as shown in Figure 6. Total number of elements used for simulations were 9,035,891 for fluid domain and 707,072 for solid domain. There were 8 prism layers on the walls to capture the boundary layer. After testing for several different sizes of first prism layer, it was found that 0.005 mm was the most appropriate first layer thickness to obtain required y+ value around 1 for this given input data.

Calculations were done using HPC cluster with 32 cores. Total wall clock time for one steady state proposed CFD case was about 1.48 hours. Convergence criteria for maximum residuals was specified as 1E-4. Besides the momentum residuals, mass and energy imbalances were also observed. Both mass and energy imbalances were below 0.1%. The full transient CHT case took about 62.4 hours to finish one cycle. The time step was varied dynamically as the case was running to shorten the calculation time. The average time step size was about 0.2 sec.

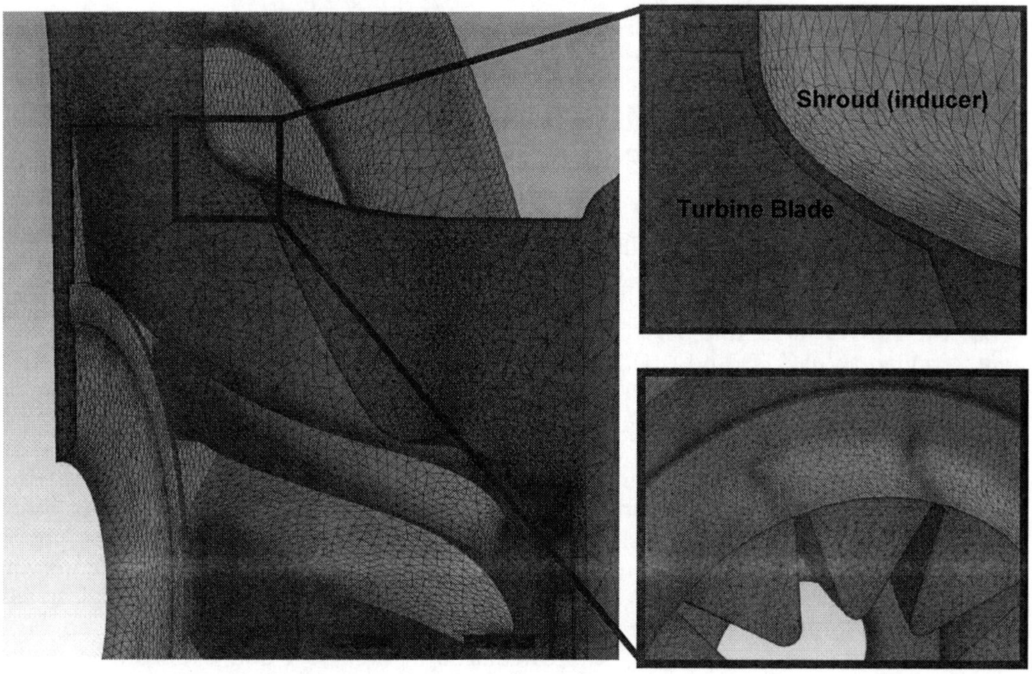

Figure 6 - Details of the fluid mesh around the turbine blade

The next step after proposed CFD simulation was to process thermal FEA. Thermal FEA was processed in the Ansys Workbench environment. Once the turbine housing was meshed, the proposed HTC along with the wall adjacent temperature were mapped onto the turbine housing as shown in Figure 7-A. To mimic the gas stand test, both the HTC and bulk temperature were linearly ramped in 70 seconds from idle to rated power rated condition as shown in Figure 7-B. Similarly, the cool down curve was simulated with a ramp down of the HTC and the temperature within 100 sec from rated power to idle condition. Overall, two cycles were simulated. The temperature results of second cycle were saved and compared with the measured thermocouple data (see Section 5). The bearing housing was also included in the thermal calculation because it acts as a heat sink at the flange interface between the turbine housing and bearing housing. In thermal simulation, the contact between the turbine housing and bearing housing is simulated as bonded contact with a thermal contact conductance coefficient of 3000 W/(m^{2}K). Inverse of this property is called thermal contact resistance. Cooling of the bearing housing, provided by oil flow, is simulated using heat transfer coefficients and sink temperature along the oil wetted surfaces. These thermal boundary values were developed by BorgWarner Inc. using experimental measurements.

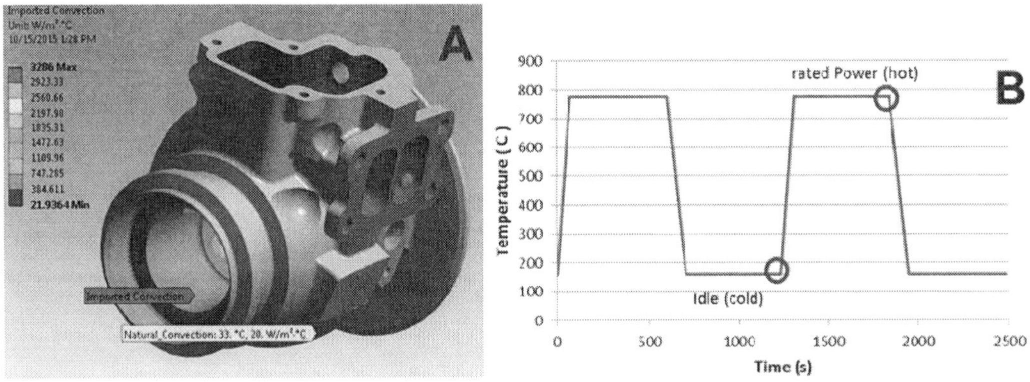

Figure 7 - A) Mapping the HTC and Bulk temperature obtained from CFD onto the turbine housing and B) Inlet input temperatures for two cycles

4.2 Case 2

The second test case chosen was a turbine housing of a small passenger car application from one of our customers. Both cases, CHT and proposed method, did not have the bearing housing in the calculations. In order to test the proposed method, inclusion of the bearing housing is not necessary. The key for the comparisons is to keep the interface boundaries in both calculations, in CHT and FEA, as identical. In this case 2, the flange of the TH (interface to bearing housing) was kept as adiabatic. Similar to previous case, a constant convective thermal boundary condition was specified on the outside wall boundary of the TH (HTC=20 W/m^2K, T_{air}=32 C) for both CHT and the FEA cases.

The second case did not have thermocouple measurements. Therefore, the thermal predictions of FEA using proposed methodology were correlated with a full transient CHT simulation. This means that there were two CFD runs, one including the turbine housing in the calculations and one without the turbine housing. The CFD and FEA processes for this case were the same as the previous case except for the cycle definition. Instead of calculating 2 cycles, a step input was applied. Inlet exhaust temperature was increased from 32 C to 1000 C within one second and held at that temperature for 400 seconds. See Table 2 for other input variables applied to the CFD simulations.

The meshing process of the second case was similar to case 1. The number of mesh elements used in fluid volume and solid volume were about 8,654,347 and 424,447 respectively. A steady state CFD case without including the solid volume run for about 1.39 hours using the HPC cluster with 32 cores. The full transient CFD, on the other hand, took about 20.13 hours.

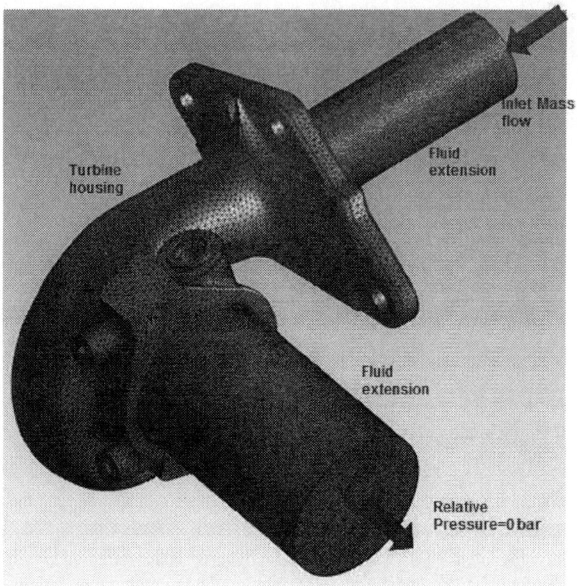

Table.2	CFD input data-Case2
Mass flow (kg/sec)	0.1
Rump up (s)	1
Wheel Rotation (rpm)	168000
Outlet Relative pressure (bar)	0

Figure 8 - CHT model of turbine housing and the input variables in Table 2

5. RESULTS and VALIDATIONS

5.1 Validation of Case 1

Typical contour plots of the thermal FE analysis are shown in Figure 9-A and 9-B. Figure 9-A is temperature distribution at idle condition, whereas Figure 9-B is the temperature distribution at rated power condition. Usually the volute, the divider wall, the tongue, and the shroud areas are heated and cooled much faster compared to other areas. The flange to the bearing housing side acts mostly as a heat sink due to the contact heat transfer. Therefore, as shown in Figure 9-A and 9-B, the flange of the turbine housing (interface to bearing housing) is the coldest area.

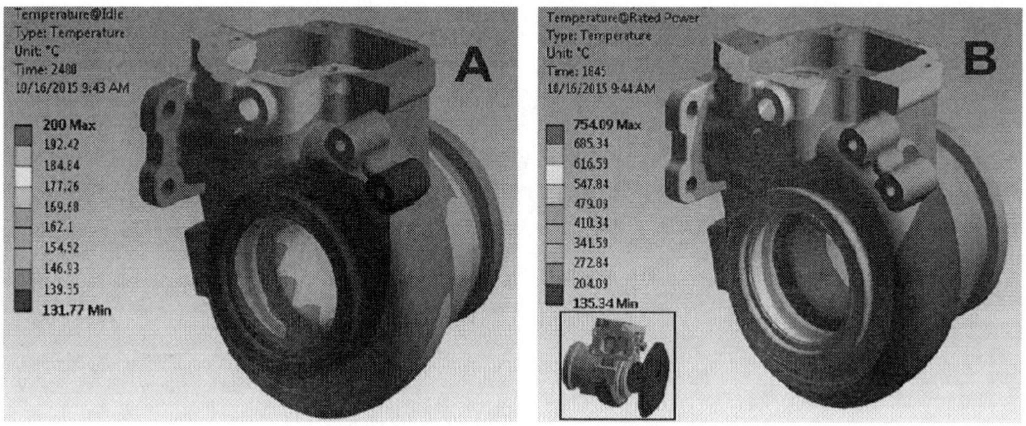

Figure 9 - Temperature contour plot at A) idle, and B) rated power condition

Figures 10-A, 10-B, and 10-C show the local comparisons between the simulations (using proposed HTC calculation and wall adjacent temperature from CFD) and measured data using thermocouples. Approximate thermocouple locations are indicated using a red dot. In reality, all thermocouples are embedded into the material. Figure 10-A shows the comparison of the transient temperatures between the experiment and calculation at the inlet divider wall. The thermocouple location in Figure 10-B is also in the divider wall but in half-way down the volute. Thermocouples are embedded 5 mm from the top into the divider wall. The thermocouple shown in Figure 10-C is drilled into the boss 5 mm from top. All three thermocouple temperatures agree well with the simulations. Calculations show consistently slightly lower temperature at the idle condition compared with the experiment. This might be related to thermal accuracy of the material data used. However, overall these results are very encouraging.

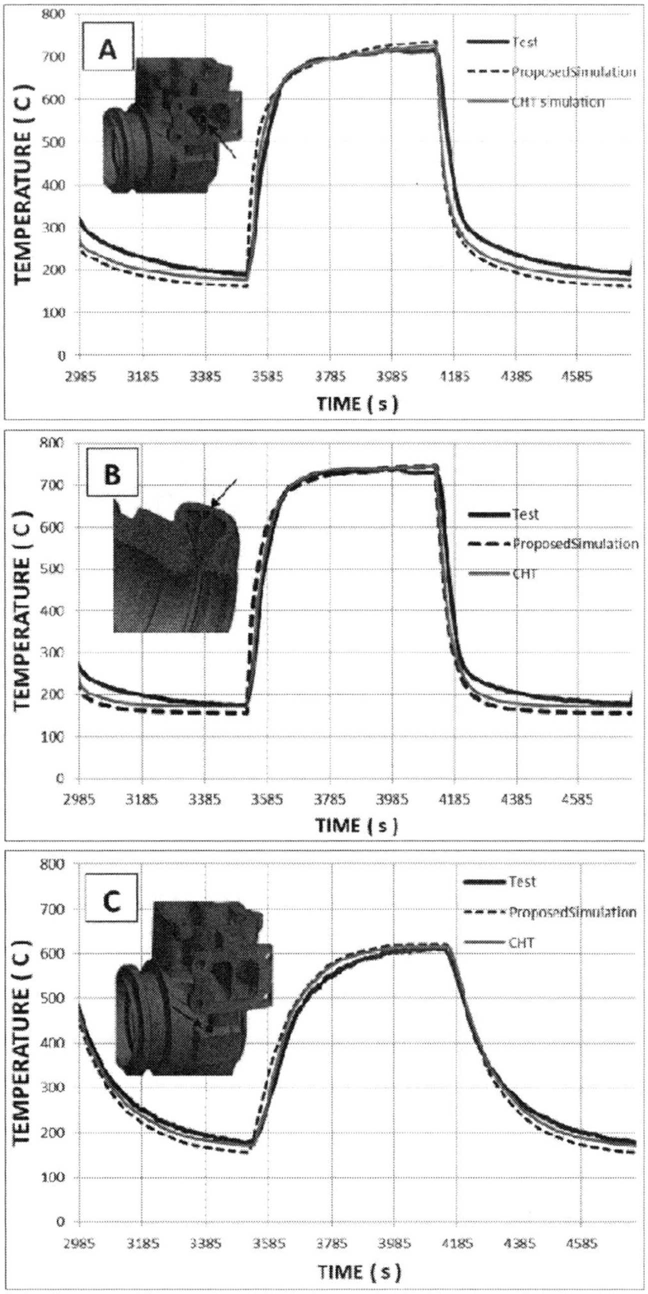

Figure 10 - Comparisons of temperature data between the simulations using the proposed method, full transient CHT and the measured data A) Inlet divider wall, B) Volute divider wall, C) Boss

5.2 Validation of Case 2

As was described in the previous section, case 2 did not have experimental measurements. Instead, the proposed methodology was validated with a full transient CHT calculation. Figure 11 compares temperature contour plots of the proposed methodology and a CHT calculation at t=80 sec. Figure 11-A is calculated using the proposed methodology and Figure 11-B is calculated using full transient CHT. Slight differences are observed at the flange (connection to manifold). However, overall temperature distribution is very similar between the two cases. Figure 12-A, 12-B, and 12-C show more detailed local comparisons between the CHT and proposed FEA method. Numerical probe locations are indicated as red dots on the part. Several different locations are selected for comparison. As shown in Figure 12-A, 12-B, and 12-C, the correlation between the FEA using proposed HTC and bulk temperature and the CHT is very good. Probe 1, which is at the base of the flange, shows slight under prediction of the temperature compared with the full CHT for the proposed method. Generally, these differences will not make a significant impact when one calculates thermal stresses and strains to calculate damage on the part.

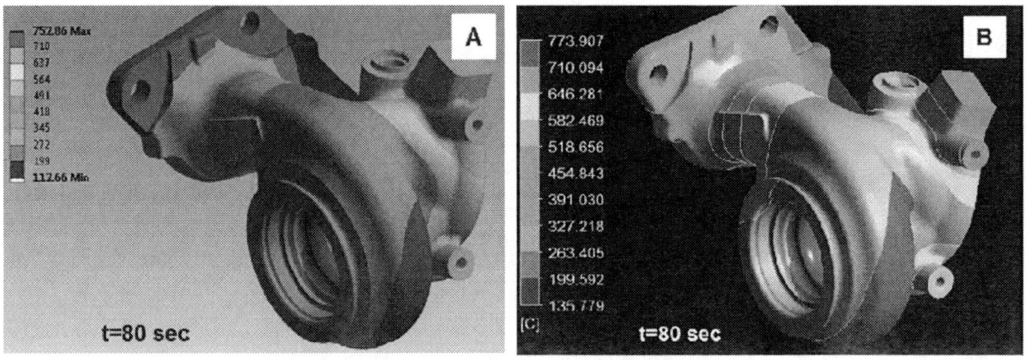

Figure 11 - Contour of temperature A) Calculated using FEA, B) Calculated using CHT

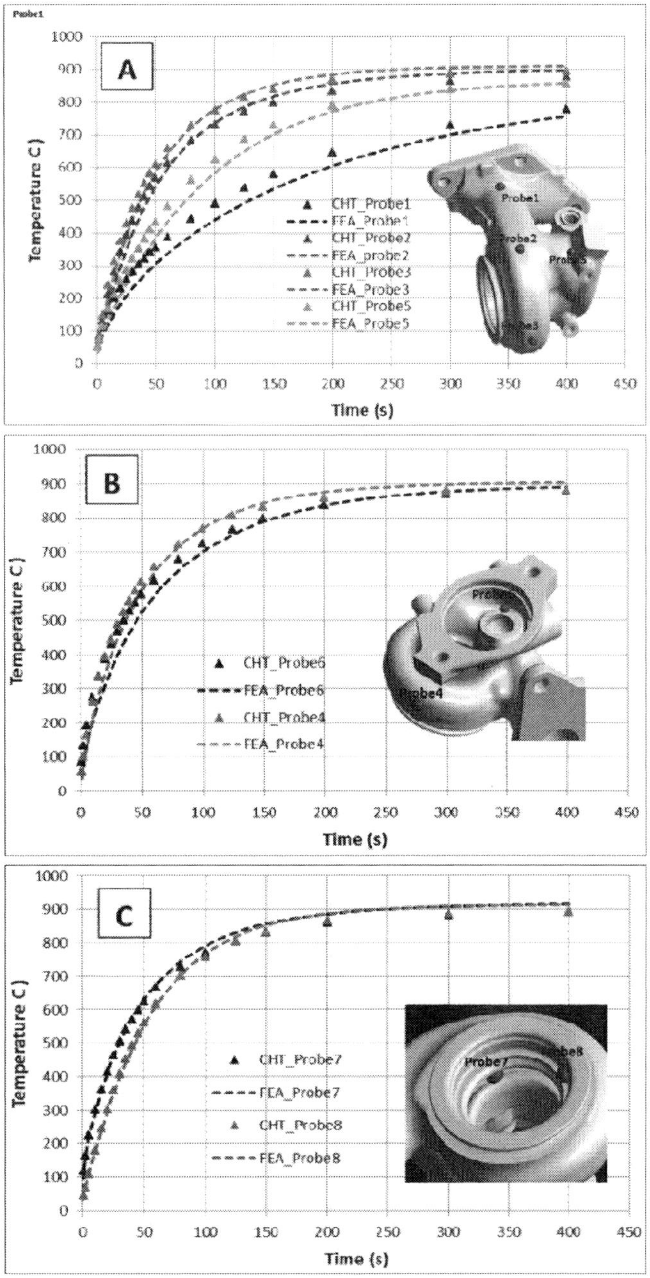

Figure 12 - A, B, C Local transient thermal responses of proposed approach (FEA) and the CHT approach

6. CONCLUSIONS

The HTC calculated from the CFD code, CFX, showed that depending on the turbulence model used (SST vs. K-ε), the output of HTC can vary significantly. Therefore the HTC provided by the CFD code is not advised to be used for further thermal processes. The source for the inaccurate HTC, by the CFD code, is due to usage of y^+ of the first element next to the wall. The proposed heat transfer coefficient calculations, using y^+ defined in the overlap turbulence region (log-law region), provides accurate thermal simulations. The proposed HTC methodology can be applied to both SST and K-e turbulence methods to correct HTC for further use.

Using the proposed method, the simulated temperature response of a commercial diesel turbine housing correlated well with the test data (see Figure 10). There are slight under predictions during the idle phase. The reasons could be related to the discrepancy between the gas stand input and the numerical input. The first reason could be due to differences in the ramp rate of mass flow and temperature. An additional reason could be the accuracy of the temperature-dependent conductivity data used in the numerical simulation.

The temperature response of the second case of a small passenger car turbine housing also correlated well using the proposed method (see Figure 12). Due to lack of testing data for this case, the proposed method was compared with full transient CHT simulation data. Slight under predictions of the proposed method at probe1 could be improved by varying the y^+ value, picked in the turbulence region, for the proposed HTC calculation.

In addition to accuracy and consistency, the proposed approach, as explained in section 3, offers significant savings in time and resources during turbine housing design iterations. Depending on the cycle time, full transient CHT may take several days to finish one thermal cycle calculations. The proposed method, on the other hand, will take a couple of hours to finish the calculation for one full thermal cycle.

REFERENCES

1. J. L. Chaboche, Cyclic Viscoplastic Constitutive Equations. Part I: A Thermodynamically Consistent Formulation. In: J Appl Mechanics 60 (1993), pp 813-821.
2. R. Neu and H. Sehitoglu, Thermo-mechanical Fatigue, Oxidation and Creep, Part I Experiments, Part II: Life Prediction, Metal Transactions, 20A, pp 1755-1783.
3. F. Längler, H. Alexanoglu, T. Mao, and A. Scholz: Validation of a phenomenological lifetime estimation approach for its application on turbine housing of exhaust turbocharger, 9th international conference on Turbo chargers and Turbocharging, Institution of Mechanical Engineers, May 19-20, 2010 London, pp 193-205.
4. F. Ahdad, M. Groskreutz, W. HuiMing, TMF design optimization for automotive turbochargers turbine housings, Proceedings of GT2007 ASME Turbo Expo 2007: Power for Land, SEA and Air, May 14-17, 2007, Montreal, Canada.
5. D. Bohn, T. Heuer, K Kusterer, Conjugate Flow and heat Transfer Investigation of a Turbo Charger, 2005, ASME Journal of Engineering for Gas Turbines and Power Vol.127, pp 663-669.
6. C. O. Brandenburg, K. Shoghi, M. Gugau, F. Kruse: On the influence of thermal boundary condition on the Thermo Mechanical Analysis of turbine housing of a turbocharger. 10th International Conference on Turbochargers and turbocharging. May 2012. P83.
7. F. P. Incropera, P. David DeWitt (2007), Fundamentals of heat and Mass Transfer (6th ed.) Hoboken, Wiley. pp 490-515. ISBN 978-0-471-45728-2.
8. B. S. Petukhov, Heat Transfer and Friction in Turbulent Pipe Flow with Variable physical Properties, In Advances in Heat Transfer, vol.6, eds. J. P. Hartnett and T. F. Irvine, Academic Press, New York (1970).
9. Frederic William Dittus, Llewellyn Michael Kraus Boelter (1930), Heat Transfer in Automobile Radiators of the Tubular Type, University of California publications in Engineering, Volume 2; Volume 13 of California publications in engineering: University of California Press.
10. H. Schlichting, Boundary-layer Theory, Seventh edition, McGraw-Hill Book Company.

11. Williams Kays, Michael Crawford (19080), Convective Heat and Mass Transfer, McGraw-Hill Ryerson.
12. H. Tennekes and J. L. Lumley. *A First Course in Turbulence*. Massachusetts Institute of Technology, Cambridge, 1972.
13. F. M. White, Viscous Fluid Flow, 3rd Edition, McGraw-Hill Education.
14. D. E. Coles, E. A. Hurst, "Computing of Turbulent Boundary Layer", AFOSR-IFP Stanford Conference", Proc. 1968 Conf., vol.2, Stanford Univ., Stanford, California.
15. F.R. Menter, Zonal Two Equation k-ω Turbulence Models for Aerodynamic Flows, AIAA Paper 93-2906, 1993.
16. D.C. Wilcox, Multiscale model for turbulent flows, In AIAA 24th Aerospace Sciences Meeting. American Institute of Aeronautics and Astronautics, 1986.
17. Ansys CFX documentation, Version 16.2, 2015.
18. B. A. Kader, Temperature and concentration profiles in fully turbulent boundary layers, International Journal of Heat and Mass Transfer, 24(9):1541-1544, 1981.
19. A. F. Mills, Heat Transfer, Richard D. Irwing, Inc., 1992.
20. K. Shoghi, Influence of holding time in thermal cycle proven by state-of-the-art thermomechanical fatigue calculation in the turbine housing of a turbocharger, 11th International Conference on Turbochargers and Turbocharging, 13-14 May 2014. London, pp. 469-461.

Numerical investigation of the effect of different nozzle clearance at hub and shroud side on variable nozzle turbine performance

X Shi, S Guo, B Zhao, C Yang, C Ma, M Qi
School of Mechanical Engineering, Beijing Institute of Technology, China

ABSTRACT

It is clear that when the nozzle vanes opening is adjusted to minimum area, the increased leakage flow has greater influence on the performance of variable nozzle turbine(VNT). While working with the engine, the nozzle clearances will not remain fixed and same with those under the design condition. In this paper 3-D numerical simulations of a VNT were performed to investigate the effect of different nozzle vane clearances at hub and shroud side on the turbine performance. The steady simulation results show that the turbine efficiency can reach to maximum while the shroud clearance value is about 10% of total clearance and then decrease with the increase of nozzle shroud gap. The VNT internal flow was analyzed for two typical ratios of clearance at shroud side to total clearance. It shows that although a non-uniform flow distribution at nozzle exit for the ratio of 0.15, it reduce the flow loss on the rotor suction side compared to that for equal clearance. It can be illustrated that for the ratio of 0.5, due to the centrifugal force, the low-energy fluids near the rotor hub side will migrate from hub side toward shroud side and more loss is generated during their transportation along rotor passage. The unsteady numerical simulations of this VNT with those two typical ratios were also undertaken. The change of nozzle clearance distribution also plays an important role of shock wave near the vanes exit. Compared to equal clearance at both hub and shroud, the strength and the length of the shock wave can be increased when the ratio value is 0.15, which leads an enhanced unsteady load of turbine rotor near the leading edge. This may increase the risk of HCF failure of the turbine rotor.

1 INTRODUCTION

Variable nozzle turbocharger(VNT)can get a better match with vehicle engines by adjusting the turbine nozzle vanes throat area to achieve low fuel consumption and low emissions. In order to pivot the nozzle vanes under different engine operation conditions there must be small clearance between the vanes and the side wall (both hub and shroud side).This leads to the leakage flow passing through the nozzle clearance and therefore a significant impact on downstream turbine rotor flow.

Normally the nozzle clearance at hub and shroud side was designed same for VNT. But while working on the actual conditions the spanwise position of the nozzle vanes cannot be controlled so accurately that the two gaps at hub and shroud will not remain constant and same. Some researchers have carried out experimental investigations of the impact of different nozzle spanwise positions on the turbine performance[1,2].

But the reason why the different nozzle clearance at hub and shroud can affect the turbine performance has not been understood deeply so that how a better performance can be achieved is not clear. Therefore it is necessary to carry out investigations to understand the involved mechanism.

In this paper 3-D numerical simulation of a VNT was performed to investigate the effect of different nozzle clearance at hub and shroud side on the turbine performance based on detailed analysis of internal flow phenomenon of the turbine. The experimental measured data has not been presented and at this stage this is an analytical study.

2 DESCRIPTION OF TURBINE

The analyzed variable nozzle turbine (VNT) consists of 9 pivoting nozzle vanes and mixed flow rotor with 13 blades, as shown in Fig.1 The three-dimension Navier-Stokes equations (RANS) were solved using commercial CFD software (NUMECA).The detailed description of the numerical method and steady computation verification can be found in the paper presented at the last conference in 2014[3].

The passage flow area can be changed by rotating the vane around a pivot point. This requires a clearance between the nozzle vane tip and the sidewall to accommodate the vane rotation. Fig.2 shows the locations of the clearance. Such a gap allows space for a clearance flow between the pressure and the suction sides of the vanes, which in turn decreases dramatically the efficiency. It is the ideal condition of a minimum clearance gap, which ensures a reliable rotating vane in order to reduce the flow loss. In fact, however, this minimum gap should be increased in design process due to thermal-expansion phenomenon.

Figure 1 Geometry of the variable nozzle turbine

Under the actual working conditions the two gaps are not equal due to the manufacture error, the aerodynamic interaction and the working environment. A geometrical parameter UCR(Uneven Clearance Ratio) was defined to quantify the uncertain matching states of the two gaps, such as the following definition

$$\mathrm{UCR} = \frac{\mathrm{Gap1}}{\mathrm{Gap1} + \mathrm{Gap2}}$$

where Gap1 and Gap2 are the clearance at shroud and hub respectively.

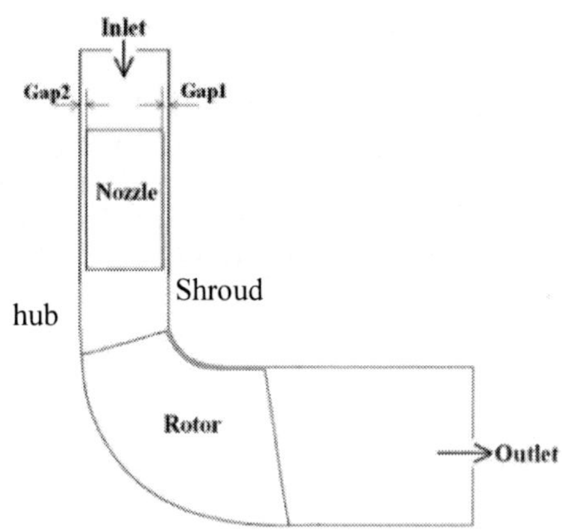

Figure 2 Illustration of the rotor, nozzle gaps, meridional view

3 ANALYSIS AND DISCUSSION

3.1 Turbine characteristic with clearance

In this section, the steady numerical simulations were undertaken and the turbine performance and internal flow were analyzed with different UCR value. Many researchers have revealed that the nozzle vane's leakage flow increases as the nozzle opening is closed small and deteriorates turbine efficiency[4-6]. So only small opening is used for the numerical simulations in this paper.

The trend of the turbine isentropic efficiency with UCR value was shown in Figure 3, which includes three kinds of inlet total pressure conditions under the same rotational speed 120000r/min. It shows that for VNT although the same clearance was designed at two ends of nozzle vanes, there is an optimum UCR value for the turbine maximum efficiency. This optimum UCR value is about 0.1 of total clearance regardless of the inlet total pressure condition. Besides, the efficiency reduces to the minimum value when the UCR value increases to 1.0. In either inlet total pressure case, the maximum difference between the maximum and minimum efficiencies reaches nearly to 6 percentage points.

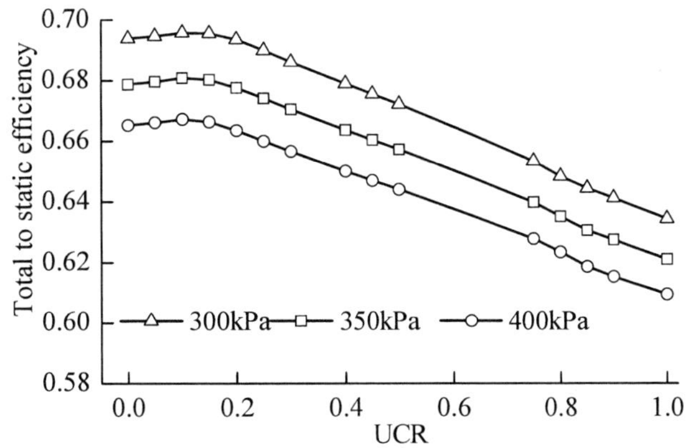

Figure 3 Total to static efficiency versus UCR value

Two typical UCR values, UCR=0.50 and UCR=0.15, are selected to analyze the physical mechanism why the turbine characteristic varies with UCR value for a specified inlet total pressure of 400kPa. The all following discussions are based on this inlet condition. Fig.4 compares the characteristics of nozzle vane's clearance leakage for two typical UCR values on the surface near the shroud. Left plots are corresponding to distribution of stream lines, and the right ones are for the entropy distribution.

The results in Fig.4 show that the UCR value decided the clearance leakage route. Compared with UCR=0.15, the clearance leakage flow route is far away from the suction side of nozzle vane when the UCR=0.50. This is due to that the clearance leakage through nozzle clearance driven by the pressure gradient between the pressure and suction side of vane increases significantly as the clearance scale increases, as shown in Fig.4. This causes more clearance leakage injects into the free stream and leads more flow loss. For example, for UCR=0.50, the entropy in the clearance leakage field is higher than that corresponding to UCR=0.15 in right plots of Fig.4.

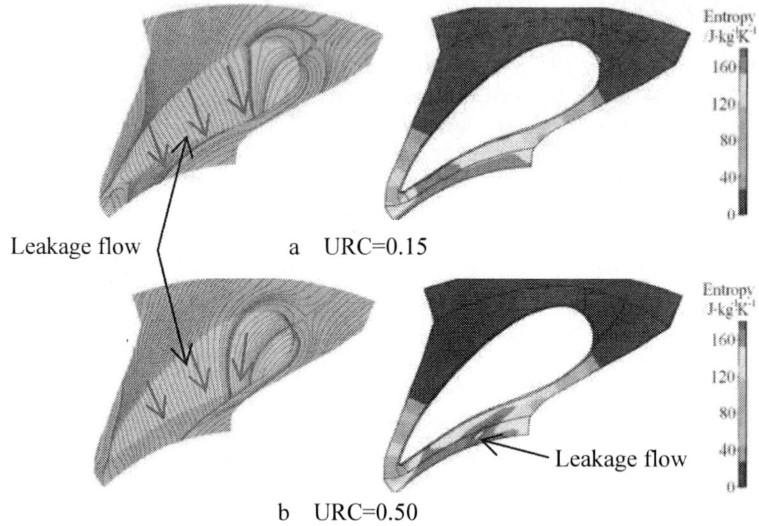

Figure 4 Characteristics of nozzle vane's clearance leakage on the surface near shroud

Clearance leakage variability, included the route and flow rate, must induce the redistribution of the flow loss at the nozzle exit. The total pressure loss coefficients, defined as equation (1), are compared for two typical UCR values in Fig.5. Fig.6 compares the density flow along spanwise direction for both the UCR values.

$$p_k = \frac{p_{01} - p_{02}}{p_{01}} \quad (1)$$

Where p_{01} and p_{02} are the circumferential average values of total pressure at nozzle inlet and outlet respectively.

The curve features with symmetric distribution when the UCR=0.50. This is because of the mass flow rate is identical on both hub and shroud side, as shown in Fig.6. This feature is broken and the total pressure loss coefficient redistributes along the spanwise direction by changing the UCR value from 0.50 to 0.15. The decreased clearance scale results in that the clearance leakage weakens and total pressure loss decreases on the shroud side. On the hub side, however, the clearance leakage is enhanced by increasing clearance scale so that the pressure loss rises. Besides, the location of the minimum pressure loss shifts toward to spanwise direction due to the redistribution of density flow, as shown in Fig.6.

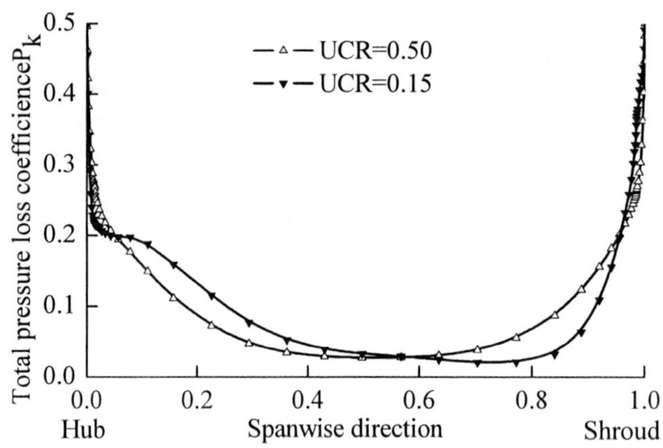

Figure 5 Comparison of total pressure loss coefficient of nozzle for two UCR values

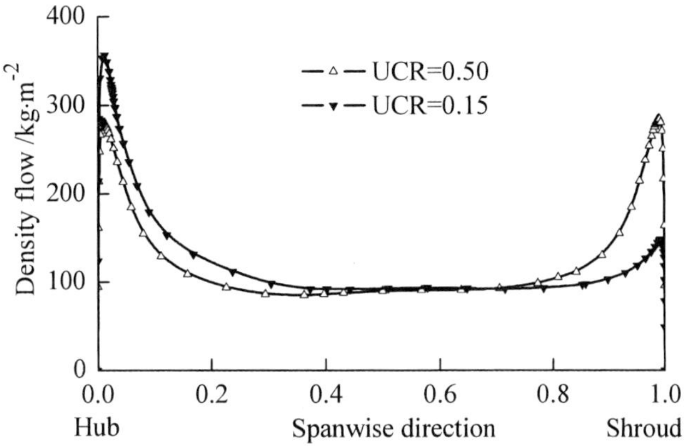

Figure 6 Spanwise distribution of the density flow at nozzle outlet for two UCR values

The change of the fluid through the nozzle clearance leads to redistribution of the flow angle at the rotor inlet. Fig.7 shows the relative flow angle at rotor inlet for two typical UCR values. For UCR=0.50, the strip with low flow angle on the hub is similar with that on the shroud while the strip is significantly increased on the hub side and the one on the shroud side is obviously decreased when UCR value is 0.15. The changes of the relative flow angle could be explained as follow: the fluid flows through nozzle clearance from the pressure side to the suction side of the nozzle and the tangential velocity of the clearance leakage is lower than that in the free stream at the nozzle's exit. As a result, the relative flow angle of the rotor decreases on the hub and shroud side. As the UCR value changes from 0.50 to 0.15 the clearance on the hub side increases which result in an enlarged area of small relative flow angle. It is obvious that the leakage flow features with high meridian velocity forms small positive even negative incidence angle. This is benefit to reduce incidence loss in the rotor inducer. It can also be seen that when the clearance on the hub side increases a part of main flow which is pushed by leakage vortices also forms a smaller range of high positive flow angle for UCR=0.15. This is one of main reasons to get an optimistic turbine performance at small UCR value.

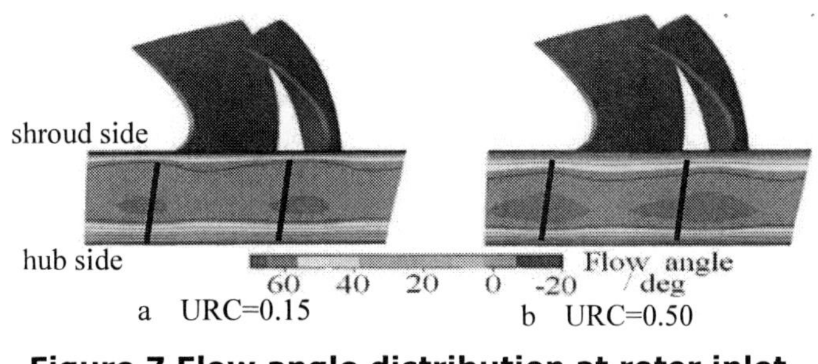

Figure 7 Flow angle distribution at rotor inlet

Fig.8 compares the transport of the low energy fluid for two UCR values. Due to the low energy fluid rotates with the rotor and could not overcome the centrifugal force. As a result, the low energy fluid begins to migrate from hub towards shroud side in the rotor passage. The result is that the low energy fluid transports along high spanwise location and it is near to shroud at the exit of the rotor. An obvious difference of flow loss induced by low energy fluid between both UCR values can be observed as well. I.e. for UCR=0.50, the low energy fluid occupies more flow passage and thus more loss is generated compared to the one when UCR=0.15. This is another main reason that the turbine performance can be changed by UCR value.

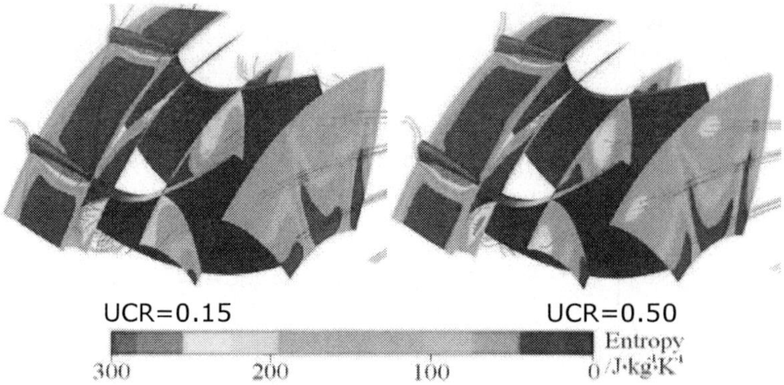

Figure 8 Entropy generation contours at streamwise planes

3.2 Shock wave variability

It is clear that when the flow area through nozzles is the minimum, a high pressure drop will generate shock waves near the trailing edge of nozzle vanes and bring a strong rotor-stator interaction. Some references indicated that the nozzle shock wave is one of the principal reasons for the rotor blade excitation[3,7,8]. In this section the unsteady numerical simulations were also carried out to investigate the impact of different clearance distribution on the shock wave characteristic and rotor unsteady load consequently.

Fig.9 compares the instantaneous distributions of circumferential Mach gradient at the exit of the nozzle for two UCR values. The strip with high Mach gradient in the mean section represents the shock wave. Compared with UCR=0.50, the strength of the shock wave is enhanced by 20% and the length is increased by 8% when the UCR=0.15. Besides, the mean position of the shock wave shifts upward (toward to shroud side) when the UCR value was changed from 0.50 to 0.15. The reason is that the low energy fluid formed by the leakage inside the nozzle has a tendency to block the gas flow, as a result the expansion ratio decrease in the region near the endwall. For UCR=0.15, the more low energy fluid is formed on the hub side and less leakage is generated on the shroud side so that the stronger shock wave shifts along the spanwise direction.

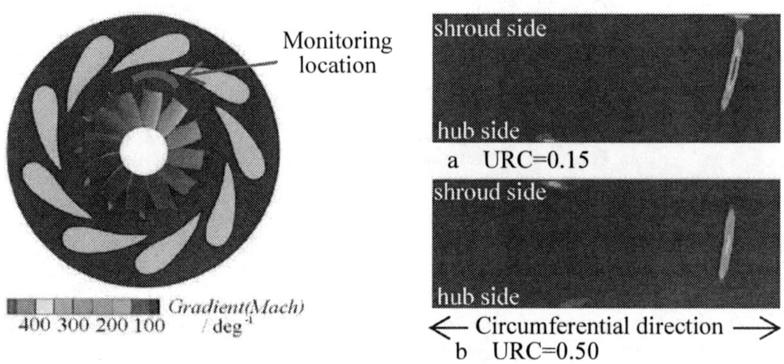

Figure 9 Comparison of the Mach gradient distribution at the exit of the nozzle, t/T=1/120

The shock wave is unsteady because there is aerodynamic interaction between the nozzle and rotor. Fig. 10 shows the time-space plot of circumferential static gradient. The vertical axis stands for the dimensionless time that is the ratio between the transient time(t) and the period of a blade passing three nozzle channels(T), and the horizontal axis is for dimensionless pitch from a nozzle vane to the next. The strip with high value represents the shock wave. Note that its location and strength is varying with time either UCR=0.50 or UCR=0.15. Moreover, for the UCR=0.15, the strength of the shock wave is enhanced compared to that corresponding to UCR=0.50.

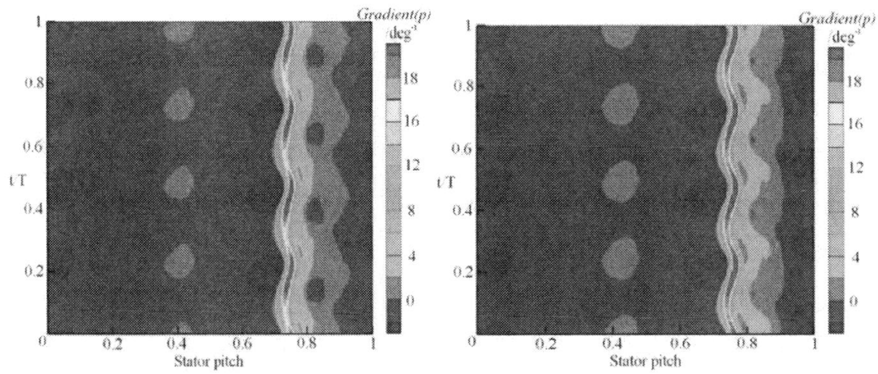

UCR=0.50 UCR=0.15

Figure 10 Time-space plot of the static pressure gradient at nozzle exit, midspan

Fig.11 shows the non-dimensional unsteady blade loading at 5% span for two UCR values. In the plots the non-dimensional rotor chord starts from the leading edge and ends at the trailing edge of a turbine wheel blade. The figures show a time history of the dimensionless static pressure difference between the pressure and suction side as the rotor passes upstream vanes. The structure of the rotor blade load is affected mainly by clearance leakage formed by nozzle clearance on the hub side. When the UCR value is changed from 0.50 to 0.15, the changing load is enhanced due to more clearance leakage on the hub side.

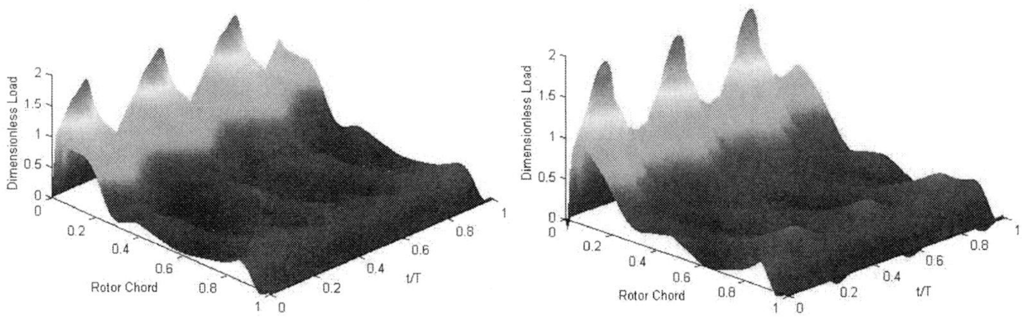

UCR=0.50 UCR=0.15

Figure 11 Time history of the dimensionless load for rotor, 5% span

Fig.12 represents the time history of the dimensionless load for rotor at 50% spanwise location. The variability of the dimensionless load is influenced by two main unsteady aerodynamic interaction phenomena: firstly shock interaction and secondly wake and potential field interaction. For the UCR=0.50, the variability in the structure of the unsteady load affected by the shock wave is not obvious while a peak can be induced by shock interaction when UCR=0.15. The emergence of the peak, induced by shock wave, will increase the risk of the rotor blade failure.

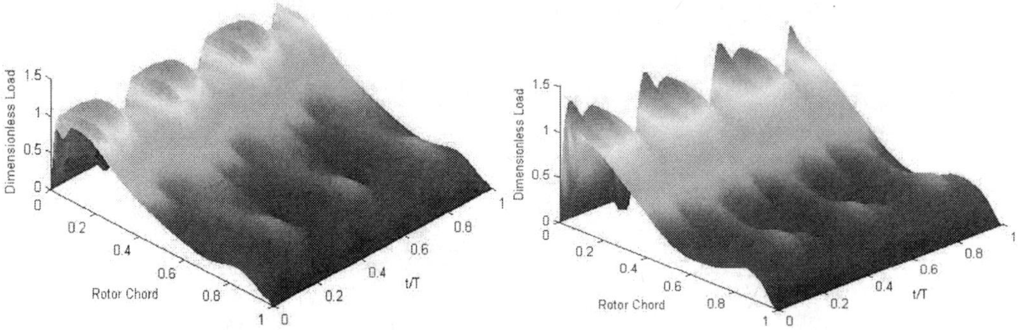

UCR=0.50 UCR=0.15

Figure 12 Time history of the dimensionless load for rotor, 50% span

Fig.13 compares the dimensionless unsteady load at 95% span for both UCR values. The comparison indicates that the changing load is reduced when the UCR value is modulated from 0.50 to 0.15 since the leakage vortex decreases as the clearance gap decreases.

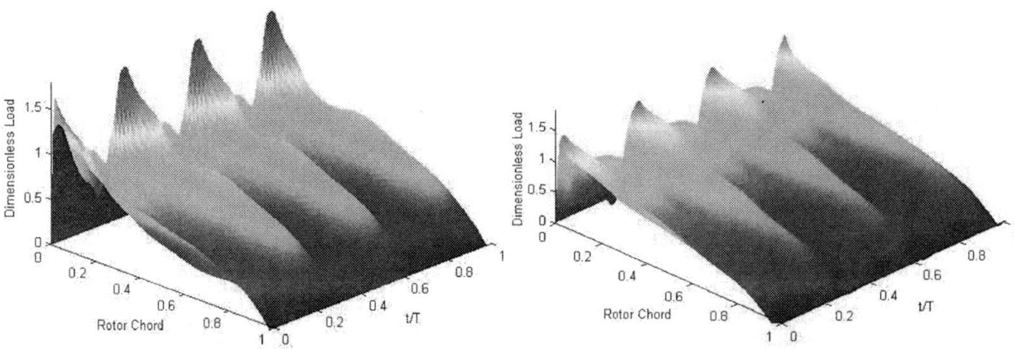

UCR=0.50 UCR=0.15

Figure 13 Time history of the dimensionless load for rotor, 95% span

4 CONCLUSION

The uncertain nozzle clearances have a significant effect on the turbine efficiency characteristic. For current research turbine, the numerically predicted range within which the turbine efficiency varies with different UCR values is about 6 percentage points. The UCR value can change the inlet condition of the rotor. For the UCR=0.15, the non-uniform flow distribution at nozzle's exit leads to a good incidence angle distribution at rotor inlet and reduce the flow loss on the suction side near leading edge compared to that when UCR=0.50. The result is that the turbine performance was improved for UCR=0.15. The UCR value decides the characteristic of the shock wave, which stands on the suction surface of the nozzle. The strength and the length of the shock wave can be increased by 20% and 8% when the UCR value is changed from 0.50 to 0.15. And in this case, a peak of unsteady load induced by stronger shock interaction presents at leading edge of the rotor blade. This increases the risk of the rotor failure.

ACKNOWLEDGMENT

This research was supported by Ford Motor Company and National Natural Science Foundation of China (NSFC, 51276018). The authors would like to acknowledge their financial support and Dr. Harold Sun of Ford for his strong technical support.

REFERENCES

(1) Walkingshaw, J.R., Spence, S.W., Thornhill, D., Ehrhard, J., "An experimental assessment of the effects of stator vane clearance location on an automotive turbocharger turbine". 10th international conference on Turbochargers and turbocharging, 2012,London, UK.
(2) Roumeas, M., Cros, S., "Aerodynamic investigation of a nozzle clearance effect on radial turbine performance". Proceedings of ASME turbo Expo 2012, GT2012-68835, Copenhagen, Denmark.
(3) Shi, X., Yang, C., Liu, Y.H., Liu, S.T., Zhao, B., "Numerical investigation on the high-cycle pressure fluctuation mechanism of VNT rotor". 11th international conference on Turbochargers and turbocharging, 2014,London, UK.
(4) Hu, L., Sun, H., Yi, J., et al, "Investigation of Nozzle Clearance Effects on a Radial Turbine: Aerodynamic Performance and Forced Response", 2013 SAE International, 2013-01-0918.
(5) Hu, L., Yang, C., Sun, H., et al, 2011, "Numerical Analysis of Nozzle Clearance's Effect on Turbine Performance", Chinese Journal of Mechanical Engineering, 2011, 24(4): 618-625.
(6) Tamaki, H., Goto, S., Unno, M., "The effect of clearance flow of variable nozzle on radial turbine performance", Proc. Of ASME Turbo Expo, GT2008-50461.
(7) Chen, H., 2006, "Turbine wheel design for Garrett advanced variable geometry turbines for commercial vehicle applications", 8th international conference on turbochargers and turbocharging, 2006, London UK.
(8) Kawakubo, T., 2010, "Unsteady rotor-stator interaction of a radial inflow turbine with variable nozzle vanes", Proc. Of ASME Turbo Expo, GT2010-23677.

SESSION 4A

BEARINGS, SEALS AND ROTORDYNAMIC DESIGN

Heat flow in a turbocharger shaft and its impact on the bearing system

T Kleine Sextro, N Lübbert, H Rätz, J R Seume
Institute for Turbomachinery and Fluid-Dynamics,
Leibniz Universitaet Hannover, Germany

ABSTRACT

Previous studies have shown that the temperature distribution in a turbocharger shaft and the resulting heat flow have a significant impact on the performance of full floating-ring journal bearings. The object of the present paper is therefore the development of a 1D heat flow model of the shaft using boundary temperatures in order to predict the shaft temperature field. This 1D heat flow model is developed based on CHT simulations and coupled to a bearing simulation. The results of the 1D heat flow model show an improved prediction of the bearing performance validated by its ring speed.

NOMENCLATURE

A	Area
c	Velocity
Ma	Mach number
n	Speed
$\dot{Q}$	Heat flow
R	Specific gas constant
T	Temperature
u	Circumferential velocity
w	Relative velocity
x	Position
α	Flow angle

Subscripts

BA	Axial bearing
BB	Bearing bushing
BC	Compressor-end bearing
BT	Turbine-end bearing
C	Compressor
T	Turbine

1 INTRODUCTION

The efficiency of modern combustion engines can be increased utilizing current downsizing concepts leading to less fuel consumption and emissions. However, downsizing also leads to a higher thermal load of the overall engine system. This affects especially the turbocharger and its bearing system. A rotors used in a passenger turbocharger is typically supported by two floating-ring journal bearings. These bearings consist of a short cylindrical ring placed between the shaft and the turbocharger housing forming two thin fluid films shown in Figure 1. This oil film is used as lubricant and is very sensitive to temperature changes at the bearing bushing and the shaft. A higher thermal load on the bearing system results in higher shear-driven friction in the oil films resulting in a power loss of the turbocharger.

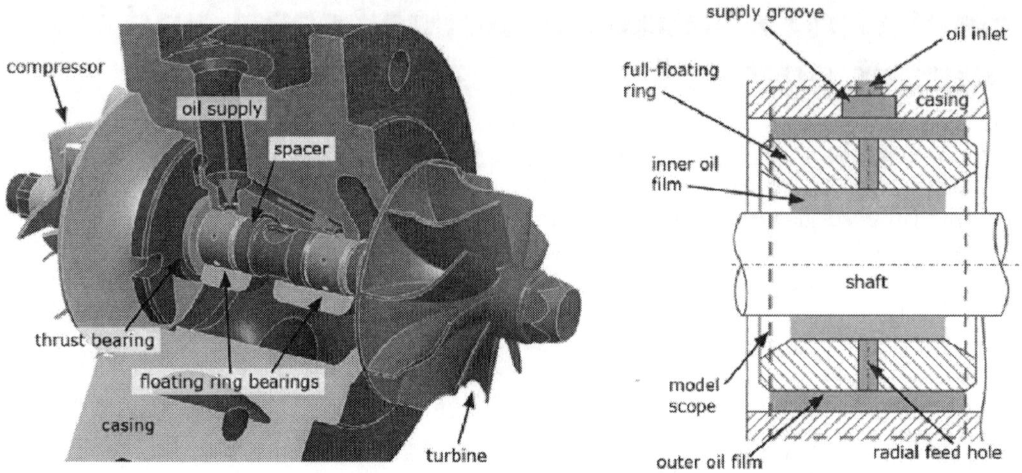

Figure 1: Bearing system of a turbocharger including a schematic of the oil films (5)

Besides the shear-driven frictional loss acting as heat source, the turbine wheel is also exposed to hot exhaust gas from the engine. For modern diesel engines, gas temperatures up to 600°C are reached and even higher temperatures for modern gasoline engines. This results in non-uniform temperature distributions for the turbocharger shaft, its housing, and bearing system. A determination of the latter is most difficult due to a variety of different heat flow paths (1) (2).

For the prediction of the bearing performance, also the lubricant viscosity and change of operating clearances in the lubrication films are important parameters. To estimate some of these parameters, a model incorporating a lumped-parameter thermal energy balance is developed by San Andrés et al. (3) (4). In a further publication, San Andreas et al. (2) presented a model for floating and semi floating ring bearings including the thermal energy transport in the ring bearing system. This model is used to predict the pressure and temperature distributions in both lubrication films. The presented thermal model does not include non-uniform temperatures in the shaft and the turbocharger housing. It assumes uniform housing and shaft surface temperatures over the whole operating range.

A thermo-hydrodynamic computational model for the prediction of relevant bearing parameters for floating ring bearings is presented by Porzig et al. (5). This model has been experimentally validated to improve the sensitivity of the bearing system for varying thermal boundary condition. By using fixed shaft temperatures instead of an adiabatic shaft approach, the prediction showed an improved agreement with the test data over a wide operating range. To get an indication of real shaft temperatures during test rig operation, sensitivity analyses were conducted for three discrete shaft speeds and shaft temperatures ranging from 100°C to 250°C in steps of 25°C. The lubrication temperature at the bearing inlet was kept constant at 70°C. The results shown in Figure 2 (right) include a prediction for an adiabatic shaft segment (gray line) and indicate a deviation of the ring-speed ratio for the adiabatic approach compared to measured data. A reference shaft temperature is obtained by the intersections of the measured ring speed ratio (orange points) with the calculated ring speed ratios for different shaft temperatures of the three shaft speeds (blue points). For the lowest shaft speed of 40,000min^{-1}, a shaft temperature of 150°C is assumed while a shaft temperature of 200°C is assumed at high shaft speeds.

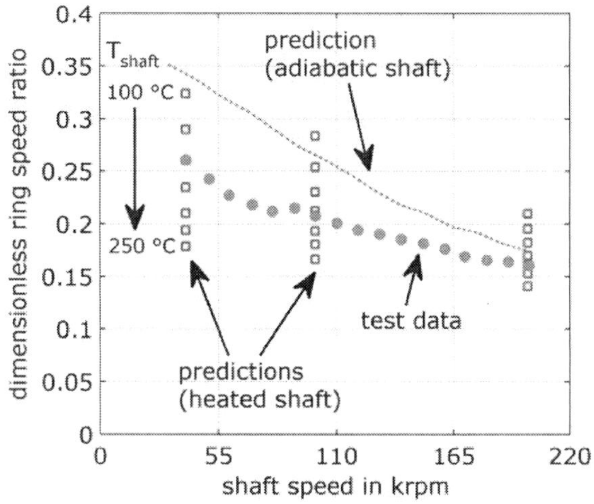

Figure 2: Predicted effect of thermal boundary conditions compared to test data for a lubricant temperature of 70°C at the bearing inlet (5)

The publications regarding the prediction of the bearing performance show that thermal boundary conditions have to be determined. Neglecting non-adiabatic influences result in an inaccurate prediction of the bearing friction losses and therefore of the overall turbocharger performance especially at part load conditions. In this paper, an approach for the computation of the heat flow in the turbocharger shaft from the hot turbine to the cold compressor and its influence on the bearing system is described. As boundary conditions for the temperature field, various heat flows are modeled.

2 METHODOLOGICAL APPROACH OF THE 1D HEAT FLOW MODEL

Reference values for the shaft temperature field are determined with CHT simulations of the entire rotor with turbine and compressor impeller. A coupled calculation of flow and solid models is utilized for a computation of the transport of energy in the form of the heat flow between fluid and solid, as shown in Figure 3. In addition, the heat flow at the bearings is taken into account which is coupled to the simulation model for hydrodynamic bearings of Porzig et al (5).

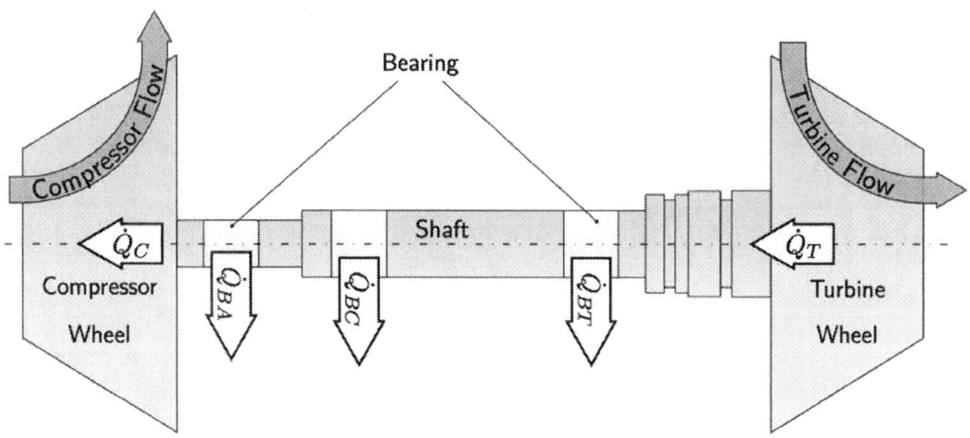

Figure 3: Sections of the 1D model for the heat flow in the solid and rotating turbocharger components

The methodological approach to develop the one-dimensional model for calculating the heat flow through the turbocharger shaft is shown in Figure 4. In the first step of method 1, the measured data obtained from Porzig et al. (5) for a single stage wastegate turbocharger is used as a boundary condition to perform CHT simulations and generate a detailed validation basis for the development of the 1D model. The CHT simulations are used to analyze the energy transport in form of heat flow through the turbocharger shaft in detail by coupling a bi-directional CFD simulation of compressor and turbine with a FEM simulation of the rotating sections consisting of the impellers and the shaft. In the second step, the CHT model is gradually replaced by modeling individual sections in order to facilitate the heat flow calculation in its complexity. Therefore, a 1D solid model of the shaft "1D Shaft Model" including an heat transfer exchange to the radial bearings is coupled with the impeller domains of the CHT simulation. Temperature boundaries of the "1D Shaft Model" are located at the intersections of the shaft and the impellers.

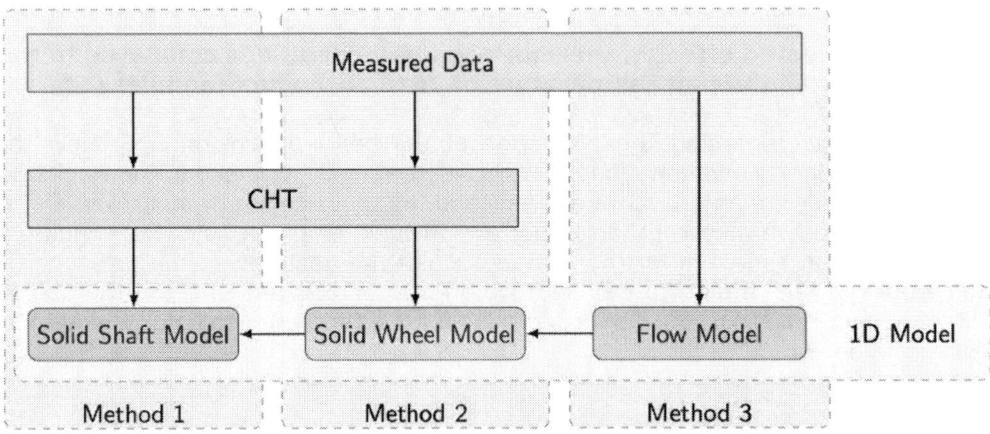

Figure 4: Methodological procedure of the 1D heat flow model based on CHT simulation and measured data

In method 2, the "1D Shaft Model" is extended by the "1D Wheel Model" which takes the heat transfer from the flow to the solid-state of the impellers into account. The models for the "1D Shaft Model" are derived from the data basis and enhanced with physical and empirical equations. The calculation of the heat flow in the "1D Wheel Model" is validated based on the CHT results. The thermodynamic boundary conditions of the flow are transferred at the interfaces of the impeller inlet and outlet sections.

An analytical approach of the flow-conducting components is achieved in the "1D Flow Model" of method 3 using 1D models of the individual components. The "1D Flow Model" uses the measured data from the test stand as boundary condition to calculate the heat flow without the need of CHT simulation. The results of the CHT simulation are used to validate the heat flow calculation by the developed "1D Flow Model".

Finally, the heat flow to the radial bearing and its impact on the bearing performance are taken into account by coupling a simulation model for hydrodynamic plain bearings (5). To validate the 1D heat flow model, the resulting speeds of the radial bearings are calculated.

3 CHT-SIMULATION OF THE COMPRESSOR AND TURBINE COUPLED WITH THE SHAFT

The numerical model of the turbocharger for the CHT simulation is shown in Figure 5 and consists of the flow domains for the compressor and turbine stage coupled with the solid domains of the impellers and the shaft. In the fluid domains of the turbine and compressor impellers, the back disks are also taken into account. The solid domains of the turbine and compressor impellers are represented individually with their specific material properties in order to calculate the specific heat transport. The shaft nut of the compressor is also considered as individual solid domain. The shaft itself is separated in several solid domains with respect to the bearing positions to include a bearing heat transfer.

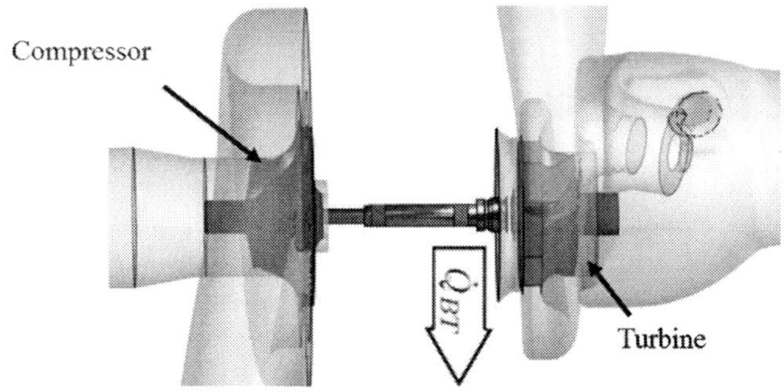

Figure 5: Coupling of fluid and solid domains for the CHT simulation

The inlet and outlet boundary conditions for the CFD simulations of the turbine and compressor are given for three chosen operating points from measured data as indicated by the red circles in the compressor and turbine run-maps in Figure 6. The operating point OP2 for a medium mass flow is used as reference point in this paper. The boundary conditions for the FEM simulation are generated based on the temperature distribution of the simulated flow at the surfaces. For the simulation of the heat transport through the solid sections, the surface of the shaft is considered to be adiabatic except for the domains at the radial bearing positions. To reduce the complexity of the simulation, only the heat transfer to the turbine-end bearing is considered. All additional heat flows of the CHT simulation are neglected. The mesh statistics of the fluid and solid domains with their y^+-values are given in Tab. 1 (left). Also the settings of the CHT simulation are shown on the right side.

Table 1: Mesh statistics (left) and settings of the CHT model (right)

Mesh Statistics		Model Settings
Number of Elements (Fluid)	19,000,000	• Stationary CFD • Coupled Heat Transfer Calculation • SST Turbulence Model • Fluid-Solid-Interface (Compressor and Turbine Impeller) • Solid-Solid-Interface (Impellers and Nut to Shaft) • Heat Transfer at Near-Turbine Bearing
Number of Elements (Solid)	700,000	
Average y^+ (Compressor)	0.73	
Average y^+ (Turbine)	0.80	
Wheel Area with $y^+ < 1$ (Compressor)	80.60 %	
Wheel Area with $y^+ < 1$ (Turbine)	69.51 %	

For the validation of the CHT simulation with the measured data, the heat transfer to the turbine-end bearing is also neglected and therefore considered to be adiabatic. For the numerical simulations of the turbocharger, ANSYS CFX 14 is used with the SST turbulence model of Menter [10] with a 2nd order discretization scheme for the CFD simulation. The results of the CHT simulation are indicated as blue dots in Figure 6 which show only a small deviation compared to the measured data for all three operating points of the compressor and turbine.

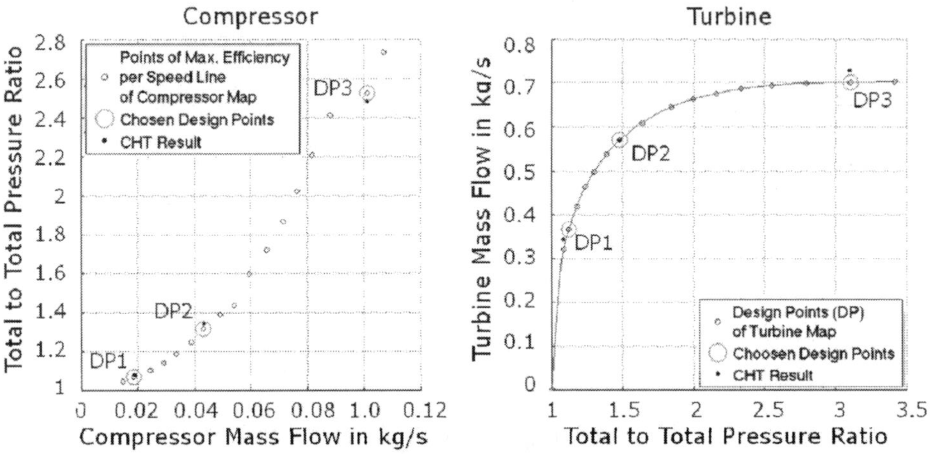

Figure 6: Run-up maps of compressor and turbine including the results of the CHT simulation for three operating points

To estimate the effect of the heat flow to the turbine-end bearing, a variation from $\dot{Q}_{BT}$ =0W (adiabatic) to $\dot{Q}_{BT}$ 120W is simulated for operating point OP2. In Figure 7, the sensitivity of the heat flow variation to the shaft temperature is shown. The shaft temperature shows a high sensitivity to the variation of heat flow at the turbine-end bearing with a temperature margin of 700K. Due to the thermal resistance of the friction welding calotte, the amount of heat flow from the turbine wheel to the shaft driven by the temperature difference is restricted. This leads to an unphysical calculation of a shaft temperature of 0K. Therefore the maximum heat flow of the turbine-end bearing is set to $\dot{Q}_{BT}$ = 120W for the further analysis.

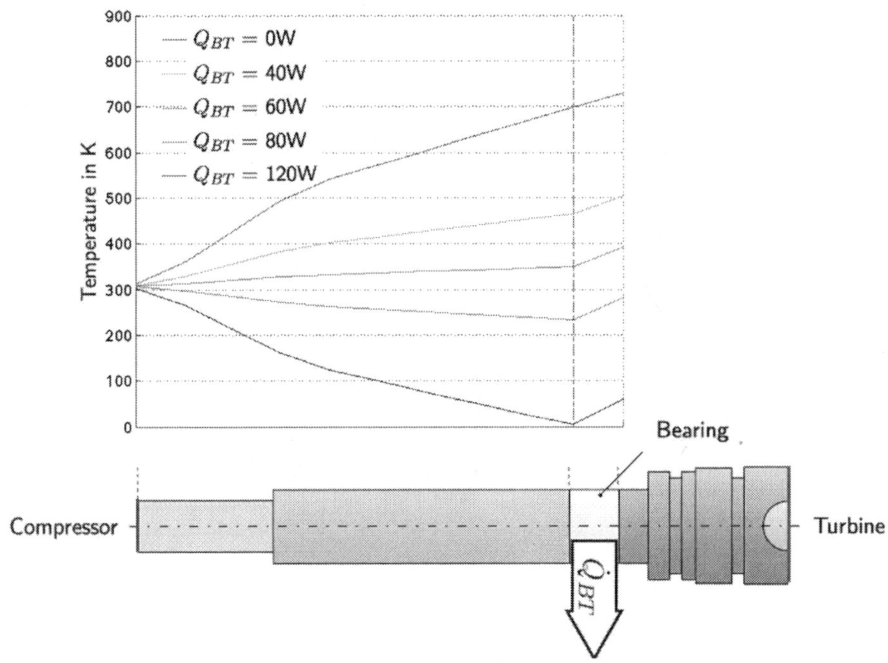

Figure 7: Sensitivity of heat flow variation at the turbine-end bearing to shaft temperature for OP2

The effect of the heat flow variation on the performance of the turbocharger compressor and turbine respectively is shown in Figure 8. The heat flow variation shows only a low sensitivity to the static flow temperatures in the compressor and turbine impellers. On the compressor side, the deviation of the static flow temperature from the inlet to the outlet of the impeller is below 1K. A slightly higher deviation can be seen on the turbine side. The deviation of the static flow temperature is 3.74K and therefore the heat flow variation does not have a significant effect on the performance of the turbine and compressor due to the relatively small shaft heat flow in comparison to the stage work.

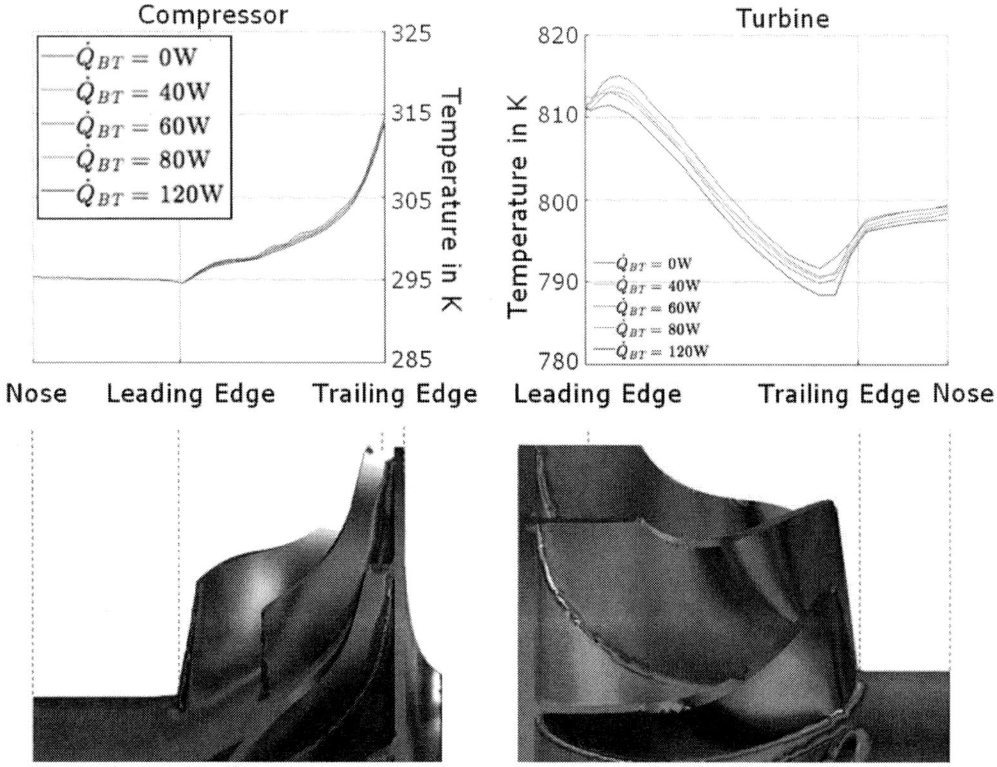

Figure 8: Impact of the heat flow at the turbine-end bearing on the temperature distribution in the compressor and turbine impeller for OP2

4 1D MODEL OF THE HEAT FLOW OF SHAFT AND IMPELLERS

The 1D models of the heat flow in the turbocharger are based on the diabatic approach by Shaaban and Seume (6). This approach determines only a heat flow between the impellers of the compressor and turbine through the shaft like in Figure 3.

A discretization of the impellers and shaft is needed for the 1D model to determine the heat flow in the solid domains as shown in Figure 9. A number of simplified assumptions are made for the 1D model having the focus on the impeller heat flow. According to the CHT simulation, diabatic effects in non-rotating sections of the turbocharger are neglected. All other shaft sections are assumed to be adiabatic. This includes also heat flow by convection and radiation to the environment as well as between the housings of the turbine and the compressor. Only the heat transfer along the surfaces of the impeller and at the turbine-end bearing is considered for the 1D model.

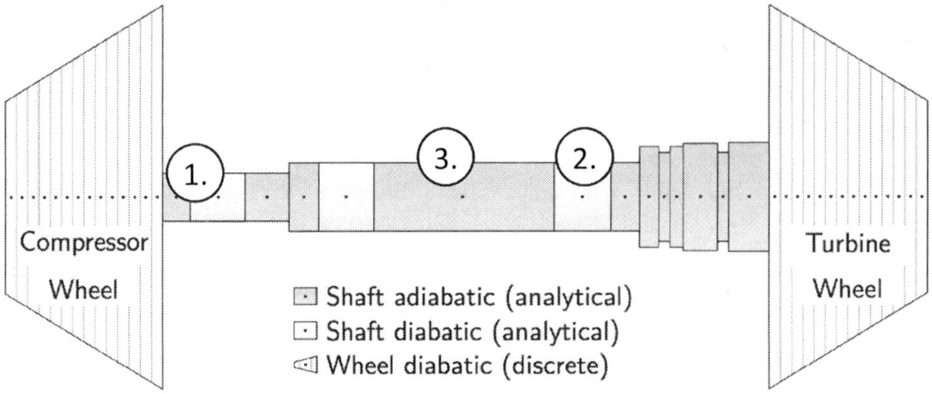

Figure 9: Discretization and simplifications of the 1D heat flow model

For the shaft, further simplifications are defined as shown in Figure 9:

1. Heat flow to the axial bearing is neglected.
2. The temperature gradient in the shaft at the radial bearing position is calculated in axial direction using an analytical solution for conduction in a rod with convection on its lateral surface.
3. The adiabatic shaft temperature is calculated from conduction only.

4.1 1D Heat Flow Model of the Turbocharger Shaft

In the first step, only the turbocharger shaft between the impellers is considered for the "1D Shaft Model" (marked by the red and blue lines in Fig. 9). The "1D Shaft Model" is coupled to the solid domains of the impellers of the CHT simulation. As boundary condition for the calculation, the temperatures of the CHT simulation are given at the intersection of the compressor and turbine impeller to the shaft. The heat flow between the impellers and the shaft itself is considered as degree of freedom of the system.

The "1D Shaft Model" is discretized as a rod with the difference method. Each node is assigned to a section with a finite thickness which is referred in following as block. For each block, the energy conservation equation is approximated by using a difference quotient and transferred into a system of algebraic equations. For the energy conservation, thermal conduction between adjacent blocks as well as a heat transfer to its respective environment is considered. A finite number of shaft blocks with an average cross-section is used for an approximation of the geometry. The blocks near the turbine end of the shaft also include the decrease of the cross-sectional area due to the friction welding volume between the shaft and the turbine impeller. The interfaces of the first and last block of the shaft are divided in half according to the approach of Taler (7) to match the temperatures of the nodes at the interfaces with the temperatures of the CHT simulation. The temperatures of the CHT simulation at the interfaces between the shaft and impeller are used as boundary conditions for the "1D Shaft Model".

To validate the "1D Shaft Model" respectively, the results of the coupled FEM simulation for the heat flow in the shaft are used. Therefore, the effect of the previous heat flow variation $\dot{Q}_{BT}$ at the turbine-end bearing on the resulting heat flows $\dot{Q}_C$ and $\dot{Q}_T$ at the intersection of the shaft to the impellers are investigated. In Figure 10, the results of the heat flow to the turbine and compressor impellers are shown for the heat flow variation at the turbine-end bearing. The heat flow at the intersection to the compressor impeller has a similar behavior for the "1D Shaft Model" and the CHT simulation with a maximum deviation of $\dot{Q}_C$ = 1.09W. Both methods show a reversal of the heat flow $\dot{Q}_C$ for an increasing heat flow $\dot{Q}_{BT}$ from the shaft to the bearing. Also for the heat flow at the interface between shaft and turbine impeller, both methods show a similar behavior with a maximum deviation of $\dot{Q}_T$ = 0.94W. In general, the "1D Shaft Model" is able to predict the resulting heat flow in a turbocharger shaft with good precision.

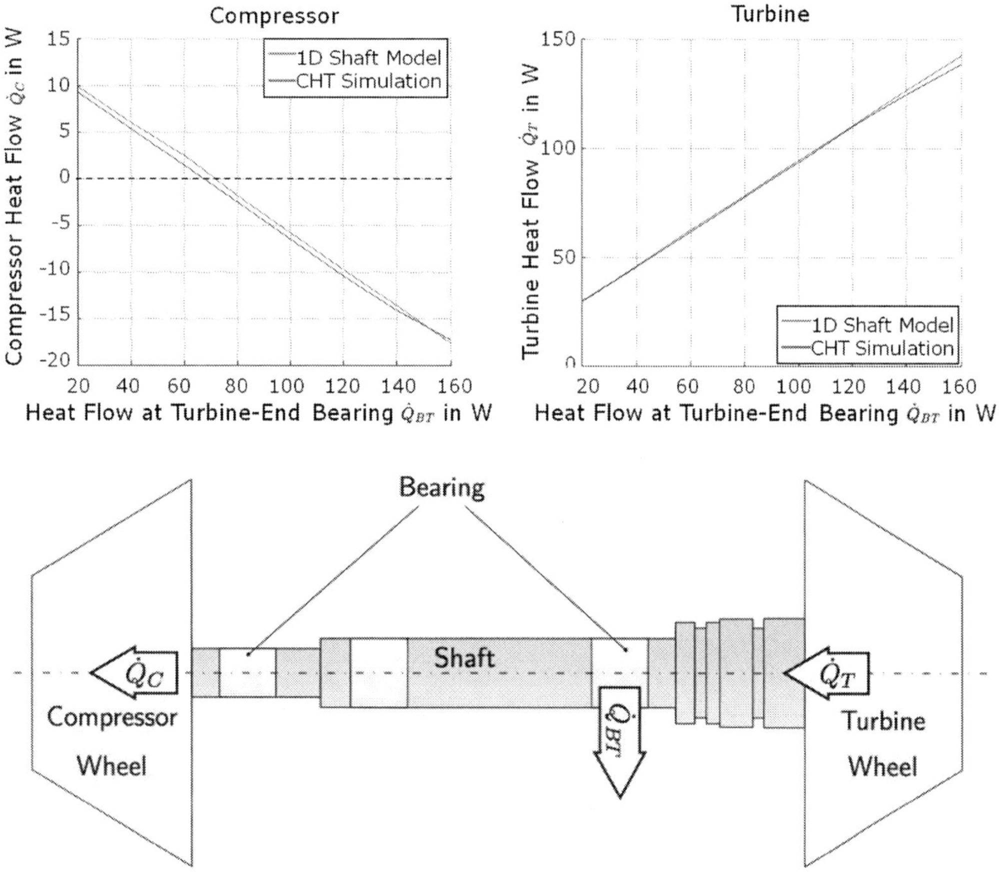

Figure 10: Results of heat flow variation at the turbine-end bearing with the 1D Shaft Model validated by CHT simulation for OP2

4.2 1D Heat Flow Model of the Impellers

In the next step, the impellers of the turbine and compressor are additionally discretised in the"1D Wheel Model" in a similar fashion to the "1D Shaft Model" (see Fig. 9). In order to calculate the local heat transfer between solid and fluid, the fluid properties at the discrete nodes on the impeller surface must be known. The properties along the flow path are approximated using a set of linear functions derived from the CHT simulations. Only the fluid state at inlet and outlet of the impellers is needed. According to Figure 11, the heat transfer is modeled for local surfaces individually using individual models to calculate the specific heat transfer coefficient. The impeller is therefore divided in three characteristic flow regions of a rotating cylinder (edge of the disk), a rotating disk (back disk and nose) (8) (9), and a flate plate (blade passage) (10).

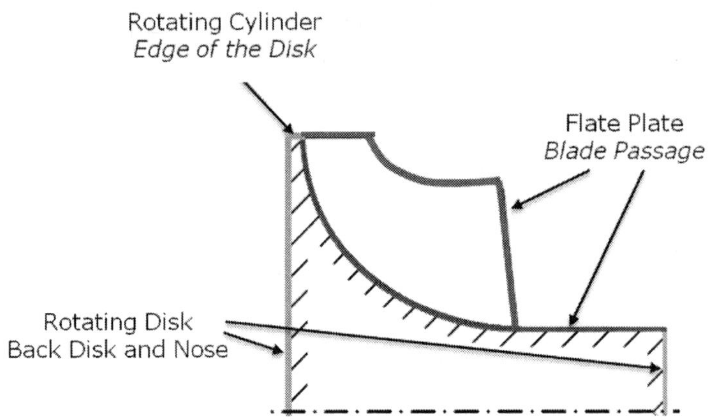

Figure 11: Local surfaces of the impeller and their individual models

For the validation of the "1D Wheel Model" and the local heat flows of the turbine and compressor impeller, the cross-sectional averaged solid temperatures of the impellers are used. The solid impeller temperatures of the CHT simulation are therefore averaged in radial direction according to the discretization of the 1D model. Figure 12 shows the solid temperature of the "1D Wheel Model" in comparison to the minimum, maximum and average temperature from the CHT simulation. The main focus of validation lies on matching of the solid temperatures at the intersections of impellers and shaft.

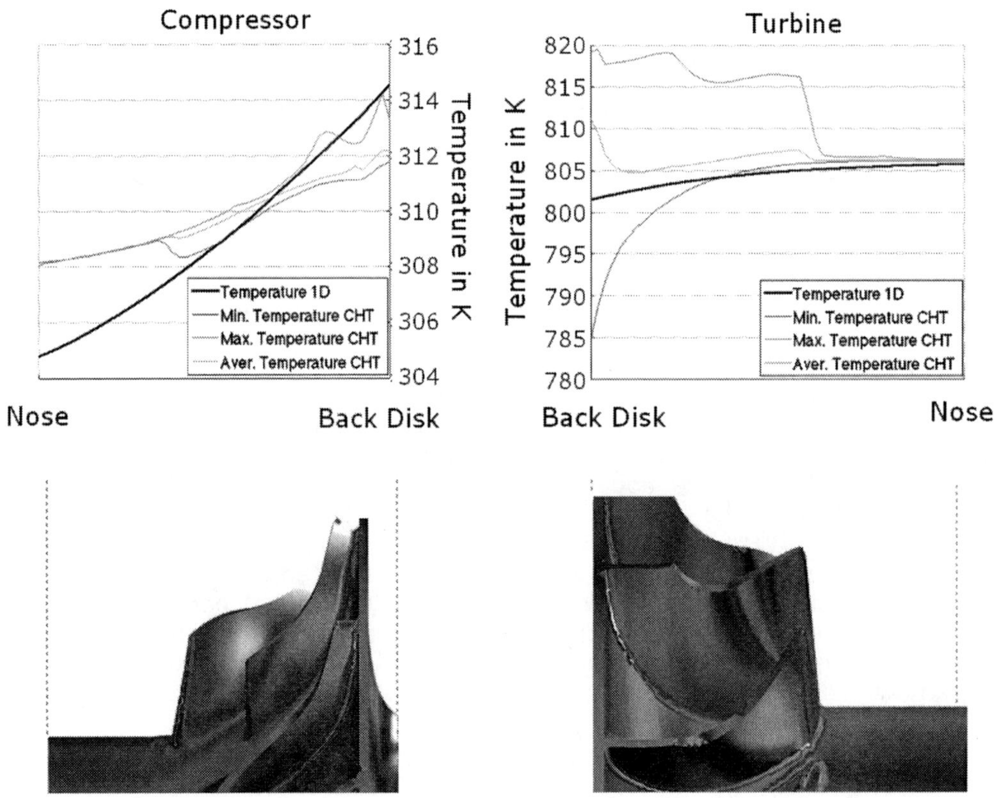

Figure 12: Temperatures of the impeller validated by the CHT simulation for OP2 and $\dot{Q}_{BT}$ = 20W

On the compressor side, the "1D Wheel Model" underestimates the average solid temperature at the nose by 3.3 K compared to the CHT simulation. In both methods, the solid temperature increases from nose to the back disc. For the CHT simulation, the slope of the solid temperature in this direction decreases by reaching the leading edge of the blade. Due to the additional surface on the blade, an additional heat flow is transferred between the fluid and solid. This effect is not shown in the "1D Wheel Model" and leads to a deviation of the solid temperature slope. At back disc, the solid temperature of the "1D Wheel Model" is overestimated leading to a temperature deviation to the CHT simulation of 2.5K. The simulated and modeled solid temperatures show the highest deviation of 8.8K at the back disc of the turbine and only a small deviation of 0.8K in the area of the shaft nut. Overall, the results of the average solid temperature for compressor and turbine show a good approximation between the CHT simulation and "1D Wheel Model" for the compressor and turbine.

4.3 1D Flow Model of the Flow Paths

In order to avoid the use of complex CHT simulations, a one-dimensional approach for the turbine and the compressor flow is used to determine the boundary conditions for the "1D Wheel Model". The "1D Flow Model" is based on the dimensionless mass flow equation of Whitfield and Baines (11) which allows the determination of the thermodynamic states at inlet and outlet of the "1D Wheel Model". For this purpose, the turbine and compressor stages are divided in characteristic components to determine the discharge state based on the inlet and outlet conditions of the turbocharger and characteristic geometrical parameters.

The dimensionless mass flow equation is based on the continuity equation for a stream tube. The stream tube is defined from position 1 with the state 1 to position 2 with the state 2. Equation 1 of the dimensionless mass flow equation used to calculate the thermodynamic sizes at state 1 and 2 based on the total values of pressure, temperature, and Mach number using the ideal gas law. This equation is solved iteratively for the Mach number Ma and the flow angle α_2.

$$\frac{\dot{m}}{p_{01}}\sqrt{\frac{RT_{01}}{\kappa}} = A_2 \cos\alpha_2 Ma_2 \left(1+\frac{\kappa-1}{2}Ma_2^2\right)^{-\frac{\kappa+1}{2(\kappa-1)}} \quad [1]$$

The dimensionless mass flow equation is used as analytical approach by neglecting the corresponding loss correlations of the flow components. The inlet and outlet conditions for the "1D Wheel Model" are calculated by the state changes in each component along the flow path of the turbine and compressor, as shown in Figure 13 for the turbine. E.g., the components inlet nozzle (1 to 2), volute (2 to 3), and ring nozzle (3 to 4) are used to calculate the inlet state of the impeller on the turbine side.

The calculation of the stream tube along the impeller is difficult due to complex, irreversible flow processes. A calculation of the state change along the impeller is therefore omitted. Instead, the outlet state of the impeller is determined with a backward stream tube based on the outlet conditions of the turbine stage. Equation 1 can also be used for a backward calculation due to the fact that the stream tube is calculated as isentropic state change. Therefore the outlet diffuser in Figure 13 is used to calculate the outlet status of the turbine impeller. An analogous structure is developed for the compressor. The calculated states are utilized as boundary conditions to the "1D Wheel Model". For the inlet and outlet conditions of the turbocharger, measured data from a test stand is used.

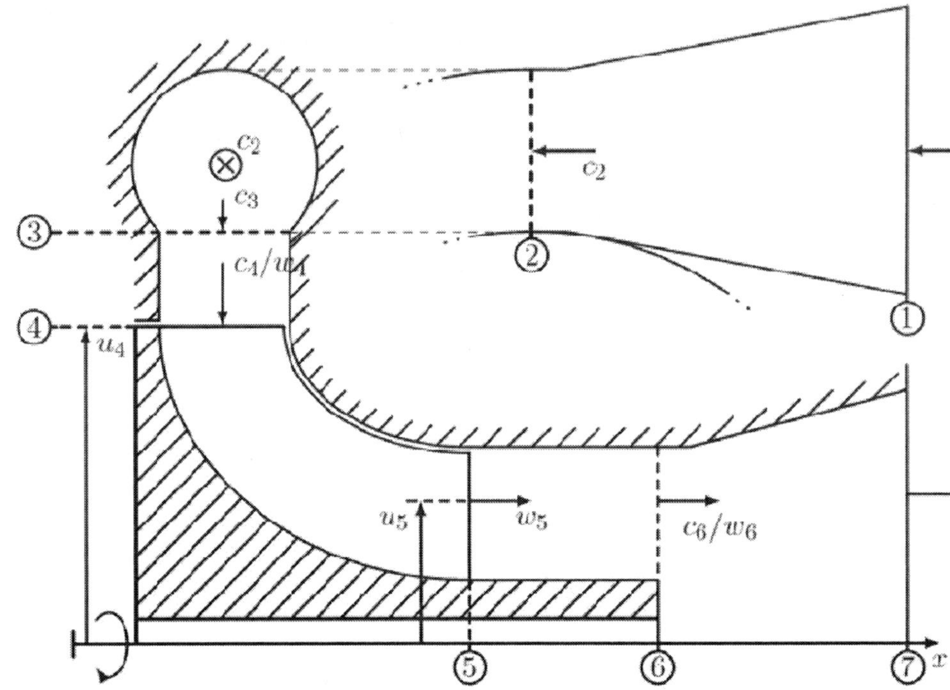

Figure 13: Components for the flow propagation of the 1D Flow Model

For the validation of the heat flow calculation based on the "1D Flow Model" with CHT simulation data, similar parameters are used as for the validation of the "1D Shaft Model". In Figure 14, the previous validation results and the results of the "1D Flow Model" and "1D Wheel Model" in combination with the "1D Shaft Model" are shown. The heat flows on the compressor and turbine side show a similar behavior for all 1D and CHT approaches. A reversal of the heat flow at the compressor side is shown for increasing bearing heat flows, with a maximum temperature deviation of 3,7W of the "1D Flow Model" approach to the CHT simulation. In general, the "1D Flow Model" and "1D Wheel Model" in combination with the "1D Shaft Model" agree well with the CHT simulation at the intersections of the compressor and turbine impellers to the shaft.

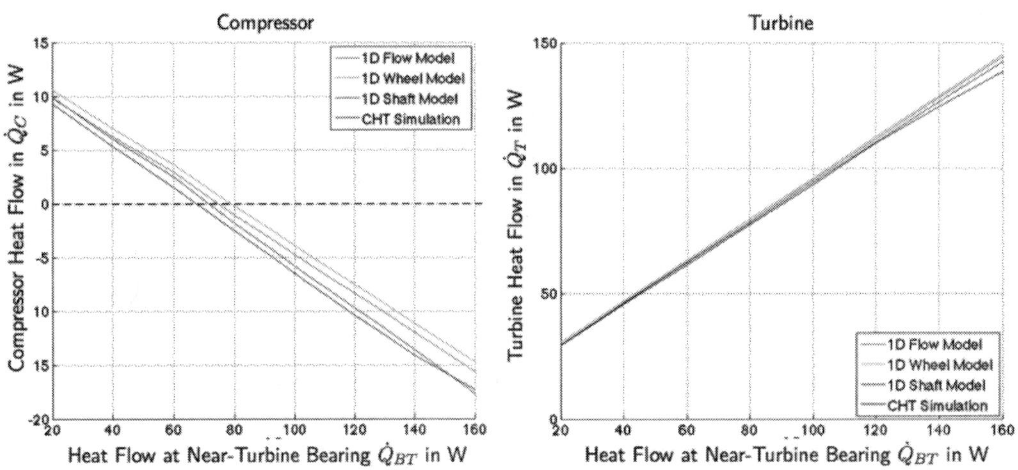

Figure 14: Results of heat flow variation at the turbine-end bearing from the 1D Shaft Model for OP2

4.4 Coupling of the bearing simulation model to the 1D Heat Flow Model

In order to improve the calculation of bearing parameters, the 1D heat flow model is coupled to the simulation model for hydrodynamic bearings (5) to account for the heat exchange of the lubrication oil and bearings, as shown in Figure 15. The simulation model for hydrodynamic bearings is coupled to 1D heat flow model with the temperature and heat flow at the turbine-end bearing. The bearing model also uses the measured temperature at the bearing bushing as second inlet condition.

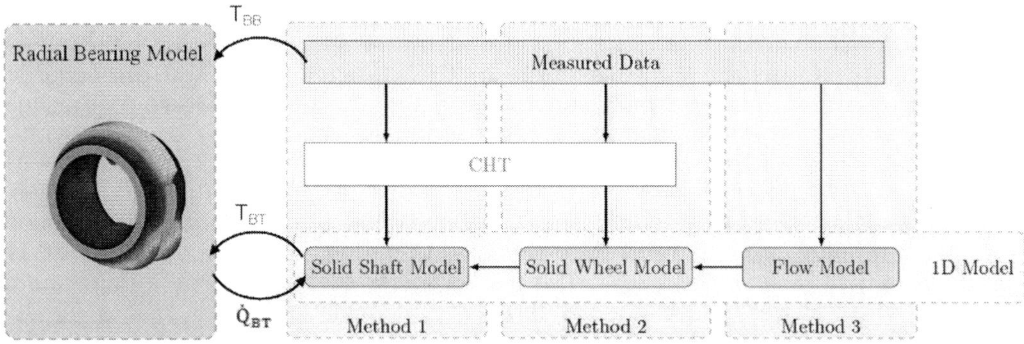

Figure 15: Coupling of 1D Flow Model with the Radial Bearing Model to predict the radial bearing speed

The results of the calculated and measured ring speed are shown in Figure 16. The calculated ring speeds of the radial bearing show a qualitatively good agreement of the characteristics for the compressor-end and turbine-end bearing with the measured speeds. Quantitatively, a nearly constant offset over the entire speed range can be seen. The ring speed drop for the compressor-end bearing at higher shaft speeds cannot be reproduced by this approach due to model simplifications, for example assuming adiabatic sections, neglecting the influence of the thrust bearing as well as rotor dynamic influences. A deeper rotordynamic analysis based on the measurements presented in this paper is given by Vetter (12). In summary, the ring speed gradient over shaft speed is predicted with a good precision using an one-dimensional diabatic approach for turbine and compressor impellers and shaft bearings.

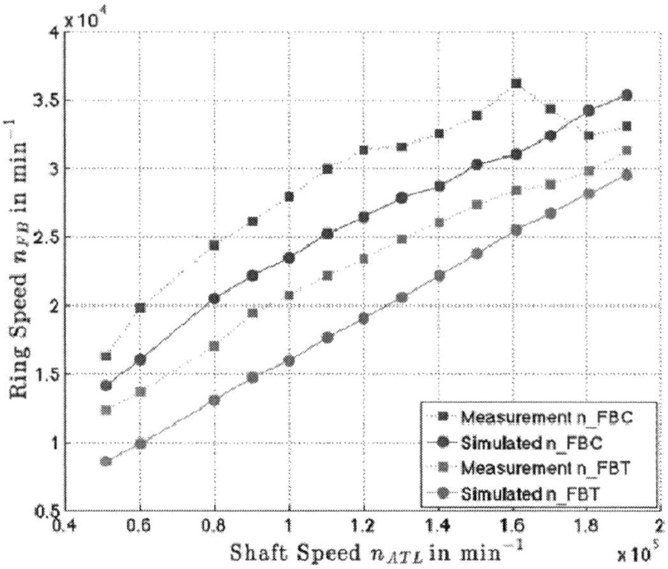

Figure 16: Results of the ring speeds of the radial bearings using the 1D heat flow model coupled with the bearing model

5 SUMMARY AND CONCLUSIONS

Previous studies have shown the need to improve the estimation of the boundary conditions for bearing systems used in automotive turbochargers in order to achieve a better performance prediction for the bearings. Especially the shaft temperature was identified as an important impact factor on the bearing performance.

This paper therefore presents a one-dimensional approach of the heat flow in a turbocharger shaft to calculate the shaft temperature and estimate its influence on the bearing performance. For this purpose, CHT simulations of the turbocharger were performed and show a high sensitivity of the shaft temperature to the heat flow in the shaft. The CHT model was then gradually replaced by three individual 1D approach in order to develop a simplified 1D heat flow model. The individual 1D approaches, "1D Shaft Model", "1D Wheel Model" and "1D Flow Model" were validated against the CHT results. The "1D Shaft Model" predicts the heat flow in the shaft including a heat transfer to the radial bearings within a deviation of 2W for bearing heat flows up to 120W. The heat transfer calculation in the impellers of the "1D Wheel Model" shows a solid temperature deviation of up to 8.8K on the turbine side between modeling and CHT results.

With the "1D Flow Model", an analytical approach was developed to calculate the states of the fluid along the flow paths of the compressor and turbine in order to determine the heat flow based on measured data only. The developed 1D heat flow model agrees well with the CHT simulation of the turbocharger. The final coupling of the 1D heat flow model to the bearing model reduces the model's degrees of freedom by an iterative calculation of the bearing heat flow and shaft temperature. The overall prediction of the bearing performance shows a qualitatively good agreement by comparing the predicted ring speed with measured data.

The next step of the development of the 1D heat flow model is its extension to predicting the heat flow to the axial bearing. For this purpose the heat flow in the bearing housing will be modeled based on CHT simulation.

6 ACKNOWLEDGEMENTS

The authors would like to acknowledge the support by the Research Association for Combustion Engines e.V. (FVV), which have been made this investigation possible.and in particular the chair of the industrial advisory group of this project, Uwe Tomm. Thanks are also to ANSYS for providing CFX and ICEM CFD in an academic license. The authors also acknowledge the contribution of the Regionales Rechenzentrum für Niedersachsen (RRZN) which provided the computing resources.

7 REFERENCE LIST

(1) Baines, N., Wygant, K. D., Dris, A., (2012); "The Analysis of Heat Transfer in Automotive Turbochargers", J. of Engineering for Gas Turbines and Power, 132(4), 042301

(2) San Andrés, L., Barbarie, V. C., Bhattacharya, A., Gjika, K., (2012); "On the Effect of Thermal Energy Transport to the Performance of (Semi) Floating Ring Bearing Systems for Automotive Turbochargers", J. of Engineering for Gas Turbines and Power, 134(10), 102507

(3) San Andrés, L., Rivadeneira, J. C., Chinta, M., Gjika, K., LaRue, G. (2007); "Nonlinear Rotordynamics of Automotive Turbochargers: Predictions and Comparisons to Test Data", J. of Engineering for Gas Turbines and Power, 129(4), pp. 488-493
(4) San Andrés, L., and Kerth, J. (2004); "Thermal Effects on the Performance of Floating Ring Bearings for Turbochargers", Proc. Inst. Mech. Eng., Part J: J. Eng. Tribol. 218, pp. 1–14
(5) Porzig, D., Raetz, H., Schwarze, H., Seume, J. R. (2014); "Thermal analysis of small high-speed floating-ring journal bearings", In 11th International Conference on Turbochargers and Turbocharging: 13-14 May 2014: 13-14 May 2014 (p. 421). Elsevier
(6) Shaaban, S., and Seume, J. R. (2012); "Impact of Turbocharger Non-Adiabatic Operation on Engine Volumetric Efficiency and Turbo Lag." International Journal of Rotating Machinery 2012
(7) Taler J. und Duda P. (2006); "Solving Direct and Inverse Heat Conduction Problems", Springer-Verlag Berlin Heidelberg
(8) Shevchuk I.V. (2009); "Rotating disk heat transfer in a fluid swirling as a forced vortex", Heat Mass Transfer, Vol. 41, pp. 1112-1121
(9) Shevchuk I.V. (2009); "Convective Heat and Mass Transfer in Rotating Disk Systems", Springer-Verlag Heidelberg Dordrecht London New York
(10) Baehr H.D. und Kabelak S. (2012); "Thermodynamik", Springer-Verlag Berlin Heidelberg
(11) Whitfield A. und Baines N.C. (1990); "Design of radial turbomachines", Longman Group, UK
(12) Vetter, D., Hagemann, T., Schwarze, H. (2014); "Predictions for run-up procedures of automotive turbochargers with floating-ring bearings including thermal effects and different bearing setups", Proc. of the 11th Int. Conf. on Turbochargers and Turbocharging, IMechE, London, GB

Experimental and numerical analysis of sealing ring of turbocharger

C C Ma, Z Huang, B Jin, H Zhang
School of Mechanical Engineering, Beijing Institute of Technology, China
Y Qu, Y Zheng
Skydragon Turbocharger Manufacturing Co., Ltd, China
S H Huang
Hunan Tyen Machinery Co., Ltd, China

ABSTRACT

A device for measuring the contact pressure between the sealing ring and the bore was developed. The contact pressure of the whole circle of a sealing ring was measured and analyzed. Based on finite element method, the modal frequency and modal shape of the sealing ring under contact pressure were obtained. Then numerical investigations of the sealing ring vibration behavior in the frequency range of external excitation were carried out. The results show that the radial response of the sealing ring under specific exciting frequency may lead to the leakage of oil.

1 INTRODUCTION

Seal devices are used in turbochargers both to prevent leakage of gases from the compressor housing and turbine housing into the bearing housing and to prevent leakage of oil from the bearing housing into the compressor housing and turbine housing. If there is leakage of exhaust gas from the turbine housing into the bearing housing, the oil may be contaminated by the exhaust gas, resulting in deterioration of the bearing lubrication, and therefore the bearing cannot have a long-term and reliable operation. If there is leakage of compressed air from the compressor housing into the bearing housing, it will cause loss in compressor performance, especially significant when the mass flow of the compressor is smaller. On the other hand, if there is leakage of oil from the bearing housing into the compressor housing or turbine housing, deterioration of emissions and turbocharger performance degradation will likely occur. The engine will not work properly and it will result in unwanted emissions as the oil flow with the outlet air of compressor into the engine. Furthermore, the lubricating oil flowing through the seals at backside of the impeller so that the compressor acts just as a pump, a part of the useful power consumed in pumping oil, which makes compressor efficiency even lower.

The commonly used type of seals in turbochargers are metal sealing rings similar to the piston ring seals used in engines. Figure 1 shows the sealing ring position at both the compressor and turbine ends. This type of sealing rings work efficiently with two sealing surfaces. One is the outer circumferential surface of seal ring in a radial direction, named sealing surface 1 as show in Figure 2. Another one is side face of the sealing ring in an axial direction, named sealing surface 2. A very small amount of oil Leakage at the sealing surface 2 is permitted, which can offer lubrication between sealing ring and the axis. Since the piston ring type seal ring is stationary relative to the turbocharger housing, the leak at sealing surface 1 is not needed for lubrication. While a certain amount of leakage is allowed at surface 2, leakage at sealing surface 1 must be reduced or avoided.

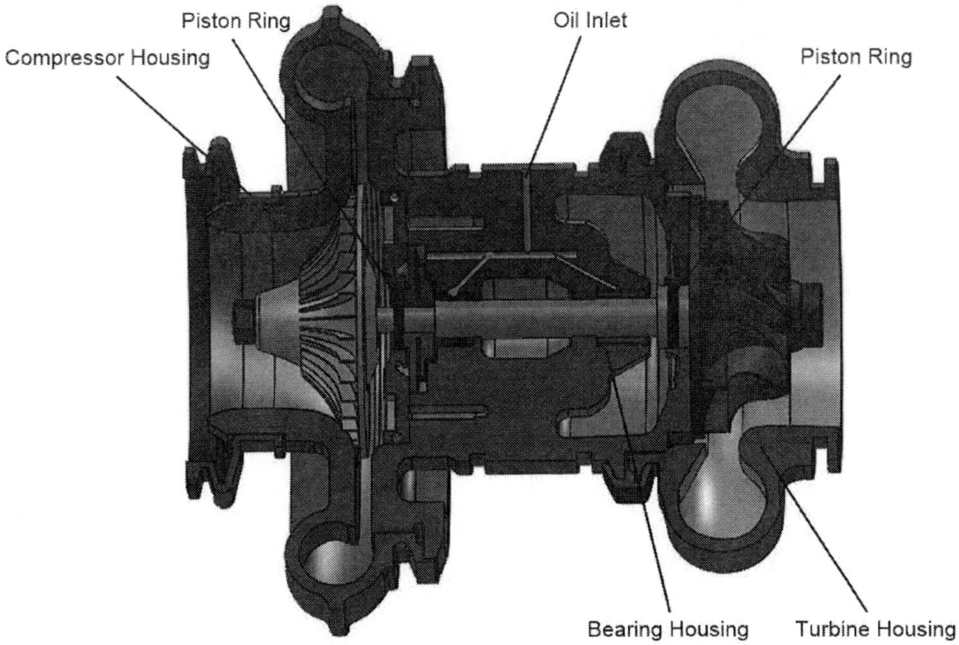

Figure 1: Sealing rings position in turbocharger

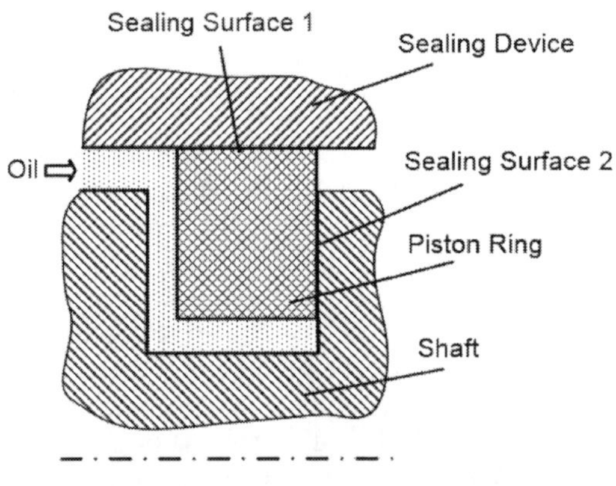

Figure 2: Sealing surface of sealing ring

Face deformation is a significant factor leading to poor sealing performance [1,2]. In actual operation, due to random external excitation caused by the changes in vehicle working conditions and manufacturing tolerances of transmission parts, seal ring deformation will change the oil film thickness of the sealing faces and result in the leakage of oil [3, 4]. Recent research on the sealing ring based on the assumption of process stability [5-8]. However, little research to date has focused on the response of sealing ring to external excitation. In this paper, the contact pressure of the sealing ring was measured by a self-developed device at first. Then a pre-stressed modal analysis was applied to calculate the frequencies and mode shapes of the seal ring under the contact pressure. For studying the harmonic resonance of the sealing ring in disturbed state, a mode-superposition method was applied to calculate the response character of the sealing ring.

2 MEASUREMENT OF CONTACT FORCE DISTRIBUTION

A sealing ring seals the oil and air through inherent pressure between the surfaces mentioned above, which is achieved by the elastic force. Generally, a sealing ring is manufactured into an original perfect circle, and then, it is cut and a gap of size m of working status is obtained, as shown in Figure 3. It is hot formed by means of heat treatment after putting a wedge-shaped insert of size M into the gap of the circle. Using this manufacturing process technique, a sealing ring of non-working status (or free status) with non-circular shape is obtained. When the non-circular ring is compressed and put into the bore with the same size and shape of the original perfect circle, a theoretic perfect sealing surface 1 (Figure 2) can be obtained with a gap of size m.

An important design consideration of sealing ring is contact pressure between sealing ring and the bore of seal device. Contact pressure is usually dependent on the elasticity of the sealing ring material, and the difference between the sizes of M and m (Figure 3).

Even if there are no manufacturing tolerances, the distribution of contact pressure at arbitrary point of cylinder circumference (φ) may be uneven when the sealing ring is mounted to the bore. Because the sealing ring achieved by this kind of manufacturing process technique has a non-circular shape in free status, the magnitude of the elastic stress within sealing ring at arbitrary position is uneven when the ring is compressed in working status.

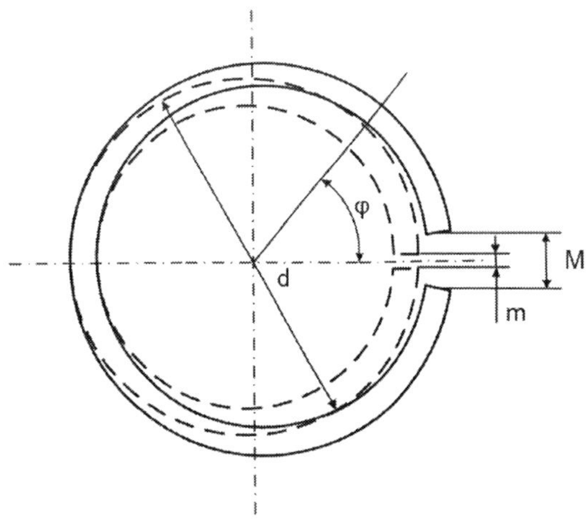

Figure 3: Sealing ring in free and compressed states

In order to measure the pressure distribution on surface 1 of a sealing ring, a device was developed. The schematic of the device used for measurement of contact pressure distribution is presented in Figure 4. Mandrels attached with strain gages generate radial force to make the sealing ring in a compressed state. The strain experienced by the mandrels is transferred to the strain gage, which responds with a linear change in electrical resistance. A bridge configuration with a voltage excitation source are used to measure the small changes in resistance. After amplifying and filtering, the output voltage of the bridge is converted into strain value by MCU and display in a digital display.

Construction of device for the measurement of ring contact pressure distribution is presented in Figure 5. Along the circumference of fixing frame which encompasses the tested sealing ring, there are six holes through which mandrels loads radially the ring. A sleeve is used for generation of radial force to push the sealing ring in compressed state. Six values of contact pressure can be obtained each time. By rotating the shaft of the device, strain data can be measured at any positions on the sealing ring.

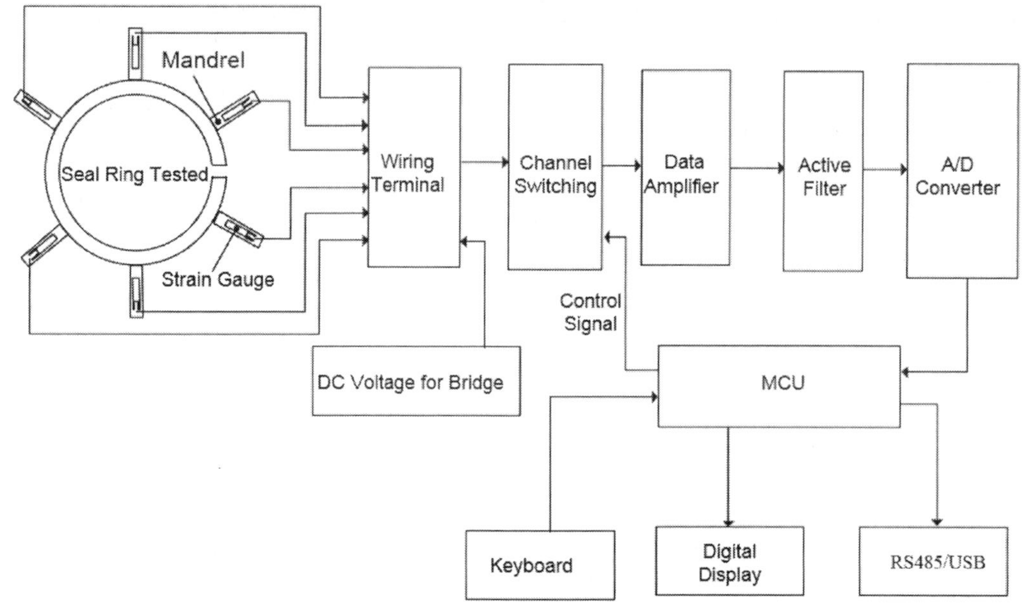

Figure 4: Schematic of self-developed device for measuring the contact pressure

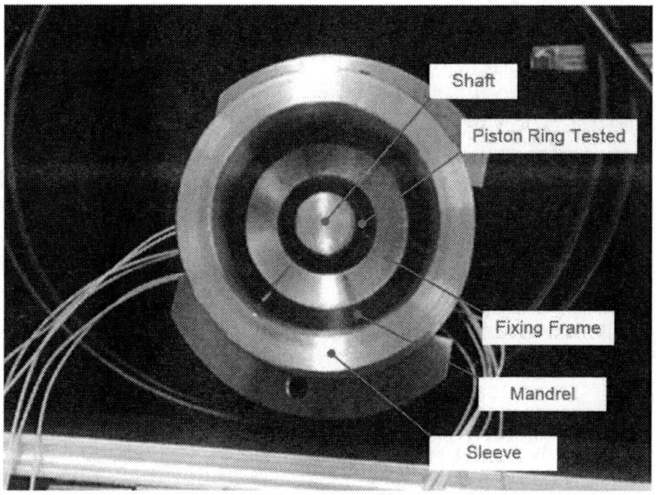

Figure 5: Construction of testing device

Before measuring, the testing device was calibrated by a standard press. Figure 6 shows the non-uniform distribution of circumferential contact pressure of sealing ring which was measured by this device. The contact pressure near ring gap ($\varphi = 0°, 360°$) has the minimum value. In long axial position of the sealing ring ($\varphi = 90°, 270°$), the contact pressure reach the maximum value. The low contact pressure at the vicinity of gap can lead to the incomplete contact of ring face and the bore of sealing device. The circumferential slits are easy to occur there, which lead to the leakage of oil.

To assess the testing results, the commercial code ANSYS was used as the numerical analysis tool for calculating the contact pressure. The numerical model has an ideal geometry; thus, it has no manufacturing tolerances. The calculated results were also plotted in figure 6. There is a good agreement between the experiment and numerical analysis. Figure 7 presents the contour of contact pressure on sealing surface 1 when the ring is compressed in the bore of seal device. The contact pressure distribution of

ANSYS calculated results is similar to that of experiments. The uneven stress within the sealing ring under working status is showed in figure 8, which leads to the uneven distribution of contact pressure.

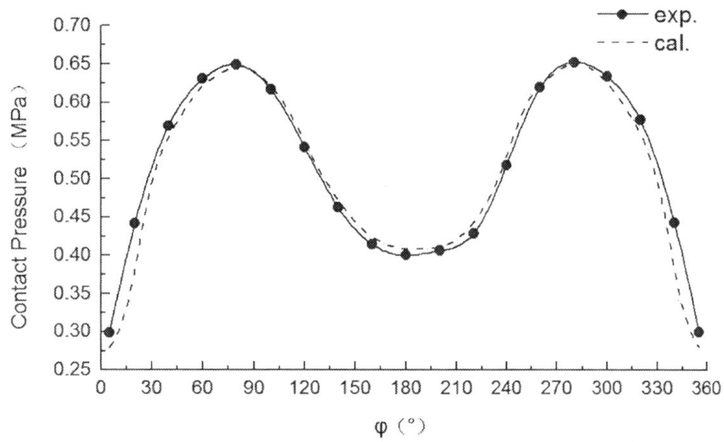

Figure 6: Distribution of circumferential contact pressure of sealing ring

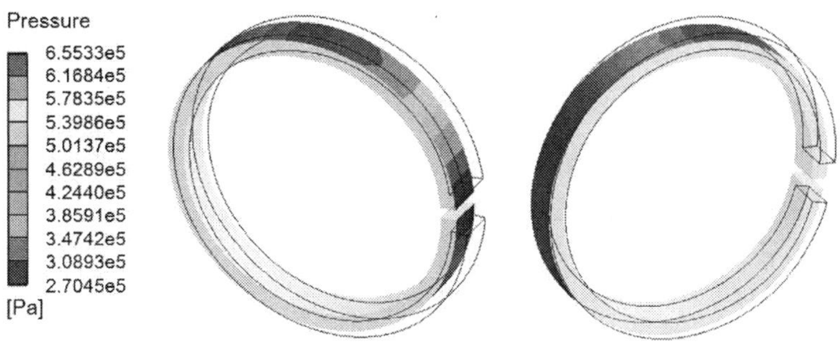

Figure 7: Contours of contact pressure

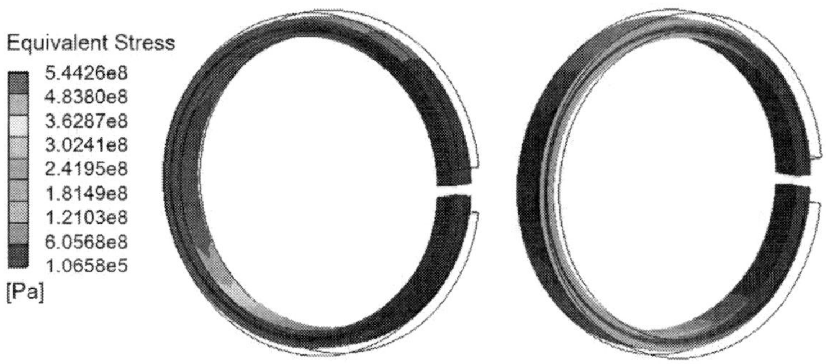

Figure 8: Contours of stress

3 MODAL ANALYSIS

The response frequencies can be changed by the stress stiffening because of the fact that the stress state would influence the values of the stiffness matrix. Therefore a pre-stressed modal analysis was performed to calculate the frequencies of the sealing ring under contact pressure illustrated in the above section. The Block Lanczos method was applied for searching eigenfrequencies as extracting the modes.

In our previous studies, the excitation information has been collected from vibration sensors mounted on the bearing housing of a turbocharger which is connected to an engine. The vibration data including X, Y, and Z directions was measured while the engine was running from idle speed to the maximum speed. Depending on the experimental data, the excitation frequency range from 10.8Hz to 22039.2Hz. The first 7 modal frequencies covering the working frequency range of the turbocharger was extracted to determine the vibration characteristics of the sealing ring. The value of the natural frequencies is presented at table 1.

Table 1: Results of the Modal Frequency

Mode	1	2	3	4	5	6	7
Frequency（Hz）	2040	5741	8936	9489	10640	14340	28788

The first 6 mode shapes of the sealing ring are shown in Figure 9. Mode shape 7 is not included here since the modal natural frequency has exceeded the range of external excitation. The maximum magnitudes of the sealing ring deformation region can be identified from the mode shape results. The deformation direction of mode 1, mode 2, and mode 3, mode 6 is axial which could lead to slits in sealing surface 2. The deformation direction of mode 4, mode 5 is radial and this could lead to slits in sealing surface 1. All these slits in sealing surface may cause the leakage of oil.

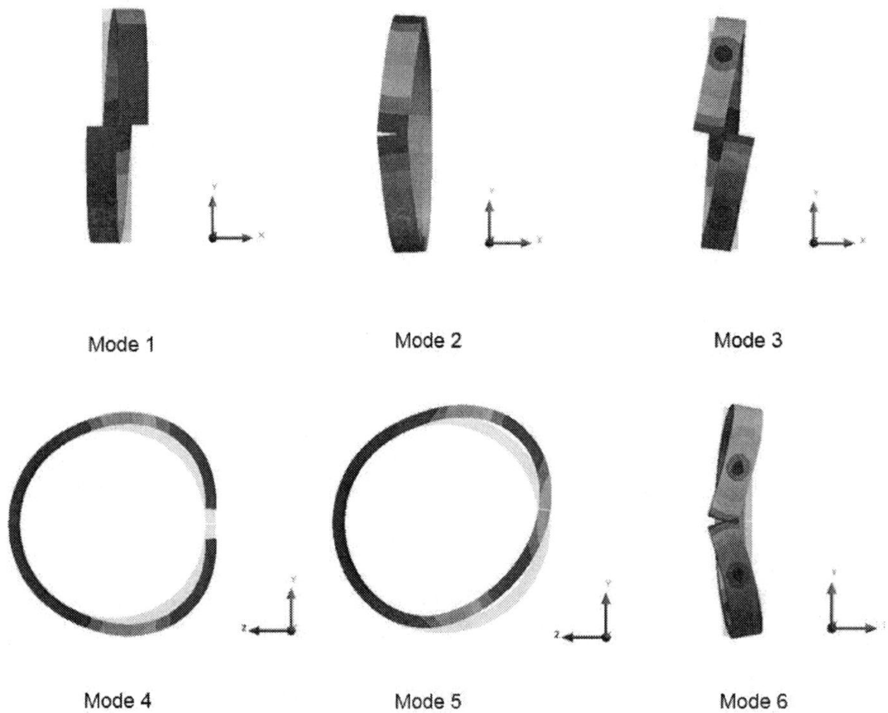

Figure 9: Mode shape of the sealing ring

4 HARMONIC RESPONSE ANALYSIS

A harmonic analysis was performed by summing factored mode shapes obtained from the modal analysis to calculate the response of the sealing ring. Sustained cyclic loads was applied on the contact surface of the sealing ring contacting with the bore of sealing device. Rayleigh damping was applied to account for dissipative mechanisms in the dynamic analyses. Material Damping in which the damping force is assumed proportional to the displacement was incorporated to analyze the hysteretic behavior due to internal material friction. Friction damping against the surface was neglected in this study. Considering the modal frequency and the working vibration frequency range of turbocharger, a sinusoidal varying loads were set sweeping across a range of frequencies from 1000Hz to 20000Hz.

The frequency response of radial displacement is presented in Figure 10. The amplitude is displacement of the maximum directional deformation region on the excited sealing ring compared with the original undeformed model. Peak response was reached at the excitation frequency of 9489Hz, which corresponds to the natural frequency of the modal 4. The maximum amplitude of the sealing ring response to the excitation in radial direction is 35μm. The frequency response of axial displacement is presented in Figure 11. Axial resonance occurs when the imposed frequency approaches the natural frequency of mode 2, mode 3 and mode 6. The maximum displacement of axial response occur at mode 2 and the amplitude is 0.75μm.

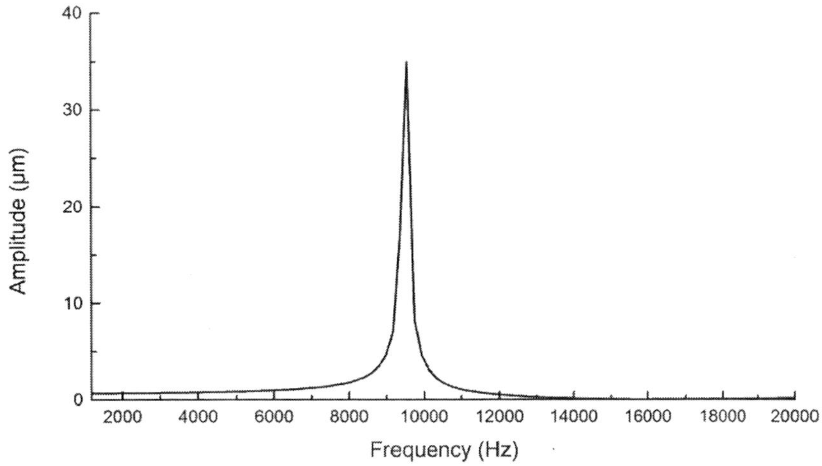

Figure 10: Frequency response of radial displacement

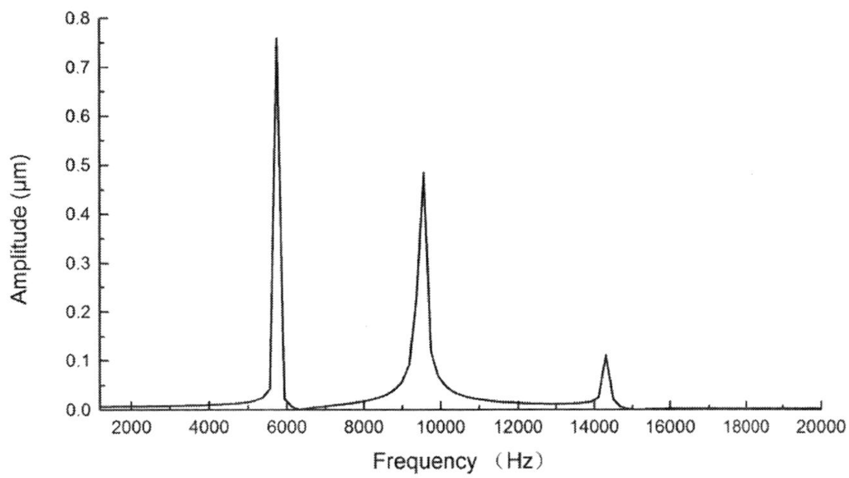

Figure 11: Frequency response of axial displacement

As aforementioned, the deformation of mode shape 4 could lead to the slits on sealing surface 1 and the deformation of mode shape 2 could lead to the slits on sealing surface 2. However, maximum frequency response of radial displacement is two orders of magnitude larger than that of axial displacement. Therefore the natural frequency of mode 4 is the critical resonance frequency of the sealing ring and the external excitation at this frequency should be avoided. Although the axial displacement is relatively small, one conjecture is that there would be a pump effect that the oil is transported to the turbine housing or the compressor housing by the axial vibration of the sealing ring.

This study is the first step to enhance the understanding of forced response behavior of sealing ring. A number of potential limitations still need to be considered in the harmonic response analysis. Future work already underway includes the temperature effects and effects of manufacturing tolerances on sealing performance.

5 CONCLUSION

With an objective to study the oil leakage problems in turbocharger, the experimental and numerical studies are conducted on the piston type seal ring commonly used in turbocharger. The main results are summarized as follows:

A self-developed device was design for the precise measurement of non-uniform distribution of circumferential contact pressure on sealing ring. It is found that the contact pressure near ring gap have the minimum value. In long axial position of the sealing ring the contact pressure reach the maximum value. The contact pressure distribution of test results is in good agreement with the numerical calculation.

The usage of modal analysis on the sealing ring vibration character shows that the axial deformation direction is presented at mode 1, mode 2, and mode 3, mode 6. The radial deformation direction is presented at mode 4 and mode 5.

The behavior of the sealing ring response to periodic external excitations was investigate. The harmonic analysis indicates that the maximum frequency response of radial displacement is two orders of magnitude larger than that of axial displacement. The natural frequency of mode 4 is the critical resonance frequency of the sealing ring and the external excitation at this frequency should be avoided.

In conclusion, the results encourage further investigation in including the influences of external excitation during the sealing performance improvement of turbocharger. The next design optimization of the sealing ring is pushed to combine a more detailed factors with the verification of the dynamic response in order to achieve an even better sealing performance sealing ring.

ACKNOWLEDGMENT

This research was supported by National Natural Science Foundation of China (NSFC, 51375048).

REFERENCE LIST

(1) Green I., 2002, "A transient dynamic analysis of mechanical seals including asperity contact and face deformation", Tribology Transactions, vol.45, 3: pp.284-293
(2) Salant R F, Cao B. ,2005, "Unsteady analysis of a mechanical seal using Duhamel s method", Journal of Tribology, vol.127, 3: pp.623-631
(3) Kaneko S, Tanaka T, Abe S., et al, 2004, "A study on squeeze films between porous rubber surface and rigid surface: analysis based on the viscoelastic continuum model", Journal of Tribology, vol.126, 7: pp.19-27

(4) Zhu Hanhua, Yan Xinping, Liu Zhenglin, et al, 2007, "Research on the influence of rotation speed upon rigidity of lubricant oil film and vibration of propeller shaft", Ship & Ocean Engineering, vol.36, 4: pp.83-85
(5) Zhou Dan, Sheng Xuanyu, 2008, "Finite element contact analysis of O ring sealing structure in horizontal channel", Nuclear Power Engineering, vol.36, 4: pp.22-24
(6) Sekiguchi S, Suzuki H, Ryper R G, et al, 2009, "Development of a new seal material to reduce energy losses in vehicle transmissions", Sealing Technology, vol.11: pp.8-11
(7) Wang Yuming, Yang Huixia, Wang Jianli, et al, 2009, "Theoretical Analyses and Field Applications of Gas -Film Lubricated Mechanical Face Seals with Herringbone Spiral Grooves", Tribology Transactions, vol.52: pp.800-806
(8) Wang Yuming, Liu Wei, Liu Ying, 2011, "Current research and developing trends on non-contacting mechanical seals", Hydraulics Pneumatics & Seals, vol.2: pp.29-33

The effect of oil film instability on power losses prediction of a turbocharger rotor-fully floating ring bearing system

L Tian, M Wakelin, C Lancaster, M Lindsay
Cummins Turbo Technologies, UK

ABSTRACT

This paper proposes a new method for calculating fully floating ring bearing (FFRB) power losses using the output from advanced nonlinear rotordynamic transient simulation. In comparison with the extensively used *static method* in previous studies, the results given by both methods agree well with each other prior to the onset of a phenomenon termed *Critical Limit Cycle Oscillation* (*CLC Oscillation*). However, at the onset of *CLC Oscillation*, the power losses calculated by the new *transient method* showed a significant increase as compared to the static method, which illustrates that oil film instability should be carefully considered in FFRB power loss calculations.

1 INTRODUCTION

By utilising the exhaust energy otherwise wasted to raise the amount of air intake, turbochargers (TCs) are nowadays extensively employed as a key element to boost the power of internal combustion engines. In the meantime, the engine overall efficiency can also be improved, as a turbocharged engine can be downsized without compromising the power output as compared to a naturally aspirated engine. In order to meet the increasingly stringent emission regulations and continuous demands for higher performance, the overall efficiency of the TC is required to be further optimised. Apart from improving the efficiency of the compressor and turbine, recently, the power losses or mechanical efficiency of TC bearing systems have also interested both TC and engine manufacturers. The reason is that a TC bearing system consuming less power can transmit more power from the turbine to the compressor, which will therefore lead to a faster acceleration response under transient conditions, and to reduced fuel consumption under steady state conditions with lowered engine back pressure (1). Among all bearing types available to the TC, hydrodynamic fluid film bearings are currently the predominant bearing technology and will remain dominant in the foreseeable future, mainly due to their low cost for mass production and proven reliability. The TC hydrodynamic bearing system typically consists of two fully floating ring bearings (FFRBs) or a single piece semi-floating ring bearing (SFRB) to support the radial dynamic load, and a thrust bearing to withstand the resultant axial load given by the unbalanced aerodynamic forces on the surface of both wheels. Both types of radial bearings incorporate two active layers of oil film, though the main structural difference between them is that a FFRB is free to rotate typically at 10% to 30% of the TC rotor speed (higher ring speed was expected but cannot be realised due to the lower oil viscosity given by the hotter inner film), while a SFRB is kept from rotating by an anti-rotation pin to form a squeeze film damper at the outer film. The rotation of a FFRB is seen as an advantage over a SFRB to reduce the bearing friction, as it gives reduced relative speed of the shaft to the bearing at the inner film (2,3).

So far, considerable effort has been devoted to analytically and experimentally quantifying the power losses of the TC hydrodynamic bearing system (4–11). In those studies, a SFRB is the mostly used radial bearing type. As a SFRB is prevented from rotating, the interaction between the inner film and outer film pressure field is disregarded in those studies. It is observed that the power losses of the entire bearing system are found significant in comparison with the total turbine power especially under low engine load conditions (9). Additionally, it is demonstrated that the thrust bearing can occupy more than half of the overall bearing power losses (8,11). Considering the influence of oil inlet conditions, it is seen that, under the same rotor speed, higher oil inlet temperature (OIT) can directly lead to a reduction of bearing power loss due to lower viscosity (8). By contrast, increasing oil inlet pressure (OIP) results in an increase in bearing power losses, as the increased oil flow gives enhanced cooling, which renders lower oil temperature and higher viscosity within the bearing clearance (8). The theoretical models used to derive power losses given by the radial bearing in those studies are developed from a linear point of view (4-6,9), which usually assumes that the oil film within the bearing clearance is stable. On this assumption, the dynamic loading given by the rotor unbalance is the only excitation resource and will only cause the rotor to orbit around the static equilibrium with a small amplitude and with a single frequency synchronous with the rotor speed, which means the static equilibrium is reasonably representative of the mostly dynamically loaded TC rotor bearing system. Therefore, only the static load of rotor weight is utilised to calculate the bearing static equilibrium, from which the position of the geometry centre of shaft journal relative to the geometry centre of bearing shell, i.e. the bearing eccentricity, is determined, and accordingly the bearing power losses are calculated. Ref. (4) shows that the power losses of radial bearings are greatly dependent on bearing eccentricity. In this paper, this method is termed *static method*.

However, those simplified assumptions deviate significantly from one important fact. Both theoretical and experimental rotordynamic studies show the dynamics of TC rotor-FFRB/SFRB system is highly nonlinear as a result of dominant sub-synchronous vibrations in addition to the synchronous response. Those sub-synchronous vibrations originate from the onset of oil film instability and usually have frequencies lower than the rotor rotational frequency with irrational orders of 0.1X to 0.5X. Consequently, significantly increased bearing eccentricities are seen in recent simulation work in contrast to the *static method* (12–24). Thanks to the significant improvement of computational power, nonlinear transient rotordynamic simulation reveals that the inner and outer films of a FFRB can individually become unstable and cause the rotor whirling in different mode shapes (17,20,21). Additionally, the FFRB rotational speed, linked directly to the FFRB power loss calculations, is also influenced especially at the onset of oil film instability (20,21). Further, the dynamics of a TC rotor-FFRB system shows great sensitivity to physical parameters, e.g., bearing and rotor structural parameters, unbalance distribution across the rotor and oil inlet conditions especially OIP (17,20,21,25,26). Those influential parameters can sometimes lead to an unfavourable phenomenon termed *Total Instability* in Ref. (25,27) or *Critical Limit Cycle Oscillation* (*CLC Oscillation*) in Ref. (18,21,22), referring to the situation in which the inner film and outer film simultaneously become unstable, and the frequencies of the resultant sub-synchronous vibrations corresponding to each film are synchronised. In this paper, *Critical Limit Cycle Oscillation* (*CLC Oscillation*) is adopted. The onset of *CLC*

Oscillation is normally accompanied by excessive rotor motion and tremendously high bearing eccentricities close to unity, which can eventually lead to TC failure (24). Unfortunately, the effect of those complex rotordynamic characteristics of a TC rotor-FFRB system on bearing eccentricity and FFRB rotational speed cannot be investigated using the *static method* developed in previous studies. As a result, the bearing power losses predicted by the *static method* can be inaccurate and incomplete.

In order to fully understand the power losses of a TC rotor-FFRB system, this paper proposes a novel method of calculating the FFRB power losses by utilising the bearing eccentricity data output from nonlinear transient rotordynamic simulation, which is termed *transient method*. The simulations were carried out with an in-house developed code in the MATLAB® environment, and its correctness is proven by good correlation to the results of shaft motion tests. Accordingly, the effects of different types of sub-synchronous vibrations on the power loss calculation are investigated in detail, and comparisons between the *transient method* and the *static method* are made. It is shown that by considering the effect of oil film instability, the calculated bearing power losses are significantly increased at the onset of *CLC Oscillation*. The rotordynamic performance of a rotor-FFRB system plays a significant role in determining the bearing system power losses.

This paper is structured as follows. The model set-up of the investigated TC rotor-FFRB system and the simulation method are briefly described in Section 2. The results of nonlinear transient rotordynamic simulation are compared to the shaft motion test results in Section 3. In Section 4, the predicted results of FFRB power losses using both methods are given and a comparison between them is made. Finally, conclusions are reached in Section 5.

2 MODEL AND HARDWARE DESCRIPTIONS

The investigated TC rotor-FFRB system as shown in Fig. 1 is modelled through the method as detailed in Ref. (19–21). As illustrated in Fig. 2, the rotor assembly is discretised into 22 Timoshenko beam sections using the finite element method (FEM) developed in the toolbox according to Ref. (28). The mass and stiffness contributions given by the target nut for shaft motion measurement (node 1 to 5), oil slinger (node 9 to 11) and thrust collar (node 11 to 13) are included. The mass and inertia properties of the compressor are attached to the shaft from node 6 to 8 by equally sectioning it to three portions according to its overall axial length, while the turbine wheel is modelled as part of the shaft by six beam sections. The black arrows in Fig. 1 show the location and phase angle of unbalance. In shaft motion test, a pair of eddy current displacement sensors are placed in the compressor cover to point to the position of node 2 on the shaft in the model. Some physical parameters and oil inlet conditions of the investigated rotor-FFRB system are listed in Table 1.

The coupling of FFRB forces to the FEM model is at the heart of the model. In nonlinear transient rotordynamic simulation, the instantaneous bearing forces are required for each time step of the simulation, which demands high reliability of the bearing model to offer good stability as well as sufficient accuracy. Due to the trade-off between accuracy and calculation time, usually a short bearing approximation is used to derive closed form bearing force equations (19–21), since an excellent balance between simulation time and accuracy can be achieved for plain bore type FFRBs. However, the effect of more complex bearing geometries, e.g. bearing circumferential groove, and oil inlet conditions, e.g., OIP, cannot be modelled by short bearing approximation. In this case, the solution of oil film force will have to resort to numerical methods, such as FEM (17,25,27) and Finite

Difference Method (FDM), to discretise the full Reynolds Equation for the pressure distribution of the oil film. In this paper, the bearing model is extensively developed through FDM, and OIP and the bearing circumferential groove are fed into the bearing model as boundary conditions. An example of calculated oil film pressure distribution is illustrated in Fig. 3, which depicts the pressure field of half of the bearing land for each film.

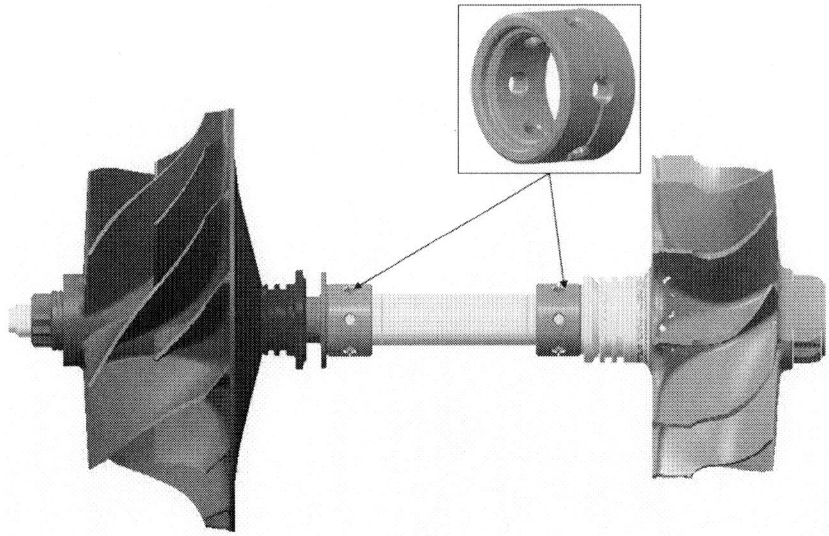

Figure 1: CAD model of the studied TC-FFRB system.

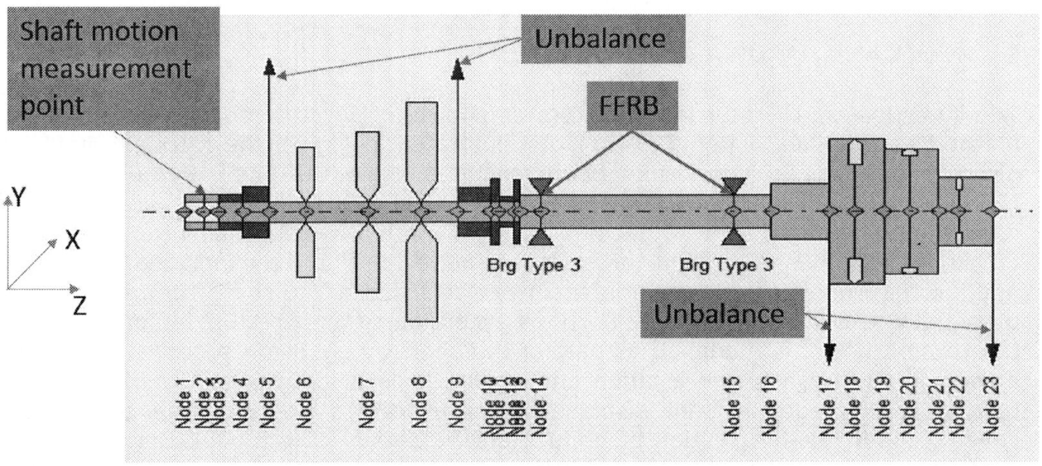

Figure 2: TC rotor-FFRB FEM model.

Table 1 TC rotor-FFRB system physical parameters and oil inlet conditions

Parameters	Value
Rotor Mass	778 gram
Rotor Length	190 mm
Oil grade	SAE 20
OIP	2e5 Pa
OIT	105 °C
FFRB inner/outer radial clearance	8 μm/32 μm
FFRB inner/outer length	6.5 mm/9.3 mm

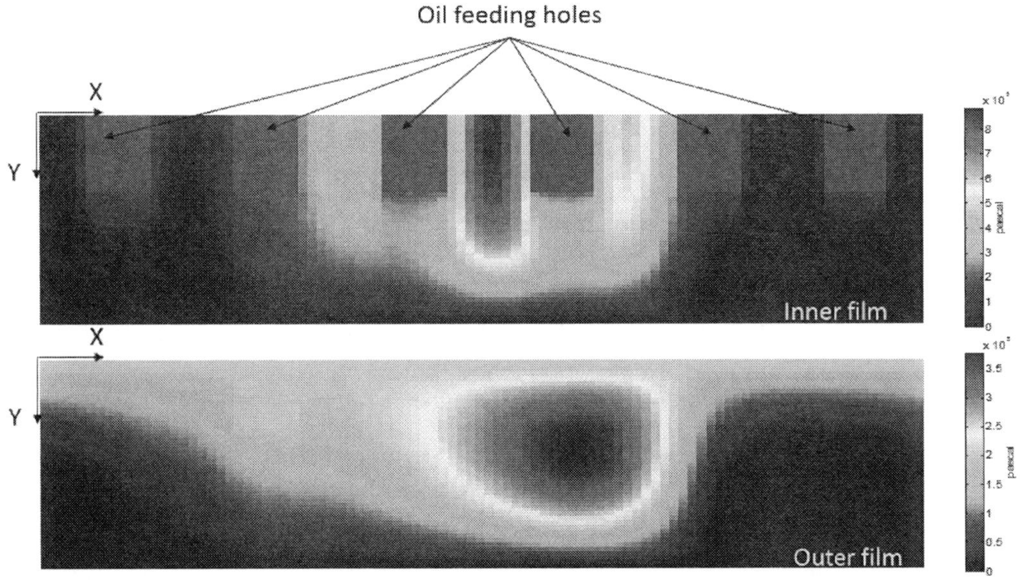

Figure 3: An example of the oil film pressure field calculated by the FDM bearing model over half of the bearing land; X is pointing to the unwrapped bearing circumferential direction, while Y is pointing to the bearing rotational axis direction; the following colour pattern corresponds to the ascending order of pressure magnitude: dark blue, light blue, cyan, yellow, orange and red.

In order to capture the effect of the thermal boundary condition, the developed FDM model is firstly applied to build a lumped thermal model following a similar approach to Ref. (29) to predict the effective oil viscosity for each bearing clearance and bearing clearance change due to differential thermal expansion coefficients. The lumped thermal model is essentially a *static method* model, which solves a series of nonlinear equations considering both the static force balance and heat flow balance between the involved components. The predicted hot bearing clearances and oil viscosity are then input into the nonlinear transient rotordynamic simulation. Additionally, output from the lumped thermal model, such as bearing eccentricities and floating ring speed, are utilised to calculate the FFRB power losses, which give the base results of the *static method* for comparison with the results given by the *transient method*.

For both the *static method* and the *transient method*, the bearing frictional torque on the shaft journal is calculated from the oil film pressure field predicted by the FDM bearing model:

$$T_j = \int_0^{L_i} \int_0^{2\pi R_{ri}} \left\{ + \frac{h_i}{2} \frac{\partial P_i}{\partial x} + \frac{\mu_i}{h_i} (\Omega_j - \Omega_r) R_{ri} \right\} R_{ri} dx dy \qquad (1)$$

where T_j is the frictional torque on the shaft journal. R_{ri}, L_i represent the inner radius and the inner length of the FFRB, respectively. h_i denotes the oil film thickness. P_i is the inner film pressure. μ_i denotes the effective viscosity of the FFRB inner film. Ω_j and Ω_r denote rotor speed and FFRB speed, respectively. x and y denote the bearing coordinate system according to Fig. 3.

The FFRB power losses P_{loss} are obtained from

$$P_{loss} = T_j \Omega_j \qquad (2)$$

3 ROTORDYNAMIC SIMULATION COMPARED TO SHAFT MOTION MEASUREMENTS

Figure 4(a) and (b) show the results of a run-up simulation and the corresponding test in 2D waterfall diagrams, respectively. Shaft motion testing was conducted on a hot gas stand, which burns diesel to meet the target turbine inlet temperature (TIT) and to achieve the designed top running speed, i.e., 123 krpm (in this instance), subject to the stress limit of the compressor wheel. The TIT was kept at 550 °C during the shaft motion measurement, which enables the rotor to rotate within the speed range from 44 krpm to 123 krpm, as shown in Fig. 4b. As for the rotordynamic simulation, the start speed is set to 10 krpm, following which a linear speed ramp-up procedure is adopted until 130 krpm. Note that the STFT (Short Time Fourier Transformation) used to produce the waterfall plot slightly shortens the simulation speed range. In Fig. 4(a), the results within the speed range from 12 krpm to 127 krpm are plotted. In all waterfall diagrams presented in this section, the amplitude of the frequency components is normalised against the maximum amplitude of rotor displacement. The following colour pattern corresponds to the ascending order of response magnitude: dark blue, light blue, cyan, yellow, orange and red.

It is seen that the main characteristics of the test results are well replicated in the simulation. A rich sub-synchronous response is identified throughout both the simulated and tested speed range. As the rotor speed increases, the simulation results depicted in Fig. 4(a) show that the dominant sub-synchronous response changes with the following pattern and characteristics:

1. up to 23 krpm: Sub 1, inner oil whirl
    a. High eccentricities of both inner film are seen in Fig. 5(a)-(b). Bearing eccentricity is a non-dimensional parameter calculated from the ratio of the distance between the geometry centres of shaft journal and bearing shell to bearing radial clearance. For further information, the reader is referred to Ref. (19,25).
    b. Fig. 6(a) shows the rotor mode shape is *conical*. Rotor mode shape plot shows, for a frequency component on the waterfall diagram, the relative position of different locations across the TC rotor. For further information, the reader is referred to Ref. (20,21).
    c. FFRBs are *out-of-phase* and their orbits are approximately 90 degree behind the rotor orbits.

2. from 23 krpm to 95 krpm: Sub 2, inner oil whirl
    a. High eccentricities of both inner films are shown in Fig. 5(a)-(b).
    b. Fig. 6(b) indicates the rotor mode shape is *cylindrical*.
    c. FFRBs are *in-phase* and their orbits are approximately 90 degree behind the rotor orbits.

3. from 95 krpm to 120 krpm: Sub 3, outer oil whirl
    a. The onset speed of Sub 3 is much earlier than 95 krpm, yet it does not dominate the sub-synchronous response until the disappearance of Sub 2 at 95 krpm.
    b. Fig. 6(c) shows the rotor mode shape is *conical*.
    c. FFRBs are *in-phase* and their orbits are behind the rotor orbits by more than 90 degrees.

4. above 120 krpm: Sub 3 switching to *CLC Oscillation*
    a. The transition from Sub 3 to *CLC Oscillation* increases the sub-synchronous frequency.

b. The mean values of the bearing eccentricities of inner and outer films of both ends are considerably increased.
c. Rotor mode shape is *conical* as shown in Fig. 6(d).
d. FFRBs are *out-of-phase* and their orbits are behind the rotor orbits by approximately 90 degrees.

Comparing Fig. 4(a) with Fig. 4(b), the main differences between the simulation and test results are:

- The onset speed of *CLC Oscillation* in test is approximately 115 krpm, which is lower than that found in the simulation.
- The magnitude of the first order response in the test is higher than that in the simulation.

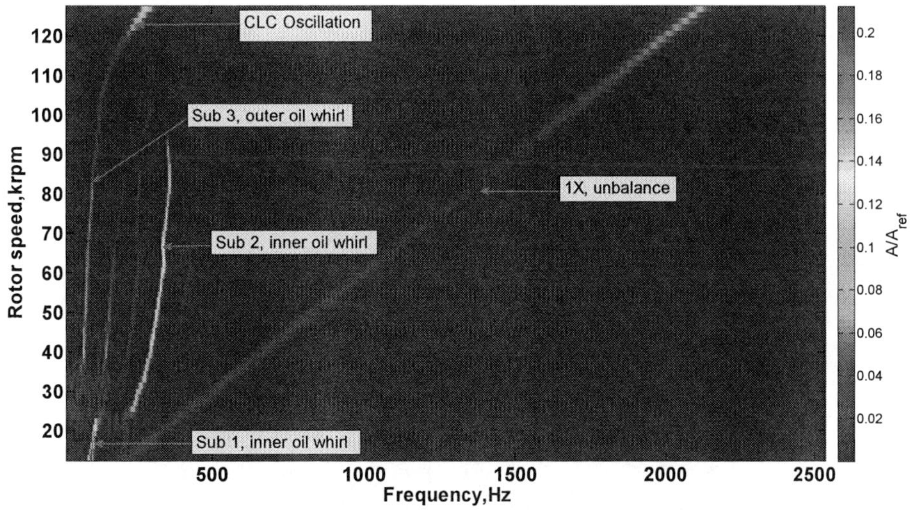

(a) Run-up simulation results, from 12 krpm to 127 krpm

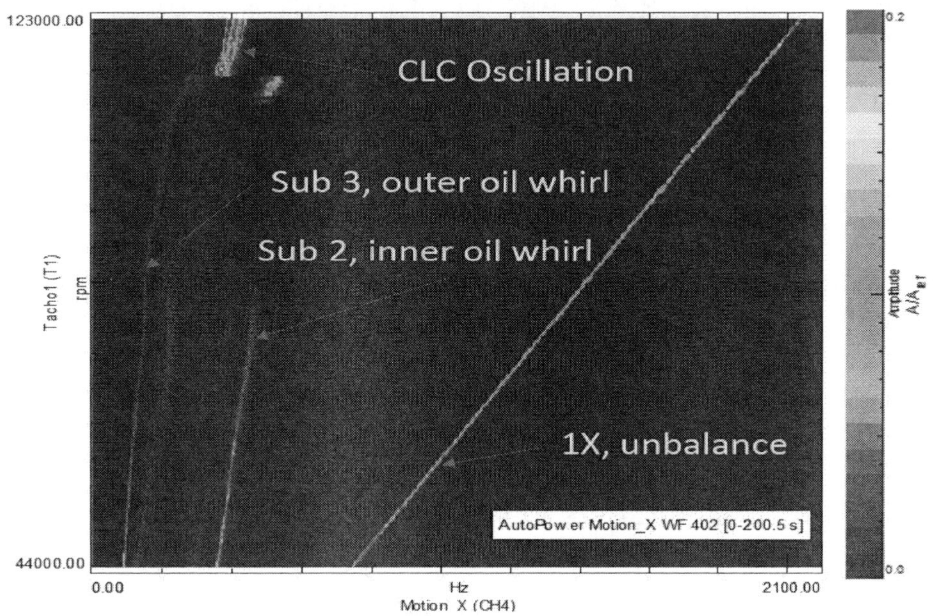

(b) Ramp-up test results, 44 krpm to 123 krpm

Figure 4: 2D waterfall diagrams of the displacement of node 2 in X; simulation in comparison with test results.

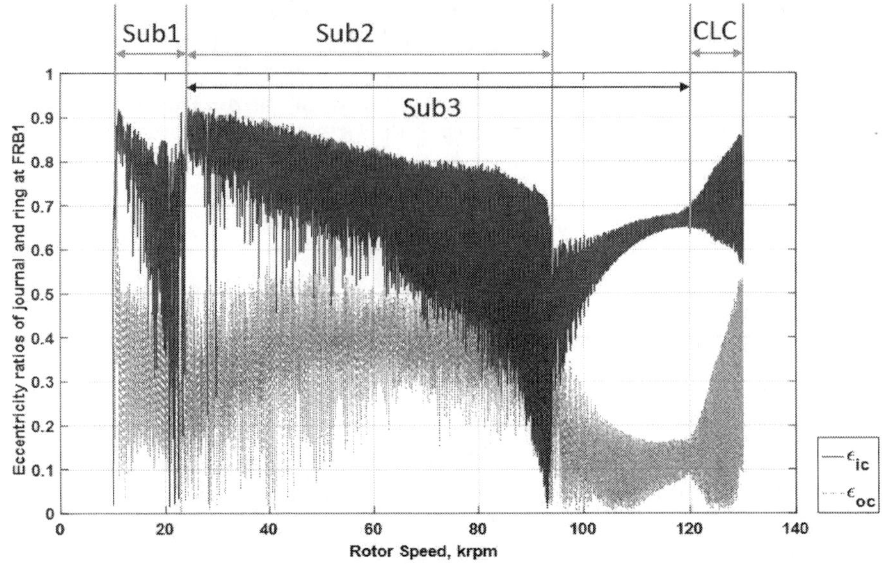

(a) Compressor end bearing

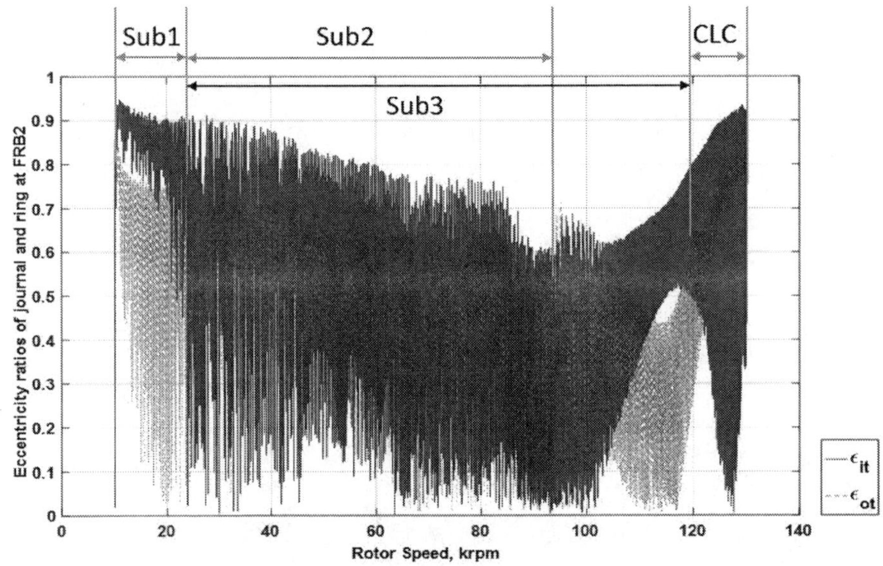

(b) Turbine end bearing

Figure 5: Bearing eccentricity plot; subscript i denotes inner film; subscript o denotes outer film; subscript c denotes compressor end bearing; subscript t denotes turbine end bearing.

One of the potential reasons for the two differences is that the exact unbalance distribution across the rotor is unknown, whereas the unbalance distribution is known as a significant influential factor for the TC rotor response (17,21). The tested TC rotor assembly is only low speed balanced to the component balance limit with a 180 degree phase difference between the compressor and turbine ends in a rigid state prior to shaft motion test. Accordingly, in simulation, the component balance limit is applied to relevant component balance planes with a 180 degree phase difference between both wheels, as shown in Fig. 2, which means that the used unbalance input in the simulation can deviate from the actual equivalent unbalance loading under high speed, since the low speed balancing procedure is not representative of the flexible nature of the TC rotor under much higher operating speed.

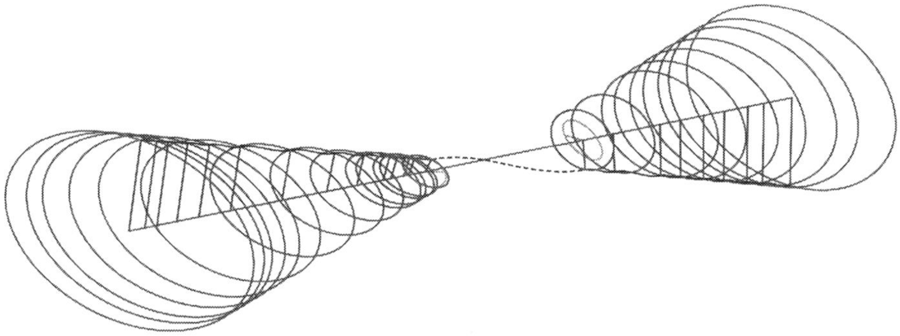

(a) Sub 1, 103 Hz at 16 krpm

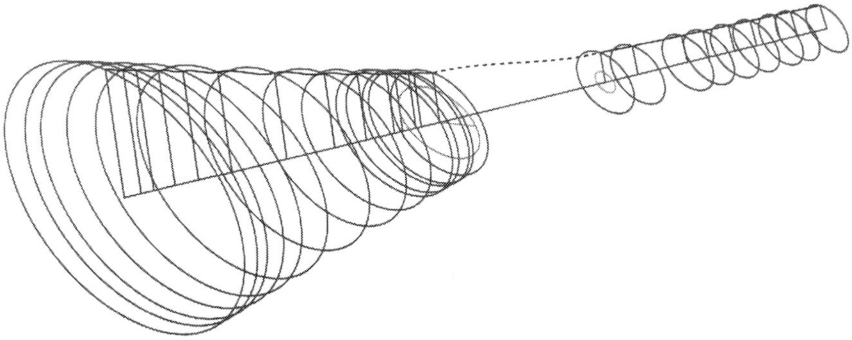

(b) Sub 2, 317 Hz at 50 krpm

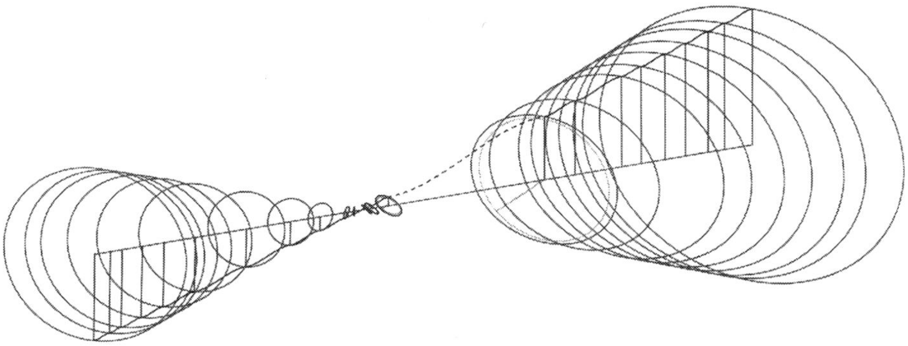

(c) Sub 3, 103 Hz at 81 krpm

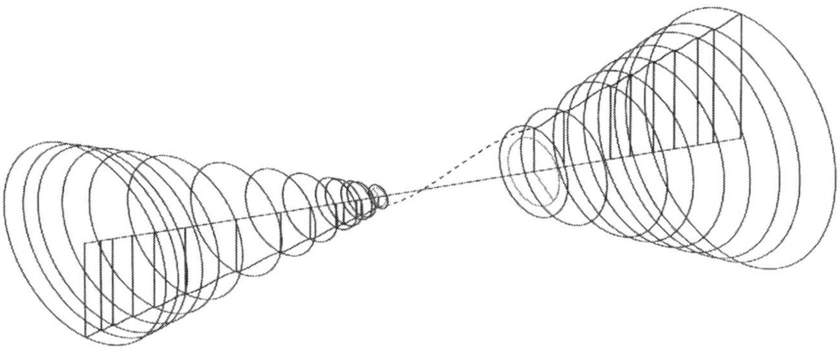

(d) CLC Oscillation, 278 Hz at 127 krpm

Figure 6: Filtered mode shapes of different sub-synchronous components; the compressor end is on the left hand side; blue circles show the relative motion magnitude of each node across the rotor, while red circles indicate the mode of FFRB; the radial lines show the angular positions of all nodes and FFRBs at a certain moment.

An additional simulation given in Fig. 7 under the condition of in-phase unbalance (meaning that there is no angular phase difference of the applied four unbalance vectors) shows a much earlier onset of *CLC Oscillation* below 100 krpm, as compared to the out-of-phase simulation results given in Fig. 4(a). Additional differences which can also be evidently seen are:

- Sub 2 is significantly suppressed and disappears at 62 krpm.
- The rotor response is stable from 62 krpm to 97 krpm.
- Sub 3 is not observed throughout the speed range.

Another reason for the difference in the magnitude of the first order response is that an inevitable error is introduced during shaft motion measurement due to the fact that the shaft motion probe holder is not exactly aligned to the bearing housing axis, which leaves an initial shaft centre line eccentricity.

With good correlation to the test, the simulation results show deep insight into the test results by providing additional useful information unavailable to test, for example, the bearing eccentricity plot as shown in Fig. 5 and ring speed ratio (see Fig. 10), which defines the exact bearing dynamics and is critical to accurate power loss prediction.

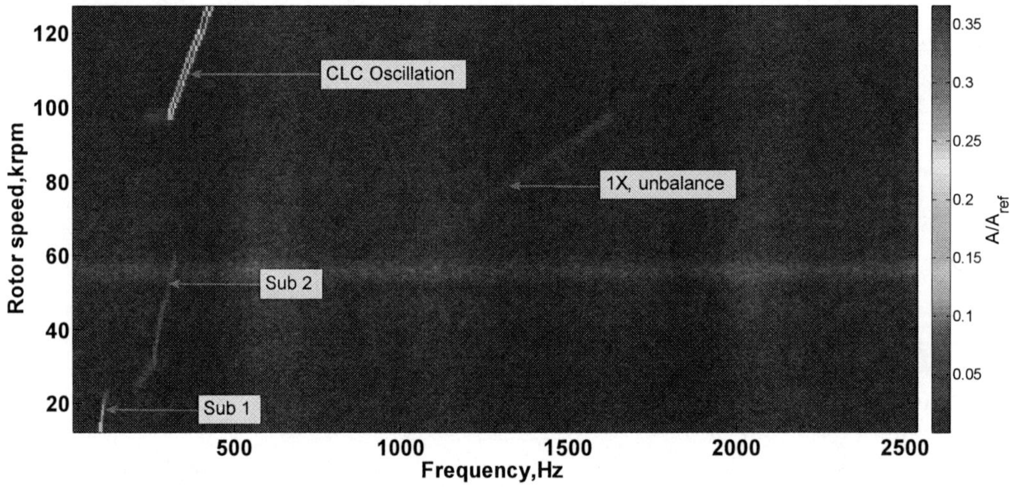

Figure 7: Simulated 2D waterfall diagram of the displacement of node 2 in X under the condition of in-phase unbalance.

4 RESULTS OF POWER LOSSES CALCULATION AND DISCUSSION

In order to extensively evaluate the effect of sub-synchronous vibrations on the power losses of FFRB, the nonlinear transient simulation results under both in-phase and out-of-phase unbalance conditions are utilised for the power losses calculation. The in-phase and out-of-phase results are presented in Fig. 8 and Fig. 9, respectively, with the *static method* predicted results. For a better interpretation of the results, a median filter is applied to smooth the raw signal of power losses calculated by the *transient method*, as the black line on top of the transient data shown in Fig. 8-9.

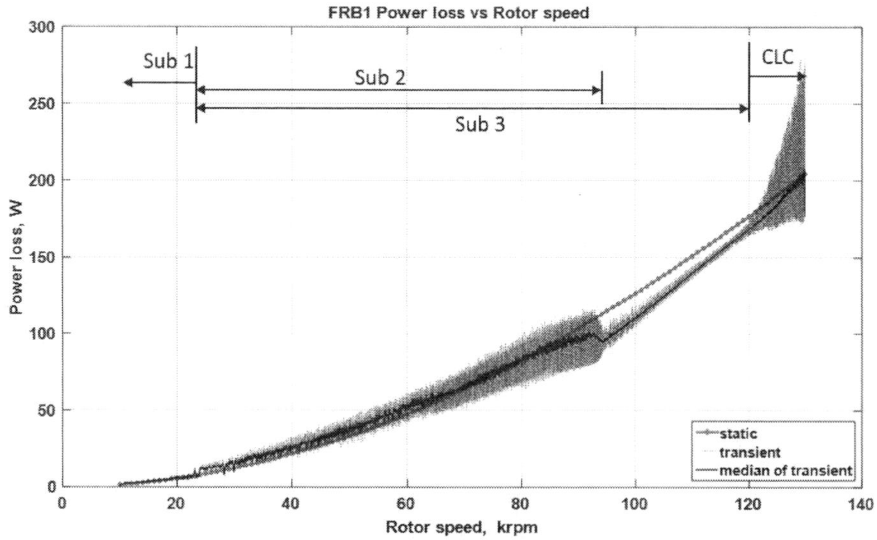

(a) Compressor end

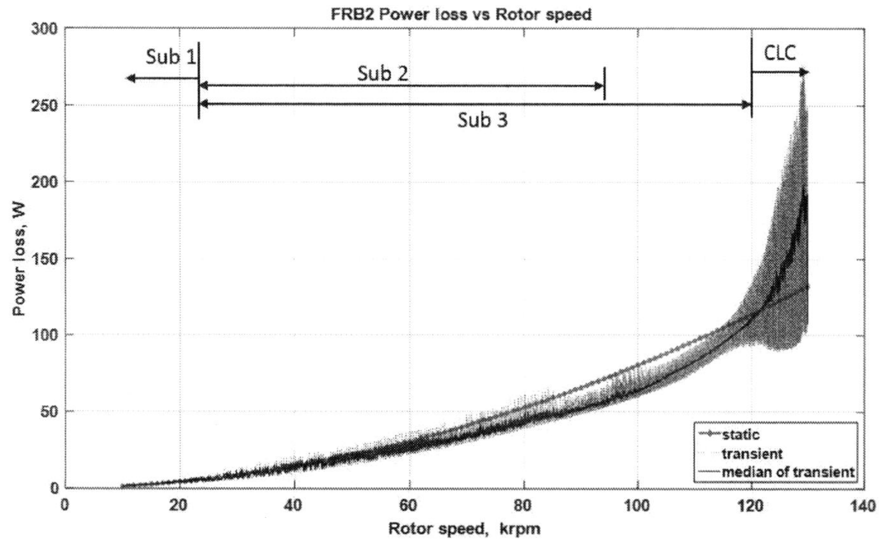

(b) Turbine end

Figure 8: Comparison of power losses prediction between the static and transient method; the transient method uses the nonlinear transient simulation results under out-of-phase unbalance condition.

Under out-of-phase unbalance conditions, as illustrated in Fig. 8, the comparison of the predicted results between the *static* and *transient methods* shows:

- For the compressor end bearing FRB 1, up to 95 krpm where Sub 2 terminates, the two methods generally agree well with each other, since the higher eccentricity given by transient method is counterbalanced by its higher FFRB speed ratio. From 95 krpm to 120 krpm, the transient results are lower than the static results, mainly due to the reduced inner eccentricity and higher speed ratio given by the transient rotordynamic simulation (see Fig. 10). Above 120 krpm, the rise of inner eccentricity leads to increased power losses, which peaks at 279 W at 130 krpm, as compared to 204 W of the static results, although the median value of transient results closely matches that of the static results.

- For the turbine end bearing FRB 2, good agreement is seen between the two methods up to 60 krpm. From 60 krpm to 120 krpm, the power losses of transient results are lower than those of static results. However, the onset of *CLC Oscillation* significantly increases the power losses. At 130 krpm, the median value of the transient results is 185 W, which is 40% higher than the static results.

Under in-phase unbalance conditions, as shown in Fig. 9, the results provided by the two methods match well with each other until the onset of *CLC Oscillation*, which lead to a dramatic increase of the power losses of both bearings. At 130 krpm, the median value of the *transient method* is approximately 350 W for FRB 1 and 274 W for FRB 2, compared to 204 W for FRB 1 and 132 W for FRB 2 from the *static method*. The total power losses of both FFRBs from the *transient method* are 85% higher than the results given by the *static method*.

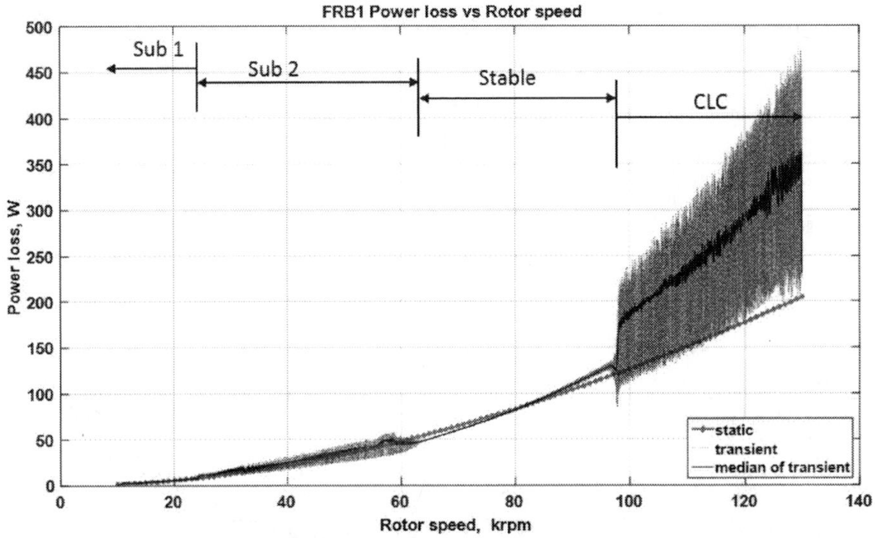

(a) Compressor end

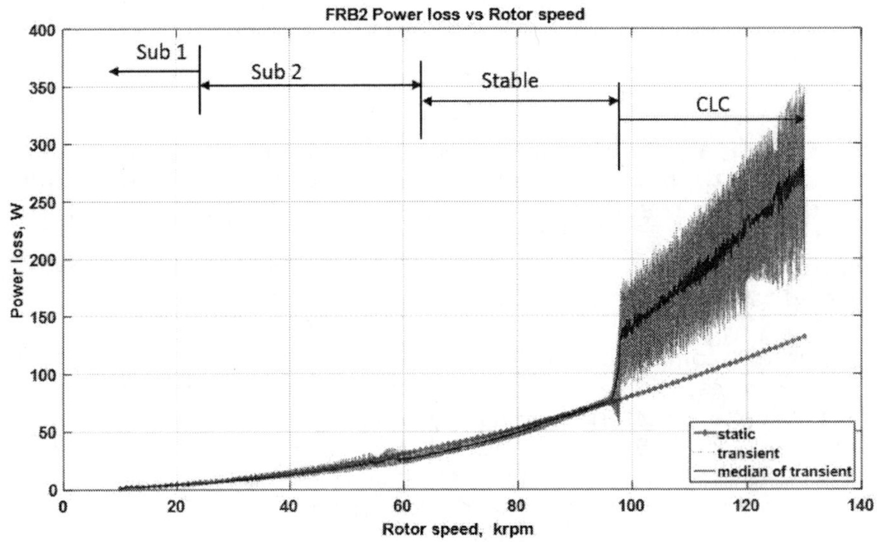

(b) Turbine end

Figure 9: Comparison of power losses prediction between the static and transient method; the transient method uses the nonlinear transient simulation results under in-phase unbalance condition.

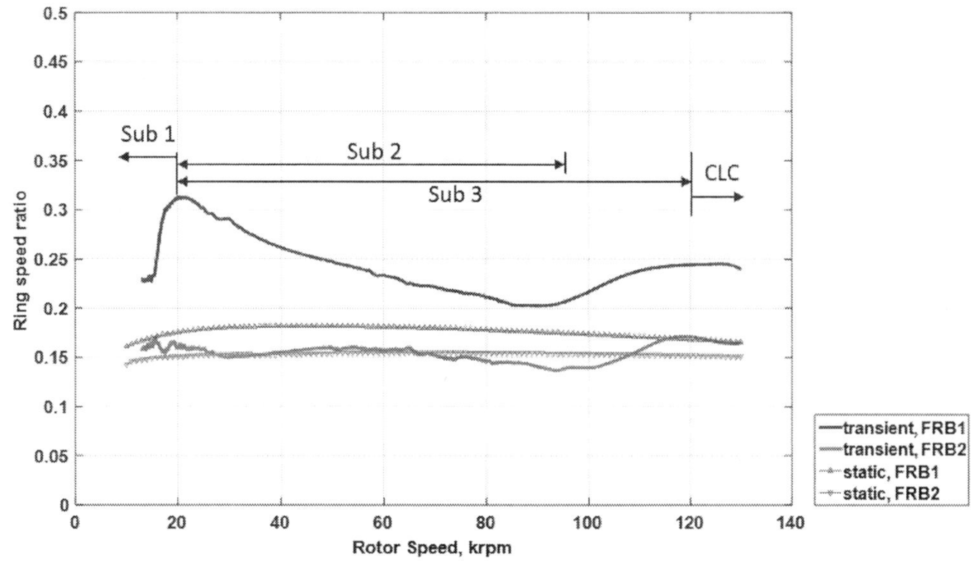

Figure 10: Predicted FFRB speed ratio to TC rotor speed by both transient and static methods under out-of-phase unbalances; FRB 1 denotes the compressor end FFRB; FRB 2 denotes the turbine end FFRB.

5 CONCLUSION

This paper advances a novel method of calculating FFRB power losses by utilising the bearing eccentricities and FFRB speed ratio output from nonlinear transient rotordynamic simulation. The new method, which is termed *transient method* in this paper, can fully take into account the effect of oil film instability, which is commonly seen on TCs equipped with FFRBs and is overlooked by the *static method* adopted by many previous studies with relation to TC power losses prediction. Comparisons between the two methods under both out-of-phase and in-phase unbalance conditions show that the results given by the *static method* generally agree well with the *transient method* predicted results prior to the onset of *CLC Oscillation*. By contrast, the occurrence of *CLC Oscillation* dramatically raises the power losses of FFRB due to excessive bearing inner eccentricities. At the top speed of the considered speed range, the median value of the power losses of both FFRBs given by the *transient method* under in-phase unbalance is 85% higher than that calculated from the *static method*. Under out-of-phase unbalance condition at top speed, the median value of the predicted power losses by the *transient method* is 40% higher than those given by the *static method* at the turbine end bearing. The rotordynamic performance of a TC rotor-FFRB system is critical to the power losses of the FFRB, and its effect on the mechanical efficiency of FFRBs should be carefully considered in the FFRB design process. In other words, optimising the TC rotor-FFRB system in terms of rotordynamic performance is essential for reducing the power losses of FFRBs. Using the developed advanced tool to carry out rotordynamic performance optimisation and FFRB power losses optimisation will be the topic of a future paper.

REFERENCE LIST

1. Burke RD, Zhang N, Duda T, Brace CJ, Akehurst S, Capon G, et al. Experimental and analytical investigation of implementing a ball bearing turbocharger on a production diesel engine. In: 11th International Conference on Turbochargers and Turbocharging. London. 2014.

2. Trippett RJ. Measured and predicted friction in floating-ring bearings. SAE Technical Paper 860075; 1986.
3. Shi F, Deng D. An Analysis for floating bearings in a turbocharger. SAE Technical Paper 2011-01-0375; 2011.
4. Lamquin T, Gjika K. Power losses identification on turbocharger hydrodynamic bearing systems: test and prediction. In: ASME Turbo Expo 2009: Power for Land, Sea, and Air. American Society of Mechanical Engineers. p. 153–62.
5. Deligant M, Podevin P, Vidal F, Tyminski W, Guilain S, Lahjaily H. 3D thermal steady-state CFD analysis of power friction losses in a turbocharger's journal bearing and comparison with finite difference method and experimentation. 12th EAEC. Bratislava.
6. Deligant M, Podevin P, Descombes G, Vidal F, Lefebvre A, Lamquin T. Computational Fluid Dynamics Calculations of Turbocharger's Bearing Losses. SAE Int J Engines. 2010;3(2010-01-1537):103–14.
7. Payri F, Serrano JR, Olmeda P, Paez A, Vidal F. Experimental methodology to characterize mechanical losses in small turbochargers. In: ASME Turbo Expo 2010: Power for Land, Sea, and Air. American Society of Mechanical Engineers; 2010. p. 413–23.
8. Deligant M, Podevin P, Descombes G. Experimental identification of turbocharger mechanical friction losses. Energy. 2012;39(1):388–94.
9. Serrano JR, Olmeda P, Tiseira A, García-Cuevas LM, Lefebvre A. Theoretical and experimental study of mechanical losses in automotive turbochargers. Energy. 2013;55:888–98.
10. Serrano JR, Olmeda P, Tiseira A, García-Cuevas LM, Lefebvre A. Importance of mechanical losses modeling in the performance prediction of radial turbochargers under pulsating flow conditions. SAE Technical Paper; 2013.
11. Hoepke B, Uhlmann T, Pischinger S, Lueddecke B, Filsinger D. Analysis of Thrust Bearing Impact on Friction Losses in Automotive Turbochargers. J Eng Gas Turbines Power. 2015;137(8):082507.
12. Li C-H. Dynamics of rotor bearing systems supported by floating ring bearings. J Tribol. 1982;104(4):469–76.
13. Holt C, San Andrés L, Sahay S, Tang P, La Rue G, Gjika K. Test Response of a Turbocharger Supported on Floating Ring Bearings: Part I—Assessment of Subsynchronous Motions. In: ASME 2003 International Design Engineering Technical Conferences and Computers and Information in Engineering Conference. American Society of Mechanical Engineers; 2003. p. 969–74.
14. Holt C, San Andrés L, Sahay S, Tang P, La Rue G, Gjika K. Test Response of a Turbocharger Supported on Floating Ring Bearings: Part II—Comparisons to Nonlinear Rotordynamic Predictions. In: ASME 2003 International Design Engineering Technical Conferences and Computers and Information in Engineering Conference. American Society of Mechanical Engineers; 2003. p. 975–80.
15. Holt C, Andres LS, Sahay S, Tang P, Rue GL, Gjika K. Test Response and Nonlinear Analysis of a Turbocharger Supported on Floating Ring Bearings. J Vib Acoust. 2005;127(2):107–15.
16. Schweizer B, Sievert M. Nonlinear oscillations of automotive turbocharger turbines. J Sound Vib. 2009;321(3):955–75.
17. Schweizer B. Dynamics and stability of turbocharger rotors. Arch Appl Mech. 2010;80(9):1017–43.
18. Boyaci A, Seemann W, Proppe C. Bifurcation analysis of a turbocharger rotor supported by floating ring bearings. In: IUTAM Symposium on Emerging Trends in Rotor Dynamics. Springer; 2011. p. 335–47.
19. Tian L, Wang WJ, Peng ZJ. Dynamic behaviours of a full floating ring bearing supported turbocharger rotor with engine excitation. J Sound Vib. 2011;330(20):4851–74.

20. Tian L, Wang WJ, Peng ZJ. Effects of bearing outer clearance on the dynamic behaviours of the full floating ring bearing supported turbocharger rotor. Mech Syst Signal Process. 2012;31:155–75.
21. Tian L, Wang WJ, Peng ZJ. Nonlinear effects of unbalance in the rotor-floating ring bearing system of turbochargers. Mech Syst Signal Process. 2013;34(1):298–320.
22. Boyaci A, Schweizer B. Nonlinear Oscillations of High-Speed Rotor Systems in Semi-Floating Ring Bearings. In 2015. p. 845–54.
23. Koutsovasilis P, Driot N, Lu D, Schweizer B. Quantification of sub-synchronous vibrations for turbocharger rotors with full-floating ring bearings. Arch Appl Mech. 2015;85(4):481–502.
24. Knoll IG, Proppe IC, Koch D-IR, Backckhaus IK, Boyaci D-IA. Run-up of turbocharger rotors in nonlinearly modelled floating bush bearings. MTZ Worldw. 2010;71(4):50–5.
25. Schweizer B. Oil whirl, oil whip and whirl/whip synchronization occurring in rotor systems with full-floating ring bearings. Nonlinear Dyn. 2009;57(4):509–32.
26. San Andres L, Rivadeneira JC, Gjika K, Groves C, LaRue G. Rotordynamics of small turbochargers supported on floating ring bearings—highlights in bearing analysis and experimental validation. J Tribol. 2007;129(2):391–7.
27. Schweizer B. Total instability of turbocharger rotors—physical explanation of the dynamic failure of rotors with full-floating ring bearings. J Sound Vib. 2009;328(1):156–90.
28. Friswell MI. Dynamics of rotating machines. Cambridge University Press; 2010.
29. San Andres L, Kerth J. Thermal effects on the performance of floating ring bearings for turbochargers. Proc Inst Mech Eng Part J J Eng Tribol. 2004;218(5):437–50.

SESSION 4B

MATERIALS, HOUSING, BURST CONTAINMENT, EGR AND FATIGUE LIFE

Explicit dynamic finite element simulation of turbocharger containment and wheel burst

L Wang, D M Eastwood
Applied Mechanics, Cummins Turbo Technologies, UK

ABSTRACT

Turbocharger containment is the capability of a turbine or compressor housing to contain a burst wheel. It is essential to ensure the product safety and to prevent catastrophic accidents. In recent years, Cummins Turbo Technologies developed an explicit Finite Element Analysis (FEA) based approach to simulate the turbocharger containment, aiming to minimise the cost associated with testing, optimise the product design and shorten the new concept development cycle. In this study, a strain rate and temperature dependent plasticity material model with fracture modelling capability is employed to model the turbine housing material behaviour. In addition, a methodology of optimising the weakening slot design to burst a turbine wheel at any specified speed is presented.

Keywords: Explicit Dynamic FEA, ANSYS Autodyn, turbocharger, containment, wheel burst

NOMENCLATURE

A	Initial yield stress
B	Hardening constant
C	Strain rate constant
I	Internal element force
M	Mass matrix
m	Thermal softening exponent
n	Hardening exponent
P	External applied force
T	Current temperature
u	Displacement
Δt	Time increment
$\dot{u}$	Velocity
$\ddot{u}$	Acceleration
T_H	Homologous temperature
T_{melt}	Melting temperature
T_{room}	Room temperature
ε_p	Effective plastic strain
ε_p^*	Plastic strain rate
σ	Stress

1. INTRODUCTION

Turbocharger containment is the capability of a turbine or compressor housing to contain a burst wheel. Since the blade tip speed of a wheel can reach 600 *m/s*, it is critical to ensure the product safety and to prevent catastrophic accidents. In the turbocharger industry experimental testing has been widely used to investigate turbocharger containment, due to its "straightforward" nature and its capability to take material variation as well as geometrical variation into account. Physical testing inevitably costs a significant amount of time and money, however. In recent years, Cummins Turbo Technologies has developed a Finite Element Analysis (FEA) based approach to simulate turbocharger containment, aiming to minimise the cost associated with testing, optimise the product design and shorten the new concept development cycle.

Until now, few papers were published on the topic of numerical simulation of turbocharger containment. Memhard et al [1] and Zheng Jing [2] have done a significant amount of work on material testing, aiming to develop the material plasticity and failure models at various strain rates and temperatures for the turbine housing materials. Very limited work was presented, however, regarding the simulation techniques of turbocharger containment. Winter et al [3] did some pioneering work on the FEA simulation of compressor containment, where relatively good correlation was shown between FEA and testing. Two of the major drawbacks of this paper were: firstly no failure criteria were used on the housing material, hence it may be very challenging to quantify the failure of a housing; secondly, the impeller was built as three parts to represent the three fragments in testing. This may result in difficulty of using this FEA model to predict the burst speed and modelling crack propagation. Most recently, Ramamoorthy et al [4] developed a method to predict the wheel burst speed and an impressive correlation (2% in speed) to test was achieved. However, their criterion to verify the housing containment may be limited.

In this paper, the explicit FEA solution method has been chosen to simulate the containment process, because it is more robust and efficient in modelling problems that involve high nonlinearities, high speed impact and material fracture. A strain rate and temperature dependent plasticity material model with fracture modelling capability is employed to model the turbine housing material behaviour. In addition, a methodology of optimising the weakening slot design to burst a turbine wheel at any specified speed is presented. This approach has been implemented in Cummins Turbo Technologies and good correlation is achieved.

2. FINITE ELEMENT SOLUTION METHOD

Finite Element solution methods are generally resolved into the implicit method and the explicit method [5]. The implicit FEA method iterates to find the approximate static equilibrium at the end of each load increment. It controls the increment by a convergence criterion throughout the simulation. For a highly non-linear problem, a large number of iterations have to be carried out before finding the equilibrium. Thus the global stiffness matrix has to be assembled and inverted many times during the analysis. Therefore, the computation is extremely expensive and memory requirements are also very high. It is difficult to predict how long it will take to solve the problem or even if convergence can be achieved. Thus the implicit method is preferable to analyse problems under static and simple loading conditions.

The explicit method determines a solution by advancing the kinematic state from one time increment to the next, without iteration. The explicit solution method uses a diagonal mass matrix to solve for the accelerations and there are no convergence checks. Therefore it is more robust and efficient for complicated problems, such as dynamic events, nonlinear behaviours, and complex contact conditions. However, in order to obtain accurate results, the time increment has to be extremely small, which ensures that the acceleration through the time increment is nearly constant. Therefore it typically requires many thousands of increments. At the beginning of the time increment (*t*), based on the dynamic equilibrium equation [6-7]:

$$P - I = M\ddot{u} \tag{1}$$

The nodal accelerations ($\ddot{u}$) are calculated as:

$$\ddot{u}\,|_{(t)} = (M)^{-1}(P - I)\,|_{(t)} \tag{2}$$

where *M* is the nodal mass matrix, *P* is the vector of externally applied force and *I* is the vector of internally induced element force. The acceleration of any node is completely determined by the mass and the net force acting on it. Through time the accelerations are integrated using the central difference rule, by which the change of the velocity ($\dot{u}$) is calculated from equation (3), assuming that the acceleration is constant:

$$\dot{u}\,|_{(t+\Delta t/2)} = \dot{u}\,|_{(t-\Delta t/2)} + \frac{\Delta t|_{(t+\Delta t)} + \Delta t|_{(\Delta t)}}{2}\ddot{u}\,|_{(t)} \tag{3}$$

The velocities are integrated through time and added to the displacement (*u*) at the beginning of the increment to calculate the displacements at the end of the increment:

$$u\,|_{(t+\Delta t)} = u\,|_{(t)} + \Delta t\,|_{(t+\Delta t)}\dot{u}\,|_{(t+\Delta t/2)} \tag{4}$$

The element strain increments $d\varepsilon$ is calculated from the strain rate, and then the stresses are obtained from the constitutive equations:

$$\sigma_{(t+\Delta t)} = f\left(\sigma_{(t)}, d\varepsilon\right) \tag{5}$$

In this study, the Explicit FEA solution method provided by commercial FEA software - ANSYS Autodyn has been chosen to analyse the wheel burst and housing containment of turbochargers.

3. SIMULATION OF TURBINE WHEEL BURST

In the turbine containment testing there are two failure modes of the turbine wheels, i.e. hub burst and blade detachment [8-9]. To provoke the blade detachment failure, it is required to cut a weakening slot (either angular or straight) on the backface of a turbine wheel [9]. However, trial-and-error is sometimes required in designing the correct weakening slot, to burst the wheel at its target burst speed. Testing inevitably costs a considerable amount of money and takes time. The objective of this study is to develop an analysis-led-testing method by optimising the weakening slot design.

In the FEA models, it is assumed that

1) Aerodynamics load has no effect on wheel failure and housing containment.
2) The turbine wheel and housing have uniform density and precise geometry.

3) Materials are isotropic, homogeneous and defect-free.
4) Material failure is independent with respect to the types of loadings, strain rate and temperature.

An angular velocity of 218750 rpm, which is the target burst speed of this wheel as per specification of [9], is applied to the wheel only FEA model. Due to the complexity of the geometry tetrahedral elements (first order, explicit) are used to mesh the turbine wheel resulting in 1.8 million elements, as shown in Figure 1.

The elastic-plastic properties of Inconel 713C, taken from a uniaxial tensile test, are used to simulate the material behaviour of the turbine wheel under the high centrifugal force. A plastic strain based material failure criterion is applied to model the ductile material failure of the wheel burst. It is essential that the failure criterion does not over-predict or under-predict the burst speed for a specified weakening slot design. The failure criterion is achieved by employing the workflow as shown in Figure 2. The comparison between test and simulation is shown in Figure 3.

Figure 1 Meshing of the weakened turbine wheel FE model

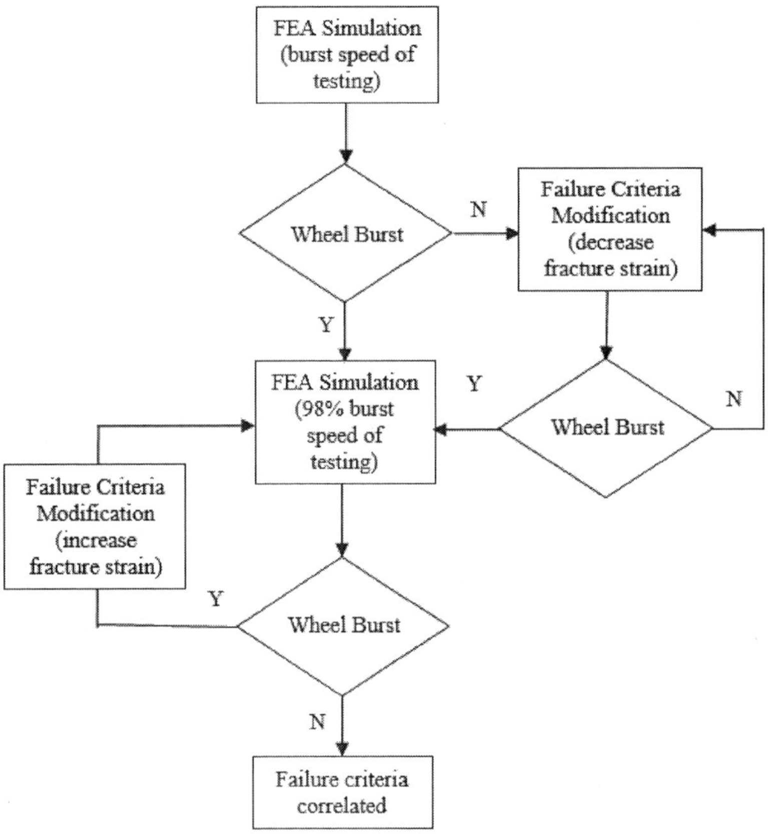

Figure 2 Workflow of achieving the material failure criterion

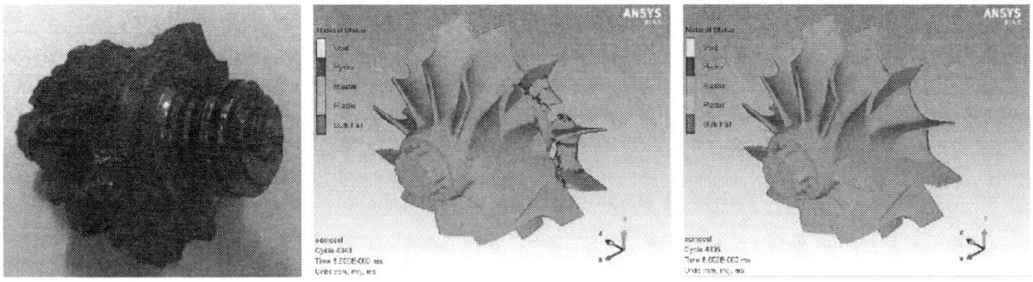

(a) Test part of burst speed (b) FE model of burst speed (c) FE model of 98% burst speed

Figure 3 Comparison between test and simulation of wheel burst

4. RESPONSE SURFACE ANALYSIS OF WEAKENING SLOT DESIGN

Response Surface Method (RSM) is a collection of mathematical and statistical techniques for modelling and analysing problems where several variables (input factors) influence the response (output factors) and the objective is to optimise the response [10]. A number of input factors, as shown in Table 1 and Figure 4, could affect the turbine wheel burst process. In this study, a RSM analysis of the wheel burst FEA with two factors, i.e. slot depth and slot Outer Diameter (OD), and three levels (coded as -1, 0 and 1) has been carried out. The lower and upper limits of the inputs vary by approximately 10% from the baseline. For all the FEA runs, the speed is set as the target burst speed of this wheel, the slot angle and slot width remain the same.

Table 1 Input factors and levels in RSM analysis of wheel burst FEA

Inputs	Lower limit (-1)	Baseline (0)	Upper limit (1)
Slot Depth (mm)	3.4	3.8	4.2
Slot OD (mm)	25	28	31
Wheel Speed (rpm)		218750	
Slot Angle (Degree)		89.5	
Slot Width (mm)		1.5	

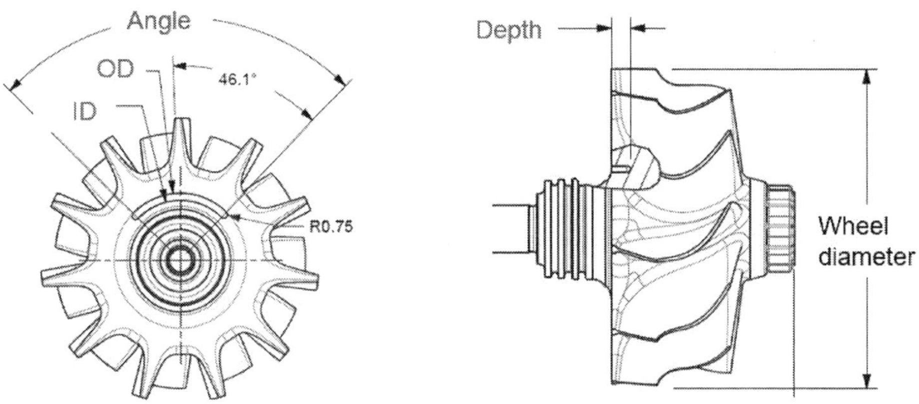

Figure 4 Geometrical factors of weakening slot

The Central Composite Design (Face Centred) method [10] is applied to generate the RSM runs, as shown in Table 2. The nominal maximum plastic strain, which determines whether wheel burst takes place, is studied as the output of this RSM analysis. If the nominal plastic strain is equal or greater than 1, then wheel burst (material failure) occurs.

Table 2 RSM runs of wheel burst FEA

StdOrder	RunOrder	PtType	Blocks	Slot Depth	Slot OD	Burst (Y/N)	Nominal Max. Plastic Strain
2	1	1	1	4.2	25	N	0.96
4	2	1	1	4.2	31	Y	2.10
8	3	-1	1	3.8	31	Y	1.50
1	4	1	1	3.4	25	N	0.71
6	5	-1	1	4.2	28	Y	1.35
3	6	1	1	3.4	31	Y	1.00
7	7	-1	1	3.8	25	N	0.82
9	8	0	1	3.8	28	Y	1.03
5	9	-1	1	3.4	28	N	0.84

The statistics software Minitab has been used to analyse the effects of inputs to the outputs. According to the 3D scatterplot and main effects plot, as shown in Figure 5 and 6, high plastic strain (wheel burst) tends to occur if increasing the OD (outer diameter) or increasing depth of the weakening slot. Furthermore, in this study no interaction between slot depth and slot OD exists, as shown in Figure 7.

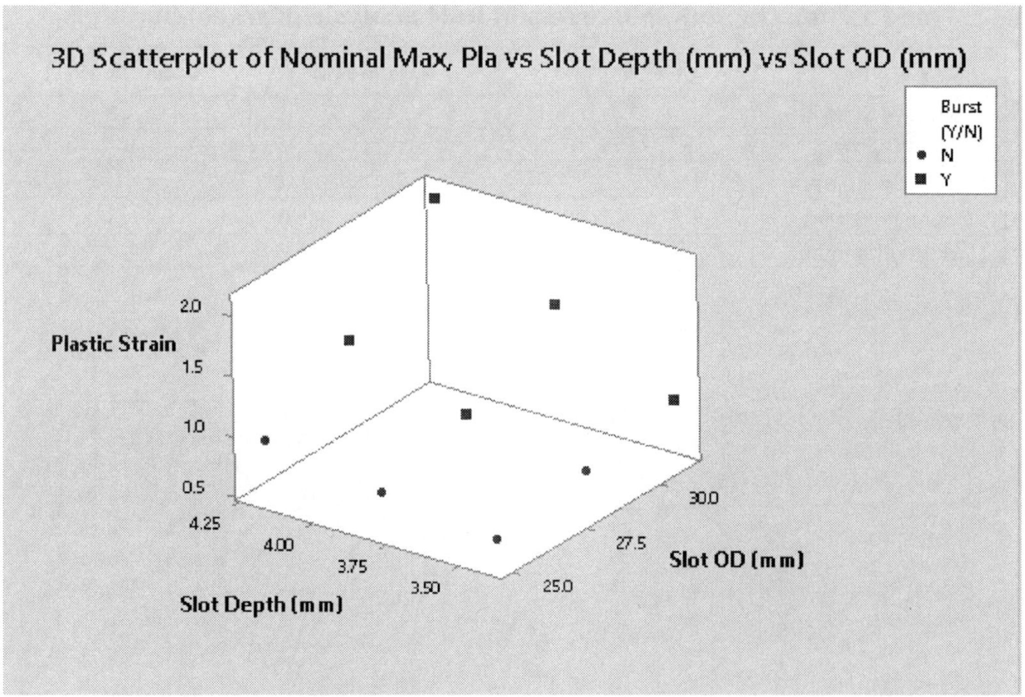

Figure 5 3D scatterplot of nominal plastic strain vs slot depth and slot OD

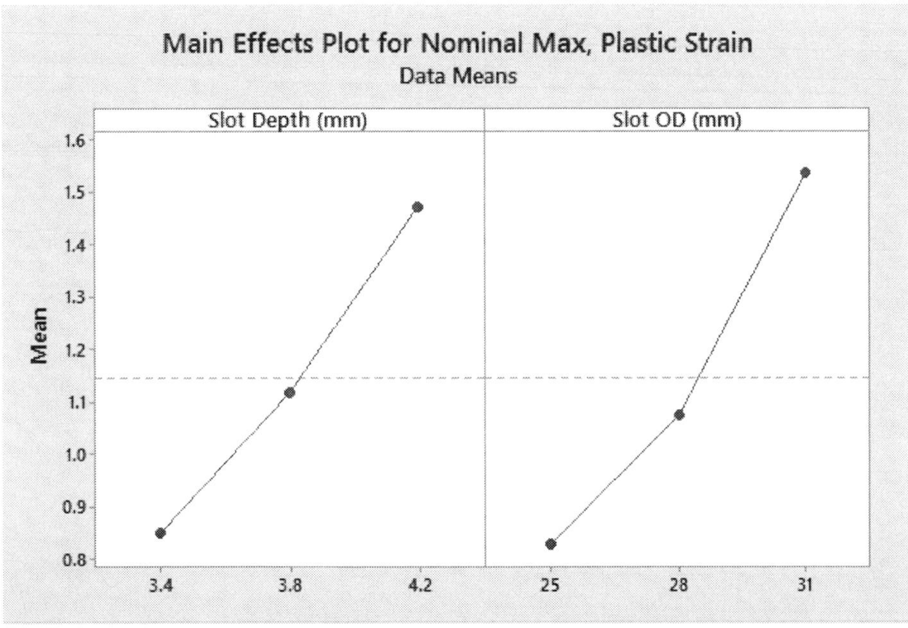

Figure 6 Main effects plot for nominal plastic strain

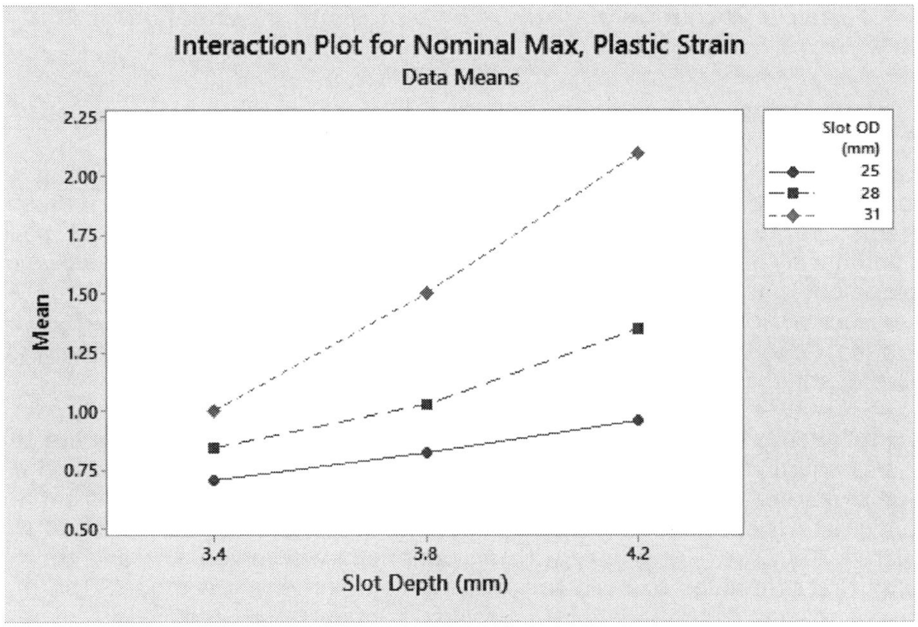

Figure 7 Interaction plot for nominal plastic strain

In the process of optimising the weakening slot, it is essential to find a weakening slot design which can burst the wheel at exactly the target burst speed. It is believed that if the OD or the depth of the slot is set to be too large, the wheel would burst at a speed lower than the target speed and vice versa. Figure 8 illustrates a workflow of optimising the slot depth for a given slot OD and target burst speed. Based on the capability of the actual cutting tool, a minimum slot depth increment, i.e. x, should be used, aiming to find the "boundary" between burst and burst-free of a turbine wheel.

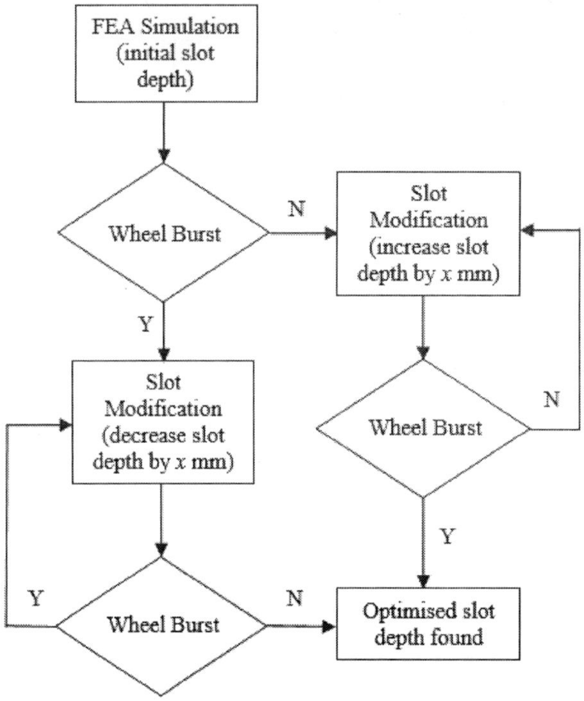

Figure 8 Workflow of optimising slot depth in wheel burst FEA

5. SIMULATION OF TURBINE HOUSING CONTAINMENT

Containment is the capability of a turbine or compressor housing to contain a burst wheel. It is critical to ensure the product safety and to prevent catastrophic accidents. In Cummins Turbo Technologies, all new products must pass the containment testing before being released to the market. To assist the new concept development, shorten the design cycles and decrease the testing cost, an explicit FEA based technique of housing containment has been developed, by virtually evaluating and comparing various housing concepts before carrying out any physical testing.

The simulation process includes two sequential analyses: 1) Steady-state thermal analysis, where the temperature distribution of components is analysed and used to take the material thermal softening effects into account in 2) Transient explicit dynamic analysis, in which wheel burst as well as housing containment are simulated. The wheel geometry with an optimised weakening slot and the turbine housing are assembled to build the housing containment FEA model. The mesh of which is shown in Figure 9, where 5.1 million elements are used.

The target burst speed is applied to the weakened wheel as the initial condition of the FEA model. Impact from wheel fragments to the housing is modelled by using the penalty formulation of the trajectory contact approach [11], where a contact event is detected, if the trajectory of a node and a face intersects during one computation cycle. A local penalty force is calculated to push the node back to the face. Equal and opposite forces are calculated on the nodes of the face in order to conserve linear and angular momentum. Frictional effects are considered and a frictional coefficient of 0.2 is assumed between the wheel and housing.

Figure 9 Meshing of housing containment FEA model

In this study, the reduction of elastic modulus due to elevated temperature is neglected because plastic deformation and material fracture dominate in the containment process. The Johnson Cook strength model [12] is used to model the plasticity behaviour of the ductile SiMo iron housing, which is subjected to large strain, high strain rate and high temperature. The yield stress is defined as:

$$Y = \left[A + B(\varepsilon_p)^n\right]\left[1 + C\ln(\varepsilon_p^*)\right]\left[1 - (T_H)^m\right] \quad (6)$$

where ε_p is the effective plastic strain, ε_p^* is the plastic strain rate; T_H is the homologous temperature, which is defined by

$$T_H = \frac{T - T_{room}}{T_{melt} - T_{room}} \quad (7)$$

There are five material constants in the Johnson Cook model: A is the initial yield stress, B is the hardening constant, n is the hardening exponent, C is the strain rate constant, and m is thermal softening exponent.

To simulate the material failure there are a number of material fracture modes [13-15], most of which, however, require comprehensive testing. In this study, a relatively simple plastic strain based criterion is used to model the housing material failure and good correlation is shown between FEA and testing. Element erosion [6] is employed to simulate the penetration of the housing due to the high speed impact from the wheel fragments, as shown in Figure 10. It is clear that the blades detach from the wheel at the region near the weakening slot, gradually impacting the housing. The wheel fragments reach speed zero at the end of the impact and the housing successfully contains the wheel fragments in this case study.

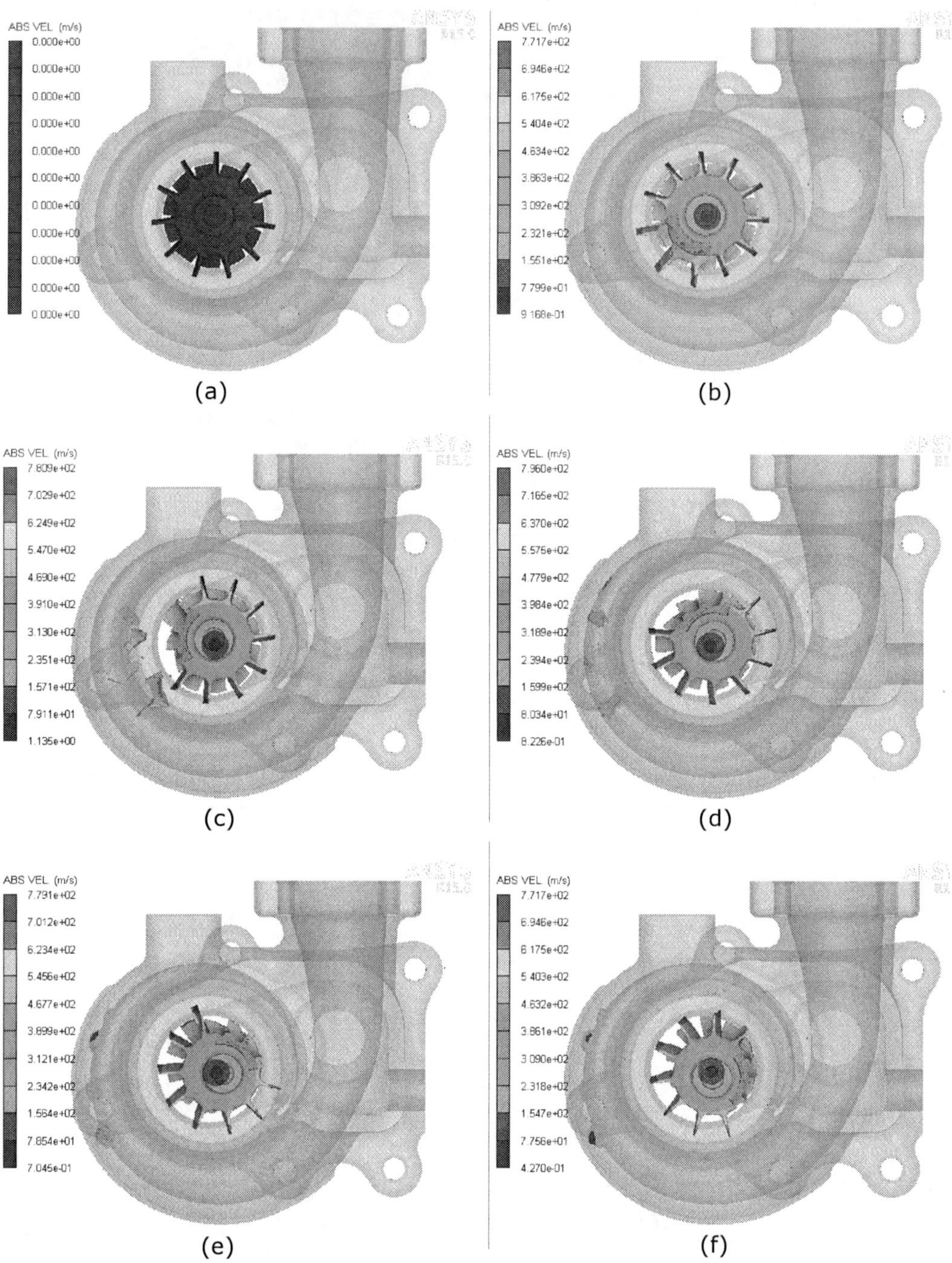

Figure 10 Wheel velocity plots of the housing containment FEA model

6. COMPARISON BETWEEN TEST AND SIMUALTION

The explicit FEA method of modelling the wheel burst and housing containment is applied to other case studies, where good correlation between test and simulation has been achieved. Figure 11 compares the wheel burst FEA with test by using the straight cut weakening slots. Again it demonstrates the importance of the weakening slot design. In the case study 3 shown in Figure 11(a), all three blades detach from the hub. Nevertheless, by using the slot design of case study 4, as shown in Figure 11(b), the middle blade remains intact, indicating that the slot depth may not be deep enough to result in the detachment of all three blades. Figure 12(a) and Figure 12(b) illustrate the correlation in terms of the damage at the inner wall and inlet of two turbine housings, respectively. Furthermore, the technique of turbine housing containment FEA has been applied to the compressor containment with Aluminum impeller and ductile iron housing. Figure 13 showcases the correlation of the fractured impeller from the compressor containment test and FEA simulation.

(a) Case study 3

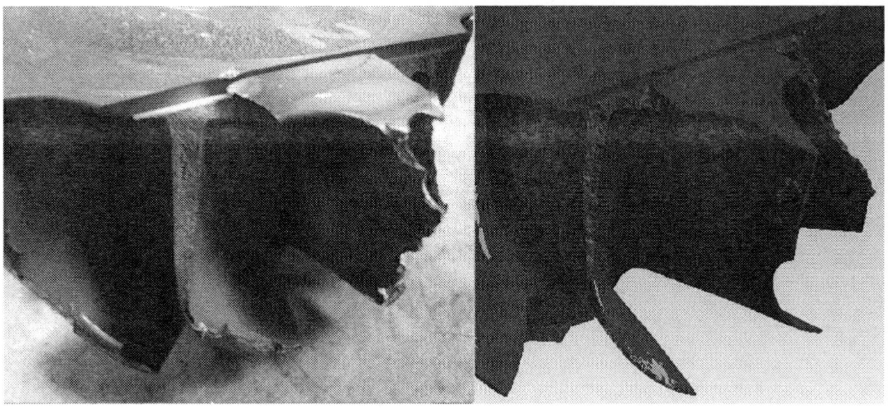

(b) Case study 4

Figure 11 Comparison of turbine wheel burst test and FEA

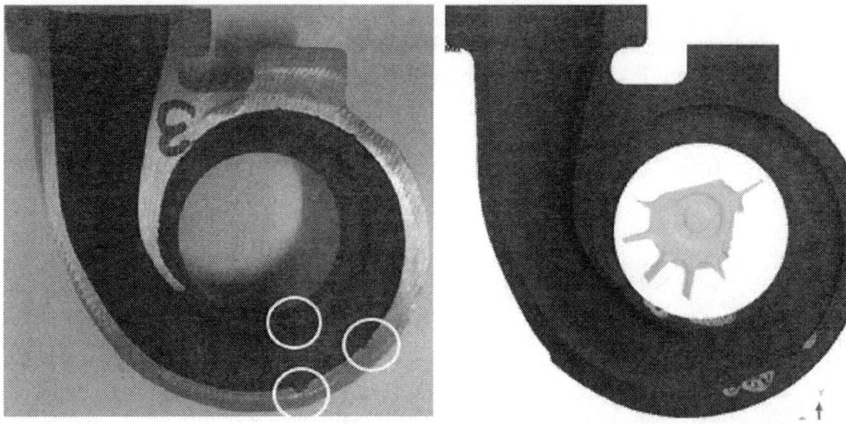

(a) Case study 5

(b) Case study 6

Figure 12 Comparison of turbine housing containment test and FEA

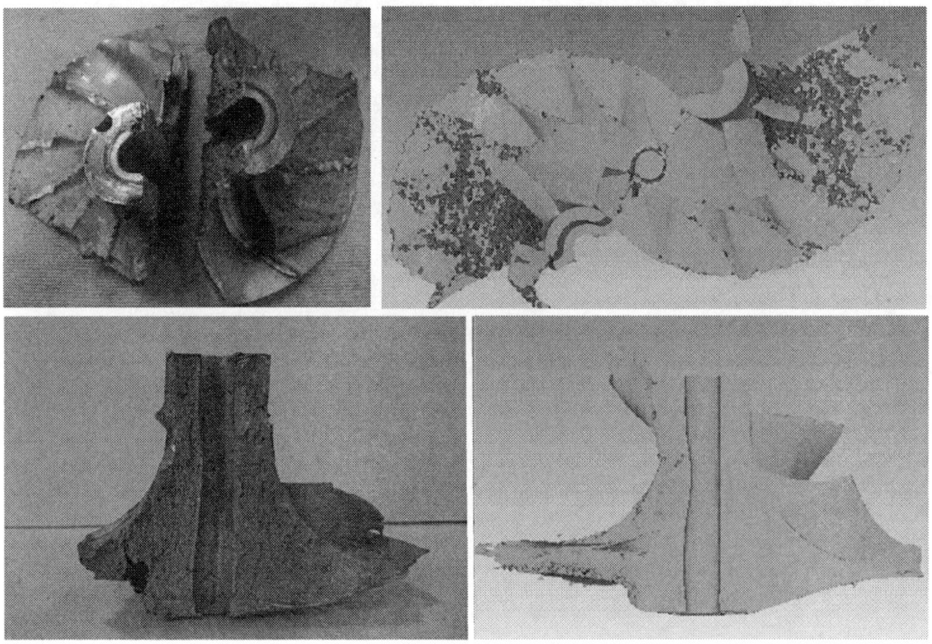

Figure 13 Comparison of impeller burst in compressor containment test and FEA

7. CONCLUSION

In this paper two major aspects of the turbine containment, i.e. wheel burst and housing containment, have been simulated by explicit FEA; which demonstrates excellent correlation with testing. Based on this study, the following conclusions may be drawn:

a) Wheel burst tends to occur if increasing the outer diameter or depth of the weakening slot.

b) In the process of optimising the weakening slot, it is essential to find a weakening slot design which can burst the wheel at exactly the target burst speed. It is believed that if the outer diameter or the depth of the slot is set to be too large, the wheel would burst at a speed lower than the target speed and vice versa.

c) The Johnson Cook strength model is suitable to simulate the plastic behaviour of turbine housing, which is subjected to large strain, high strain rate and high temperature in the containment simulation.

d) A plastic strain based failure criterion combined with element erosion technique may be capable of modelling the ductile material failures in turbo containment FEA.

ACKNOWLEDGEMENTS

The authors would like to thank Cummins Turbo Technologies to allow publishing this paper. Our sincere thanks also go to Mr. Owen A Ryder, Mr. Jeffrey D Jones and Mr. Simon J Tooley who provided many valuable suggestions and proof read this paper. We would like to acknowledge the consistent support from other team members of the Applied Mechanics department.

REFERENCE LIST

1) Dieter Memhard, Florence Andrieux, Dong-Zhi Sun, R Häcker (2011) Development and verification of a material model for prediction of containment safety of exhaust turbochargers, 8th European LS-DYNA Users Conference, Strasbourg.
2) Zheng Jing (2013) Prediction of failure of cast iron with dynamic loading and high temperature, Materials Science and Engineering A, 566, 71-81.
3) Thomas Winter, Armin Huß, Heiko Beck (2007) Simulation of containment-tests of fast spinning rotors by explicit FEM, 6th European LS-DYNA Users' Conference, Gothenburg.
4) J M Ramamoorthy, S S Parikh, S Pandian, P S Kasthuri Rangan (2014) Containment simulation and validation of turbocharger housing design. 11th International Conference on Turbochargers and Turbocharging, London.
5) F J Harewood, P E McHugh (2007) Comparison of the implicit and explicit finite element methods using crystal plasticity. Computational Materials Science, 39, 481-494.
6) Explicit Dynamics Theory Guide, ANSYS Mechanical User's Guide, ANSYS documentation, version 15, 2014.
7) Explicit Dynamic Analysis, Abaqus Analysis User's Guide, Abaqus documentation, version 6.14, 2014.
8) White Paper No. 2, Burst & Containment: Ensuring Turbocharger Safety, Garret by Honeywell.
9) Containment policy, Cummins Turbo Technologies, HOL-01-06-02.

10) D C Montgomery (2009) Design and analysis of experiments. John Wiley & Sons.
11) Body Interactions in Explicit Dynamics Analyses, Setting Connections, ANSYS Mechanical User's Guide, ANSYS documentation, version 15, 2014.
12) G R Johnson and W H Cook (1983) A constitutive model and data for metals subjected to large strains, high strain rates and high temperatures, Proc. 7th Int. Symp. on Ballistics, 541-547, The Hague, Netherland.
13) G R Johnson and W H Cook (1985) Fracture characteristics of three metals subjected to various strains, strain rates, temperatures and pressures, Engineering Fracture Mechanics, 21, 31-48.
14) Y Bao and T Wierzbicki (2004) On fracture locus in the equivalent strain and stress triaxiality space, International Journal of Material Sciences, 46, 81-98.
15) X Teng and T Wierzbicki (2006) Evaluation of six fracture models in high velocity perforation, Engineering Fracture Mechanics, 73, 1653-1678.

An improved approach to HCF development for vaneless turbine stages

W Smith, C Wilkins, W Magalhaes, B Sirakov
Honeywell Turbo Technologies, USA
D Gouravaraju
Honeywell Technology Solutions, India
Z Neterda
Honeywell Technology Solutions, Czech Republic

ABSTRACT

Turbochargers have become a key element of modern engines with their increasing demands to improve performance and efficiency as well as to enable the compliance with emissions legislation. To support these trends in engine development turbocharger turbines have been required to meet ever increasing aerodynamic performance requirements, resulting in more elaborate designs, whilst also being subject to ever more demanding service conditions in terms of rotational speed, pressures and gas temperatures. All turbocharger turbine stages whether vaned or vaneless are subject to High Cycle Fatigue (HCF) driven by vibratory loading. In all cases the vibratory loading occurs when the harmonic of the frequency of the unsteady aerodynamic loading matches one of the blade natural frequencies. This unsteady aerodynamic loading is due to the interaction between the stationary and rotating parts in the turbine stage. In the case of a vane less turbine the stationary part of the stage contributing to the interaction is the scroll or volute specifically the tongue region where it comes closest to the passing blades creating non uniformity. When this unsteady aerodynamic loading reaches a sufficient level, and the blade is in resonance, it can cause excessive mechanical loading resulting in the formation of cracks in the blades and ultimately failure. Since HCF can be very challenging to detect during the development of new turbine stages this type of failure can potentially occur late in the engine development cycle or even in the field.

This paper highlights the steps that have been taken to improve the design, analysis, measurement and verification of HCF in vane less turbine stages, the intent being to improve both the HCF robustness and maintain performance. In making these improvements the impact on the aerodynamic performance of the turbine stages was also strongly considered to ensure that improvements made for HCF were not made at the expense of performance. The investigation of these new design options required developments in analytical capabilities, specifically 3D unsteady CFD, in order to make the necessary iterations with adequate fidelity in a practical timeframe. To verify the analysis techniques comprehensive experimental measurement of blade strain levels across multiple designs was also conducted and the results compared with analysis. Finally endurance testing was conducted to confirm that the measured improvements in HCF resulted in longer (and sufficient) service life of the turbine stage. Estimation of the required service life was accomplished using the process presented by Kulkarni et al (IMechE 2010).

Based on the results of this work it has been possible to propose new design and analysis practices that ensure future turbine stage designs are robust to HCF without any impediment to the continual improvement of aerodynamic efficiency. These practices have been validated through experimental and endurance testing of real world parts and will facilitate the design of future generations of turbine stage designs with state of the art performance that are able to withstand the increasingly severe service conditions required by modern engines.

NOMENCLATURE

HCF	High Cycle Fatigue
CFD	Computational Fluid Dynamics
A/R	Volute **A**rea divided by centroid **R**adius
(X)EO, M(Y)	X harmonic of Y vibratory mode shape

1. INTRODUCTION / BACKGROUND

Turbine HCF is one of the key areas of challenge for the turbomachinery industry in general and more specifically to the turbocharger business. Turbochargers in particular have some unique aspects with respect to HCF, most notably the wide range of different operating conditions that they can experience over an equally variable range of operator duty cycles. Unlike aerospace type applications which typically have a somewhat consistent mission profile, and one that is understood in advance of the design process, turbochargers can be used in a variety of very different ways depending on the end application as well as the particular region of operation and the operator. One other somewhat unique aspect of turbocharger turbines with respect to HCF is the use of vane less stators / housings compared with the typical fixed and variable vaned stator arrangements more commonly seen in aerospace, industrial and other turbomachinery application types. This combination of the variable usage cycle and the vane less stator arrangement led to the study that was undertaken and is reported in this paper.

For vane less turbine arrangements the most typical HCF failures observed are for the 1st blade mode (exducer flapping mode). 2nd mode (inducer flapping) has also been observed on occasion but since it is less common it won't be discussed in this work. Figure 1 shows a typical 1st Mode blade failure where a piece of the blade exducer is visibly missing. This failure is compared to the modal displacement and modal strain plots for the 1st Mode and a good agreement is found confirming this as the most likely failure mode. Since we are discussing vane less turbine stages the order of excitation for this failure is not related to a nozzle vane count order but rather a harmonic of the 1st order and is related to the single tongue in the turbine housing (see fig 2 showing a typical Campbell plot for a vane less radial turbine stage). Industry practice is to design the 1st mode frequency of the blade such that the 4th harmonic of the 1st mode (4EO) is avoided within the expected operating range of the turbine stage (i.e. within the max speed and max operating temperature). Since 4EO M1 is avoided the lowest order within the operating range becomes 5EO M1 and this is typically responsible for most of the observed HCF failures from vane less turbine stages. However failures have also been observed at lower speeds corresponding to 6EO M1. In cases where blades are designed to avoid 5EO M1 then 6EO M1 would then occur at higher turbo speeds and pressures and would consequently be expected to have a similar risk of failure as 5EO M1, this leads us to the conclusion that simple mode avoidance approaches, whether they be 4th or 5th order avoidance, are no longer sufficient to avoid potential HCF issues.

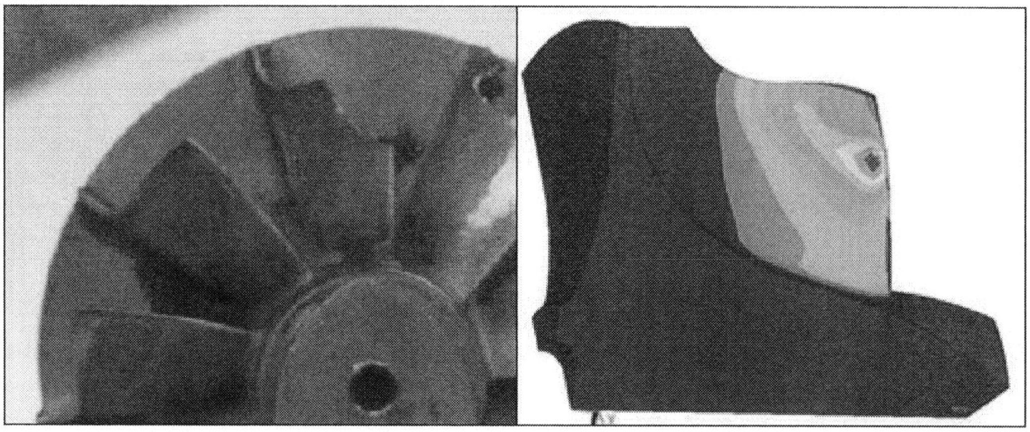

Figure 1 – Typical HCF mode 1 failure compared to modal stress analysis

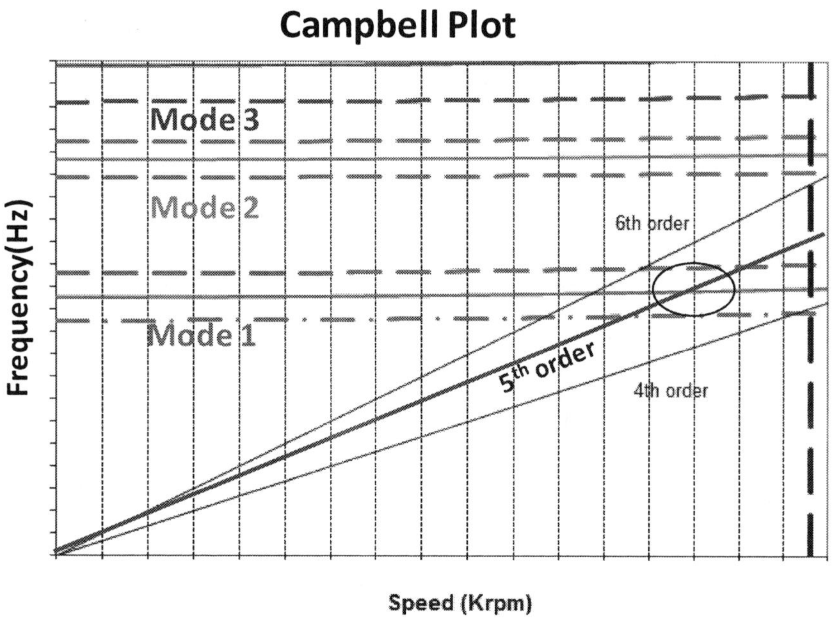

Figure 2 – Campbell plot showing critical interactions

As stated above the mechanism for the HCF failure is the interaction between the turbine blades and the tongue of the vane less turbine housing which creates a cyclic aerodynamic force that acts on the blades as they rotate. This interaction is shown in the below figure 3 and will be discussed in more detail in section 2.

When the frequency (or harmonic of the frequency) of this forcing function matches the modal frequency of the blade then HCF loading occurs. The structure of the turbine blade will then experience a mechanical loading, depending on the mechanical design of the blade (shape, thickness and thickness distribution) this loading will then result in a stress distribution. If the localized stress level and number of cycles are great enough a crack will initiate ultimately leading to blade failure.

Figure 3 – Mach number close to turbine housing tongue

1.1 Approaches to deal with HCF

There are essentially 3 ways to address HCF in the design of a vane less turbine stage:

1. Avoid the interaction - increase blade modal frequency or reduce rotational speed limit in order to avoid the critical Mode / order
2. Improve the mechanical design of the blade as a structure to reduce the local stress during HCF loading dropping it below the threshold for failure in the expected exposure to cycles
3. Reduce the strength of the alternating aerodynamic loading

All three of the above mechanisms were investigated to understand the potential benefits and drawbacks.

For the first approach the reduction of speed limit is an obvious and effective way to prevent an HCF issue. However looking at a typical Campbell plot (figure 2) it can be seen that the level of speed limit reduction required would not be practical for most applications resulting in an unacceptable reduction in performance potential for the engine or a complete lack of altitude capability. This leaves the possibility to increase the blade frequencies during the design process or perform some post treatment to the blades by machining (so called clipping or trim reduction). While this increase in blade frequency is possible it would bring significant compromises to the design with respect to both performance and centrifugal stress as the blades thickness would be increased. Initial experimental investigation also indicates that the magnitude of the alternating strain on turbine blades is typically more a function of aerodynamic boundary conditions (see figure 4 showing measured strain level with fixed expansion ratio and varied backpressure for different for different housing designs). It follows that if we remove 5EO M1 from the speed range we typically replace it with a 6EO M1 at similar boundary conditions and thus little change in strain amplitude.

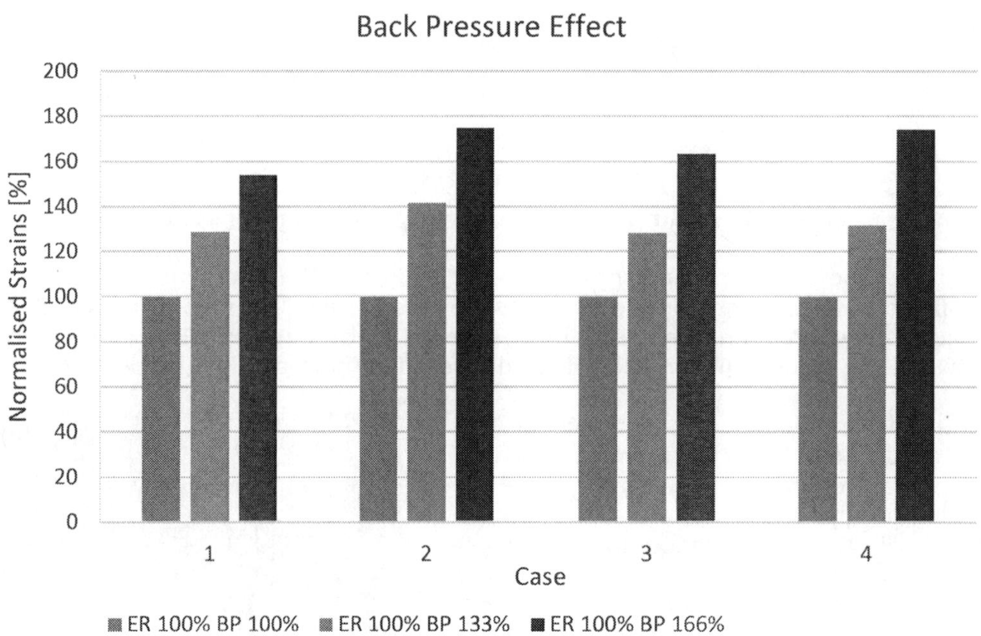

Figure 4 – Strain impact with back pressure (BP) change for constant expansion ratio (ER)

For the second approach several studies have been undertaken that won't be discussed in full detail here. In this case there is also a requirement to redesign the blades, typically to be somewhat thicker but also to re-optimize the thickness distribution to reduce any stress concentrations. Figure 5 shows an example of this approach and its impact on the 1st Mode stress distribution in the blade. This type of approach has also been studied experimentally and found, in certain cases, to be quite effective. This improvement is shown on the Goodman diagrams in figure 6 comparing design A and design B of a given wheel. However as with the blade frequency increase approach this method tends to result in thicker blades which can often result in a performance impact (see figure 7) and in other cases there may also be an unacceptable increase in the centrifugal stresses resulting from the thicker blades.

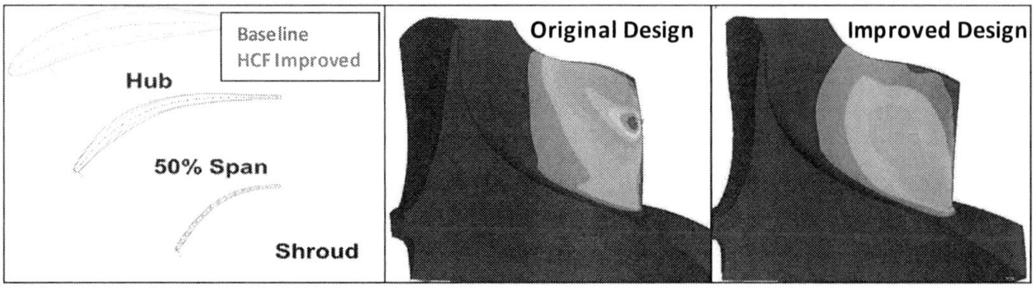

Figure 5 – Blade thickness modification and impact on stress distribution

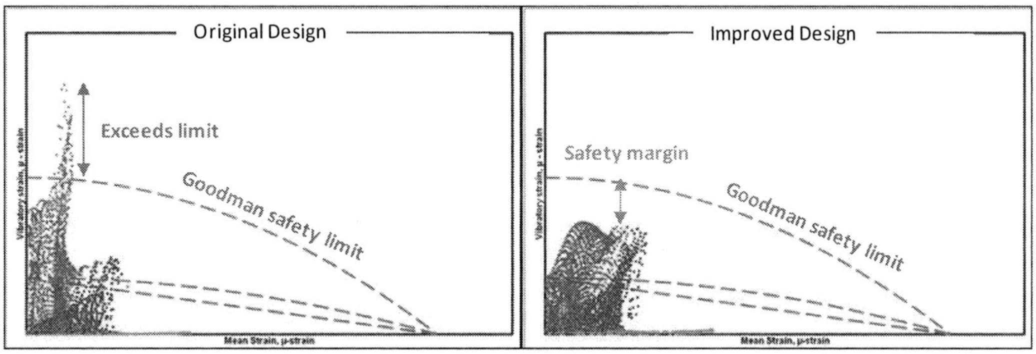

Figure 6 – Goodman diagrams for baseline and improved turbine stage

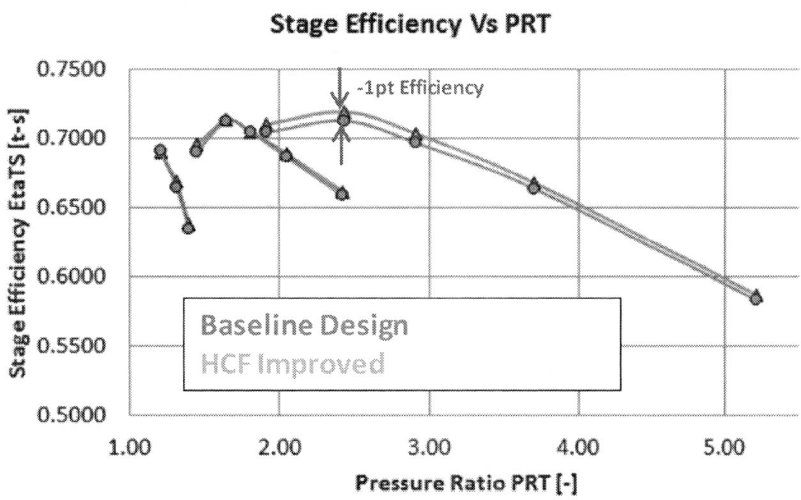

Figure 7 – Performance impact of mechanically redesigned turbine stage

The third approach is the one that is the main focus of this work. In this case the aerodynamic shape of the turbine blades is largely set by the performance requirements of the turbine stage so the main area of focus is on the turbine housing design itself which to a large extent determines the aerodynamic loading.

2. PLAN OF WORK

In order to proceed further on this final topic a plan of work was put together as follows:

1. Identify a suitable turbine stage to use for analytical and experimental evaluation
2. Perform baseline HCF analysis on the stage to understand the detailed flow mechanisms
3. Identify the key design parameters and create a family of volutes to study them
4. Perform HCF and performance analysis on the range of volutes
5. Down select key volutes to be manufactured and conduct experimental evaluation of performance and HCF
6. Compare experimental results to analytical
7. Confirm the best configuration in terms of HCF / performance tradeoff
8. Re evaluate performance / HCF on best design and confirm actual life increase
9. Validate overall findings on an additional turbine stage size/design

3. INITIAL ANALYSIS AND UNDERSTANDING OF AERODYNAMIC LOADING

As discussed in the previous section it was identified that the design of the turbine housing and specifically the area around the tongue is very critical to the alternating aerodynamic loading that contributes to HCF. In order to progress further it was necessary to get a more detailed understanding of the flow mechanisms occurring in the volute/rotor.

In this section we will first introduce the turbine stage that was used as the basis of this study. A 39mm, 11 bladed radial turbine stage was selected, this is a modern design considered to have good aerodynamic performance. For the purposes of this study the turbine housing of this stage was simplified to remove elements not directly under study, for example the waste gate passage was removed and the outlet was simplified and made axis-symmetric. While both the outlet and the waste gate can, if designed improperly, impact HCF they were not the focus of this study and so were removed to isolate the volute effects. The turbine design is shown below in figure 8.

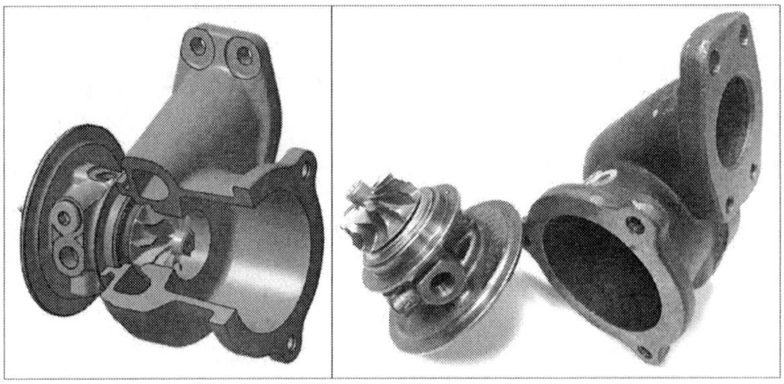

Figure 8 – Turbine stage model and hardware

3.1 Analysis process overview

Two types of analysis were performed in order to assess the various turbine volute configurations:

1. Traditional 'steady' CFD for performance assessment using commercial CFD tools
2. Full unsteady CFD analysis linked to structural analysis for HCF assessment using a proprietary Honeywell process

In the case of 1. The analysis was performed using a commercial CFD code on a full stage (full volute plus single blade passage) using a mixing plane approach. This analysis and its results won't be further discussed in this paper in the interests of space and are superseded by the experimental performance results which are published in section 6.

For case 2. Above the overall process is laid out as follows:

1. First blade vibration characteristics are analyzed using modal analysis. Modal Analysis → Blade natural frequencies/mode shapes
2. Create a Campbell diagram to identify the critical resonance speeds/ conditions for excitation orders (EO) and modes
3. Forcing function i.e. unsteady pressure forces acting on the blade surface is calculated at these critical conditions using 3D unsteady simulations
4. Unsteady pressure forces on the blade are then decomposed to each EO by performing Fourier series decomposition
5. Harmonic analysis is performed specifying the decomposed pressures as an external force of excitation. One way coupling is used between fluid and structure by specifying the total damping ratio. Finally resulting blade displacements are converted into strains

3.2 Details of 3D unsteady CFD simulations

The turbine stage is modeled with single scroll and full annulus wheel. The housing is meshed with 2.65 million hexahedral cells, whereas the single passage wheel consists of 1.25 million cells, example meshes are shown in figure 9 (the wheel has 11 blades). In this case the 5EO M1 response is analyzed as this is the lowest harmonic of Mode 1 present within the operating range of the turbo and has the highest response.

It is important to note that for a vane less turbine stage it is necessary to model the full volute and to perform full unsteady analysis. While this approach is computationally demanding it is not possible to further simplify without missing key elements of the mechanism.

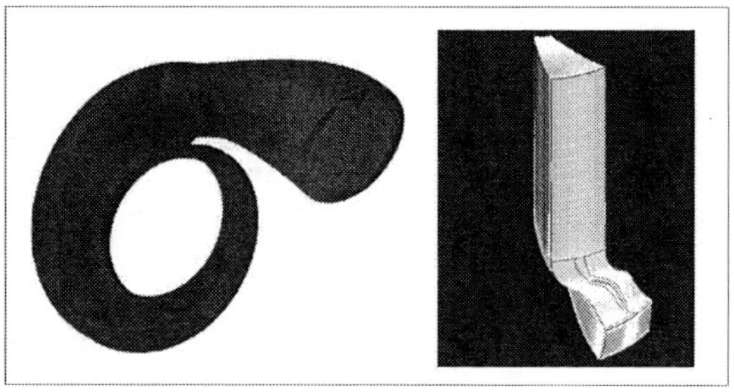

Figure 9 – Examples of volute and wheel meshes

Initially steady state simulations are performed as follows:

- A mixing plane approach is used to model the housing-rotor interface.
- Steady state results are used for initialization of the unsteady simulations to reduce convergence time.
- For the transient housing-rotor interface a sliding plane approach is used.
- Wheel rotational speed corresponding to 5EO M1 interaction from Campbell diagram is specified.

- Constant Total pressure and Temperatures are specified at the inlet and constant back pressure is specified at the wheel exit (values are based on expected in service conditions at the turbo speed corresponding to 5EO M1 interaction under evaluation).
- Spalart-Allmaras turbulence model is used.
- 500 time steps are used for one rotor revolution and simulation is run for a total 12 rotor revolutions to ensure the convergence of higher harmonics of the pressure field.

3.3 Understanding of flow mechanisms

The first objective of the analytical study was to better understand the flow mechanics that lead to the unsteady loading on the turbine blades. For this purpose the baseline turbine volute case was used and a detailed review of the pressure and Mach number of the flow approaching, and in, the blade channels was undertaken.

It can be observed that as the blades of the turbine rotor approach the tongue of the housing during their rotation there is an interaction that results in an acceleration of the flow (increasing Mach No) resulting in a drop in static pressure. This interaction (drop in static pressure) results in strong local flow accelerations and pressure gradients and a resulting significant decrease in the mass flow rate through the passing blade channel. Then as the blade continues to rotate and passes the closest point to the trailing edge of the tongue it passes through the tongue wake and then experiences a rapid increase in pressure and the full flow rate returns to the blade channel. This process is detailed in figures 10, 11 and 12 below.

There are several key conclusions that can be made at this stage:

1. The tongue in the turbine volute creates a very significant non uniformity in the flow entering the turbine rotor
2. There is a very strong interaction between the rotating blade and the volute with the flow field changing substantially as the blade rotates indicating that both rotor and volute need to be modeled together in an unsteady manner to accurately model the physics
3. The important interaction between the rotor blades and the volute begins well before the blade reaches the closest point (the tip) of the tongue

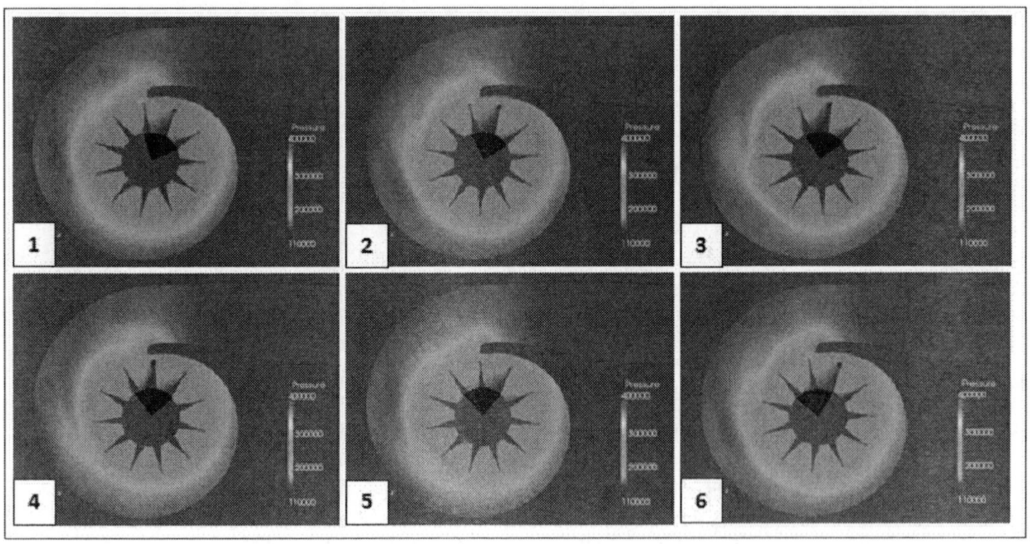

Figure 10 – Static pressure in volute/rotor at different time steps

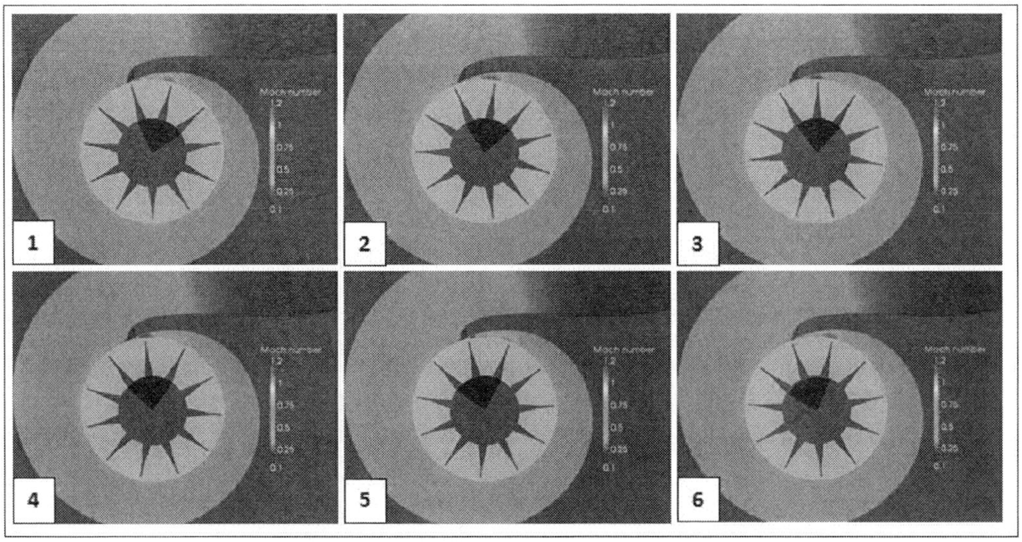

Figure 11 – Mach number in volute at different time steps

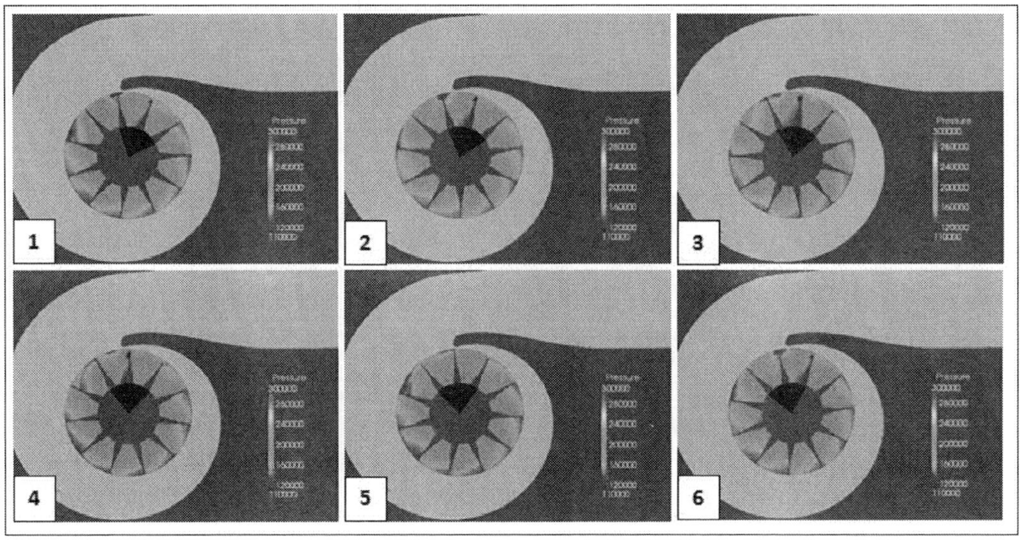

Figure 12 – Pressure in rotor at different time steps

4. VANELESS VOLUTE DESIGN PARAMETERS

In this section some of the detailed design features of the turbine housing are discussed. The goal was to identify these key features and then to create a range of designs that could be evaluated both analytically and experimentally with the ultimate goal being to understand the influence of each parameter and their interactions on both HCF loading and performance of the stage.

Based on the analytical results from the baseline volute case, and the understanding generated, a series of key geometric parameters for the turbine volute and tongue were identified. These were the parameters that were judged to be the most potentially impactful to the unsteady aerodynamic loading.

7 key parameters were identified and the majority are outlined in figure 13.

1. Turbine housing A/R
2. Gap or ratio between wheel inducer OD and closest point of tongue
3. Evolution of volute cross sectional area as a function of angle around volute
4. Angle of trailing edge of tongue from radial
5. Tongue thickness
6. Volute cross sectional aspect ratio
7. Post processing of tongue to 'machine back' along its centerline (potential remedial action for casting)

Clearly these parameters are somewhat interconnected and could also be defined in a different manner but for practical reasons and to align with standard design features they will be defined in this manner for the purposes of this study.

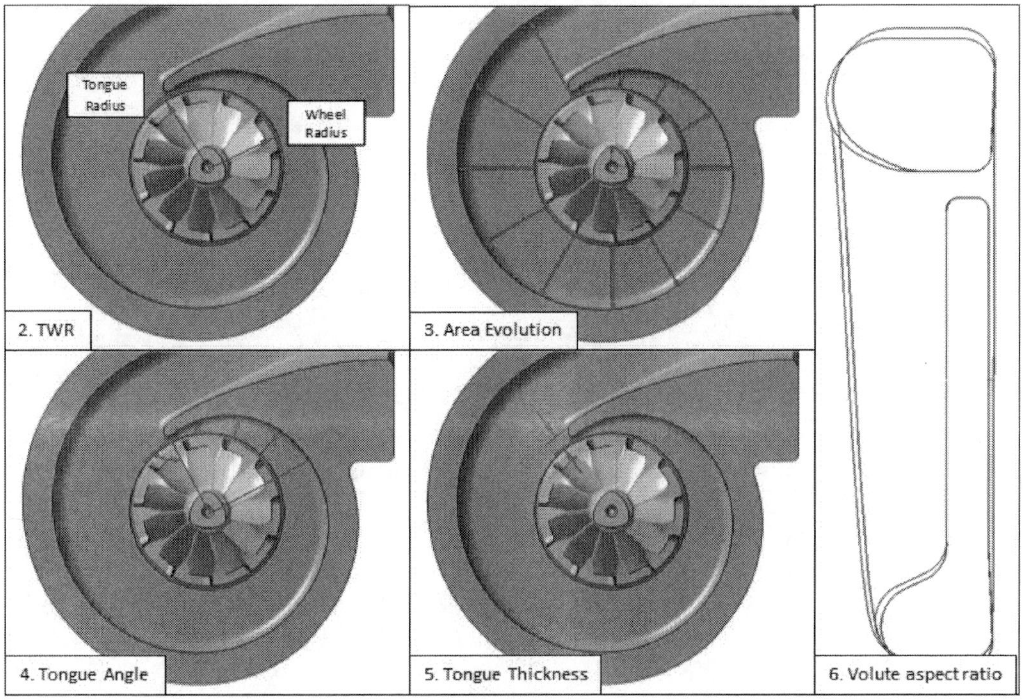

Figure 13

Based on the above parameters a series of 22 different turbine gas passages were designed that were intended to be used for analysis and testing. The list of different gas passage configurations and their associated non dimensionalised attributes is shown in figure 14 below.

Configuration	Parameter						
	1	2	3	4	5	6	7
1	M	2	1	3	1	N	N
2	M	3	1	3	1	N	N
3	M	4	1	3	1	N	N
4	M	6	1	3	1	N	N
5	M	3	1	3	1	N	Y
6	M	4	1	3	1	N	Y
7	M	3	1	3	0.6	N	N
8	M	3	1	4	1	N	N
9	M	3	1	1	1	N	N
10	M	3	2	3	1	N	N
11	M	3	1	3	1	Y	N
12	H	3	1	3	1	N	N
13	L	3	1	3	1	N	N
14	M	2	1	2	1	N	N
15	M	3	1	2	1	N	N
16	M	4	1	2	1	N	N
17	M	6	1	2	1	N	N
18	M	7	1	2	1	N	N
19	M	5	1	2	1	N	N
20	M	5	1	3	1	N	N
21	M	5	1	2	0.6	N	N
22	M	1	2	3	1	N	N

Figure 14 – L = Low, M = Mid, H = High, Y = Included, N = Not Applicable, 1 2 3 etc = variations

5. ANALYTICAL STUDY OF VOLUTE DESIGN PARAMETERS

In a similar manner to the baseline case the various different volutes listed in Figure 14 were analyzed and the resulting maximum wheel strain values and detailed flow field were reviewed in detail. It should be noted that all of these volutes were designed to match the flow capacity and efficiency characteristic of the baseline volute. While the performance CFD conducted won't be discussed in detail here this performance consistency was confirmed and the experimental performance test results will be covered in section 6.

Below in figure 15 is a summary of the output HCF strain amplitude for the different analysis cases run. In this plot the cases are shown in descending order of strain amplitude and it can be seen that the normalized amplitude varied from the max of 100% down to the minimum value of 28% indicating that the best case configuration showed reduction in strain amplitude of 72% which is very significant and greater than the improvement noted in section 1.1 related to the rotor mechanical optimization approach.

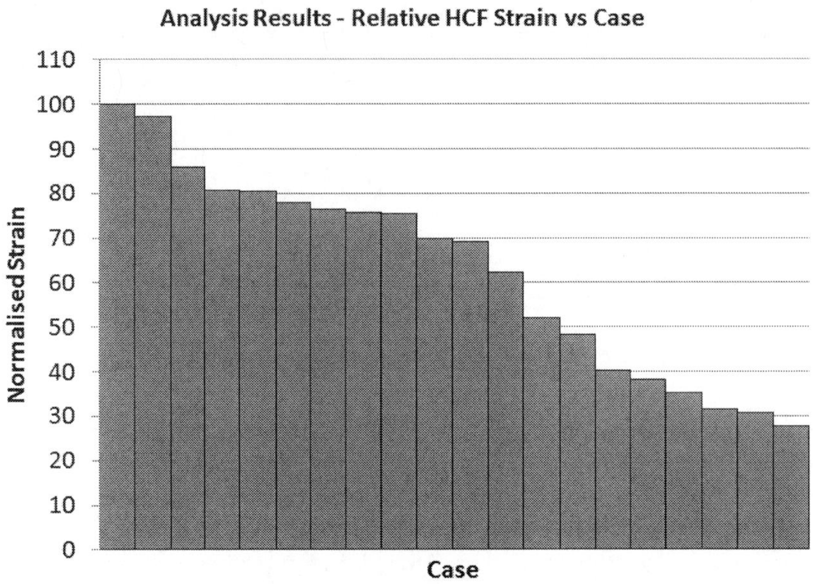

Figure 15

The cases above can then be broken down into subsets in order to understand the effect from the different volute design parameters identified in section 4. The plots for the respective volute design parameters are shown in figure 16 below. From the 7 parameters that were identified it can be noted that two of them demonstrate a very clear trend with significant reductions in strain amplitude. Two parameters does not show any clear trend with respect to strain amplitude and then the other parameters show an impact but weaker than the first two.

A summary of the impacts would be as follows:

Parameter A → no trend
Parameter B → strong trend, larger value results in lower strain
Parameter C → limited trend, keeping larger reduces strain amplitude
Parameter D → limited trend, smaller value reduces strain amplitude
Parameter E → limited trend, smaller results in lower strain amplitude
Parameter F → very little or no trend visible
Parameter G → strong trend, larger value reduces the strain amplitude

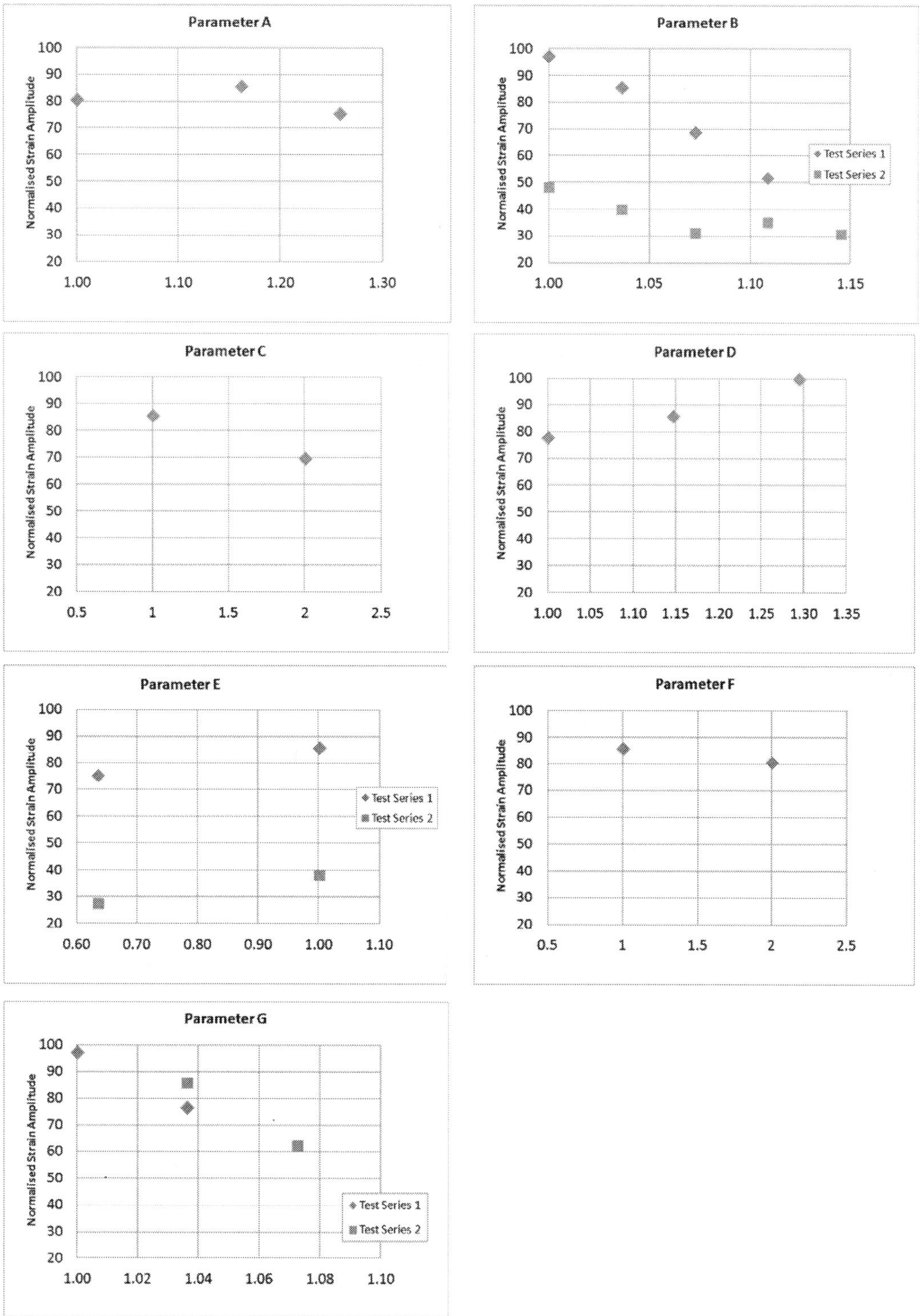

Figure 16

6. EXPERIMENTAL TESTING OF VOLUTE DESIGN PARAMETERS

In order to verify the analytical HCF results a number of the turbine volutes were manufactured and tested, two types of testing were conducted:

1. Performance testing on a hot gas stand to determine impact of parameters on mass flow rate and efficiency
2. Experimental evaluation of impact on blade strains using non contact strain measurement

The full details of the performance test process is well known in the industry and need not be further detailed here. Equally the comparison between analytical and experimental data for performance is also beyond the scope of this paper therefore in this case experimental performance data only will be reported, it can be stated though that there was generally good agreement between the expected and measured performance levels of the various configurations.

Regarding the HCF part of the testing a little more detail is provided although this process has also been widely discussed and detailed in the past [references].

The blade vibration response is measured by widely used approach in the aero industry called Blade Tip Timing method. This system can be used to measure the blade deflection amplitudes during resonance. Dynamic stresses generated in the blades can be determined from the deflections and subsequently blade durability can be predicted under resonance. The system consists of a set of case mounted probes that accurately determine the arrival time of a blade tip at a particular case location. The arrival times of each blade are converted to deflections by knowing the rotational velocity and the radius of the measurements. Multiple Probes at a common axial location are used to determine the amplitude and phase of a blade vibration. Once the amplitude, frequency and nodal diameter are determined, the dynamic blade deflection can be converted to stress from mode shape information obtained by finite element analysis.

Some pictures of the Blade Tip Timing test setup are shown in figure 17 below. In this case the testing was conducted uniquely on the hot gas stand (similar to the stand used for performance evaluation) however testing on engine has also been conducted on similar turbine stages in the past. This engine vs gas stand testing will not be discussed in detail in this work but the comparative results to date confirm that the gas stand is a relevant test method.

It should also be noted that this testing was particularly demanding due to the small size of the tested turbine (39mm wheel diameter) in combination with the high speeds and temperatures used. Success in testing small turbines under these conditions is the result of many refinements to the process and also requires extremely careful preparation and measurement of the parts used for testing.

Figure 17 – Blade tip timing setup on gas stand

A comprehensive test study was completed to investigate the effect of different turbine housing parameters on the alternating strains. Two wheels of the same design were tested to provide consistent results in term of the strain behavior for given design and the same test conditions. In fact while the magnitude of the strain amplitude varied between the two wheels the trend of results between the two was identical confirming that improvements made to the volute in reducing strain are likely to be transferrable between different rotors.

A summary of the results themselves (averaged across the 2 wheels tested and normalized to the baseline) are shown in the table below (fig. 18). Comparing back to figure 15 from the analysis it can be noted that similar reductions in HCF strain amplitude were achieved in the tested hardware as was expected from the analytical results. In the case of the experimental results in fact even bigger reductions in HCF strain were achieved with reductions of around 85% being recorded compared with the baseline housing.

Looking at these overall results we can also investigate the tradeoff between strain reduction and stage aerodynamic efficiency. Plotted in figure 19 are the efficiency results, et PR 2.20) added to the strain results for each of the tested cases. From this figure we can clearly see that there is little variation in efficiency for the different cases. To highlight this further the full gas stand efficiency results for the highest and lowest strain cases are plotted in figure 20. Here it can clearly be seen, once again, that there is no visible detriment in aerodynamic efficiency for turbine housings that give very significant reductions in HCF strain amplitude. Indeed in this case the performance appears to have actually improved somewhat. While the increase in performance may be partially due to normal part to part and testing variation it is also likely due to the reduction in the undesirable non uniformity in flow and unsteady effects around the tongue. So from this we can conclude that it is not necessary to make any sacrifice in performance to improve HCF but equally performance degradation can occur if the correct aerodynamic features are not well preserved.

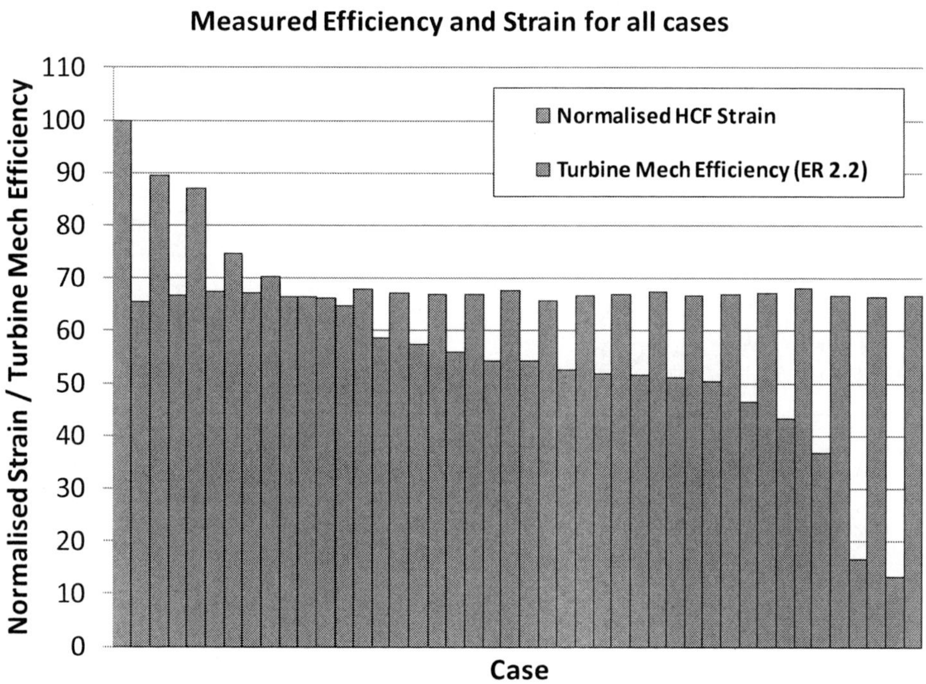

Figure 18 – Measured strain and efficiency at expansion ratio = 2.2

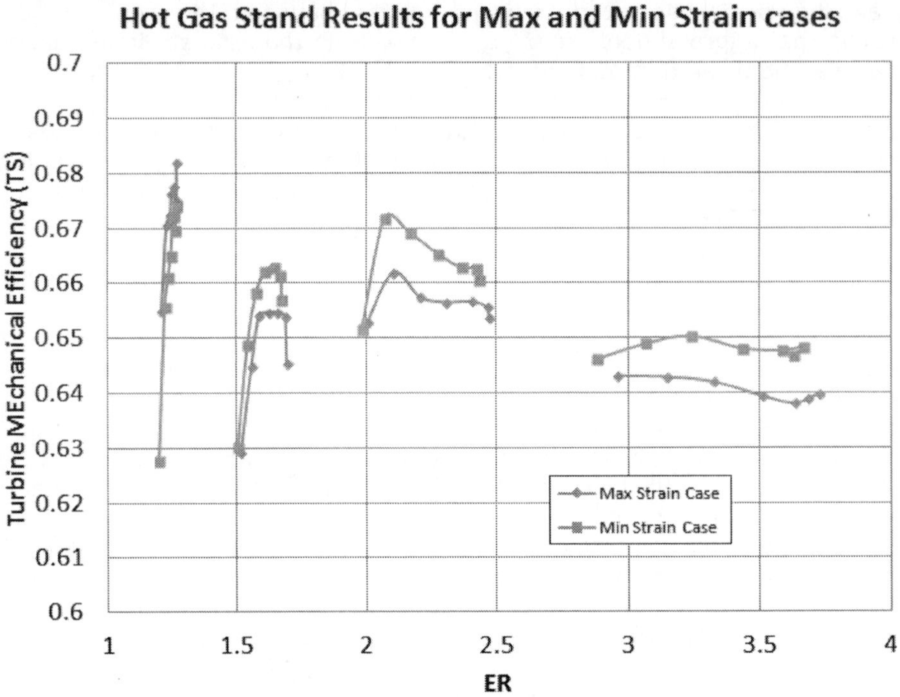

Figure 19

Now taking a closer inspection of the specific HCF measurements they can be compared to the analytical results for the first series of cases. The relative strain amplitude results are plotted below in fig 20 comparing the predicted vs measured strain for one of the parameters tested. While many of the cases show good agreement between analysis and test (particularly the extreme high and low cases) there are also a number that do not.

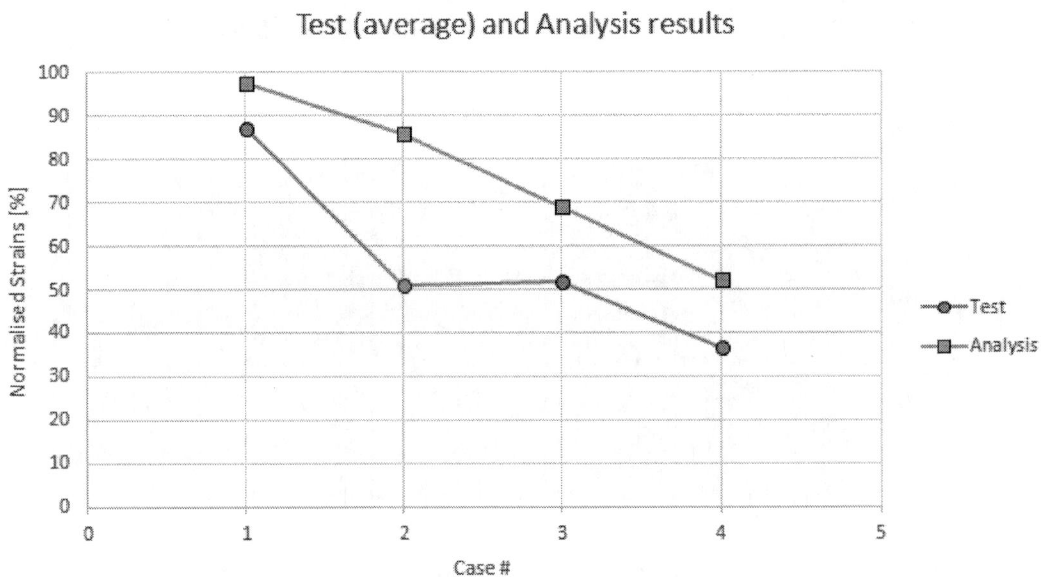

Figure 20

The lack of ideal correlation between the analytical and experimental results could potentially be attributed to multiple different factors:

1. Error or insufficient sophistication of analytical model
2. Uncertainty in strain measurement method
3. Differences between analyzed (nominal) geometry and geometry of physical test parts

It is beyond the scope of this paper to discuss this at length but due to the relatively small size of the test parts a combination of factors 2 and 3 are believed to be the dominant root causes.

7. OTHER CONSIDERATIONS

In addition to the factors discussed in detail in this work there are a number of other aspects of turbine HCF that are important to understand well. These are factors that primarily result from how the parts are manufactured. While these won't be discussed at length they are important to acknowledge.

The turbine volute which is the focus of this work is formed by a sand casting process and is subject to geometry and surface finish variation that can have a significant impact on the resulting unsteady aerodynamics of the stage. It is therefore important to understand the sensitivity of the design to these variations and implement good controls as well as ensure that the nominal design is set in a way that is robust to the expected level of variation.

The turbine wheel which is typically manufactured via an investment casting process is also subject to geometry variations that although smaller in magnitude than the housing can be just as critical. It is therefore important to have a good understanding and control of the variation through good production controls, measurement approaches, 3D scanning and reverse engineering of real geometry. In addition to the geometry resulting from the manufacturing process there is also variability in the material itself, particularly the grain structure which can vary from part to part and also between different foundry parameters and processes.

The above mentioned geometry and microstructural variation (in combination with design features of the rotor and overall stage) can result in varying levels of mistuning (blade to blade frequency variation) which can result in different levels of strain amplification for an otherwise identical turbine stage. A typical example of blade mistuning can be seen below in figure 21 which shows the typical output from a blade tip timing test showing the different responses of the blades of a single rotor which vary in both frequency and amplitude.

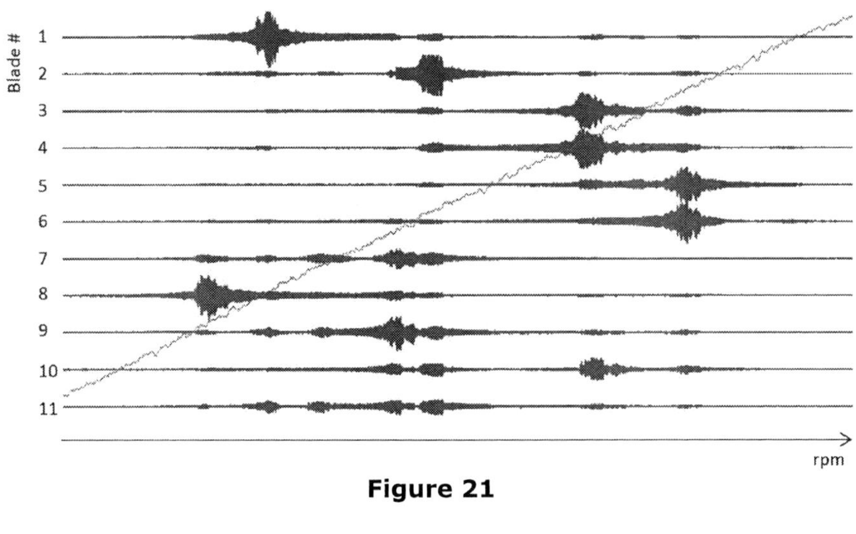

Figure 21

Based on these considerations that result in additional variation in the final HCF strain level for real world parts it highlights further the importance of achieving the lowest possible level of strain at the design stage.

8. OVERALL RESULTS SUMMARY

Summarizing the findings from both the HCF analysis and testing it can clearly be demonstrated that the turbine housing volute design has a very significant impact on HCF strain. By taking the understanding from the HCF analysis and applying it to the design principles of the turbine volute it is then possible to make very significant reductions to the HCF strain experienced by the turbine wheel (up to 85% reduction as shown in the previous section). This confirms that the turbine volute design is the most significant factor in the HCF performance of a vane less turbine stage.

By also taking into account the measured performance for the different cases a tradeoff curve between stage aerodynamic performance and HCF can also be established. When plotting measured aerodynamic efficiency vs HCF strain reduction (see figure 22 below) it can clearly be seen, once again, that the changes made to reduce HCF strain do not have any negative impact on performance. This is in contrast to the other methods for HCF improvement (mode avoidance, wheel mechanical upgrade) which do typically result in some level of performance degradation.

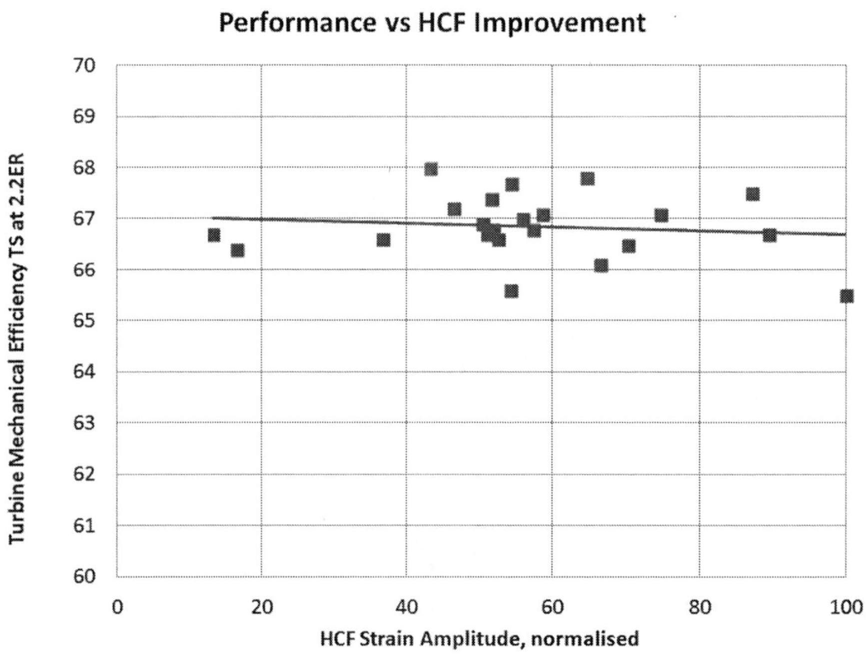

Figure 22

In order to confirm and cross check these results some of the best practices derived in terms of volute design for HCF were applied to a completely different (and larger) turbine stage that had been experiencing HCF failures during some extreme durability testing. This additional turbine stage comprised a 44mm radial turbine and a single scroll vane less turbine housing.

In this case the baseline turbine stage had the same basic volute design features as the baseline stage for the detailed study presented on the 39mm turbine wheel.

The turbine housing volute was then redesigned taking into account the impact of the different parameters highlighted in section 5.

Both baseline and improved turbine housing volutes were tested back to back over 3 different tests:

1. Performance testing on hot gas stand
2. Strain measurement using Blade Tip Timing
3. Extended resonance dwell testing to failure (killer testing)

The results of the back to back performance testing are shown in figure 23. It can clearly be seen from these results that the impact of the volute design changes on performance was negligible and the differences are within part to part geometry and measurement variations. The transparency of these performance results was also confirmed through engine testing which is not detailed here.

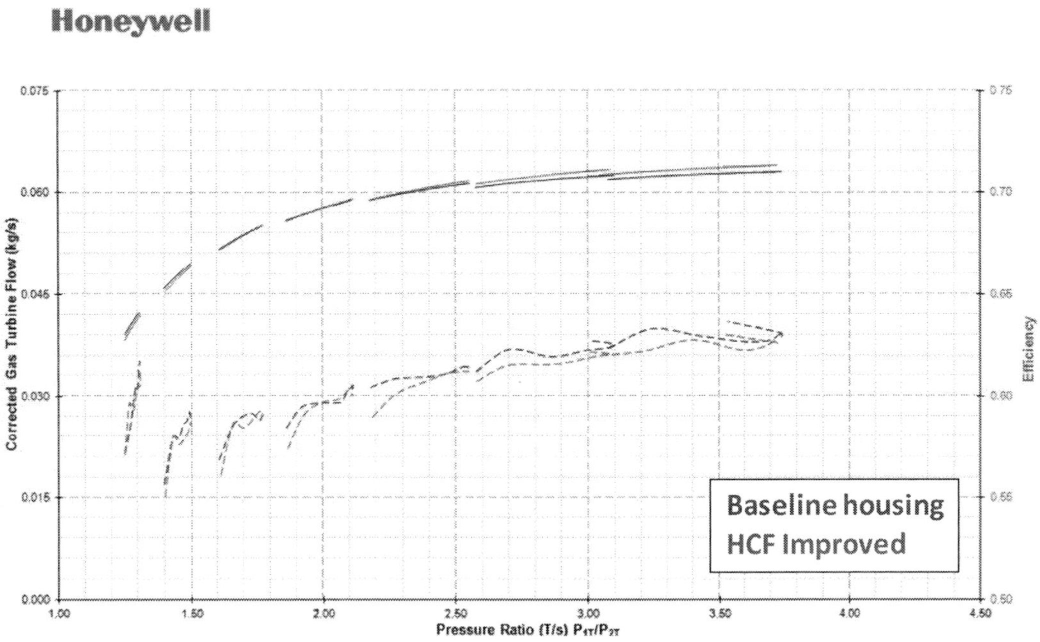

Figure 23

The results from the HCF measurement (figure 24) also confirm the findings from the main study indicating a significant reduction in the measured HCF strain. In this case the reduction was approximately 60%.

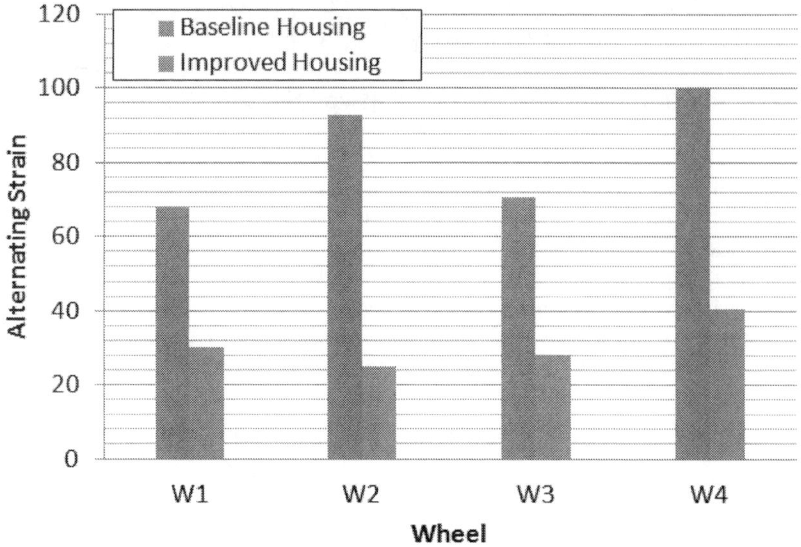

Figure 24

Finally the endurance testing was conducted in order to confirm that the reduced measured strain did indeed result in a longer life. For this test the highest responding blade is held in or close to resonance until failure occurs or large enough number of cycles to ensure adequate life is reached. The results of this testing are summarized in Table 1, where it can be seen that the baseline housing resulted in a cracked blade after 375,000 cycles whereas with the optimized housing the test was stopped after >200M cycles without failure (more than 600 X the cycle to failure of the baseline case). This result confirms that the changes made to the turbine housing were quite effective in not only demonstrating reduced strain measurement but that this also translated into very considerable life improvement for the turbine rotor.

Table 1

Housing	Normalised Strain	Cycles tested	Result
Baseline	1	375,000	Blade cracked
Improved	0.4	>200,000,000	No crack

9. WRAP-UP AND CONCLUSIONS

In summary we have identified several approaches in order to reduce HCF strain in vane less turbine stages. Due to the increasing demands placed on turbocharger turbines in terms of speeds, pressure and temperatures combined with the variability in real world parts it is important to bring HCF strain to the lowest possible level. While mode avoidance and wheel mechanical design upgrade can be effective in certain cases (especially in cases where wheels were not well designed) they bring significant drawbacks, especially in terms of performance. Conversely carefully addressing the turbine housing / volute offers a much larger opportunity for HCF reduction without any need to incur a performance penalty.

However the tools and capabilities required to address the volute design are considerable. Some of the key challenges being to both analyze and experimentally verify the HCF strain level of small high speed, high temperature vane less turbine stages. Honeywell has recognized this demand with significant investments in both analytical and testing capability and capacity to address these needs. In this work this point has been systematically proven through a range of testing and analysis and the knowledge gained is being broadly applied across Honeywell's full range of vane less turbine stages to prepare them for the future requirements of the industry.

REFERENCES

1. Ken NAKANO, Hiroaki HATTORI and Yutaka HIRATA; "Study on Blade Forced Vibration Response of Radial Inflow Turbine"; Proceedings of International Gas Turbine Congress, 2003 Tokya, IGTC2003Tokya TS-053
2. S T Kitson, D C Clay, D H Brown, R O Evans, D M Eastwood, P K Tootill; "Improving Analysis Capability in Order to Reduce Turbine HCF", 8th International Conference on TurboChargers and TurboCharging, 2006, Pg: 261-271
3. Avadhoot Kulkarni, Gerry LaRue; "Vibratory Response Characterization of a Radial Turbine Wheel for Automotive TurboCharger Applications"; Proceedings of ASME Turbo Expo 2008, GT2008-51355
4. Tom Heuer, Marc Gugau, Achim Klein and Paul Anschel; "An Analytical Approach to Support High Cycle Fatigue Validation for TurboCharger Turbine Stages; Proceedings of ASME Turbo Expo 2008, GT2008-50764
5. Tomoki Kawakubo; "Unsteady Rotor-Stator Interaction of a Radial-Inflow Turbine with Variable Nozzle Vanes; Proceedings of ASME Turbo Expo 2010; GT2010-23677
6. R.J. Miller, R.W. Moss, R.W. Ainsworth and N.W. Harvey, "Wake, shock and Potential Field Interactions in a 1.5 Stage Turbine – Part-I : Vane-Rotor and Rotor – Vane Interaction; Journal of TurboMachinery, ASME, Vol. 125, 2003
7. S T Kitson, D C Clay, D H Brown, R O Evans, D M Eastwood and P K Toothill "Improving Analysis Capability in order to Reduce Turbine HCF" **CONFERENCE DETAILS TBC**

The development of a long route EGR turbocharger for commercial engine applications

M Burkinshaw, U Mukhtar, L Sullivan
Materials Engineering, Cummins Turbo Technologies, UK

ABSTRACT

The development of a turbocharger for long route exhaust gas recirculation (LR EGR) commercial engine applications has been conducted using a multifaceted approach to testing and analysis. The complexities and difficulties of generating a LR EGR turbocharger for commercial engine applications are discussed. Untreated and surface treated aluminium alloy impeller and compressor cover material concepts were characterised in terms of their durability and consistency of performance in LR EGR environments. The range of surface treatments evaluated included anodising, plating and polymeric-based coatings. Substrates were subject to particle erosion and condensate corrosion test methodologies, which were developed using data acquired from literature and on-application experience. Particle size and density were considered, whilst condensates of numerous pH values were employed during corrosion testing. Aerodynamic performance, fatigue strength and thermal capability of the untreated and surface treated aluminium alloys were compared and contrasted. Test samples were analysed using high resolution microscopy, chemical analysis and interferometry. The impeller and compressor cover material concepts with the most superior overall LR EGR performance were identified through the use of a comprehensive scoring matrix.

1. INTRODUCTION

Worldwide emissions legislation is becoming more stringent with respect to the composition of the exhaust gas produced by the internal combustion engine (ICE) (1; 2). Whilst diesel engines provide low fuel consumption and relatively low emissions of carbon dioxide (3), such engines intrinsically produce nitric oxides (NO_x) because of the high oxygen content and peak temperature within the combustion cylinder (3).

Exhaust gas recirculation (EGR) technologies decrease NO_x emissions produced by an ICE as a consequence of reducing the temperature within the combustion cylinder (4; 5). Conventionally, short route (SR) EGR is utilised within commercial engine applications, but interest in long route (LR) EGR is rapidly growing. This is as a result of a number of key benefits provided by LR EGR transport; most notably, a reduction in NOx emissions due to a decreased charge temperature (5), a more homogeneous mixture of EGR gas and intake air (6) and a decrease in pumping loss (5).

However, there are two main disadvantages of LR EGR with respect to the turbocharger. Firstly, the turbocharger impeller is subjected to increased levels of wear through impingement by solid particles contained within the EGR gas (5). Secondly, the compressor stage of the turbomachine experiences an increase in corrosion as a result of an acidic environment generated by condensate formation (5). Therefore, in order to grow market share with regards to LR EGR within the commercial engine sector, it is vital that turbocharger manufacturers provide solutions which mitigate the detrimental erosion and corrosion environments found within the compressor stage of a LR EGR engine system.

The intention of this article is to present the methods and cutting-edge technology by which a LR EGR turbocharger has been developed for use within a commercial engine application. A wide range of surface engineering treatments have been applied to the turbocharger compressor stage (impeller and compressor cover) in order to assess their impact on both wear and corrosion resistance within LR EGR engine systems. The limitations of the surface engineering treatments have been determined in terms of tribological, corrosion, aerodynamic and fatigue performance using a comprehensive suite of test and analysis methods. The most appropriate solutions for the compressor stage of a commercial engine LR EGR turbocharger have been determined.

2. EXPERIMENTAL

2.1. The turbocharger compressor stage

The turbocharger compressor stage (Figure 1) comprises the impeller (Figure 2) and the compressor cover.

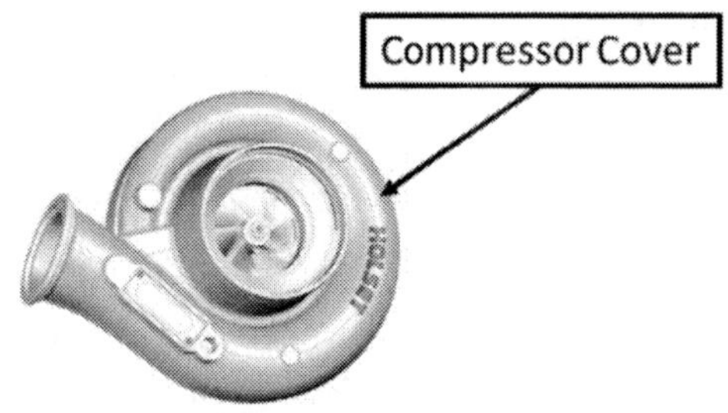

Figure 1: Turbocharger compressor stage

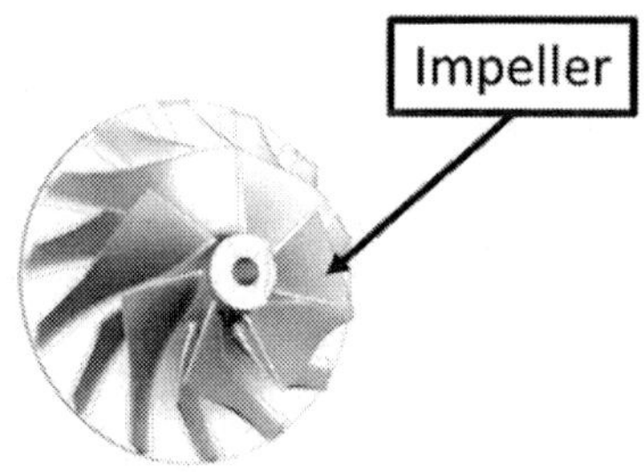

Figure 2: Turbocharger impeller

2.2. Materials

All turbocharger test materials were provided by Cummins Turbo Technologies and are listed in Table 1.

Table 1: Test materials

Component	Base Material	Production Method
Impeller	Aluminium Alloy A	Machined from solid (MFS). Heat treated and aged.
Compressor Cover	Aluminium Alloy B	Cast. Heat treated.
	Aluminium Alloy C	

Numerous surface engineering concepts were identified for use in conjunction with the impeller and compressor cover, respectively; these are listed in Table 2. The base material onto which the surface treatments were applied are also stated.

Table 2: Surface engineering concepts for LR EGR turbochargers

Component	Base Material	Surface Engineering Treatment Concept	Visual Appearance	Average Surface Engineering Treatment Thickness (μm)
Impeller	Aluminium Alloy A	Anodise A	Green	12
		Plating A	Dark grey	13
		Spray A	Black	50
Compressor Cover	Aluminium Alloy B	Anodise B	Light grey	10
		Anodise C	Dark grey	20
		Sealant	Transparent	< 1
		Plating B	Dark grey	40
		Spray B	Blue	35

Test coupons were developed in order to replicate the microstructure, mechanical properties and topography of the deck of the impeller and compressor cover volute, respectively. The dimensions of the two types of test coupon are stated in Table 3.

Table 3: Test coupons

Test Coupon	Area being Replicated	Dimensions (W x D x H) (mm)
Impeller	Deck	16.5 x 5.5 x 55.0
Compressor Cover	Volute	15.0 x 12.0 x 30.0

The details of the two types of turbocharger utilised throughout this study can be found in Table 4.

Table 4: Test turbochargers

Turbocharger Nomenclature	Impeller Diameter (mm)	Turbine Diameter (mm)
HE400VG	85	76
HE500WG	108	86

2.3. Methods

Figure 3 shows the methodology employed in this study to evaluate the concepts stated in Table 1 and Table 2 in terms of LR EGR performance.

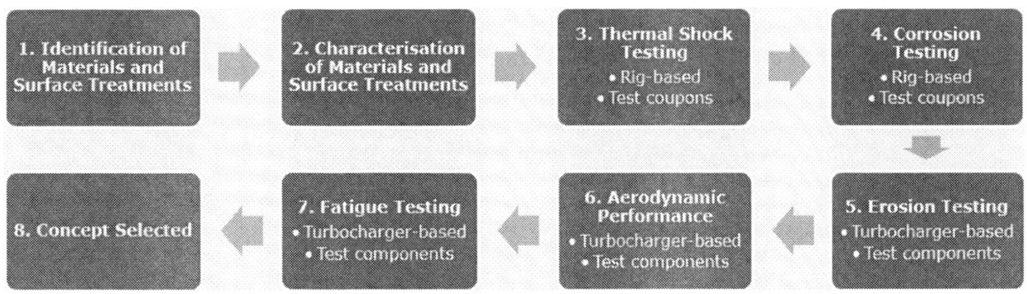

Figure 3: Approach to testing and analysis

2.3.1. Adhesion testing

Adhesion testing was conducted according to VDI 3198, using a Rockwell C type indentation method. Using test coupons, 3 off repeats per surface treatment were completed. The resultant damage to the substrate was ranked according to a scale documented in VDI 3198 (7); H1 indicated the most superior adhesion performance, whilst H6 highlighted the worst adhesion behaviour.

2.3.2. *Thermal shock*

Thermal shock testing was conducted by subjecting 3 off test coupons per surface treatment to a temperature of 180 °C for 45 minutes. Subsequently, the coupons were placed into a water bath at a temperature of 10 °C for 15 minutes. The process was repeated 15 times. All test coupons were submerged in acetone and ultrasonically cleaned for 10 minutes prior to each test.

2.3.3. *Corrosion testing*

Corrosion testing was conducted using a modified version of VDA 230-214, in which over a period of 5 days, 3 off test coupons per surface treatment type were cyclically subjected to the test plan detailed in Table 5. Prior to each test, the samples were ultrasonically cleaned in acetone for 10 minutes.

Table 5: Condensate cyclic corrosion test

Section of Test	Temperature	Time at Section
Vapour	50 °C	16 hours
Semi-immersion		6 hours
Drying	80 °C	2 hours

All test samples were subjected to corrosion testing using condensate A stated in Table 6. However, further corrosion testing was conducted using concepts which had demonstrated excellent performance in conjunction with condensate A; such concepts were subjected to condensates B-D also.

Table 6: Test condensate types

Condensate Nomenclature	pH	Acid Type
A	3.5	Organic
B	4.5	
C	2.5	Inorganic
D	1.2	

2.3.4. *Erosion testing*

Erosion testing of the impeller concepts was conducted using a Cummins Turbo Technologies HE400VG (Table 4) turbocharger which was installed in a turbocharger test cell and run at a constant speed of 110 krpm. Test media of known particle size, chemistry and density was inserted into the compressor stage whilst the turbocharger was operational. A test impeller was subjected to 25 g of test media E1 (Table 7) at a mass flow rate of 2.5 g / s to understand comparative wear performance. 3 off repeats per impeller concept were conducted.

Table 7: Erosion test media

Erosion Media Nomenclature	Media Type	Particle Size (μm)	Mass Flow Rate (g / s)
E1	Arizona Sand	20 – 200	1.2
E2		300 – 500	0.9
E3		20 – 50	1.0
E4		0 - 20	1.8

In order to understand the effect of particle size on wear behaviour, additional testing was conducted on the impeller concept which provided the most superior wear performance from media E1 testing. Such further testing was conducted using the media stated in Table 7. Mass flow rate was adjusted to consider the density of the different test media and is shown in Table 7. The total mass of erosion media used in each test was 25 g.

2.3.5. Aerodynamic performance

Aerodynamic performance of the impeller surface engineering concepts stated in Table 2 was determined using a Cummins Turbo Technologies HE400VG (Table 4) turbocharger which was installed into a state of the art turbocharger aerodynamic performance test cell. One turbocharger unit was used to obtain the aerodynamic performance data from all of the concepts, with 3 off repeats per surface treatment completed. The turbocharger was run at the 100 krpm, 110 krpm and 120 krpm speed lines across the documented sales map width of the impeller.

2.3.6. Fatigue testing

Low cycle fatigue (LCF) testing was conducted using the impeller surface treatment which had shown the greatest contextual overall performance after step 6 in Figure 3, when compared against the concepts stated in Table 2. Testing was completed using two Cummins Turbo Technologies HE500WG (Table 4) turbochargers, both of which contained surface treated 108 mm diameter impellers. The turbochargers operated in pairs and cyclically increased and decreased between two speed points over an extended period of time. A total of 8 surface treated impellers per concept were subjected to low cycle fatigue testing.

2.3.7. Evaluation of substrates

Surfaces of interest were examined using a Philips™ XL30 Scanning Electron Microscope (SEM) to obtain high magnification images, whereas low magnification images were obtained using a Leica® DM LM microscope. The thickness of the surface treatments were determined using the scanning electron microscope. A Bruker Contour GT-KA interferometer was used to obtain topographical information from substrates of interest. Such a device had a lateral and vertical resolution of 2.2 μm and < 0.1 nm, respectively.

3. RESULTS

3.1. Surface treatment concepts before testing

3.1.1. Impeller test coupons

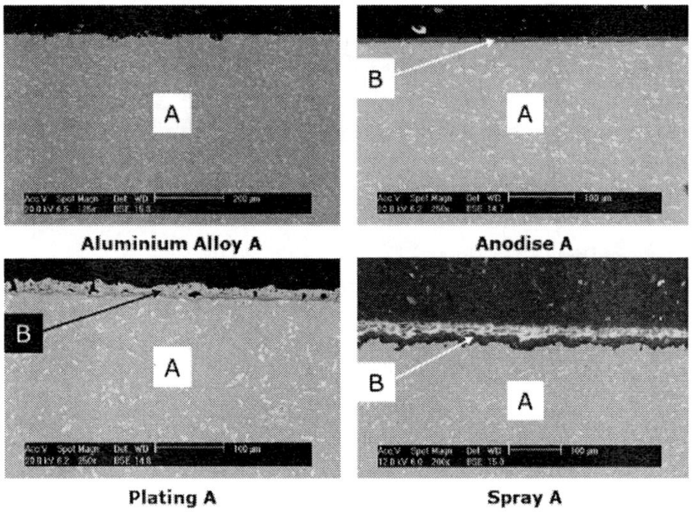

Figure 4: Impeller concepts (A = base material & B = surface treatment)

Cross sections of the impeller base alloy and surface treatment concepts are shown in Figure 4. Anodise A possessed a uniform thickness, with minimal porosity. Plating A was non-uniform in thickness and contained numerous pores of varying dimensions throughout the structure. Spray A contained no porosity and the outer layer of the surface treatment was flat.

3.1.2. Compressor cover test coupons

Cross sections of the compressor cover concepts are highlighted in Figure 5. The microstructure of the two aluminium alloys appeared very similar. The sealant could not be identified at the comparative magnifications shown herein. Anodise B possessed a more uniform thickness compared to Anodise C and both concepts contained minimal porosity. Plating B contained two distinct layers; a light grey lower and a dark grey outer. Both sections contained extensive porosity and were non-uniform in thickness. The thickness of Spray B was variable but not porous.

The topography of Aluminium Alloy B, Spray B and Plating B concepts are shown in Figure 6. There was a marked difference in the height of asperities and valleys identified on the surface of Aluminium Alloy B, whereas minimal height variations were observed on Spray B and Plating B concepts, with the surfaces appearing flat.

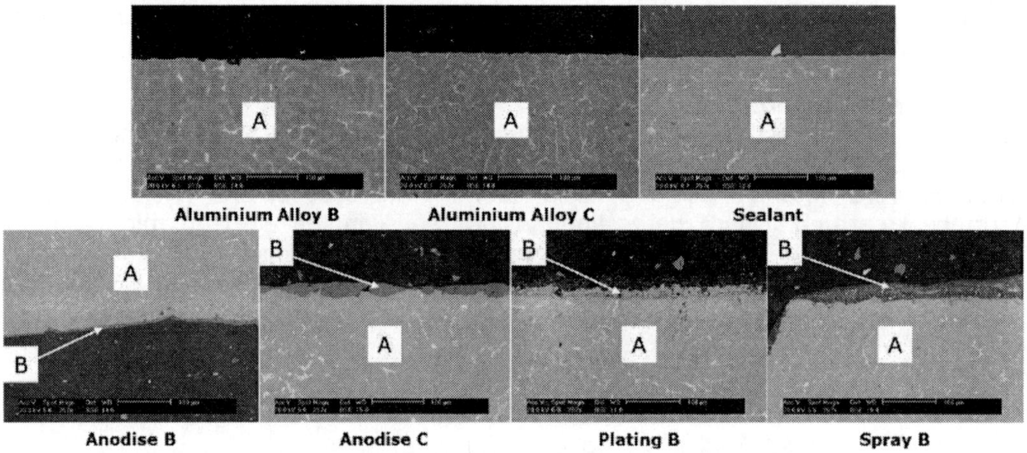

Figure 5: Compressor cover concepts
(A = base material & B = surface treatment)

Figure 6: Topographical images of compressor cover concepts

3.2. Adhesion testing

Optical images of typical indentations acquired from adhesion testing of the surface treatment concepts are presented in Figure 7 and Figure 8, respectively.

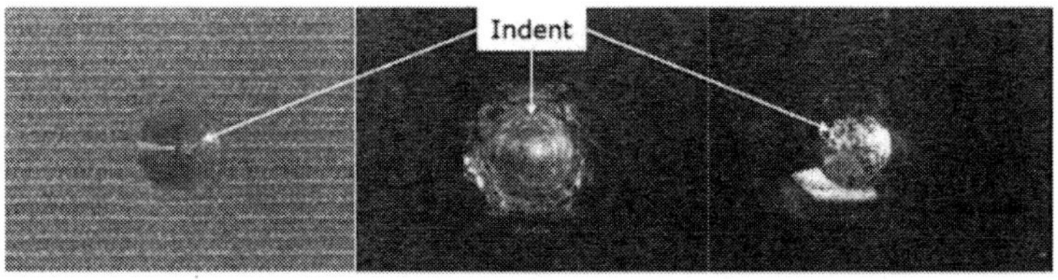

Figure 7: Adhesion test results for impeller concepts

Anodise A, B & C, Plating B, Sealant and Spray B possessed an overall rating of H1 as no microcracks or delamination were observed. However, delamination was identified on Plating A and Spray A samples; these were scored H2 and H4 accordingly.

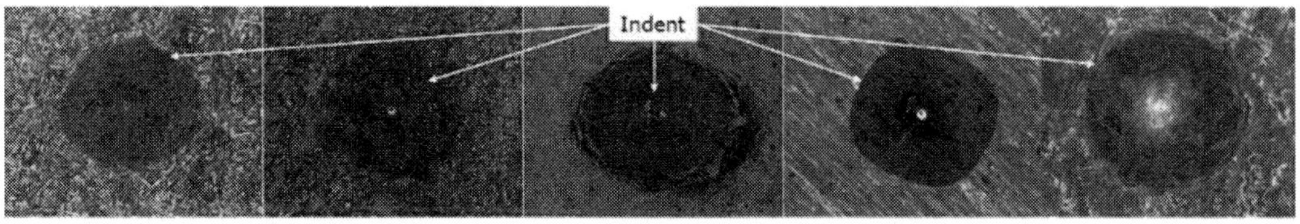

Figure 8: Adhesion test results for compressor cover concepts

3.3. Thermal shock

Delamination or crack propagation was not observed from any of the impeller or compressor cover concepts after thermal shock testing. Furthermore, there was no variation in adhesion characteristics identified on any of the surface treatments post thermal shock testing; Figure 9 highlights the typical results obtained from the compressor cover concepts.

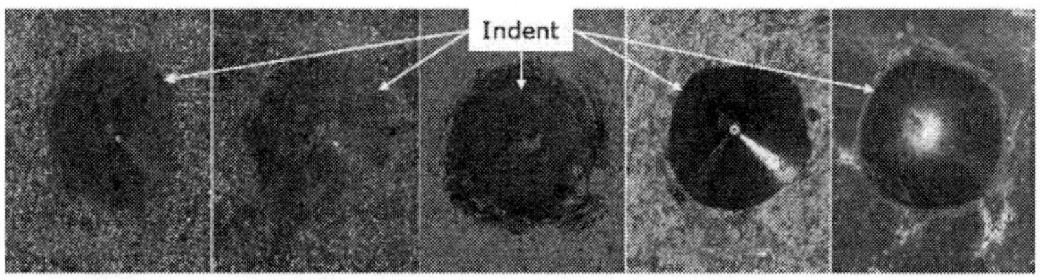

Figure 9: Post thermal shock adhesion test results for compressor cover concepts

3.4. Corrosion testing

3.4.1. Condensate A

Representative corrosion behaviour of the impeller and compressor cover concepts, when subjected to a 3.5 pH condensate environment, are presented in Figure 10 and Figure 11.

Aluminium Alloy A and Spray A had experienced significant corrosion, with numerous pits and cracks generated sub-surface. There was some corrosion cracking identified sub-surface on Plating A. No corrosion was observed with regards to Anodise A.

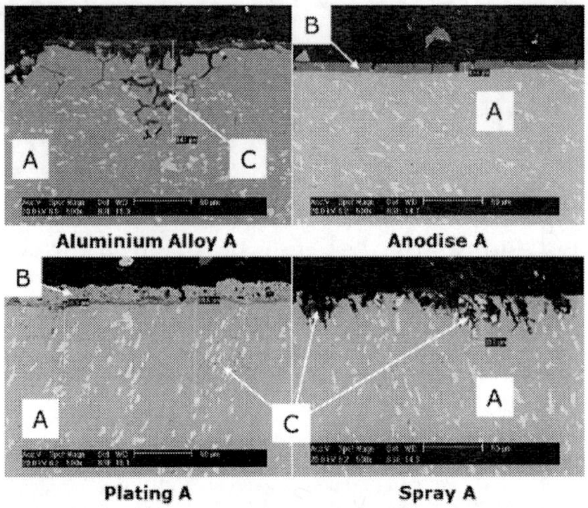

Figure 10: Corrosion behaviour of impeller concepts (A = base material, B = surface treatment & C = corrosion)

A lower volume of corrosion was identified on Aluminium Alloy C, when compared to Aluminium Alloy B. The Sealant and Anodise B concepts had experienced comparable levels of corrosion. Extensive corrosion was observed on Anodise C. There was no evidence of corrosion with regards to Plating B or Spray B.

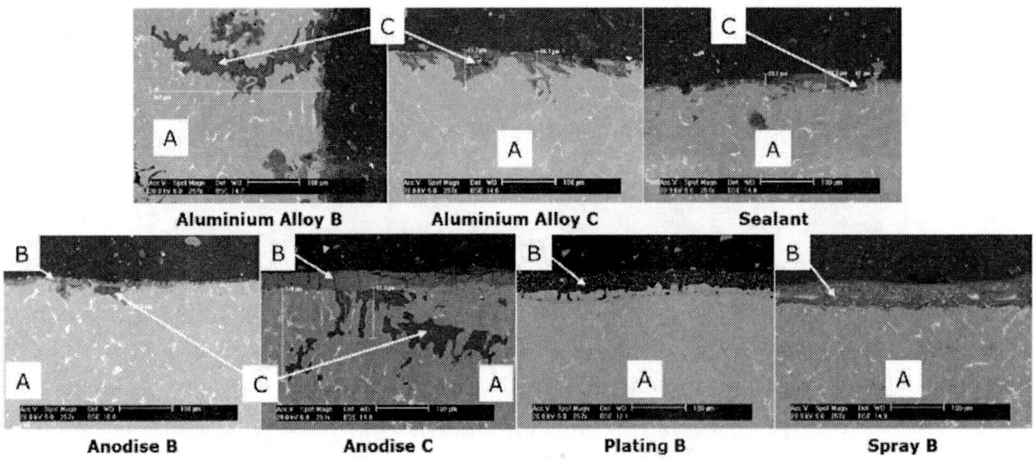

Figure 11: Corrosion behaviour of compressor cover concepts (A = base material, B = surface treatment & C = corrosion)

3.4.2. Condensates B, C & D

As described in section 2.3.3, further corrosion testing was performed on those concepts which possessed the most superior corrosion behaviour when subjected to condensate A.

Figure 12 presents the corrosion behaviour of Aluminium Alloy A and Anodise A when subjected to three condensates of varying pH level. Aluminium Alloy A experienced corrosion at all pH levels, with the greatest volume of corrosion observed at 1.2 pH. The surface treatment in Anodise A had been removed from the substrate during the test at 1.2 pH. Corrosion of the surface treatment was evident at 2.5 pH in the form of pitting, with limited sub-surface cracking observed.

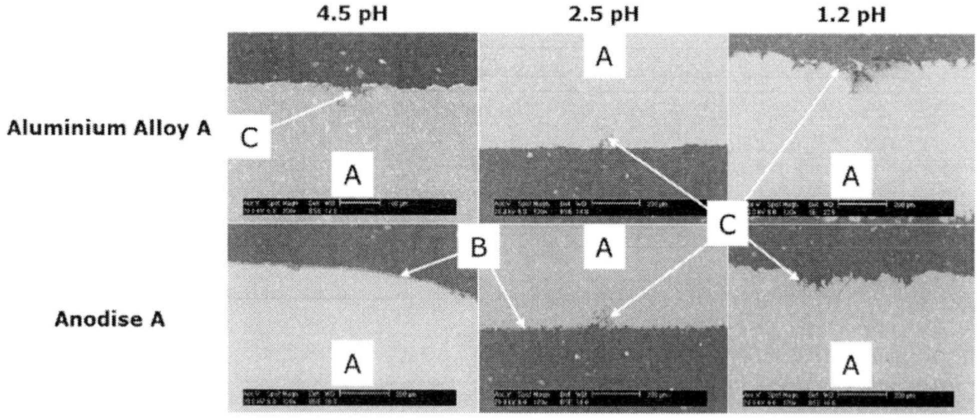

Figure 12: Corrosion behaviour of impeller concepts at various pH levels (A = base material, B = surface treatment & C = corrosion)

As can be observed in Figure 13, when subjected to condensates B, C & D, Aluminium Alloy B experienced greater levels of corrosion compared to Aluminium Alloy C. There was no corrosion identified in the base material on Plating B or Spray B concepts.

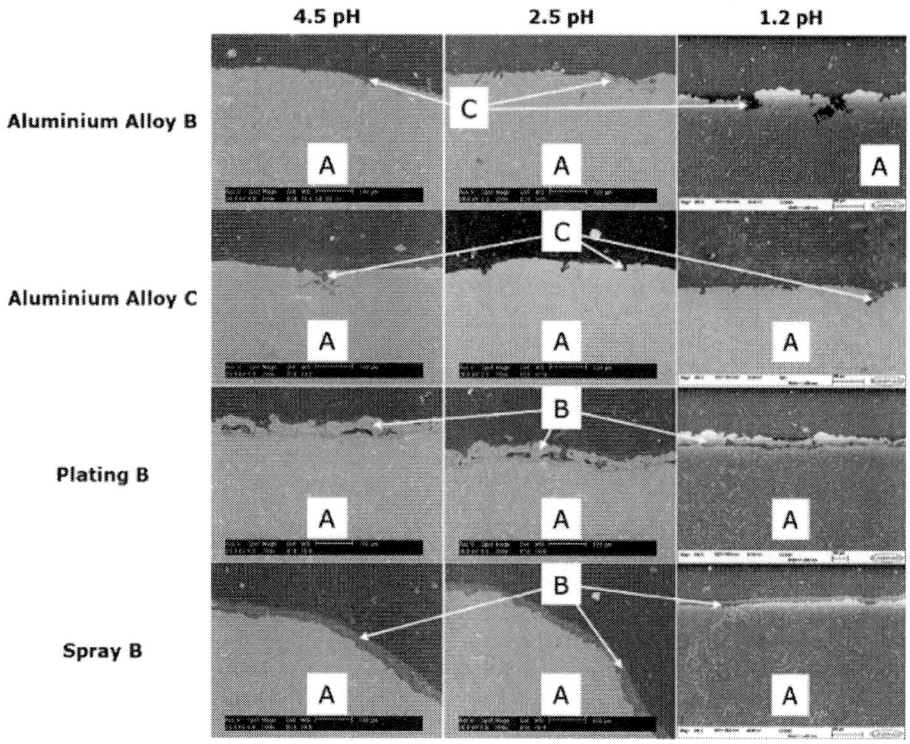

Figure 13: Corrosion behaviour of compressor cover concepts at various pH levels (A = base material, B = surface treatment & C = corrosion)

3.5. Erosion testing

Figure 14 highlights the wear regions on the pressure surface of the surface treated impellers after being subjected to erosion testing using media E1. The lowest volume of wear was observed on Anodise A, when compared against Plating A and Spray A concepts.

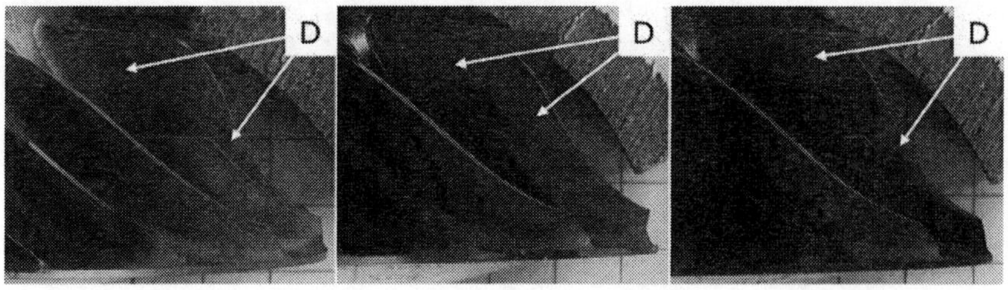

Figure 14: Post erosion test images of impeller concepts (D = wear region)

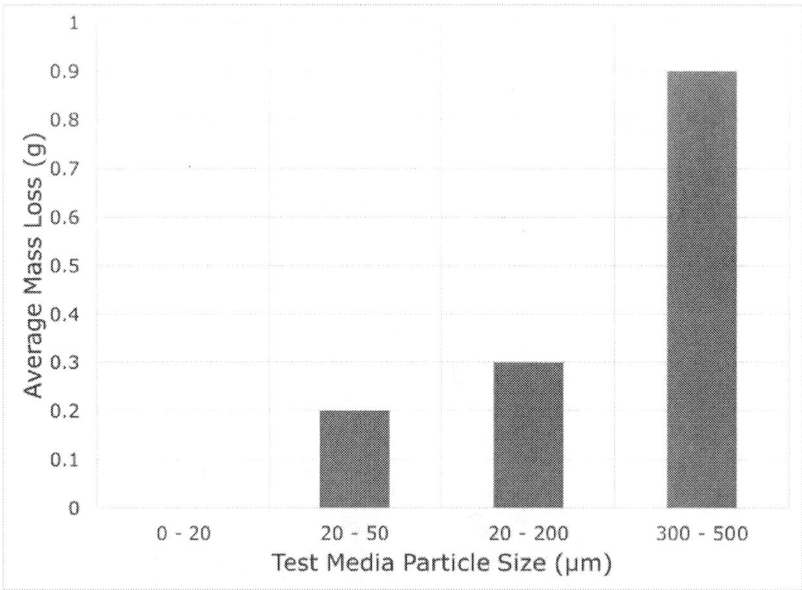

Figure 15: Mass loss with respect to test media particle size

Further erosion testing was performed using the Anodise A concept and the four respective test media and mass flow rates stated in Table 7. The mass loss from the Anodise A concept after erosion testing is presented in Figure 15. The mass of material removed from the impeller increased with particle size.

3.6. Aerodynamic performance

The aerodynamic performance of the impeller concepts is stated in Figure 16. The performance of Anodise A and Plating A surface treatments were comparable to Aluminium Alloy A. There was a marked decrease in the pressure ratio achieved by Spray A compared to Aluminium Alloy A on the 120 krpm speed line.

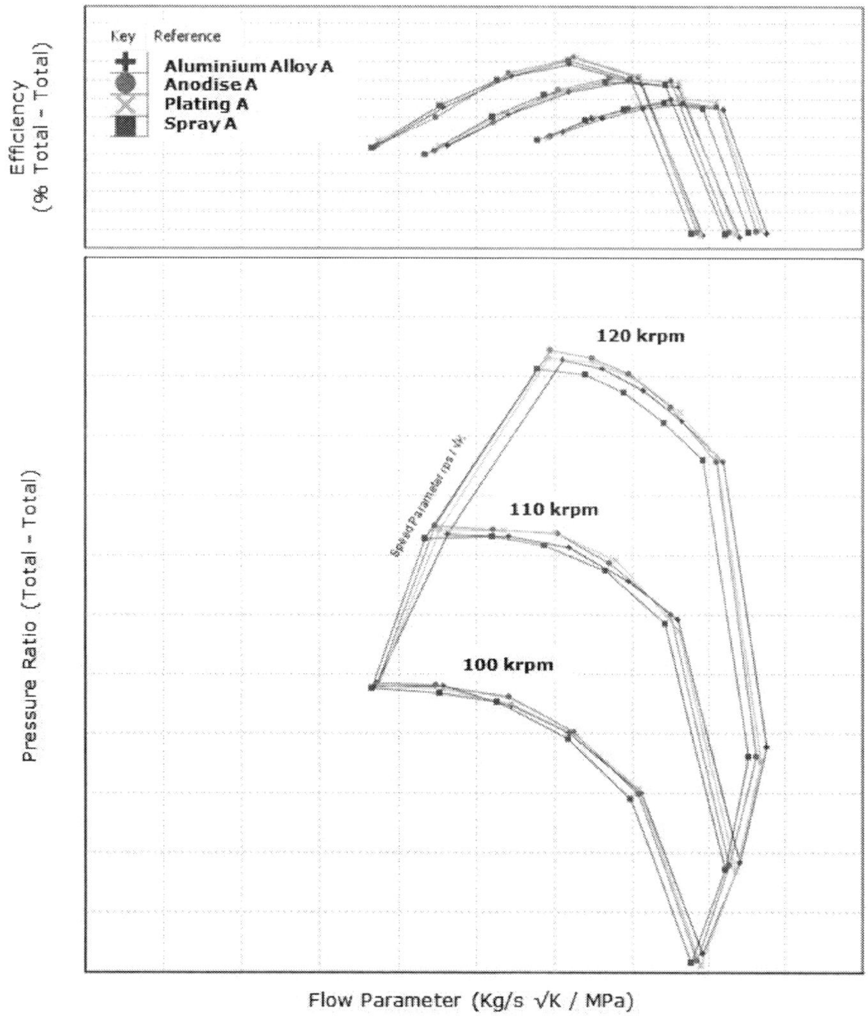

Figure 16: Compressor performance map for impeller concepts

3.7. Fatigue testing

When applied to a machined from solid aluminium impeller, Anodise A possessed acceptable fatigue performance.

4. DISCUSSION

The application of LR EGR to commercial engine turbochargers is a more complex problem compared to passenger car product. This is because of the more stringent durability targets of a commercial internal combustion engine system, which, in the context of LR EGR, demands that any applied surface treatments possess low wear rates, excellent corrosion resistance and acceptable fatigue performance over a long in-service time period.

Therefore, the testing methodologies employed in this article and presented in Figure 3, were designed to evaluate the capabilities of surface treatments in aggressive LR EGR environments. Furthermore, experiments were designed such that they delivered an accelerated test cycle; a common technique used in the subject matters of tribology (8) and corrosion (9). The accelerated test approach minimised testing costs, significantly improved repeatability and enhanced data accuracy, when compared against conventional system-level testing techniques.

Adhesion characteristics and thermal shock behaviour can provide an indication as to the long term durability of a surface treatment. Indeed, although the techniques were performed on static substrates in this study, acquired data can be used to infer performance during a dynamic test. Hence, a surface treatment would have a greater propensity to delaminate if subject to high stress conditions, should it possess poor adhesion characteristics.

Although there are numerous reported methods of determining surface treatment adhesion characteristics (10; 7; 11), the methodology employed herein afforded a macro-scale visual representation of adhesion behaviour and a direct performance comparison between concepts. Furthermore, such an approach to adhesion testing can be easily and efficiently used to measure product quality when in service. Referring to sections 3.2 and 3.3 of this article, it is evident that the majority of the surface treatment concepts under evaluation possessed excellent durability, with the exceptions of Plating A and Spray A. The inferior delamination behaviour of the latter concepts would not be appropriate for a LR EGR turbocharger application and can be accredited to the respective bonding characteristics to the underlying aluminium alloy (8).

Due to the acidic nature of the condensate-based environment generated by the LR EGR ICE (5), it is imperative that an appropriate level of corrosion resistance is possessed by the turbocharger compressor stage components. Indeed, as noted by Montemor (12), "coatings designed for corrosion protection must offer an effective physical barrier". Referring to section 3.1 and Figure 10-Figure 13, it is evident that a dense microstructure, optimal chemistry and minimal porosity contained within a surface treatment were critical to minimising condensate interaction with the base alloy and therefore achieving excellent corrosion resistance. Consequently, Anodise A possessed far superior corrosion resistance compared to its impeller concept counterparts. Although the limitations of the surface treatment were identified in section 3.4, it is evident that without such an effective layer applied to the impeller, a significantly higher turbocharger failure rate in LR EGR environments as a result of corrosion would be expected.

Whilst Spray B and Plating B compressor cover concepts provided exceptional corrosion behaviour, the cost of the aforementioned concepts was considerably greater than Aluminium Alloy C, which provided superior corrosion performance compared to its non-surface treated Alloy B counterpart. Therefore, even though Spray B and Plating B concepts were identified as the most appropriate solutions for LR EGR compressor covers, further work is recommended because the compressor cover can tolerate a greater volume of corrosion prior to failure compared to the impeller. Hence, a concept cost and performance benefit study is required when applying LR EGR turbocharger technology to commercial engines.

Although there are varying forms of LR EGR turbocharger-based erosion testing conducted by authors within the industry (13; 14), the overall consensus is that impeller surface treatments must possess excellent wear resistance. This is because of impingement of the impeller as a result of solid particles (5) and water droplets (13) in LR EGR applications. Referring to section 3.5, it can be inferred that Anodise A possessed the greatest wear resistance of the impeller concepts evaluated herein as a result of its microstructure and intrinsic mechanical properties. As has been observed in literature (15), particle size influenced the volume of wear witnessed on test substrates.

Due to the synergistic effects of corrosion and wear (16), a surface treated impeller subjected to LR EGR environments must possess sufficient tribocorrosion performance to minimise the rate at which damage is experienced by the impeller. Hence, further work is envisaged within this topic area insofar as results from individual erosion and corrosion testing will be compared and contrasted to on-application real world performance. In addition, the development of a lab-scale tribocorrosion test is recommended in order to gain greater understanding of the topic using a cost-efficient, repeatable and accurate experiment.

The adoption of LR EGR turbocharger technology within the industry would be hindered if the aerodynamic performance of the impeller surface treatment concepts were not analogous to Aluminium Alloy A. Referring to Figure 16, it was apparent that whereas Anodise A and Plating A concepts were comparable to the performance of the base substrate, Spray B possessed inferior aerodynamic behaviour as a result of its significant thickness (Table 2). Further work would be required to identify the critical surface treatment thickness at which aerodynamic performance was transparent to an untreated impeller.

Anodise A was selected as the impeller surface treatment suitable for LR EGR applications, as a result of the consistently excellent performance exhibited by the concept. A visual comparison between Anodise A and the untreated baseline impeller are shown in Figure 17. Although the fatigue performance of Anodise A was acceptable (section 3.7), it is important to understand that fatigue behaviour is a critical consideration for a turbocharger impeller, especially so in a commercial engine system. Thus, even though it would be possible to increase the thickness of the anodising layer to improve corrosion and erosion resistance, the fatigue strength of the substrate would be subsequently impaired (17; 18). Therefore, it is imperative that the environmental conditions and duty cycle of a given LR EGR application are evaluated prior to any change in surface treatment thickness.

Figure 17: Aluminium Alloy A and Anodise A impeller concepts

5. CONCLUSIONS

The research discussed herein has identified technologies which act as enablers for utilising turbochargers on LR EGR commercial internal combustion engines.

The multifaceted and comprehensive approach to developing a LR EGR turbocharger for commercial internal combustion engines has been presented. A wide range of testing and analysis techniques were utilised in order to characterise and evaluate surface engineering concepts which would enhance the capabilities of the turbocharger impeller and compressor cover. The long term durability of the surface treatments was adjudged using adhesion and thermal shock testing, whilst compatibility to LR EGR environments was determined using corrosion and erosion experiments. Compared to conventional engine-based tests, the approach to testing employed in this study improved data accuracy and repeatability, whilst reducing cost.

Anodise A was selected as the surface engineering solution for LR EGR impellers as a result of its non-porous and uniform microstructure, together with excellent adhesion characteristics. The characteristics of Anodise A ensured minimal interaction between condensate and base alloy to provide excellent corrosion resistance. Moreover, the surface treatment afforded superior wear resistance compared to the alternative concepts under evaluation, together with possessing comparable aerodynamic performance to the baseline impeller and acceptable fatigue behaviour.

Further work was envisaged to develop greater understanding of the capabilities of Anodise A. Indeed, correlation of erosion and corrosion testing results to on-application performance was suggested, together with generating a lab-scale tribocorrosion experiment. Corrosion and erosion behaviour of Anodise A may be enhanced through an increase in surface treatment thickness. However, performance and fatigue testing would be necessary to determine the critical thickness at which no detrimental behaviour was observed.

Two surface engineering concepts, Spray B and Plating B, were identified as the most appropriate to protect the compressor cover in LR EGR environments, since they provided excellent corrosion behaviour in particular. However, Aluminium Alloy C provided enhanced performance compared to the conventionally used Aluminium Alloy B. It was therefore recommended that a performance and benefit study is completed in order to determine whether an alternative base material, rather than a surface treatment, is required for specific LR EGR applications.

6. REFERENCE LIST

1. *REGULATION (EC) No 595/2009 OF THE EUROPEAN PARLIAMENT AND OF THE COUNCIL.* **European Union.** 2009, Official Journal of the European Union.
2. **United States Environmental Protection Agency.** Heavy-Duty Highway Compression-Ignition Engines And Urban Buses -- Exhaust Emission Standards. *Emission Standards Reference Guide.* [Online] [Cited: 6 October 2015.] http://www3.epa.gov/otaq/standards/heavy-duty/hdci-exhaust.htm.
3. *1D simulation of a turbocharged Diesel engine with comparison of short and long EGR route solutions.* **Cornolti, L, et al.** 2013, Applied Energy, Vol. 111, pp. 1-15.
4. *Analysis of different exhaust gas recirculation architectures for passenger car diesel engines.* **Millo, F, Giacominetto, P F and Bernardi, M G.** 2012, Applied Energy, Vol. 98, pp. 79-91.
5. *Effects of Low Pressure EGR on Transient Air System Performance and Emissions for Low Temperature Diesel Combustion.* **Heuwetter, D, et al.** 2011, SAE paper, 2011-24-0062.
6. *Appliance of High EGR Rates With a Short and Long Route EGR System on a Heavy Duty Diesel Engine.* **van Aken, M, Willems, F and Jan de Jong, D.** 2007, SAE paper, 2007-01-0906.
7. *The VDI 3198 indentation test evaluation of a reliable qualitative control for layered compounds.* **Vidakis, N, Antoniadis, A and Bilalis, N.** 2003, Journal of Materials Processing Technology, Vols. 143-144, pp. 481-485.
8. **Holmberg, K and Matthews, A.** Coating Characterization and Evaluation. *Coatings Tribology: Properties, Mechanisms, Techniques and Applications in Surface Engineering.* s.l. : Elsevier, 2009, pp. 319-362.
9. Quality control testing of coatings using accelerated corrosion tests. **[ed.] V E Carter.** *Corrosion Testing for Metal Finishing.* s.l. : Elsevier, 2013, pp. 8-10.
10. **ASTM.** *Standard Test Method for Pull-Off Strength of Coatings Using Portable Adhesion Testers.* s.l. : ASTM International, 2009. D4541-09e1.

11. *Tribological and mechanical properties of Cr2N–11Ag-coatings deposited on Cr-V ledeburitic steel.* **Bilek, P, et al.** 2015, Wear, Vols. 340-341, pp. 47-52.
12. *Functional and smart coatings for corrosion protection: A review of recent advances.* **Montemor, M F.** 2014, Surface and Coatings Technology, Vol. 258, pp. 17-37.
13. *Effect of Water Droplets Caused by Low Pressure EGR on Spinnign Compressor Wheels.* **Karstadt, S, et al.** 2014. 19th Supercharging Conference Dresden.
14. **Munz, S, et al.** Turbocharger for emission concepts with low-pressure-end exhaust-gas recirculation. [Online] [Cited: 20 October 2015.] http://www.turbos.bwauto.com/tools/download.aspx?t=document&r=510&d=599.
15. *Determination of erosion rate characteristic for particles with size distributions in the low stokes number range.* **Solnordal, C B, et al.** 1-2, 2013, Wear, Vol. 305, pp. 205-215.
16. *Tribocorrosion of stainless steel in sulfuric acid: Identification of corrosion–wear components and effect of contact area.* **Diomidis, N, et al.** 1-2, 2010, Wear, Vol. 269, pp. 93-103.
17. *Effect of anodized layer thickness on fatigue behaviour of magnesium alloy.* **Khan, S A, et al.** 1-2, 2008, Materials Science and Engineering A, Vol. 474, pp. 261-269.
18. *Effect of hard anodize thickness on the fatigue of AA6061 and C355 aluminium.* **Rateick Jr, R G, Binkowski, T C and Boray, B C.** 15, 1996, Journal of Materials Science Letters, Vol. 15, pp. 1321-1323.

Development of a high temperature turbocharger for heavy duty applications

A Sullivan[a], D H Brown[a], D M Eastwood[a], J Green[a], A Dombrovsky[b], F J Arnau[b]
[a] Cummins Turbo Technologies, UK
[b] CMT-Motores Térmicos, Universitat Politècnica de València, Spain

ABSTRACT

Alternative fuels for heavy duty applications are beginning to become mainstream as fleet operators seek to reduce fuel consumption and CO2 emissions. This is leading to changing requirements for turbocharging and pushing the limits of traditional technologies. A duty cycle based approach has been used to develop fixed geometry and wastegated turbocharger technology for elevated temperature heavy duty applications. Numerous logged duty cycles with a variety of fuels have been combined with a novel lumped parameter model that predicts component temperature and was subsequently used to evaluate component performance at elevated temperature. Materials have been validated using innovative test techniques for components such as the turbine housing and wastegate mechanism which simulate the operational mechanisms whilst also accelerating the test. Simulation of new test methodology has enhanced test development and subsequent test results have been used to perform a robust model calibration for new materials. Duty cycle analysis has been implemented along with a strain based Thermo-Mechanical Fatigue analysis to enable estimation of life in service.

1. INTRODUCTION

Continuous reduction in permissible CO2 and NOx levels and incremental fuel economy targets in commercial vehicles has led to growth in the global natural gas fuel market [1]. Both emissions targets and alternate fuels along with changes in fuel strategy has led to increasing exhaust gas temperatures which has required a reassessment of current materials, methods and tools [2].

Cummins Turbo Technologies has a long history of supplying turbochargers for natural gas applications to the global market, however these applications tended to be lean burn which have similar characteristics to the more traditional diesel market. Stoichiometric natural gas engines are becoming increasingly popular in the commercial vehicle heavy duty market however this mode of operation has significant impact on the turbocharger requirement.

In order to ensure cost effective, robust, and dependable product a duty cycle based approach has been used to study the impact of stoichiometric natural gas operation. Numerous duty cycles have been logged on replicated customer routes with vehicles configured with either diesel or natural gas engines. Where the component metal temperature is unavailable, a novel lumped capacitance heat transfer model has been taken to predict the thermal response of the turbocharger which has been subjected to the logged exhaust gas thermal history.

Lumped capacitance heat transfer models are widely used for several engineering applications. This approach is able to provide fast estimation of heat transfer phenomena occurring in a process with a minimum computational cost. The more complex the modelling of the phenomena, the closer are the results to reality. However, excessive complexity of the model requires more inputs which are sometimes difficult to obtain in engineering practice.

In the field of turbocharging research heat transfer phenomena have been extensively studied over the past few years from both an experimental [3] and theoretical [4] point of view. The lumped capacitance approach has been widely used to model heat transfer phenomena in steady state conditions as shown in [5] and [6] and in transient conditions as shown in [7]. However, this approach has been focused mainly on improving engine modelling using 1D gas dynamic codes as has been done in [8] and [9]. To date the main application of these models has been for passenger car turbochargers. In the present paper a lumped heat transfer model has been developed to estimate turbine housing metal temperatures for a given heavy duty cycle.

A more precise assessment of component durability is made possible given the measured or predicted component thermal history. Duty cycle analysis enables classification of damage cycles to be applied within a Thermo-Mechanical Fatigue (TMF) model and predict field life for new applications in development. This is implemented in an existing stress based TMF method and also a commercially available strain based TMF method nCode DesignLife. The commercial software allows prediction of component life given a known duty cycle which includes elevated exhaust gas temperature (beyond traditional limits) and also using a variety of material classifications. Whilst the stress based approach is limited in its applicability the strain based approach has been shown to correlate well with durability testing on alternative materials subjected to increased exhaust gas temperature.

Elevated turbine inlet temperature can have a detrimental effect on the life of the turbine stage components, possible reduction in TMF life is one example. Numerous turbocharger systems and components have been tested using a number of novel methods in combination with the duty cycle assessment approach. Each of the components have been analysed and several material options considered and validated against the duty cycle target. Each of the potential failure modes have been investigated. These include TMF, Low Cycle Fatigue (LCF) and High Cycle Fatigue (HCF), creep, joint relaxation and rotation (vibration), wear and corrosion.

2. TURBINE TEMPERATURE DUTY CYCLE DATA

As heavy duty applications are considering alternative fuels such as natural gas, data logs of natural gas versus diesel turbine temperature data have been compared. Figure 1 shows normalised logged Turbine Inlet Temperature (TIT) for heavy duty line haul truck applications with diesel and natural gas engines on similar routes. Figure 1 (a) shows that a diesel engine operates over a wide temperature range whereas in contrast Figure 1 (b) shows that a gas engine where temperatures are less variable. So we can say that the gas engine here experiences relatively higher peak exhaust gas temperature and more time is spent at higher temperature.

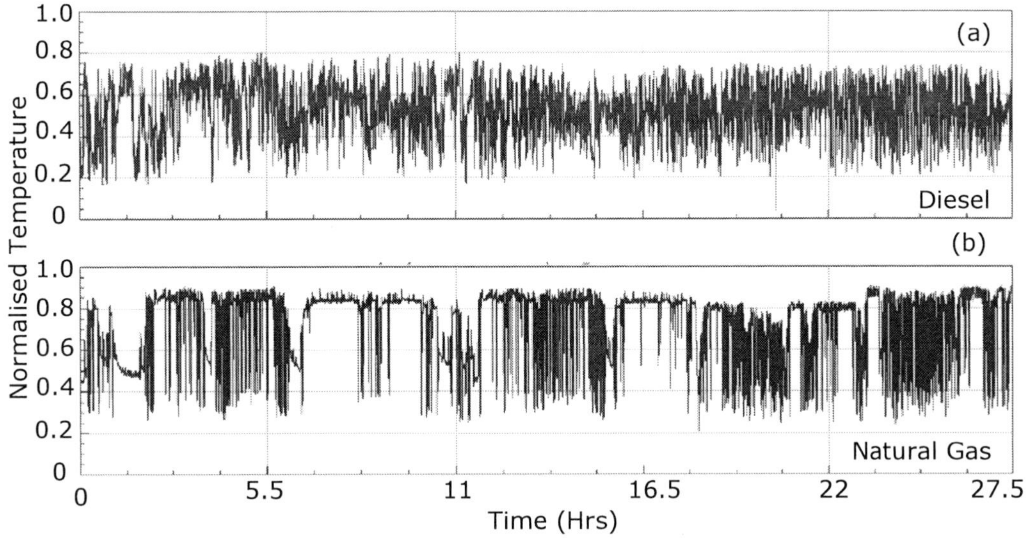

Figure 1: Normalised TIT logs for heavy duty line haul truck applications with diesel and natural gas engines.

A time at level plot of the field data shown in Figure 2 confirms this statement. The diesel log shows operation over a broad temperature range while the gas engine operates for considerably greater time at higher TIT. The next step is to account for this fundamental change in operating condition and predicting the effect on life on turbine stage components such as the turbine housing.

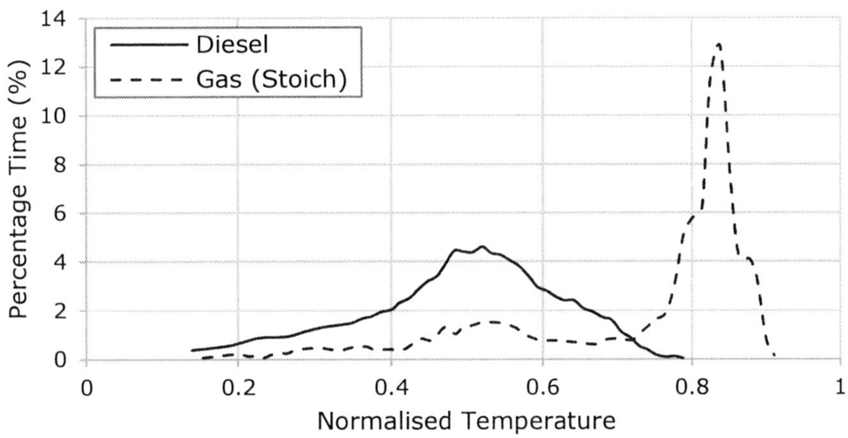

Figure 2: Time at normalised level comparison between diesel and natural gas duty cycles.

One of the challenges of predicting turbine stage component life is obtaining real world field data, either using logged gas temperatures or metal temperatures as inputs into the life analysis. Field data logs usually gather TIT information, but metal temperatures are not routinely recorded. It is therefore desirable to have a means of predicting turbine housing metal temperatures from a given gas temperature duty cycle. A method has been developed to predict metal temperatures from gas temperatures and validated using heavy duty truck data. The method is discussed in Section 3.

To provide inputs to TMF analysis for designing a turbine housing capable of higher temperature operation, the logged gas engine data was scaled. Metal temperatures were predicted from TIT and a cycle count performed to identify damaging cycles for input into TMF. Figure 3 shows the scaled TIT and the predicted turbine inlet flange metal temperatures.

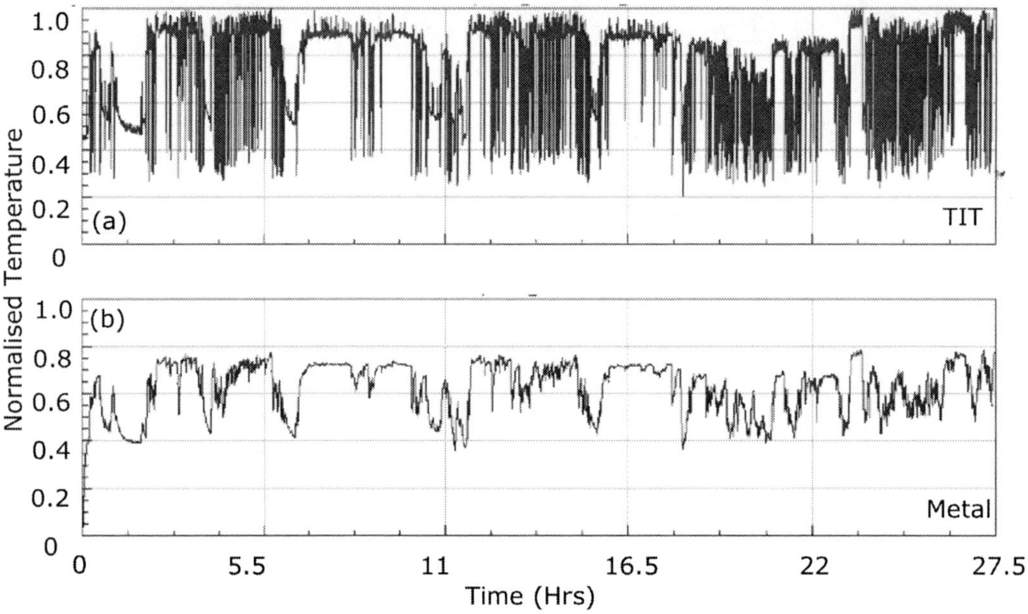

Figure 3: Use of scaling to estimate a normalised elevated temperature duty cycle.

The cycle count information was combined with the rate of change in metal temperature and used to predict TMF life of the turbine housing.

3. LUMPED CAPACITANCE HEAT TRANSFER MODEL

The lumped model developed for turbine housing is shown in Figure 4. The turbine housing (node TH) has been modelled as a single node with a lumped capacitance (C). In normal operation (hot gases in the turbine) this node losses heat to surroundings (node AMB), bearing housing (node BH) and receives heat from hot exhaust gas (node GAS). Therefore, three conductances must be determined in order to fully define the model. The first conductance (K_{conv}) represents the convective heat transfer between hot gas and turbine housing. The second conductance (K_{cond}) represents the conductive heat transfer through turbocharger parts from turbine housing to bearing housing. Finally the third conductance (K_{rad}) represents the radiation heat transfer from turbine housing to the ambient.

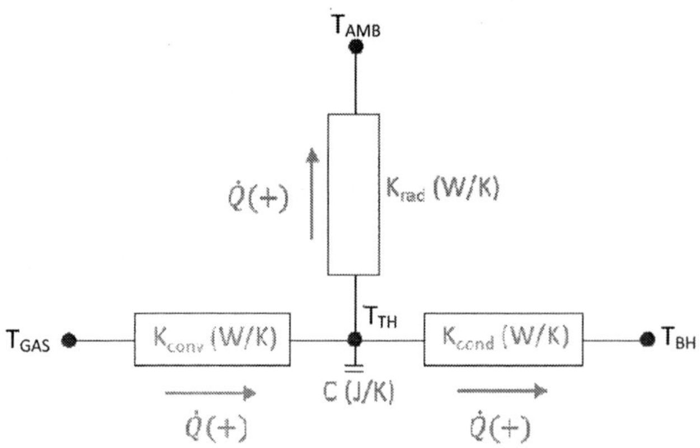

Figure 4: Lumped model.

The proposed model does not require input data regarding compressor operation or lube oil flow. Thus, the lumped model greatly simplifies the current models available in the literature [10] as the main objective is to be able to estimate turbine housing metal temperatures with as little input data as possible.

The internal energy balance for node T_{TH} ((1) can be written taking into account the previously described model. In this equation the next time step of metal temperature ($T_{TH}^{t+\Delta t}$) must be solved for each point of logged duty cycle data. The input data from logged data are turbine mass flow (to estimate convective heat transfer) and turbine inlet temperature ($T_{IT}^{t+\Delta t}$), as well as the time step of these data (Δt). The main assumption of the model is to consider a constant bearing housing temperature ($T_{BH} = 400\ K$).

$$C \cdot \frac{T_{TH}^{t+\Delta t} - T_{TH}^{t}}{\Delta t} = K \cdot \left(T_{BH} - T_{TH}^{t+\Delta t}\right) + hA_{conv} \cdot \left(T_{IT}^{t+\Delta t} - T_{TH}^{t+\Delta t}\right) - \varepsilon A_{ext}\sigma \cdot \left[\left(T_{TH}^{t+\Delta t}\right)^4 - \left(T_{AMB}^{t+\Delta t}\right)^4\right] \quad (1)$$

Where

C	=	Capacitance (J/K)	T	=	Temperature (K)
TH	=	Turbine housing	t	=	Time step
K	=	Conductance (W/K)	BH	=	Bearing Housing
h	=	Convective heat transfer coefficient (W/m^2·K)	A	=	Area (m^2)
conv	=	Convective heat transfer	T_{IT}	=	Turbine Inlet Temperature
ε	=	Emissivity (-)	ext	=	External geometry of turbine housing
σ	=	Stefan-Boltzmann constant (W/m^{2}K^4)	AMB	=	Ambient
cond	=	Conductive heat transfer	$\dot{Q}$	=	Heat flow (W)
GAS	=	Exhaust gas	rad	=	Radiative heat transfer

In order to define the different conductances, general correlations from the literature have been used. Thus, for the conductive conductance (K) the general approach from [11] is used. To use this approach it is necessary to know turbine and bearing housings external length, diameter and thermal conductivity. The estimation of the convective conductance (hA_{conv}) is based on the GAS/T correlation proposed in [12] for which it is necessary to know turbine inlet temperature (for gas properties and Prandtl and Nusselt numbers), turbine inlet port area and external diameter of the housing (to define A_{conv}) and turbine mass flow (for Reynolds number). Finally, for the definition of the radiation conductance ($\varepsilon A_{ext}\sigma$) reference [13] is used. In this work a simplified external area is defined based on turbine housing external length, diameter and an emissivity value for typical turbine housings is given. The ambient temperature can be assumed constant at 298 K if no data are available. The capacitance of turbine housing can be easily computed using the mass of the component and the specific heat capacity of the material. If these data are not available a general approach shown in [13] can be used.

Using all of the above, equation (1) is fully defined and there is only one unknown ($T_{IT}^{t+\Delta t}$) which must be solved for each point of the logged turbine inlet temperature. As the equation is implicit, an appropriate algorithm must be used to solve it.

3.1. Validity (quality)

The above described model is the most simplified approach to estimate metal temperatures from logged turbine inlet temperatures. As the whole turbine housing is lumped in a single node, there is no spatial discretization and so the results provided by the model must be seen as a first estimation of metal temperature for the whole housing.

A more complex model (with more nodes at specified spatial locations) will require determining more parameters such as conductive and convective conductances. The convective conductances determination represents a serious problem to build a reliable model as it will be necessary to perform an extensive experimental study to obtain correlations. For that reason the proposed model is based on the available lumped models for turbochargers from the literature in which turbine housing is lumped in a single node and reliable convective correlations are provided for the GAS/T heat path.

3.2. Model predictions against logged duty cycles

The validation of the model is based on logged field data for which a metal temperature has been measured at two location on the turbine housing flange. In Figure 5 a comparison between modelled and mean of measured metal temperatures is shown for part of a measured duty cycle for a natural gas application.

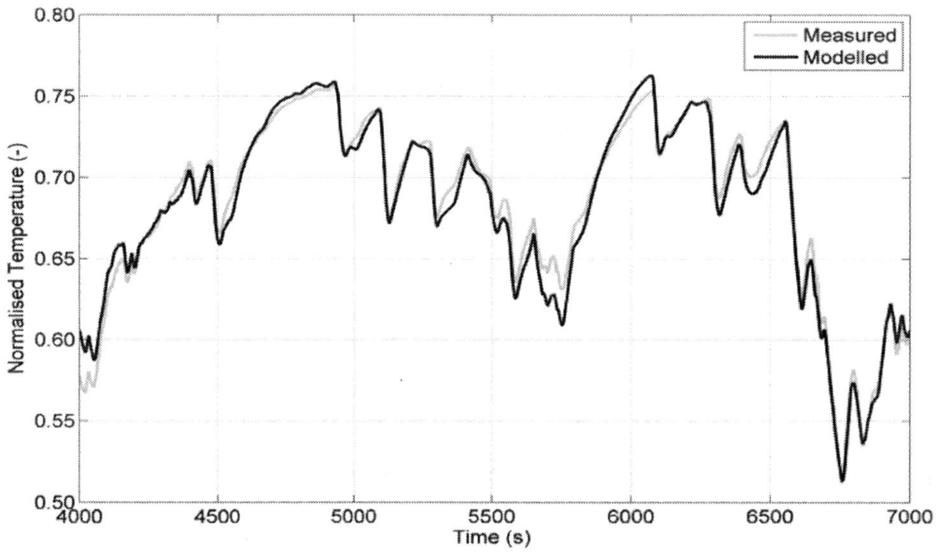

Figure 5: Comparison of normalised modelled and measured metal temperatures.

It can be observed that the model captures properly the logged metal temperature for this case with maximum relative errors of 7%. Similar results can be obtained with different turbochargers and logged cycles. It is important to remark that the model performs with good accuracy in both the slow temperature changes and in the rapid ones. This fact means that the model is quite robust and the parameters determination from the literature is appropriate.

4. TURBINE HOUSING THERMAL FATIGUE ANALYSIS

As part of the design of high temperature turbine housings, cast in materials such as Silicon Molybdenum Spheroidal Graphite Iron (SiMo), Compacted Graphite Iron (GJV) and Austenitic Spheroidal Graphite Iron (Ni-Resist), an analysis technique was required for each material to predict durability. The initial method chosen was to conduct load controlled fatigue testing on the candidate materials, with the results used to edit the existing stress life (Sn) based thermo-mechanical fatigue (TMF) macro for each material. The predicted results for each material were then compared to gas stand accelerated thermal cycle test results, on actual turbine housings assembled on turbos.

Stress and temperature results accessed by the TMF macro were calculated using a finite element model. The turbine housing geometry was imported into the Ansys FEA package from the Creo2 CAD package. The model was meshed with second order tetrahedral elements. Surface thermal boundary conditions of heat transfer coefficient and temperature were applied based on previous experience. Thermal boundary conditions were then adjusted to match metal temperature thermocouple measurements from the component thermal cycle test. Suitable measured temperature dependant mechanical and thermo-physical material properties were utilised. Once correlated transient nodal temperatures were generated to represent the component thermal cycle test, these temperatures were input as loads to run the structural model elastically with representative structural boundary conditions.

Correlation of the prediction using the existing Sn based TMF macros for new materials to component thermal cycle test was relatively poor. Ni-Resist housings were lasting longer than predicted by a factor of between 2 and 20, whereas GJV housings were failing more quickly than predicted by a factor of between 2 and 1000. FEA was predicting a large area of the Ni-Resist housing being of greater stress than temperature dependant yield strength, however, whereas the prediction for GJV was only showing a small amount of the housing being above yield strength.

A cross-functional fault tree analysis was performed to identify the reasons for poor correlation. The main reasons identified were the use of load controlled, as opposed to strain controlled specimen testing, and lack of consideration of plasticity in Finite Element Analysis (FEA) and fatigue techniques.

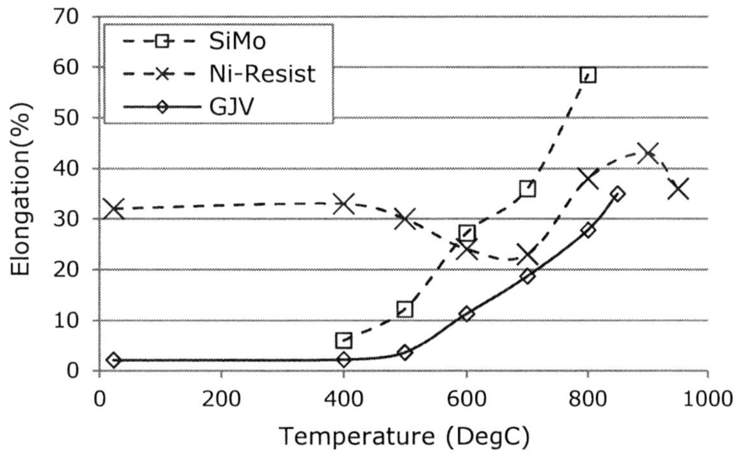

Figure 6: Comparison of ductility.

To explain the contrasting behaviour of the two materials, it is worth considering the temperature dependant elongation, which is an indication of ductility. As shown in Figure 6. GJV has an elongation of around 3% from room temperature to 450C with a slope to around 35% from 450C to 850C. In contrast Ni-Resist has an elongation of over 30% from room temperature to 400C, with a reduction to 22% at around 680C and then a slope up to 42% at 900C. At all temperatures, Ni-Resist is more ductile than GJV.

Temperature dependant strain controlled fatigue properties were obtained for the Ni-Resist housing material. The fatigue analysis was performed in nCode DesignLife using the strain-life technique with Hoffman-Seeger plasticity correction [14]. Good correlation, within a factor 2 compared to thermal cycle test at most failure locations, was achieved when the maximum temperature from the cycle was used in the fatigue calculation. Predicted cycles to failure are shown in Figure 7. The high ductility of Ni-Resist allows redistribution of stress and strain which can only be predicted when including the effects of plasticity. At even higher temperatures visco-plasticity and creep would also have to be considered.

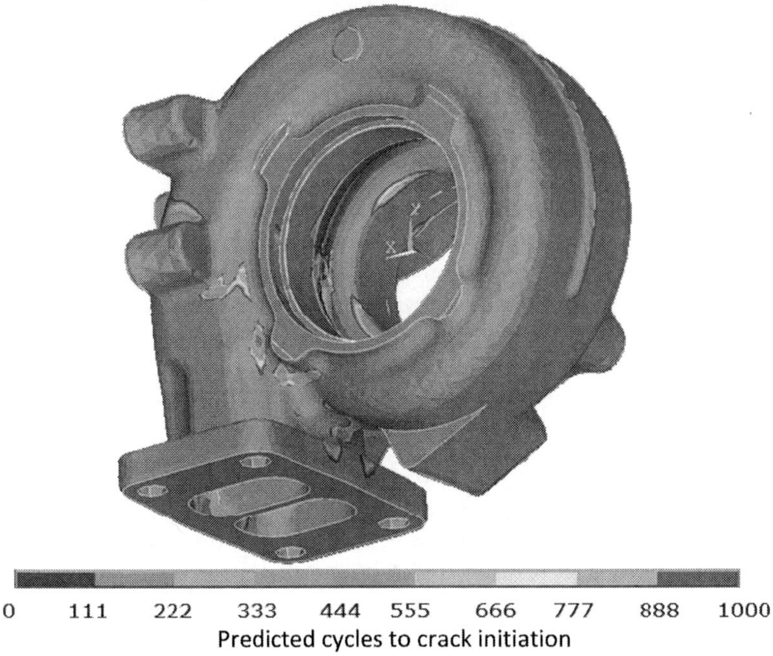

Figure 7: Predicted fatigue life plotted onto Ni-Resist turbine housing.

As GJV has very low ductility compared to Ni-Resist it does not benefit from stress re-distribution in the plastic zone and will therefore be more likely suffer from static failure when exceeding yield. It is planned, in future, to obtain strain based fatigue data for GJV to try and improve thermal fatigue prediction.

5. TURBOCHARGER SYSTEM AND COMPONENT VALIDATION

Systematic use of Design Failure Modes and Effects Analysis (DFMEA) has identified the major potential failure modes which includes TMF, LCF and HCF, creep, joint relaxation and rotation, corrosion. In order to validate a given turbocharger specification, a number of tests have been designed to validate the turbocharger system as well as specific sub components such as the turbine housing, turbine wheel, wastegate mechanism, joints and heatshield.

Duty cycle analysis has driven innovative development of application specific testing including thermal cycle testing where severe elements of known duty cycles presented in Section 2 have been combined to provide a representative thermal cycle with clear pass/fail criteria which equates to a known durability. The severe elements are highlighted in Figure 8(a) and this is translated into a repeatable thermal cycle shown in Figure 8(b) which is used in gas stand testing.

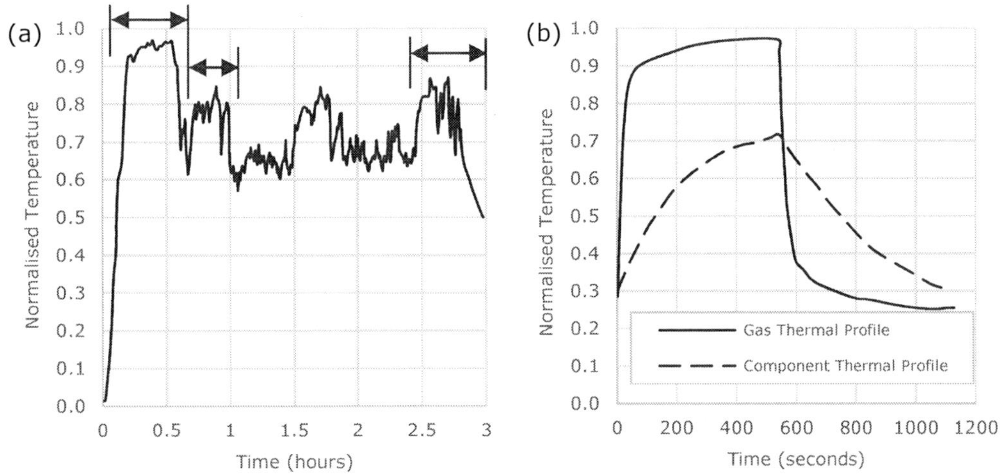

Figure 8: (a) logged duty cycle data with highlighted severe transient periods, (b) replicate thermal cycle used in product validation.

As the test thermal cycle is taken from actual duty cycle data the relative life from each cycle can be estimated to provide a clear target. Results from the gas stand show that the model predicts the high stress locations well and the cycles to crack initiation is within a factor of 2 in critical locations such as the turbine housing tongue and wastegate port. End of test images are shown in Figure 9 (a-c) of the tongue / location diameter from various materials selected for high temperature operation. It should be noted that all materials have initiated and propagated cracks however Ni-resist has a reduced severity especially in the side wall area highlighted Figure 9 (a & c). This shows good correlation in terms of low life / crack initiation to predictions in Figure 7 and also indicates the relative durability benefit provided Ni-resist materials at elevated TIT over SiMo and GJV.

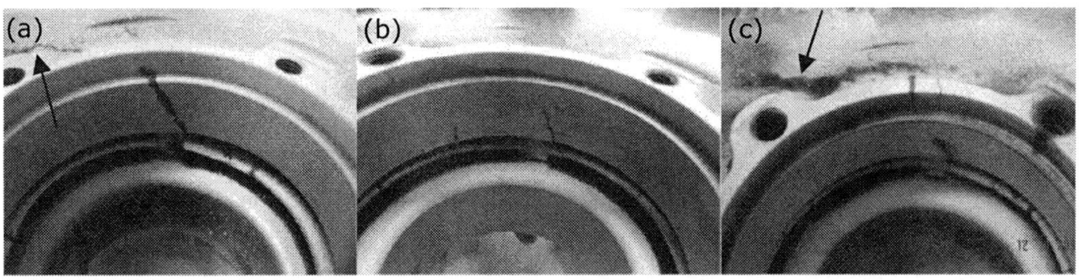

Figure 9: Example images from aggressive thermal cycle testing showing cracks developed (a) SiMo, (b) Ni-Resist, (c) GJV.

Turbocharger system level testing has been modified to account for increased running at elevated temperatures including containment, joint rotation, LCF and HCF. Particular attention is paid to joint vibration as material choices were led by finite element analysis and material thermal compatibility considerations. Vibration testing has been carried out at elevated temperature to replicate the vibration and thermal profile experienced across the joints. The arrangement is shown in Figure 10 (a) where hot gas is passed via a flexible link from a gas stand into the turbine housing which is rigidly linked to a shaker table. Figure 10 (b) shows an example using existing application limits which defines the input vibration signature and the resultant outputs from instrumentation mounted on the bearing and compressor housings. Several configurations have been designed and validated which exceed current durability limits.

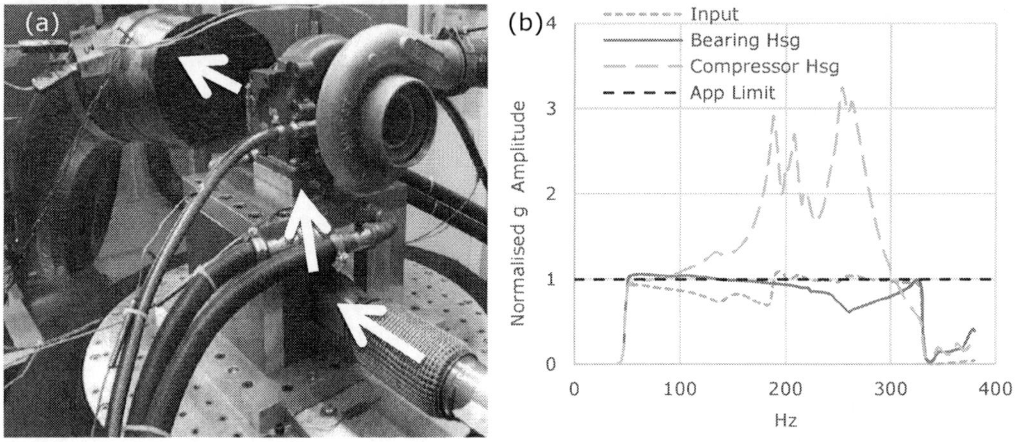

Figure 10: Vibration test fixture shown in (a) and (b) vibration profiles.

A novel approach has been used to accelerate the testing of wastegate shaft and bush wear, initial material pre selection has been carried out with use of a tribometer [15], upon selection each material pair has undergone functional testing. Functional testing consisted of four wastegate system assemblies mounted in a modular block arrangement which are pneumatically actuated as shown in Figure 11(a).

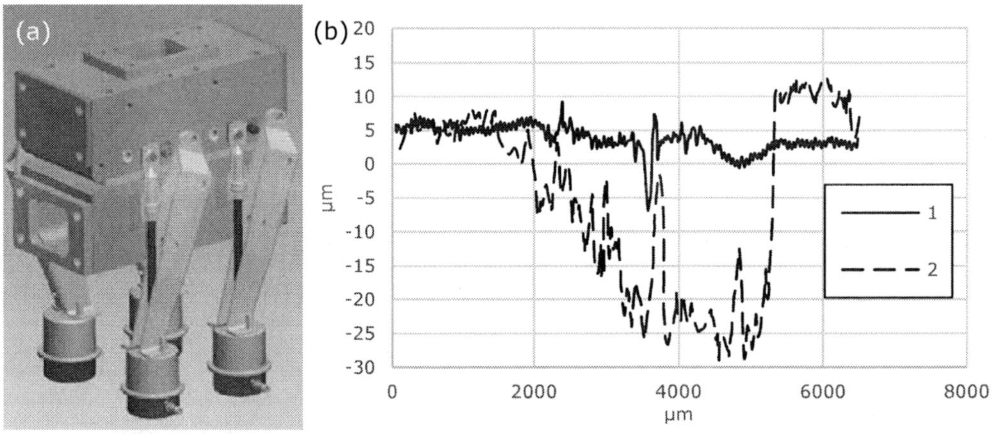

Figure 11: (a) Modular wastegate functional wear testing rig and (b) wear scar profilometer traces from functional testing 1. Cobalt alloy pairing and 2. Stainless Steel pairing.

The system has the added flexibility of being mounted on specific application engine or gas stand. Duty cycle analysis has been used to determine the target cycles to simulate 30% life wear characteristics in an exhaust gas atmosphere. Figure 11(b) shows an example of the wastegate shaft wear profiles at the end of the test which compares a traditional cobalt pair to a recently developed Austenitic/Intermetallic Stainless Steel wear pair.

Corrosion testing for exhaust gas environments has been conducted on all turbine stage components using a modified version of the VDA 230-214 [16] standard with an aggressive pH 1 and a representative exhaust gas pH 3.5 [17, 18, 19]. This test includes repeated heating cycles to maximum operation temperature specified as a baseline from current production and also the new elevated temperature. Increasing the operating temperature increases the corrosion potential in all materials but pH level has a much more significant impact. Compared to SiMo, Ni-

Resist material is clearly more resistant which is to be expected and GJV is more susceptible as shown in Figure 12.

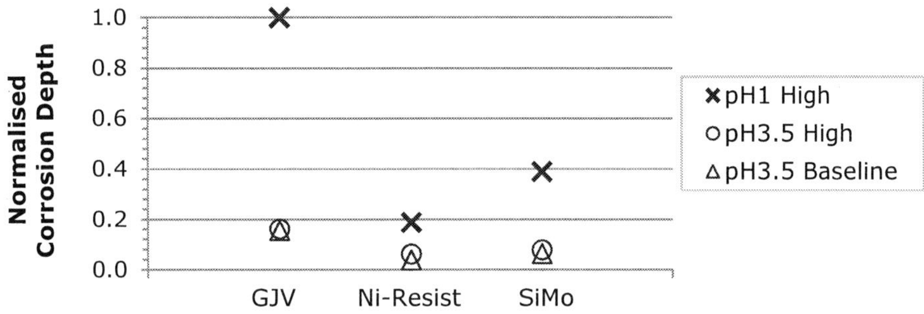

Figure 12: Corrosion depths measured from turbine housing materials after VDA [16] corrosion testing with varying pH level and temperature.

6. SUMMARY AND CONCLUSIONS

Continuous reduction in CO2 and NOx limits and incremental fuel economy targets has driven interest in alternative fuels and alternative fuel strategies. This has required an assessment of current exhaust gas temperature limits which in turn has required a reassessment of current materials, methods and tools.

We have compared data logs of natural gas versus diesel turbine inlet temperature. The diesel log shows operation over a broad temperature range whilst the gas engine operates for considerably greater time at higher TIT. This is a fundamental shift in operation and duty cycle which leads us to reconsider the boundary conditions for both analysis and testing. Here we have used a scaled duty cycle to enable new limits to be established.

A simplified lumped model has been used to estimate the metal temperature of turbine housing from a heavy duty truck exhaust gas duty cycle datalog. The model is applied to a heavy duty turbochargers and based on available work and correlations from passenger car turbochargers research literature. The proposed model shows good accuracy for a heavy duty application and is suitable to be used as a first estimation of metal temperatures. This can then be used to set the boundary conditions for more complex finite element models particularly for turbine housing TMF analysis and also other turbine stage components.

Duty cycle analysis combined with metal temperature prediction enables more detailed analysis on turbine stage components. Here a new approach has been taken for TMF analysis of the turbine housing, damage accumulation methodology and a strain based approach using nCode DesignLife software. Ductility has a major effect on fatigue life at a specific temperature or range of temperatures. Specimen testing, analysis and fatigue prediction methods should be influenced by the likely material behaviours over the temperature range of interest such as yielding, plastic behaviour and creep.

Duty cycle data has been used to generate new boundary conditions for turbine stage component testing. Turbine housing validation testing has confirmed the suitability of the new TMF modelling approach in that the duty cycle metal temperature / damage cycles and strain based fatigue calculation correlates well to component test results. New limits have been applied to containment, joint rotation, HCF, and LCF testing whilst novel techniques have been applied to joint vibration, wastegate wear testing and oxidation/corrosion testing. Validation for high temperature operation has been completed for existing component configurations and also new configurations and materials.

7. ACKNOWLEDGEMENTS

Thanks to Cummins Westport, Jon Ukpai and Anne Neville from the Institute of Engineering Thermofluids, Surfaces and Interfaces, University of Leeds, Matt Lucas and Andrew Halfpenny from HBN nCode, Sayan Bakshi and Richard Goodyear Cummins Turbo Technologies.

8. REFERENCE LIST

[1] IEA 2010, The Contribution of Natural Gas Vehicles to Sustainable Transport, International Energy, http://www.iea.org

[2] Honardar, S., Busch, H., Schnorbus, T., Severin, C. et al., "Exhaust Temperature Management for Diesel Engines Assessment of Engine Concepts and Calibration Strategies with Regard to Fuel Penalty," SAE Technical Paper 2011-24-0176, 2011, doi:10.4271/2011-24-0176.

[3] Romagnoli, A. and Martinez-Botas, R., 2013, "Heat Transfer Analysis in a Turbocharger Turbine: An Experimental and Computational Evaluation". Applied Thermal Engineering, 38, pp. 58-77.

[4] Sirakov, B. and Casey, M., 2013, "Evaluation of Heat Transfer Effects on Turbocharger Performance". ASME Journal of Turbomachinery, 135(2), 021011

[5] Cormerais, M., Chesse, P., Hetet, J.-F., 2010, "Turbocharger heat transfer modeling under steady and transient conditions", Int J Thermodyn, 12, pp. 193–202.

[6] Baines, N., Wygant, K., Dris, A., 2010, "The analysis of heat transfer in automotive turbochargers", J Eng Gas Turb Power 132:042301.

[7] Burke, R. D., Olmeda, P., Arnau, F. and Reyes-Belmonte, M., 2014, "Modelling of Turbocharger heat transfer under stationary and transient conditions". In: 11th International Conference on Turbochargers and Turbocharging, 2014-05-13 - 2014-05-14, London.

[8] Aghaali, H. and Angström, H.-E., 2013, "Temperature estimation of turbocharger working fluids and walls under different engine loads and heat transfer conditions", SAE Technical Paper 2013-24-0123.

[9] Serrano, J. R., Olmeda, P., Arnau, F. J., Dombrovsky, A., Smith, L., "Turbocharger heat transfer and mechanical losses influence in predicting engines performance by using one-dimensional simulation codes", Energy, 86, pp. 204-218.

[10] Serrano, J., Olmeda, P., Arnau, F. and Reyes-Belmonte, M., 2013, "Importance of Heat Transfer Phenomena in Small Turbochargers for Passenger Car Applications", SAE Int. J. Engines 6(2).

[11] Serrano, J., Olmeda, P., Arnau, F., and Dombrovsky, A., 2015, "General Procedure for the Determination of Heat Transfer Properties in Small Automotive Turbochargers", SAE Int. J. Engines 8(1).

[12] Serrano, J.R., Olmeda, P., Arnau, F.J., Reyes-Belmonte, M.A., Tartoussi, H., 2015, "A study on the internal convection in small turbochargers. Proposal of heat transfer convective coefficients", Applied Thermal Engineering, 89, pp. 587-599.

[13] Payri, F., Olmeda, P., Arnau, F.J., Dombrovsky, A., Smith, L., 2014, "External heat losses in small turbochargers: Model and experiments", Energy, 71, pp. 534-546.

[14] www.ncode.com/turbo

[15] M. B. D. Burkinshaw, *The High Temperature Tribological Performance of Turbocharger Wastegate Materials,* vol. Pages 289–298, 13–14 May 2014.
[16] VDA 230-214, *Resistance of metallic materials to condensate corrosion in exhaust gas-carrying components - Testing methods,* 13-Jul-2012.
[17] J. J. Garcia, *Exhaust Gas condensate Corrosion Test on Low Pressure Cooling System of Aluminum Brazed EGR, ACAC and WCAC,* SAE paper no 2012-01-1947.
[18] H. Lee, C. Jo, Y. Seungsuk, Y. Seyoung, K. Yangrae and J. Jongik, *Optimization of Dual Loop EGR of a V6 3.0 Liter Diesel Engine for CO2 Reduction,* 2013-01-0316, 4th August 2013.
[19] S. Moroz, G. Bourgoin, J. M. Luján and B. Pla, *Acidic Condensation in Low Pressure EGR Systems using Diesel and Biodiesel Fuels,* 2009-01-2805, SAE Int. J. Fuels Lubr. Volume 2 Issue 2, 2009.

SESSION 5A

NOVEL APPLICATION: EURO 7 AND ELECTRIC ASSIST PLUS E-BOOSTING AND DOWN SPEEDING, WASTE HEAT RECOVERY

Low inertia centrifugal compressor wheels: Influence of back disk cavity on aerodynamic losses and axial thrust load

T Fischer, H Rätz, M Peters, J R Seume
Leibniz Universität Hannover, Institute of Turbomachinery and Fluid Dynamics, Germany

ABSTRACT

The influences of back disk cavities on centrifugal compressor aerodynamic losses and axial thrusts are evaluated based on numerical simulations. The influences on compressor performance are found to be negligible. Nevertheless, the consideration of the back disk cavity is necessary for the prediction of axial thrusts and back disk cavity power losses. For this reason, simple analytical equations are proposed for the prediction and the results compared to the numerical simulations. The empirical results show an acceptable agreement concerning pressure prediction at the back disk. To improve the empirical solution, a new approach for the circumferential velocity prediction is introduced.

NOMENCLATURE

$\dot{V}$	volume flow	p_0	total pressure
b	channel height	PRC	compressor pressure ratio
d	diameter	r	radius
DE	diffuser exit to inlet area ratio	R_e	Reynolds number
EI	exit to inlet area ratio	s_{ax}	axial width
FACW	fraction of axial cavity width	TRIM	diameter ratio
FDFR	fraction of design flow rate	u_2	impeller tip speed
f_{fric}	friction coefficient	ΔH_s	total isentr. enthalpy change
FTR	fraction of tip radius	η	dynamic viscosity
HSR	inlet hub to shroud ratio	ρ	desity
K	rotation coefficent	φ	flow coefficient
M_{BD}	torque at back disk	ψ	work coefficient
P	power	ω	angular velocity
p	static pressure		

Subscripts

1	inlet	hb	hub
2	impeller outlet	in	cylindrical portion of BDC inlet
3	diffuser outlet	sh	shroud
BDC	back disk cavity	(t – t)	total to total

Abbreviations

BDC	back disk cavity
CW	compressor wheel
DP	design point
RANS	Reynolds-average Navier-Stokes
uRANS	unsteady Reynolds-average Navier-Stokes
WLTP	Worldwide Harmonized Light Vehicles Test Procedure
wo	without

1 INTRODUCTION

Recent political regulations for future automotive pollutant emissions (e.g. Euro 6) and the introduction of the *Worldwide Harmonized Light Vehicles Test Procedure* (WLTP), which shall replace all test procedures (except U.S.) for light vehicles as from 2017 [1], will have a huge impact on the automotive industry. To comply with the new regulations governments and automotive manufacturers are performing intense research on alternatives for the conventional internal combustion engine. Fuel cell applications are one of the most promising alternatives for the near future [2-5]. Similar to the internal combustion engine, the power density is a key feature for the size and package of this application. For this reason, fuel cells are supercharged to increase the specific performance, thereby keeping the stack size small [6]. For electrically-driven compressors used to provide boost pressure for the cathode air supply, the ramp-up time and for this purpose a low moment of inertia, as well as a high flow range are essential for compressor performance.

To meet the requirements of high flow range and low moment of inertia, high trim compressor wheel designs are one possible solution as presented by Fischer et al. [7]. They used a preliminary high trim impeller to show the potential of an optimized back disk design, regarding inertia reduction. By means of an automatic optimization procedure, an inertia reduction of 6.9 percent was achieved. However they did not investigate the influence of these back disk shapes on compressor performance and axial thrust loads.

Since electrically-driven compressors for fuel cell applications with maximum stage pressure ratios of about 2.5 - 3 [8-10] are equipped with foil respectively air bearings [11], they are particularly prone to axial overloads. Because they are not lubricated axial overloads cause higher wear from solid or mixed friction in comparison to a regular thrust bearing. To approve the suitability of a back disk optimization as developed by Fischer et al. [7] for a compressor wheel for fuel cell applications, a detailed investigation on the influences of back disk cavity shape regarding compressor performance and axial thrust load is necessary.

Fundamental research performed by various authors [12-14] showed the principle of flow and pressure fields between a stationary and a rotating disk. These investigations by means of experiments [12, 13] and direct numerical simulations (DNS) [14] were carried out for various Reynolds numbers, axial clearance to disk radius ratios and wall roughnesses, to determine the influence on disk torque, flow regime and flow transition. Based on this fundamental research, many investigations were carried out to determine the influence of back disk cavities on radial turbomachinery performance.

Based on numerical simulations of a turbocharger, Raetz et al. [15] analyzed the influence of back disk cavity and blow-by on the axial and radial impeller forces for various operating conditions. They identified the radial forces to be small compared to the axial impeller forces. They state, that the flow and pressure fields were approximately uniform in the circumferential direction. The axial thrust was mainly caused by the pressure field at the impeller back disk, which depends on the operating point and therefore on the stage pressure ratio. The influence of blow-by flow had a minor influence on the single stage axial force, but a non-negligible influence on the sum of the turbine and compressor axial forces. Sun et al. [16] concluded that blow-by also has a minor influence on the axial thrust of a centrifugal compressor impeller. Based on numerical simulations of a centrifugal compressor with and without a back disk cavity for various leakage mass flow rates and operating points, the highest difference regarding stage efficiency and pressure ratio was identified between simulations with or without a back disk cavity. They found the influence of leakage flow to be small with an inflow increased axial thrust and power consumption in the back disk cavity, while an outflow decreased both values. Only the stage pressure ratio decreases both with leakage inflow and outflow.

Several attempts on empirical correlations for the prediction of axial forces as well as pressure and flow fields have been published based on the described experimental and numerical investigations [12-14, 17-18]. All investigations were carried out for existing geometries of back disk cavities of centrifugal compressors, or for varying heights between two rotating disks. Investigations on the influence of different shapes of back disk cavity geometry on small centrifugal compressor performance are not known to the authors. For this reason, the present paper focuses on the adoption of several empirical approaches for three different three-dimensional back disk geometries presented by Fischer et al. [7] and their different back disk cavities. For this purpose, intense numerical investigations have been performed for a new high trim centrifugal compressor design developed at the Institute of Turbomachinery and Fluid Dynamics (TFD). The results of empirical correlations and the influence of the back disk cavities on centrifugal compressor performance and axial thrust load will be presented within the following chapters.

2 NUMERICAL MODEL

The numerical model is based on a preliminary high trim compressor wheel (CW) design, developed at the TFD. The compressor performance parameters at design point (DP) are summarized in Tab. 1. The geometrical parameters are given in Tab. 2 and the corresponding positions are illustrated in Fig. 1.

Table 1 – Compressor performance parameters at design point

Flow coefficient at DP	$\varphi = \dot{V}_1 / u_2 d_2^2$	0.15
Work coefficient at DP	$\psi = \Delta H_s / u_2^2$	0.5
Tip speed at DP	$u_2 = 0.5\ \omega\ d_2$	355 m/sec
Pressure ratio at DP	$PRC = p_{0,3} / p_{0,1}$	2.0

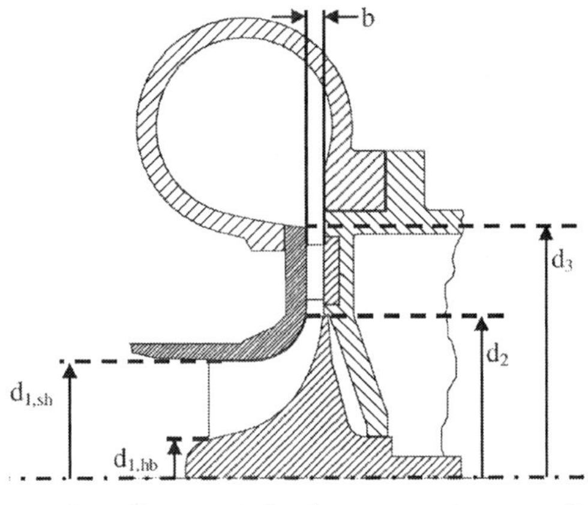

Figure 1 – Geometrical parameter positions

Table 2 – Compressor geometrical parameters

CW diameter ratio	$TRIM = d_{1,sh} / d_2$	0.8
CW exit to inlet area ratio	$EI = 4\ d_2\ b_2 / (d_{1,sh}^2 - d_{1,hb}^2)$	0.68
Diffuser exit to inlet area ratio	$DE = d_3\ b_3 / d_2\ b_2$	1.2
CW inlet hub to shroud ratio	$HSR = d_{1,hb} / d_{1,sh}$	0.3

The software used for the numerical simulations is ANSYS CFX. The numerical model consists of an inlet pipe, a compressor impeller with 6 full and 6 splitter blades, a vaneless diffuser and a collector. The collector is used instead of a real volute, to prevent asymmetric pressure distribution usually caused by the volute's tongue. Furthermore, three different back disk cavity geometries (BDC) are added to the numerical model. The fluid domain of the compressor stage is depicted in Fig. 2. Total temperature and total pressure at inlet (marked blue), and mass flow rate at outlet (marked red) are used for the numerical boundary conditions. A rotational speed is set for the impeller domain, and the wall at the impeller side of the back disk cavity fluid domain, which represents the impeller back disk. For turbulence modeling, the two-equation k-ω shear stress transport model (SST) developed by Menter [19] is used. For the prediction of complex flows with state of the art RANS simulations, the SST model provides the best agreement against experimental data, as shown by Menter [20] and Bardina et al. [21]. The transition between single domains, respectively different frames of reference is modeled with the frozen rotor approach. Wall boundaries are defined as smooth walls and cavity blow-by is neglected. All global results of the compressor stage are evaluated from the inlet of the inlet pipe to the diffuser outlet.

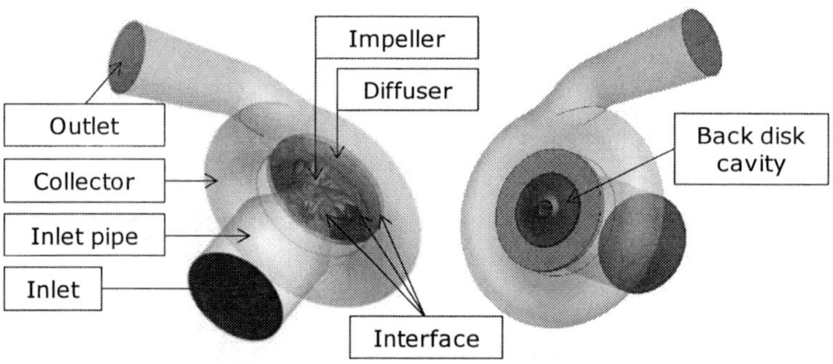

Figure 2 – Numerical model

The investigated back disk cavity geometries are depicted in Fig. 3. The three back disk geometries correspond to the geometries developed by Fischer et al. [7]. The first back disk geometry is the unchanged reference geometry. The second back disk geometry is the result of an automatic optimization procedure. The design target is a minimum moment of inertia without exceeding the required maximum stress limit at the back disk. The third back disk geometry corresponds to the manual re-design of the back disk geometry. The material thickness is reduced between 30 and 75 percent of the impeller tip diameter in order to reduce the moment of inertia. The housing geometry corresponds to the original housing geometry of the reference impeller and is the same for all three back disk cavities.

The first back disk cavity (BDC 1) is of constant gap height (0.3 mm) between housing and back disk. The second back disk cavity (BDC 2) has an increasing gap height from 0.3 to 1.0 mm. The third back disk cavity (BDC 3) consists of a constant gap height (0.3 mm) from 75 percent radius to the tip radius. Below 75 percent tip radius, the gap height increases from 0.3 to 1.5 mm at mid section and decreases to 0.3 mm at compressor shaft. In addition, a radius at the tip of the back surface is applied.

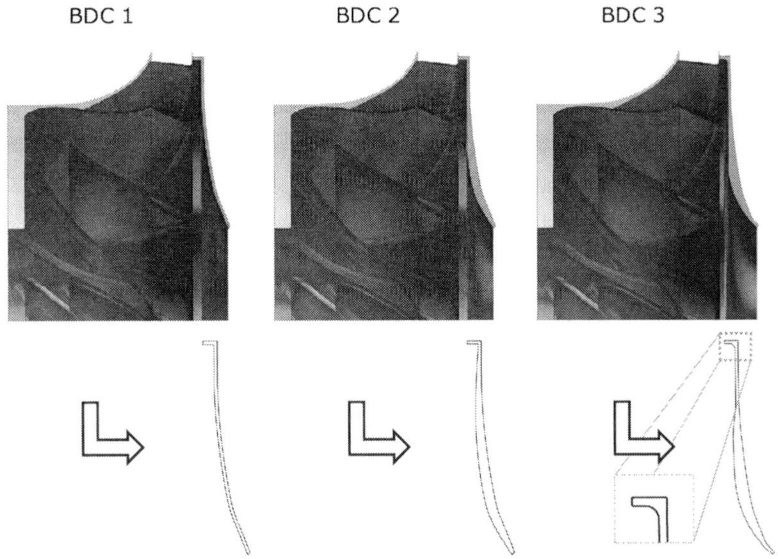

Figure 3 – Back disk cavity geometries

A full 360 degree model of the compressor stage is used for the numerical simulations. Structured meshes with H-, L- and O-grid topologies are used for the inlet pipe, the compressor impeller, the back disk cavity and the diffuser domains. The collector domain is neglected for the evaluation of stage results. For this reason, an unstructured tetrahedral mesh with additional prism layers at the wall region is used to simplify grid generation.

As no experimental data is available for the compressor stage, two grid convergence studies are performed to validate the numerical model. The first one is performed for the grid of the compressor stage and the second for the back disk cavity grid. Three grid sizes are evaluated for the compressor stage: a coarse grid with 8.5 million nodes, a medium grid with 12.6 million nodes and a fine grid with 28.6 million nodes. The results for the stage pressure ratio, isentropic efficiency and static pressure at impeller outlet (2) are shown in Fig 4 (L). Grid convergence of the back disk cavity is evaluated for three grid sizes using the medium size compressor grid. The coarse grid consists of 0.5 million nodes, the medium grid consists of 1.6 million nodes and the fine grid consists of 3.9 million nodes. The results for the axial force at back disk and back disk cavity power loss are depicted in Fig 4 (R).

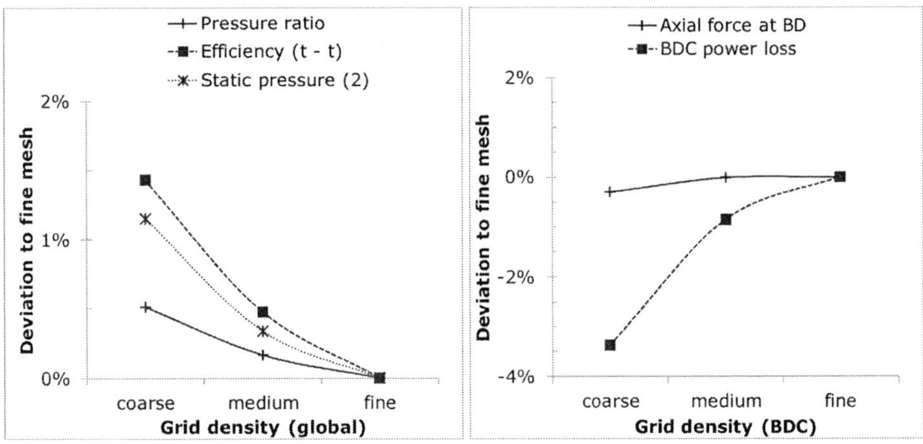

Figure 4 – Results of the grid convergence study global (L) and BDC (R)

Based on these results medium grid sizes are chosen both for the compressor stage and the back disk cavity. It is assumed that the minimal accuracy benefit of the fine grid does not outweigh the disadvantage in terms of low computational efficiency. The numbers of grid nodes and dimensionless wall distances of the chosen numerical grids are summarized in Tab. 3.

Table 3 – Numerical grid data

Total number of nodes (wo BDC)	12,616,596
Number of nodes (inlet pipe)	3,163,728
Number of nodes (impeller)	5,965,890
Number of nodes (diffuser)	2,949,192
Number of nodes (collector)	537,786
Number of nodes (BDC 1)	1,585,080
Number of nodes (BDC 2)	1,902,912
Number of nodes (BDC 3)	2,271,744
Averaged y^+ - value (inlet pipe)	3.02
Averaged y^+ - value (impeller)	1.53
Averaged y^+ - value (diffuser)	4.30
Averaged y^+ - value (BDC 1)	0.57
Averaged y^+ - value (BDC 2)	0.95
Averaged y^+ - value (BDC 3)	0.88

3 NUMERICAL RESULTS

3.1 Global results

All global results are evaluated in terms of dimensionless performance parameters to ensure comparability. These definitions are used according to Hazby et al. [22]:

Flow coefficient $\varphi = \frac{\dot{V}_1}{U_2 \cdot d_2^2}$ (1)

Work coefficient $\psi = \frac{\Delta H_s}{U_2^2}$ (2)

BDC power loss $P_{BDC} = M_{BD} \cdot \omega$ (3)

Stage-specific performance parameters like mass flow rates and shaft powers are given as a fraction of the same parameter at design point. The stage work coefficients are depicted in Fig. 5, for all numerical simulations.

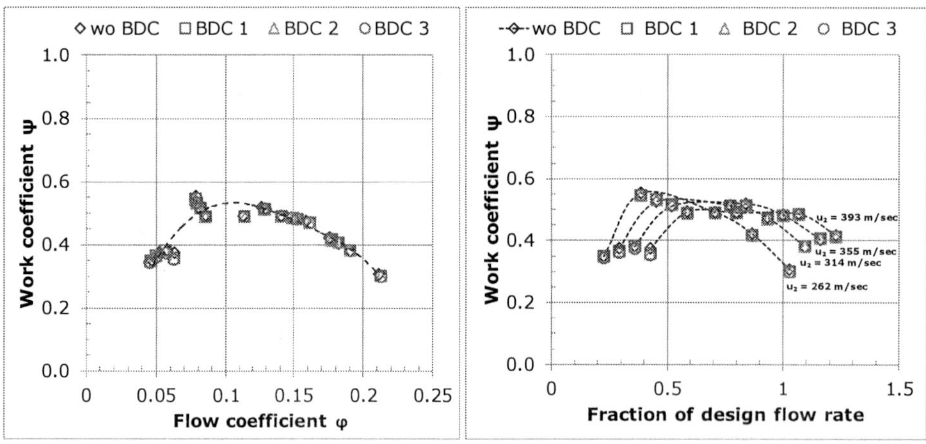

Figure 5 – Work coefficient vs. flow coefficient (L) and FDFR (R)

The corresponding stage pressure ratios are shown in Fig. 6 (L). Based on these results, no significant influence of back disk cavity geometry on stage work can be determined. A minor drop of total pressure ratio, respectively work coefficient can be observed only at high stage pressure ratios. The magnitude of the total pressure drop is summarized in Fig. 6 (R) for an impeller tip speed of 355 m/sec. It can be seen that the maximum change of stage pressure ratio is 0.8 percent at FDFR = 0.5 - 0.8, between the numerical models with and without back disk cavity.

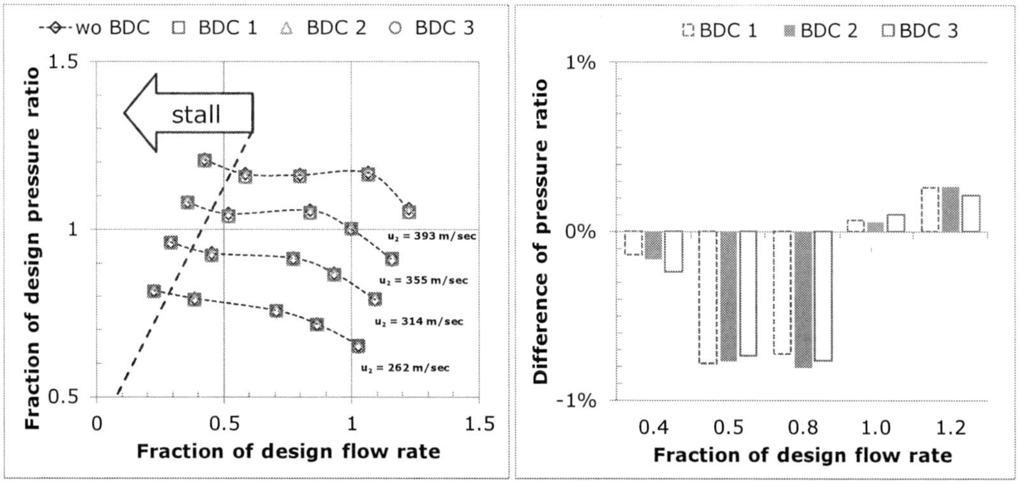

Figure 6 – Pressure ratio vs. FDFR (L) and difference of pressure ratio at design tip speed (R)

The area of the performance map declared as "stall", is defined as compressor operating points where about 30 percent of the inducer area at the shroud region is blocked by backflow. The occurring axial velocity profile is illustrated in Fig. 7. This result does not necessarily imply that a stable compressor operation is not possible, rather that steady-state numerical simulations cannot deliver accurate results due to the unsteadiness of compressor operation. Results for such operating points need to be validated by gas stand measurements or uRANS simulations. For this reason operating points with flow coefficients lower than 0.09 are not considered for the local analysis of numerical results.

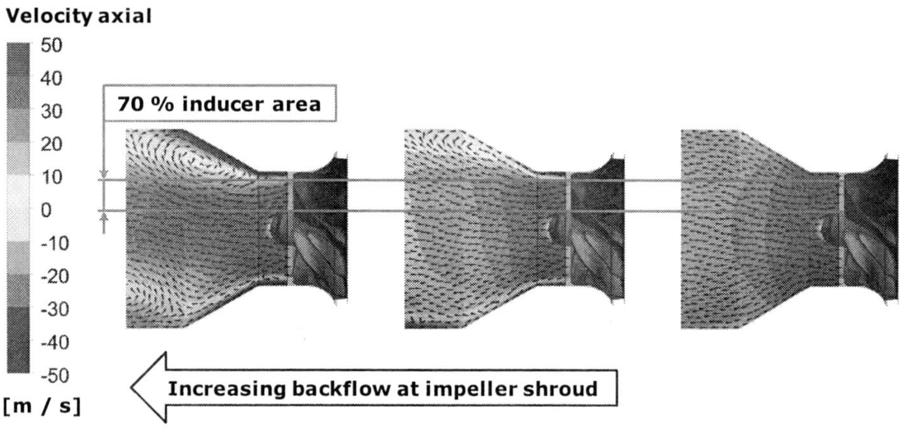

Figure 7 – Backflow (blue) at the inducer for FDFR = 0.2 (L), FDFR = 0.4 (M) and FDFR = 0.7 (R)

The impact of the back disk cavity on required compressor shaft power is small. The required shaft power generally increases for high rotational speeds and mass flow rates as depicted in Fig. 8 (L) by means of a normalized performance map. The shaft power increase due to the back disk cavity is depicted in Fig. 8 (R), at design tip speed (355 m/sec). The maximum shaft power increase adds up to 1.1 percent for BDC 1.

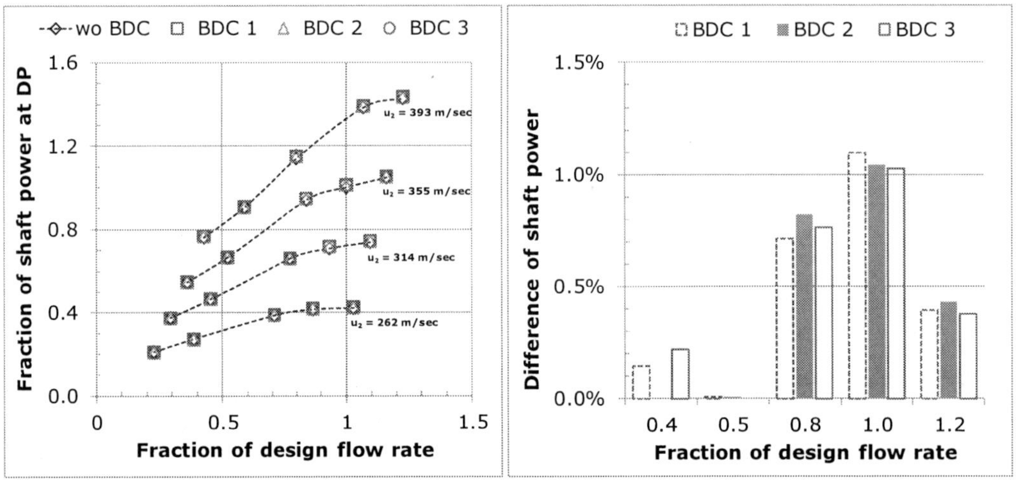

Figure 8 – Normalized shaft power vs. FDFR (L) and difference of shaft power at design tip speed (R)

The most significant influence of the back disk cavity on the numerical results of a centrifugal compressor stage is the magnitude of the predicted axial thrust loads and back disk cavity power losses. Without a back disk cavity domain, a direct evaluation of axial thrust loads from numerical simulations is not possible. With a back disk cavity domain the axial thrust loads can be predicted accurately. The influence of the back disk cavity shape on the total amount of axial thrust load and back disk cavity power loss is shown in Fig. 9.

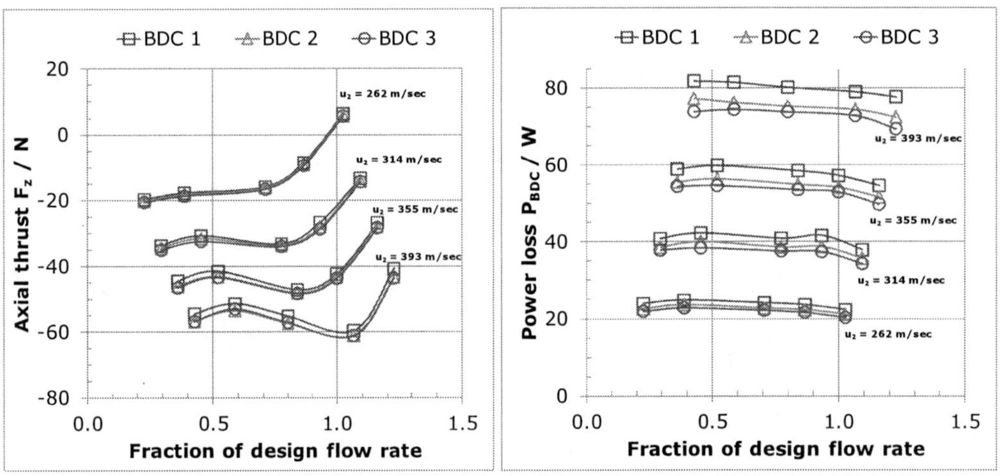

Figure 9 – Axial thrust (L) and BDC power loss (R)

Even with these relatively similar geometries the increase of predicted axial thrust loads has a magnitude of up to 5 percent between BDC 1 and 3, and an average difference of 3 percent for design tip speed (355 m/sec). The maximum back disk cavity power loss decrease of approximately 9 percent is found when comparing BDC 1 and 3. The overall reduction of power losses exceeds 5 percent for all simulated operating points at design tip speed, compared to BDC 1 (Fig. 10).

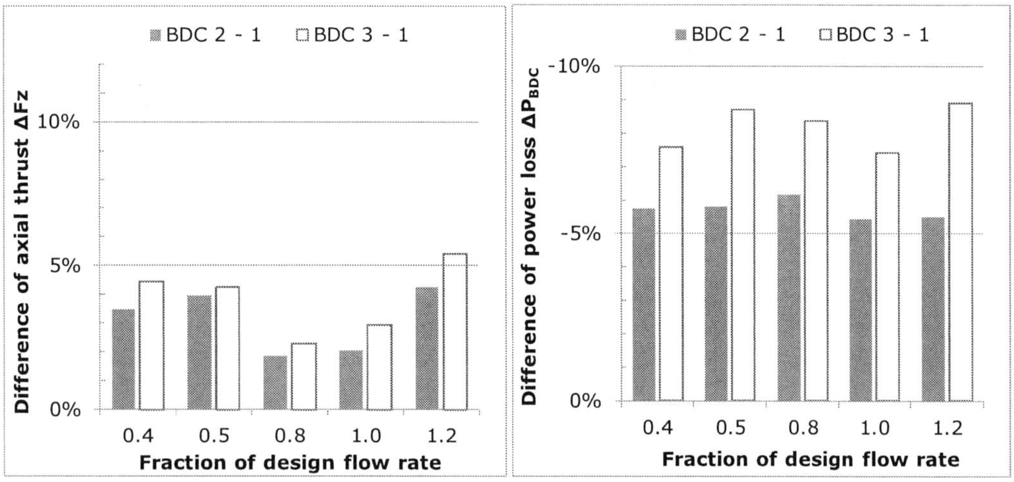

Figure 10 – Change of axial thrust (L) and BDC power loss (R) referred to BDC 1 at design tip speed

3.2 Local results

The quantification of local differences between the three back disk cavity geometries will be presented in this section. The results will be evaluated in two levels of detail. At first the distribution of thermodynamic values will be evaluated as area-averaged values at various radii. Secondly the distributions will be evaluated between the rotating back disk and the stationary housing surfaces. The positions of global and local evaluations are illustrated in Fig. 11. Twenty locations at every 0.04 normalized radius are used for the global evaluation and twenty-one axial positions at every 0.05 normalized channel width are applied for the local evaluation.

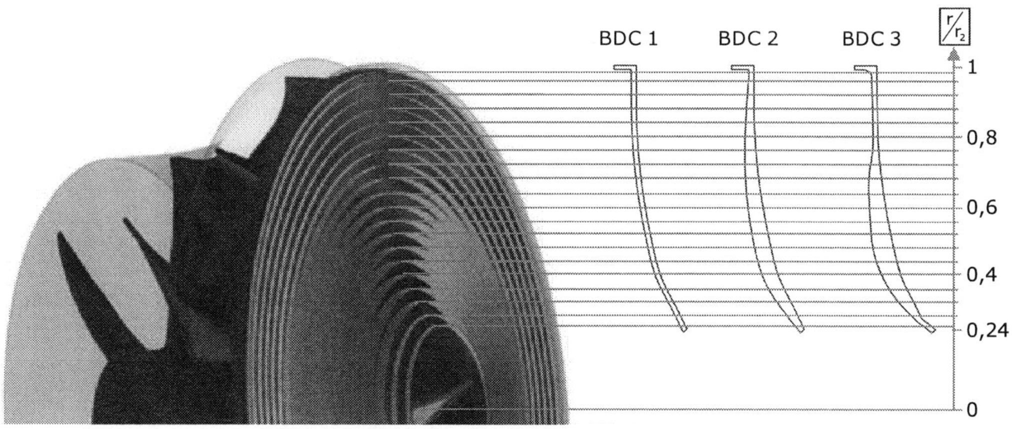

Figure 11 – Locations of evaluation

The results are presented for the design point and FDFR = 0.84 at design tip speed, because of the stable compressor operation at these points. For this reason, the influences of unsteady effects are negligible and steady state simulations have good numerical convergence. The circumferentially-averaged total and static pressure distributions at the back disk are depicted in Fig. 12, for all geometries. For both operating points the highest pressure drop occurs for BDC 1. As a consequence BDC 1 provides the lowest axial thrust load. The comparison of BDC 2 and 3 shows that BDC 3 has a lower total pressure at BDC inlet, due to the radius at impeller tip. However the continuously increasing gap width of BDC 2 results in a slightly higher static pressure drop, as illustrated in Fig. 13.

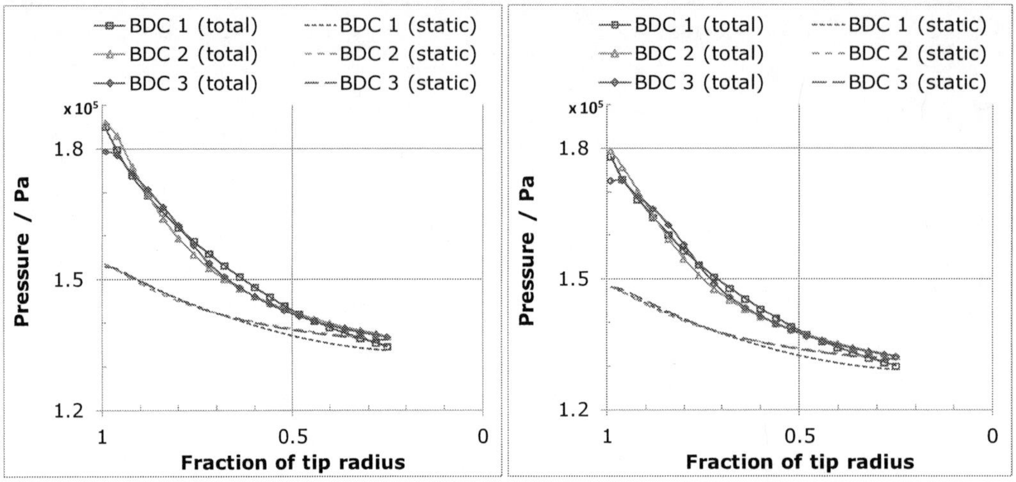

Figure 12 – BDC pressure distributions for FDFR = 0.84 (L) and DP (R)

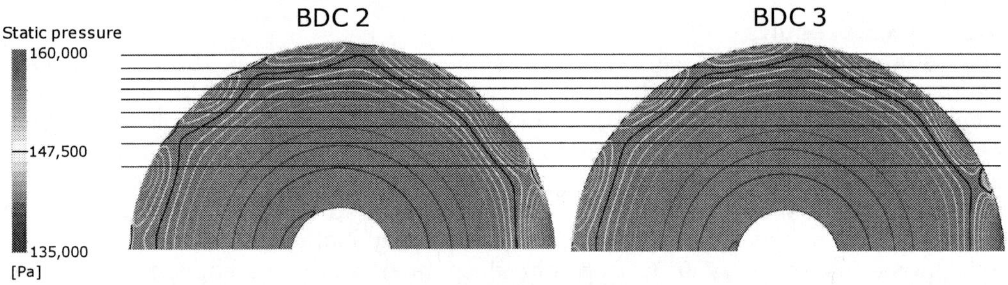

Figure 13 – Static Pressure at back disk (FDFR = 0.84)
BDC 2 (L) and BDC 3 (R)

The Pressure distribution between back disk and housing for the design point and FDFR = 0.84 are depicted in Fig. 14. For reasons of comprehensibility only positions at FTR = 0.99 (radial inlet) and at FTR = 0.52 (max. width of BDC 2 and 3) are used, and all three back disk cavities are plotted in a single diagram. The different static pressure levels are in accordance with previous results and the pressure distribution is almost constant in the axial direction.

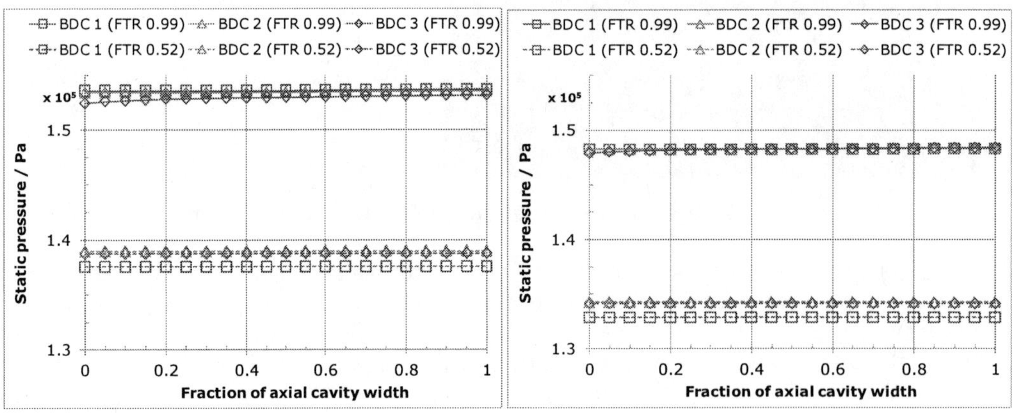

Figure 14 – Pressure distributions between back disk and housing
for FDFR = 0.84 (L) and DP (R)

The velocity distributions from back disk to the housing surface are depicted in Fig. 15, at the same radial positions. The flow regime is a blend of the regimes presented by Daly and Nece [13]: turbulent flow with separated boundary layers (s_{ax} > 0.6 mm) and turbulent flow with merged boundary layers (s_{ax} < 0.6 mm).

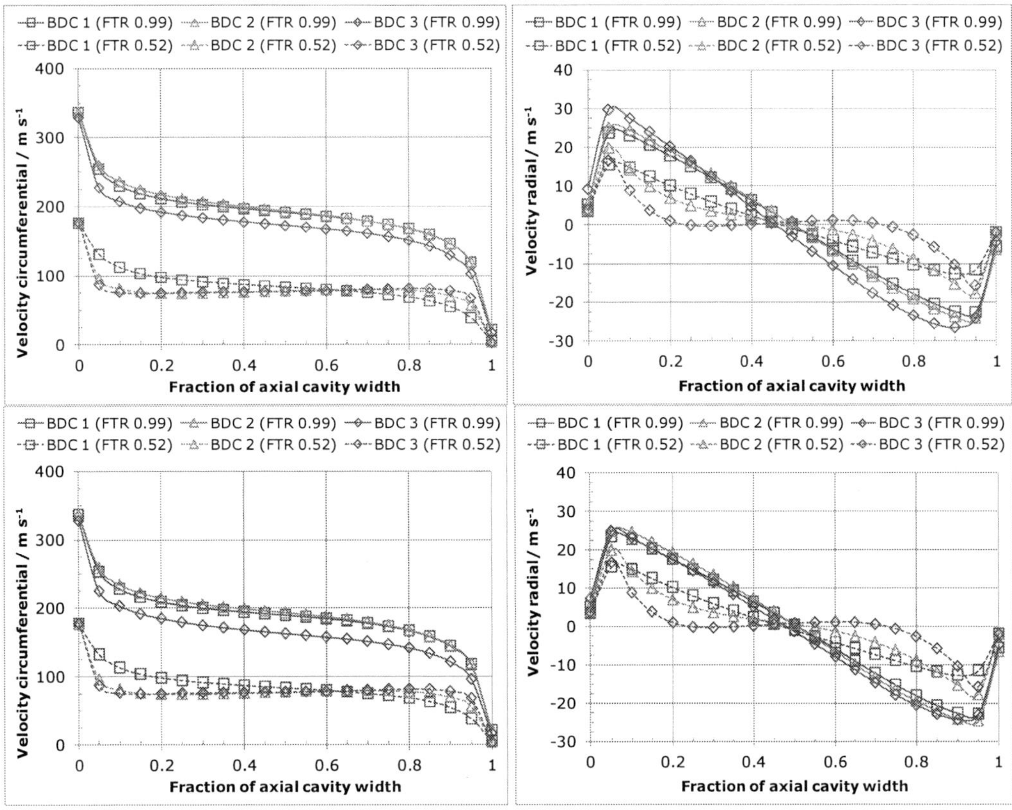

Figure 15 – Flow regime between back disk and housing; velocity circumferential (L) and velocity radial (R) (FDFR = 0.84 (T) and DP (B))

4 EMPIRICAL CORRELATIONS

As already mentioned, without a back disk cavity domain a direct prediction of axial thrust loads from numerical simulations is not possible. The influences of the back disk cavity on global numerical results of a compressor stage are however fairly negligible for the geometries investigated within this paper, as shown in the former chapter. This implies the question of whether a simple empirical correlation is capable of an accurate axial thrust load prediction, without being adjusted in advance. Such an approach significantly reduces the resources needed for numerical simulations during preliminary design of compressor stages.

The pressure distribution of the impeller back disk cavity can be calculated by the following equation:

$$\frac{dp}{dr} = \rho \cdot K^2 \cdot \omega^2 \cdot r \qquad (4)$$

The rotation coefficient K defines the ratio between angular impeller velocity ω and angular velocity of fluid rotation. The coefficient can be calculated from measured pressure distributions as follows:

$$K_{calc} = \sqrt{\frac{2\cdot(p_2-p(r))}{\rho_2\cdot\omega^2\cdot(r_2^2-r^2)}} \quad (5)$$

The rotation coefficients evaluated from CFD results for FDFR = 0.84 and the design point are depicted in Fig. 16. The values differ both for varying flow rates and different back disk cavities.

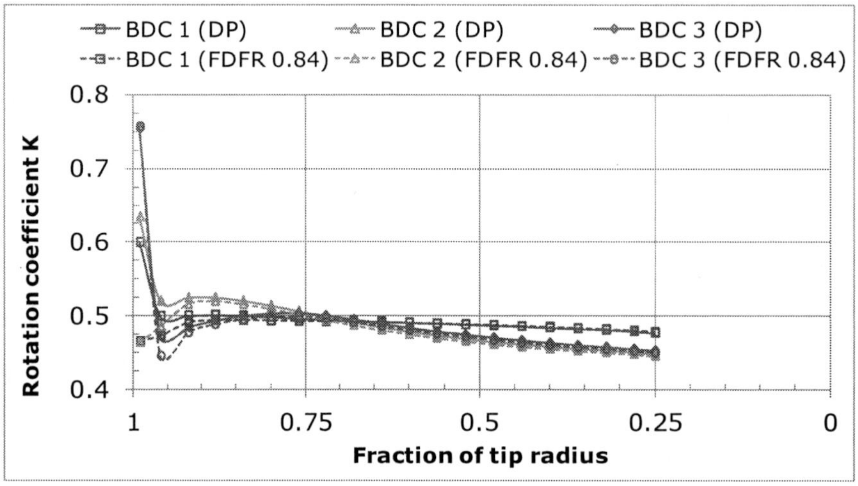

Figure 16 – Rotation coefficients from CFD simulations (Eq. 5)

There are different formulations for the empirical determination of the rotation coefficient. The following approaches are used by [17-18]:

$$K_{[17]} = 0.5 \quad (6)$$

$$K_{[18]} = \left[1+\left(\frac{r_{BDC}}{r_2}\right)^2\cdot\sqrt{\left(\frac{r_{BDC}}{r_2}+5\cdot\frac{s_{ax,in}}{r_2}\right)\cdot f_{fric}}\right]^{-1} \quad (7)$$

Based on numerical results for the three back disk cavity geometries, a new empirical correlation for the determination of the rotation coefficient is introduced:

$$K_{new}(r) = 1.4\cdot\left(R_e^{-0.75}\cdot s_{ax}^{*\ -0.5}\right)^{0.095} \quad (8)$$

The Reynolds number Re and the ratio between local cavity width and local radius s_{ax}^* are used as dimensionless parameters for geometry characterization:

$$R_e = \frac{\rho_2\cdot\omega\cdot r_2^2}{\eta_2} \quad (9)$$

$$s_{ax}^* = \frac{s_{ax,i}}{r_i} \quad (10)$$

5 COMPARISON OF CFD SIMULATIONS AND EMPIRICAL CORRELATIONS

The empirical results for the back disk cavity of constant width (BDC 1) are in very good accordance with the numerical results as shown in Fig. 17. The pressure distributions match entirely. The rotation coefficients match except for the outer radii. This difference results from the neglection of inflow pressure losses.

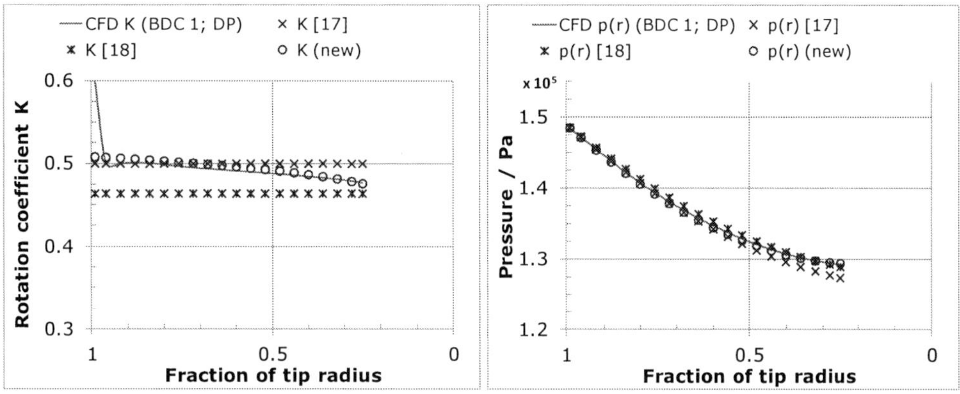

Figure 17 – Rotation coefficients (L) pressure distributions (R) (BDC 1; DP)

For the back disk cavity of varying width (BDC 2), the empirical results show good agreement with the numerical results, concerning pressure distribution (Fig. 18). The match of the pressure distributions is improved compared to the empirical correlations from literature (Eq. 6 and 7). The rotation coefficients do not match but still show a comparable trend to the numerical results.

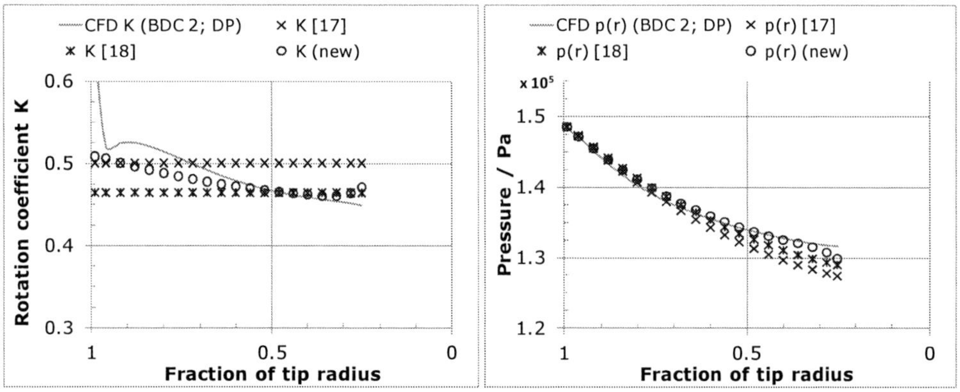

Figure 18 – Rotation coefficient (L) pressure distribution (R) (BDC 2; DP)

The results of the empirical correlations and numerical simulations are shown in Fig. 19, for the back disk cavity of varying width (BDC 3). The pressure distributions are in accordance with the numerical results. Similar to the results for BDC 2, the rotation coefficients do not match but have a similar trend to the numerical results.

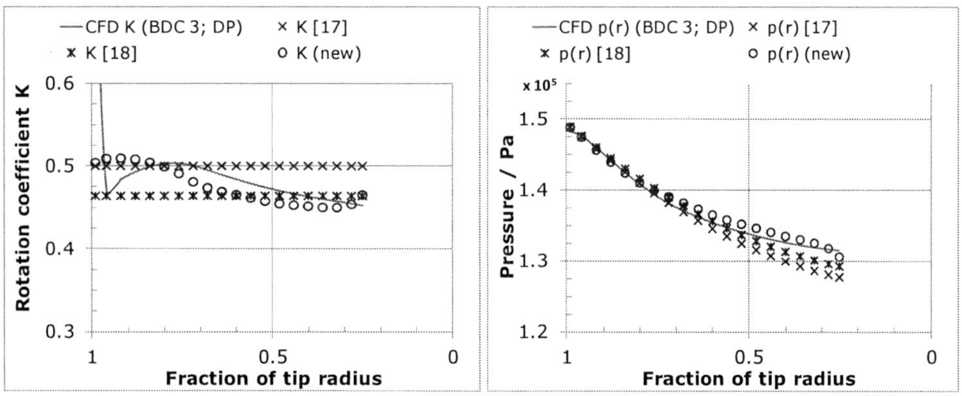

Figure 19 – Rotation coefficients (L) pressure distributions (R) (BDC 3; DP)

6 CONCLUSIONS

In the present paper, a numerical investigation of complex back disk cavity flows was presented. This research was performed to first evaluate the necessity of back disk cavity geometry consideration for numerical simulations during compressor wheel design.

The numerical model was secondly used to investigate the influences of three different back disk cavity geometries on centrifugal compressor performance. The influence on characteristic stage performance parameters was found to be minor. During preliminary design these influences can be neglected, because they were in the range of numerical uncertainty.

However it should be mentioned, that the influence of back disk cavity on stage performance tends to increase with increasing pressure ratio and rotational speed. The local influence of back disk cavity geometry on axial thrust load and back disk cavity power loss was found to be significant.

Two main conclusions can be drawn: The smaller the back disk cavity width, the smaller the area-averaged pressure at the back disk and thus axial thrust load. The larger the back disk cavity width is, the lower the back disk cavity power loss. It was also shown that a back disk cavity with a small width at the impeller tip and a large width at the shaft is advantageous concerning cavity power losses and disadvantageous concerning axial thrust loads, compared to geometries with a continuously increasing gap width.

The back disk geometries with a reduced moment of inertia were found suitable for an electrically-driven compressor because the change of axial thrust load did not exceed the capacity of a conventional thrust bearing. For the prediction of back disk pressure distributions, a new empirical correlation was introduced based on the numerical simulations. The results showed a good fit for the geometries investigated in the present paper.

Further investigations must be performed in order to extend the empirical solution to asymmetric cavity geometries. For this purpose a larger variety of back disk cavity geometries needs to be examined. Furthermore, pressure losses between impeller tip and radial cavity inflow will be investigated in further detail. Finally the influence of different volute, diffuser and impeller geometries will be investigated to extend the validity of the empirical correlation.

ACKNOWLEDGEMENTS

The authors would like to acknowledge the contribution of the Regionales Rechenzentrum für Niedersachsen (RRZN). Finally, we acknowledge the valuable suggestions of the anonymous reviewers.

REFERENCE LIST

[1] Mock, P., Kühlwein, J., Tietge, U., Franco, V., Bandivadekar, A., German, J., (2014), "The WLTP: How a new test procedure for cars will affect fuel consumption values in the EU", ICCT, working paper 214-9, 29 October, 2014
[2] Töpler, J., Lehmann, J., (2014), "Wasserstoff und Brennstoffzelle - Technologien und Marktperspektiven", Springer Vieweg, ISBN 978-3-642-37415-9, 2014

[3] Reich, G., Reppich, M., (2013), "Regenerative Energietechnik – Überblick über ausgewählte Technologien zur nachhaltigen Energieversorgung", Springer Vieweg, ISBN 978-3-8348-8614-9, 2013
[4] http://europa.eu/rapid/press-release_MEMO-13-1131_de.htm
[5] http://www.fch.europa.eu/page/who-we-are
[6] Kurzweil, P., (2013), "Brennstoffzellentechnik - Grundlagen, Komponenten, Systeme, Anwendungen", 2nd Ed., Springer Vieweg, ISBN 978-3-658-00085-1, 2013
[7] Fischer, T., Peters, M., Seume, J.R., (2015), "Low Inertia Centrifugal Compressor Wheels: Back Disk Geometry Optimization", 11th International Gas Turbine Congress, 15-20 November, Tokyo, Japan, IGTC15-0109, 2015
[8] Balasubramanian, B., Barbir, F., Neutzler, J., (1999), "Optimal Operating Temperature and Pressure of PEM Fuel Cell Systems in Automotive Applications", Preprints of Symposia, American Chemical Society, 44, 4, pp. 977-981, 1999
[9] Barbir, F., (2008), "Vehicles with Hydrogen-Air Fuel Cells", Energy Carriers and Conversion Systems with Emphasis on Hydrogen, Vol. 2, pp. 25-45, 2008
[10] Blunier, B., Miraoui, A., (2007), "Air Management in PEM Fuel Cells: State-of-the-Art and Prospectives", ACEMP '07 International Aegean Conference on Electrical Machines and Power Electronics, Bodrum, pp. 245-254, 2007
[11] Metz, D., Werner, J., Münz, S., Becker, M., (2013), "Charging System for Fuel Cell Applications", MTZ Worldwide, April 2013, Vol. 74, Issue 4, pp. 40-43, 2013
[12] Nece, R.E., Daily, J.W., (1960), "Roughness Effects on Frictional Resistance of Enclosed Rotating Disks", Trans. ASME, Journal of Basic Engineering, Vol. 82, pp. 553-562, 1960
[13] Daily, J.W., Nece, R.E., (1960), "Chamber Dimension Effects on Induced Flow and Frictional Resistance of Enclosed Rotating Disks", Trans. ASME, Journal of Basic Engineering, Vol. 82, pp. 217-232, 1960
[14] Serre, E., Tuliszka-Sznitko,E., Bontoux, P., (2004), "Coupled numerical and theoretical study of the flow transition between a rotating and a stationary disk", Physics of Fluids, Vol. 16, No. 3, pp. 688-706, 2004
[15] Raetz, H., Kammeyer, J., Natkaniec, C.K., Seume, J.R., (2011), "Numerical Investigation of Aerodynamic Radial and Axial Impeller Forces in a Turbocharger", ASME Turbo Expo, 6-10 June, Vancouver, Canada, GT2011-46360, 2011
[16] Sun, Z., Tan, C., Zhang, D., (2009), "Flow Field Structures of the Impeller Backside Cavity and its Influences on the Centrifugal Compressor", ASME Turbo Expo, 8-12 June, Orlando, Florida, USA, GT2009-59879, 2011
[17] Zhang, J., Sun, H., Hu, L., He, H., (2010), "Fault Diagnosis and Failure Prediction by Thrust Load Analysis for a Turbocharger Thrust Bearing", ASME Turbo Expo, 14-18 June, Glasgow, UK, GT2010-22320, 2010
[18] Gülich, J.F., (2003), "Disk friction losses of closed turbomachine impellers", Forschung im Ingenieurwesen 68, pp. 87-95, Springer Verlag, DOI 10.1007/s10010-003-0111-x, 2003
[19] Menter, F.R., (1994), "Two-Equation Eddy-Viscosity Turbulence Models for Engineering Applications", AIAA Journal, Vol. 32, No. 8, August 1994
[20] Menter, F.R., (2009), "Review of the shear-stress transport turbulence model experience from an industrial perspective", Int, Journal of Computational Fluid Dynamics, Vol. 23, No.4, DOI 10.1080/10618560902773387, 2009
[21] Bardina, J.E., Huang, P.G., Coakley, T.J., (1997): "Turbulence Modeling Validation, Testing, and Development", NASA-TM-110446, 1997
[22] Hazby, H. R., Casey, M. V., Numakura, R., Tamaki, H., (2014), "Design and testing of a high flow coefficient mixed flow impeller", Proceedings, 11th International Conference on Turbochargers and Turbocharging, London, pp. 55-64, 2014

Investigation of an electrically assisted turbocharger with energy recovery for a heavy duty diesel engine

A Rezk[1], J Carter[2], M L Jupp[1], A K Nickson[2], J Allport[1]
[1] Turbocharger Research Institute, University of Huddersfield, UK
[2] BorgWarner Turbo Systems, UK

ABSTRACT

An electrically assisted turbocharger is an advanced boosting technology that enhances the overall performance and reduces emissions of downsized engines. This paper investigates the effect of electric turbo assist [ETA] technology on a heavy duty, on-road diesel engine during steady state operation using a GT-Power 1-D engine simulation model. Two key findings have been observed: [1] Increasing assisting power at fixed torque, AFR and EGR rate values, leads to a reduction in pumping work, enhanced engine efficiency and reduced compressor mass flow, pressure ratio and turbocharger shaft speed. [2] Contrary to expectations, a fuel economy benefit was observed whilst assisting the turbocharger during high engine speed/full load operation and recovering energy at relatively low engine speeds and loads.

NOMENCLATURE

AFR	Air to fuel ratio [-]
B	cylinder bore [m]
CE	combustion efficiency [fraction]
D_i	Combustion state duration angle [deg]
EGR	Exhaust gas recirculation [-]
E_i	Combustion state exponent [-]
EPA	Environmental Protection Agency [-]
ETA	Electric turbo assist [-]
FGT	Fixed geometry turbine [-]
F_i	combustion state fraction [-]
GHG	Greenhouse gas [-]
htc	convective heat transfer coefficient [$wm^{-2}K^{-2}$]
ID	ignition delay angle [deg]
min	minimum
P	cylinder pressure [kPa]
P_m	motoring fluid pressure at the same angle as P [kPa]
PMEP	Pumping mean effective pressure [bar]
P_r	working fluid pressure prior to combustion [kPa]
SOI	start of ignition angle [deg]
Sp	mean piston speed [ms^{-1}]
T	cylinder temperature [K]
TAL	Time at Level [-]

TIP	Turbine inlet pressure [bar]
V_d	displacement volume [m^3]
V_r	working fluid volume prior to combustion [m^3]
VTG	Variable turbine geometry [-]
w	average cylinder gas velocity [m/s]
W	Power [kW]
WC_i	Combustion state Wiebe's constant [-]
WHSC	World heavy duty steady-state on-road cycle [-]
θ	crank shaft angle [deg]

1 INTRODUCTION

HD vehicles are the second largest contributor in terms of oil consumption and greenhouse gas emission [GHG]. They account for 26% of all the CO_2 emissions from, and fuel consumption of EU road transport with an expected increase of energy consumption and direct CO_2 emissions of almost 15% by 2030 (1). This has led to an international policy controlling HD vehicles fuel consumption and emissions. The EPA is targeting the reduction of GHG emission by 250 million metric tons and an increase in fuel efficiency of 25% during 2014-2018 (2). Engine downsizing is an emissions and GHG reduction technology that has the potential to give benefits of reduced engine cost, size, mass and friction resulting in better fuel economy. To support engine downsizing, advanced boosting technologies are required to compensate specific power output and improve the overall engine efficiency (3, 4). For down-speeded and/or down-sized engines, one form of electrical turbo assist, eTurbo™ provides an extra degree of freedom to manage the engine AFR.

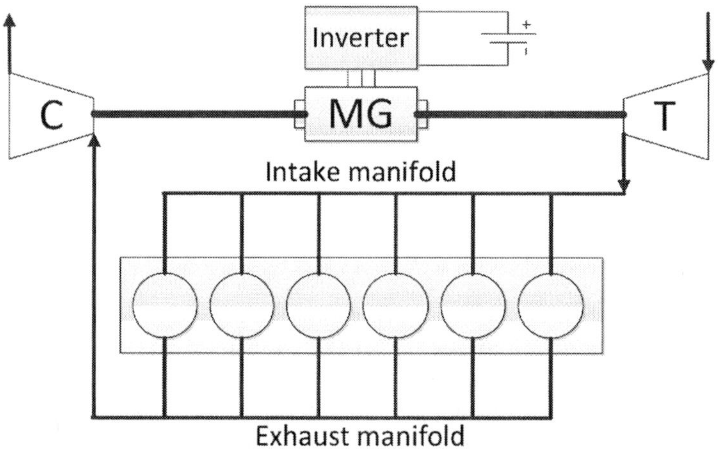

Figure 1 Schematic diagram of eTurbo unit integrated to 6 cylinder engine

eTurbo™ is a hybrid turbocharger and one of the applied ETA topologies, allowing a bidirectional energy transfer to and from the turbocharger, figure 1. This is achieved by means of a high-speed motor/generator unit mounted directly on the turbocharger shaft. Typically, the electric machine is used to assist the turbocharger at low speed where there is insufficient energy in the exhaust gas to generate the required boost, thereby improving the turbocharger transient response. At high engine speed, excess exhaust gas energy is used to generate electricity instead of being lost through opening the VTG nozzle or by wastegating (5, 6). The energy consumed during the electric assist needs to be generated and stored in the battery system during generation mode. The excess amount of the generated energy contributes to powering the engine auxiliaries and reduces the parasitic loads drawn from the engine crankshaft (7, 8).

Over the last decade, increasing interest has been noted towards ETA with different topologies to improve upon conventional turbocharger systems (9, 10). Applying ETA directly to the turbocharger shaft has been investigated for different engine applications during transient and steady state operations. Applying ETA on a fixed geometry turbine [FGT] turbocharged engine has been found to improve the steady state engine torque and fuel consumption by up to 100% and 12% respectively (11, 12). During engine acceleration, the turbo lag and hence the time required to reach the target boost has been shortened by up to 75% (11-13).

Applying ETA on a variable turbine geometry [VTG] turbocharged engine adds an extra degree of freedom and significantly affects the engine performance. The hybrid turbocharger is an efficient tool to improve the engine transient response compared with the conventional VTG turbocharger (8). Steady state simulation results of ETA coupled with a VTG revealed a decrease in engine back pressure and hence an improvement in the overall engine performance and fuel economy (7, 14, 15).

This paper extends ETA feasibility research using eTurbo™ in steady-state operation by investigating the effect of coupling a VTG turbocharger and electric assist at constant torque, AFR and EGR rates using a back to back 1-D comparative generic simulation model in GT-Power. The investigation includes the effect of using eTurbo™ over a WHSC duty cycle and the feasibility of recovering the required energy during generating mode.

2 ENGINE MODELLING

A one-dimensional generic engine simulation model in GT-Power has been developed by BorgWarner Turbo Systems. The modelled engine is a 13 litre, 6-cylinder, VTG turbocharged diesel engine with high pressure EGR. The engine is equipped with a charge air intercooler and an EGR intercooler. Such engines are commonly used in heavy duty on-road applications such as long-haul commercial vehicles and coaches, contributing 43% to transportation sector GHG. For this research, an external power module has been integrated into the existing model to simulate the electrified power exchange with the turbocharger shaft during assisting and generating modes, figure 2. The power transmission and battery charging/discharging losses are neglected at this stage of the investigation.

The in-cylinder instantaneous heat transfer is modelled using a Woschni universal model (16, 17). The Woschni convective heat transfer coefficient calculation is shown in Equation 1, where the temperature exponent is -0.5 for computational efficiency purposes (18). The average cylinder gas velocity is determined using equation 2 and the constant C_1 is determined by equation 3. The constant C_2 has a value of 0 during cylinder gas exchange and compression and 3.24E-3 during combustion and expansion.

$$htc_{Woschni} = 3.01426P^{0.8}w^{0.8}B^{0.2}T^{-0.5} \quad [1]$$

$$w = C_1\overline{S}_p + C_2\frac{V_dT_r}{P_rV_r}(P - P_m) \quad [2]$$

$$C_1 = 2.28 + 3.9 \cdot min\left(\frac{net\ mass\ flow\ into\ cylinder\ from\ valves}{Trapped\ Mass \cdot Engine\ Frequency}, 1\right) \quad [3]$$

The transient heat released during combustion is determined using Wiebe's model for every engine cycle. It is used to describe the rate of heat released during premixed, diffusion and late combustion (18, 19), Equations 3-5.

$$HR = (CE) \cdot \sum_{i=1}^{3}(F_i) \cdot \{1 - exp[(-WC_i \cdot \langle\theta - SOI - ID\rangle)^{(E_i+1)}]\} \quad [4]$$

$$WC_i = \left[\frac{D_i}{2.302^{1/(E_{i+1})} - 0.105^{1/(E_{i+1})}}\right]^{-(E_i+1)} \quad [5]$$

$$F_2 = (1 - F_1 - F_3) \quad [6]$$

Where subscript i represents the cyclic combustion states (1) premixed, (2) main diffusion and (3) late combustion. The fraction (F_i) is 0.05 during premixed state and function speed and torque during late combustion. The premixed, diffusion and late combustion exponents (E_i) are 0.7, 1 and 3 respectively. The ignition (ID) delay is a function of engine speed and torque. The engine EGR target is function of engine load and speed. The engine EGR controller manipulates the EGR valve to achieve the target level. The pressure drop across the EGR valve is a vital parameter in achieving the EGR level target.

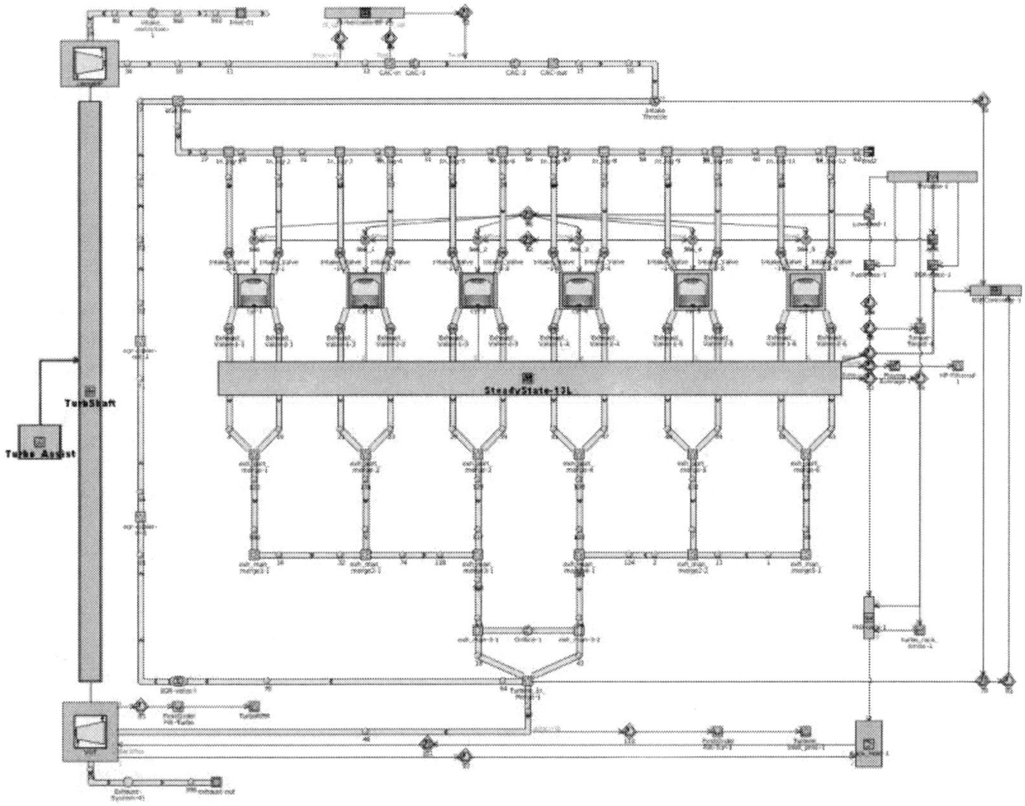

Figure 2 Engine simulation model symbolic representation

3 eTurbo™ EFFECT ON ENGINE PERFORMANCE

The engine performance was investigated by applying different ETA power levels to the turbocharger shaft. Three engine speeds have been applied with approximately constant load, AFR and EGR. Table 1 presents the operating points and the corresponding operating parameters. AFR and EGR values are based on predefined mapped values corresponding to the engine speed and load.

An unexpected key finding from this work centres on an observed reduction of turbocharger speed as the assisting power level is increased. As the ETA value is varied, the turbine nozzle adjusts its position under engine control to maintain engine torque by either reducing or increasing the turbine inlet pressure [TIP]. Adding 15 kW ETA to the turbocharger shaft at 76% and 88% normalized engine speed increased the turbine nozzle opening from 65% to 91% and 66% to 98% respectively. On the other hand, subtracting 10 kW of the turbocharger shaft reduces the turbine nozzle opening from 65% to 49% and 83% to 65% at 76% and 95% normalized engine speed.

Table 1 Operating points parameters

Normalized Engine speed [%]	Normalized Engine load [%]	EGR [%]	AFR [-]
76	100%	17	22
86	100%	17	22
95	100%	18	22

Increasing VTG nozzle opening has the effect of reducing the engine back pressure, PMEP and pumping work. The engine pumping work is the work done by the piston on the cylinder gases during the intake and exhaust strokes. The overall effect of reducing pumping work is to increase engine efficiency. The more efficient the engine, the lower the fresh air requirement to achieve the same torque at a fixed EGR and AFR resulting in a reduced turbocharger shaft speed.

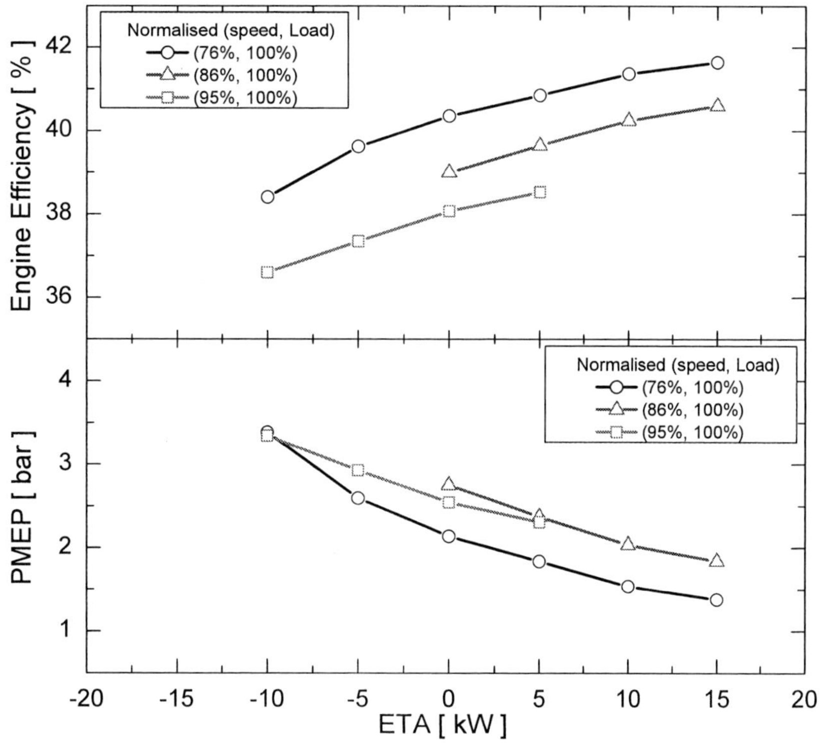

Figure 3 eTurbo™ effect of on engine efficiency

Figure 3 shows eTurbo™ effect on the engine PMEP and efficiency. Adding 15 kW ETA to the turbocharger shaft at 76% and 88% normalized engine speed reduces PMEP by 35.5% and 33% and improves the engine efficiency by 3% and 4.1% respectively. Subtracting 10 kW ETA from the turbocharger shaft at 76% and 95% normalized engine speed increases PMEP by 58% and 31.5% and decreases the engine efficiency by 5% and 4.1% respectively. Figure 4 shows the linear negative trend relationship between engine pumping work and efficiency.

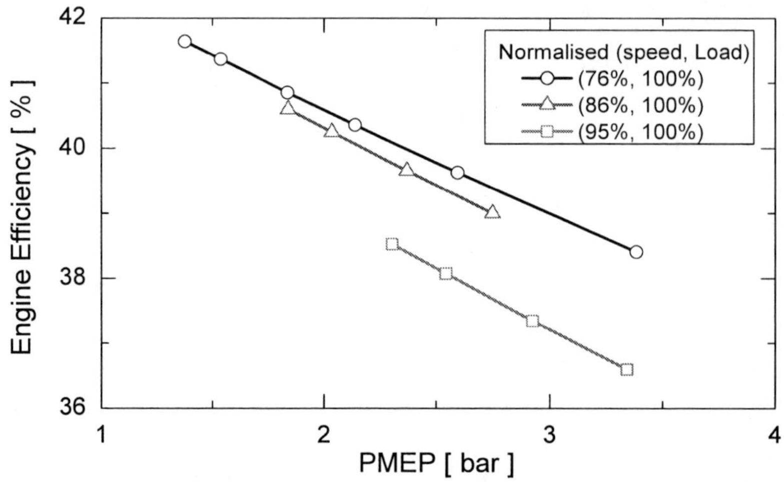

Figure 4 The relation between engine PMEP and efficiency

Figure 5 shows eTurbo™ effect on turbocharger speed, compressor pressure ratio, air flow rate and the corresponding VTG nozzle opening. Adding 15 kW ETA to the turbocharger shaft at 76% and 88% normalized engine speed reduces the turbocharger speed by 3% and 5%, compressor pressure ratio by 4.8% and 6.3% and the air mass flow rate by 4.1% and 3.9% respectively. On the contrary, subtracting 10 kW ETA from the turbocharger shaft at 76% and 95% normalized engine speed increase the turbocharger speed by 5.6% and 4.8%, compressor pressure ratio by 8.3% and 6.2% and the air mass flow rate by 6.1% and 4.1% respectively.

The above results suggest that there may be fuel consumption benefits in using VTG eTurbo™ but these benefits might be compromised in generating mode by the need to generate electricity to power the eTurbo™ motor. It could be beneficial for when using a wastegated turbocharger to limit the wastegate authority for AFR management during part load operation.

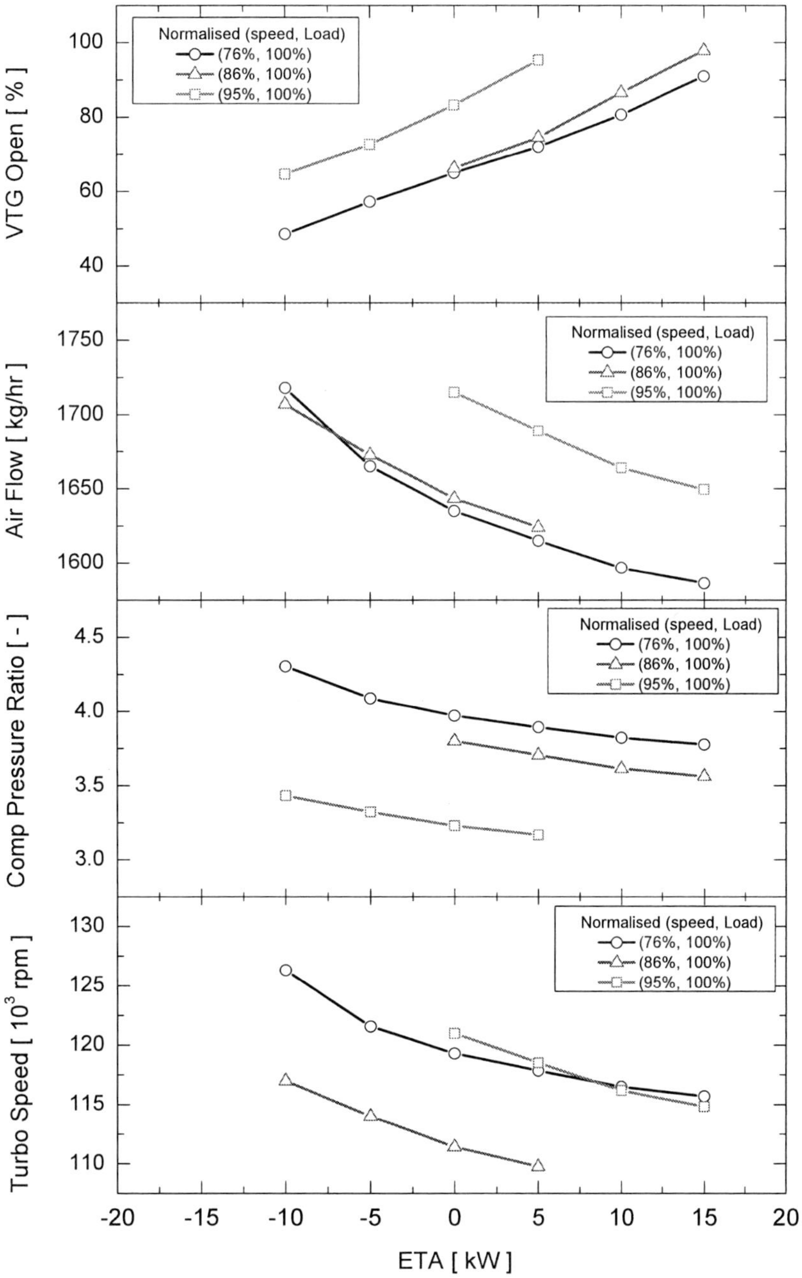

Figure 5 The effect of eTurbo™ on air flow and turbocharger speed

4 STEADY STATE DUTY CYCLE

The feasibility of using eTurbo™ is investigated over the WHSC (20). Of the 13 points specified in the WHSC, five points are available for power exchange with the turbocharger shaft, representing 17% of WHSC time-at-level [TAL], figure 6. The energy consumption is 42.5% of the total cycle. Each operating point is labelled with the percentage TAL of the duty cycle. Table 2 presents the operating parameters for the five available points. The remaining WHSC running points were not analysed because applying ETA would tend to drive the VTG beyond designated operating limits and make it impossible to maintain constant torque, AFR or EGR values.

At each power exchange point the engine speed, load and AFR are kept constant. In the simulation model the EGR rate is controlled by an EGR controller module by manipulating the recirculation line valve opening. The predefined EGR map is a function of engine speed and load. The main purpose of using EGR is to reduce the engine NO_x emissions by reducing the combustion temperature. However, for the same engine charge the more EGR fraction, the less power is produced. Generally, the EGR map varies based on engine application and the applied exhaust gas after-treatment.

Table 2 Distilled WHSC operating conditions

Index	Engine speed [%]	Engine load [%]	WHSC time [%]	Energy Consumed [%]	AFR [-]	ETA [kW]	
						From	To
A	35	100	3	4.1	25	0.0	8.0
B	45	70	4	7.1	25	-5.0	4.0
C	55	70	4	9.8	24	-5.0	0.0
D	55	100	3	9.3	23	-5.0	5.0
E	75	100	3	12.2	22	-16.0	15.5

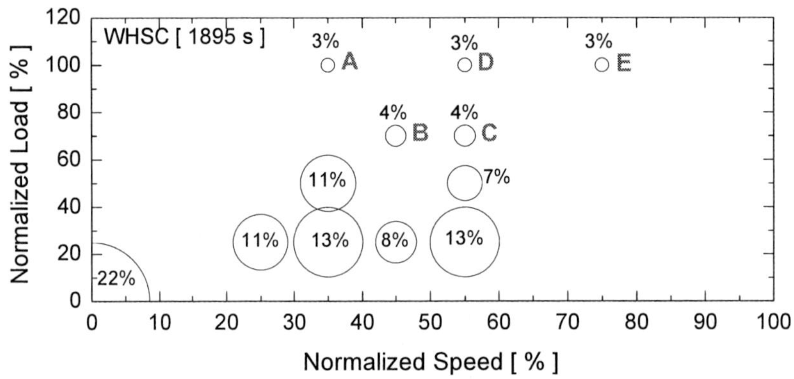

Figure 6 World heavy duty steady-state on-road cycle [WHSC]

Figures 7 and 8 show eTurbo™ effect on the VTG nozzle openings and engine BSFC at different operating conditions of distilled WHSC, each curve is labelled by EGR value. For the same operating conditions, VTG nozzle opening and BSFC change rate varies based on the EGR value. The engine fuel economy is improved as ETA value increases and vice versa by increasing the amount of power subtracted from the turbocharger shaft. By increasing the amount of the power subtracted, the turbine inlet nozzle closes to maintain the engine load and hence more engine charge is required, which results in introducing more EGR due to the increase of the pressure difference across the EGR valve, operating points B, C and D. This trend extends down to the minimum controlled nozzle opening, but not beyond. The minimum controlled nozzle openings are 43, 31, 42, 48 and 43% for operating conditions A, B, C, D and E respectively, however the minimum physical nozzle opening is 30%. Operating beyond the minimum controlled nozzle opening, the engine torque was seen to reduce. A reversed behaviour is observed by increasing ETA value, where the maximum controlled turbine nozzle opening is usually close to 100% for all operating conditions.

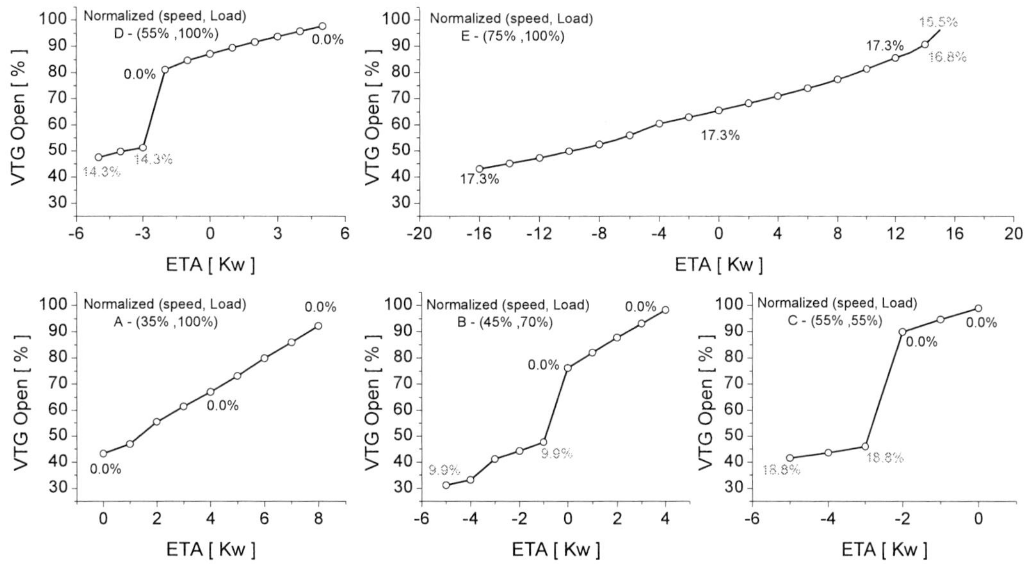

Figure 7 eTurbo™ effect on turbine nozzle opening [WHSC base]

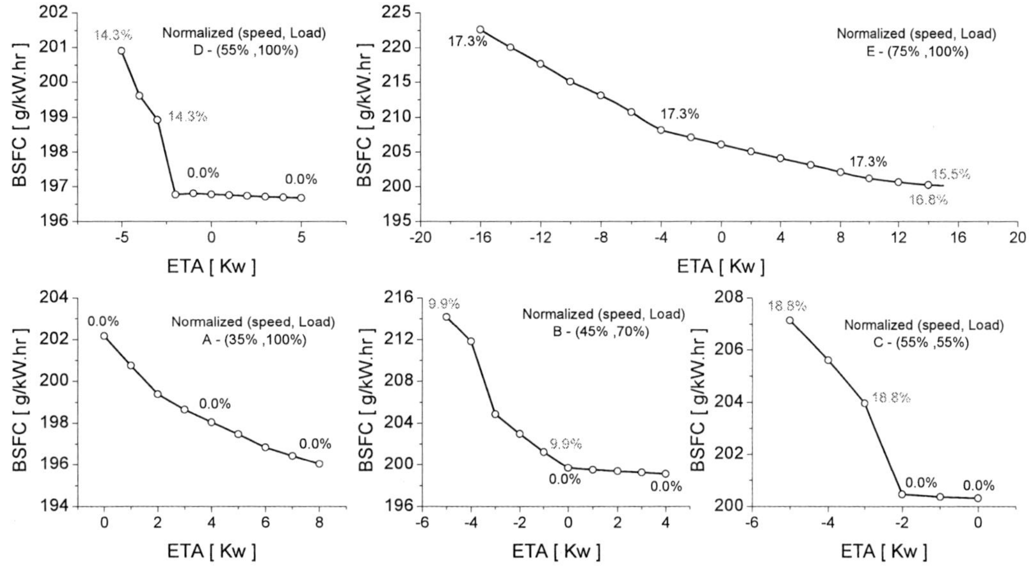

Figure 8 eTurbo™ effect on engine BSFC [WHSC base]

The key finding of applying the WHSC is that there is a benefit of assisting the turbocharger shaft at high engine speed/full load conditions. Operating point A, at a moderate engine speed/full load, is available for assist mode only, but point C [higher speed/part load] is available for generating mode only. Point B [low speed/part load], point D [moderate speed/full load] and point E [high speed/full load], are available either for generating or assist mode. Initially, it might be thought that it is beneficial to assist the turbocharger at low operating engine speeds and to generate the required energy at high speeds. The benefits, or not, of such a strategy is dependent on the amount of fuel saved and consumed during assisting and generating mode respectively, across the entire duty cycle compared with conventional VTG only operation.

In this study, two scenarios of exchanging power have been applied and Figure 9 shows the pattern of consumed and generated energy and the corresponding fuel consumption compared with conventional operation, i.e. no assist/generation:

In the first scenario, Case A, points A and E are 8kW and 12kW assisted respectively, while points B, C and D are generating 5kW each. Point E is selected for assist mode because the improvement/deficiency rate of fuel economy is higher than points B and D during assist/generating mode with respect to conventional operation. In this scenario, the energy consumed during assisting mode would be generated by the same turbocharger with 0.6% overall increase in engine fuel consumption.

The second scenario, Case B, using the same concept, energy consumed for assisting point E with 9kW ETA can be generated by subtracting 2.6kW, 2.0kW and 2.0kW from operating points B, C and D respectively resulting in a total fuel saving of 0.1%. Note: point A is not assisted because of the negligible energy saving during assisting mode. It should be noted that these findings will be made worse by taking the power conversion losses into account.

The case study results given in this paper show how ETA values for the eTurbo™ are limited by the VTG base line match. It would be reasonable to expect that the eTurbo™ could be extended beyond the current limited range with a turbocharger rematch, i.e. engine performance would benefit from a bigger compressor trim size and a re-matched VTG controller.

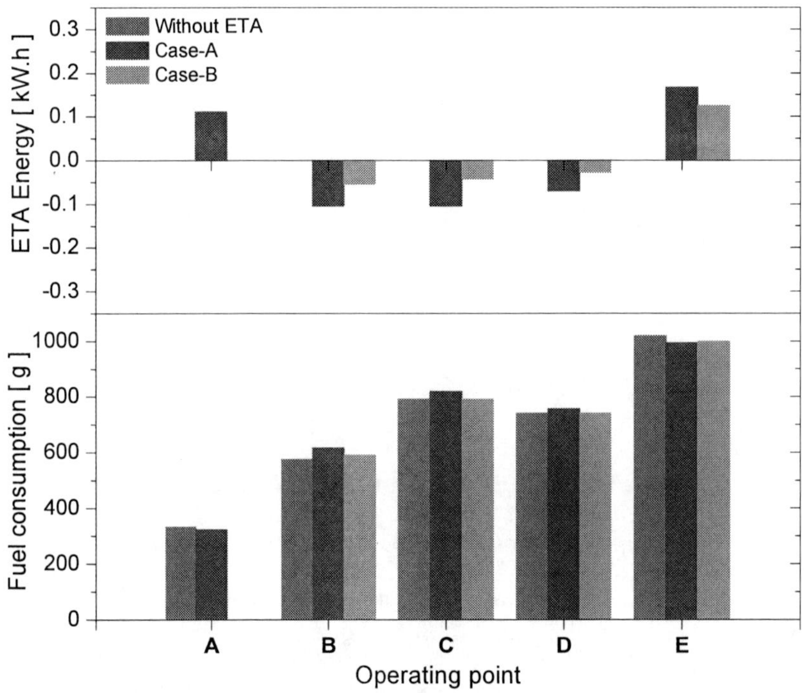

Figure 9 Energy balance & burned fuel mass consumption using ETA

5 CONCLUSION

Use of VTG ETA on heavy duty diesel engine steady state performance has been investigated. A 1-D engine simulation model in GT-Power and developed by BorgWarner Turbo Systems has been augmented by an external power module and investigated. Two key findings have been observed: [1] increasing the assisting power leads to a reduction in fuel and air mass flow requirement, compressor pressure ratio and turbocharger shaft speed, all for the same engine torque, air/fuel ratio and EGR fraction, and [2] it can be beneficial to fuel economy to assist the turbocharger during high engine speed/full load operation and to recover energy from relatively low engine speeds and loads.

Using ETA allows the VTG nozzle to open, reducing engine back pressure and pumping work whilst maintaining engine load. This results in an improvement in the overall engine efficiency and fuel economy by up to 4.1% and 3.9% respectively during the assist event itself. The energy required for turbocharger assistance at high engine speed can be recovered effectively during relatively low engine speed operation and suggests that a small overall fuel saving might be possible if the power assist/generation strategy can be optimised. There is scope to optimise AFR and EGR rates for improved fuel economy using the additional degree of freedom in the air path provided by ETA.

Future work in this area will centre on calculating energy conversion losses and optimising both assist/generating strategy and the turbocharger compressor/ turbine performance match for ETA application.

6 ACKNOWLEDGMENT

The authors would like to thank BorgWarner Turbo Systems for sponsoring this research project.

7 REFERENCES

(1) (2011) European Commission, Final Report -DG Climate Action. [online]: http://ec.europa.eu/clima/policies/transport/vehicles/docs/ec_hdv_ghg_strategy_en.pdf. Last accessed November 2015.

(2) (2010) United States Environmental Protection Agency, Office of transporatation and air quality. [online]: http://www.nhtsa.gov/staticfiles/rulemaking/pdf/cafe/CAFE_2014-18_Trucks_FactSheet-v1.pdf. Last accessed November 2015.

(3) Ikeya, N., H. Yamaguchi, K. Mitsubori & N. Kondoh, (2000) Development of advanced model of turbocharger for automotive engines. SAE Technical paper series.

(4) Godeke, H. & K. Prevedel, (2014) Hybrid turbocharger with innovative electric motor, in MTZ.

(5) Costall, A.W., R. Ivanov & T.P.F. Langley. (Year) Electric Turbo Assist: Efficient Rapid Boost for Heavy-Duty Diesel Engines. in THIESEL 2012 Conference on Thermo- and Fluid Dynamic Processes in Direct Injection Engines. of Conference.

(6) Pullen, J. & R.F. Martinez-Botas, (2001) Turbocharger motor-generator for improvement of transient performance in an internal combustion engine. Proceedings of the IMechE, Part D: Journal of Automotive Engineering. 215: p. 15.

(7) Millo, F., F. Mallamo, E. Pautasso & G.G. Mego, (2006) The potential of electric exhaust gas turbocharging for HD diesel engines SAE internatinal Technical Paper Series.

(8) Terdich, N., R.F. Martinez-Botas, A. Romagnoli & A. Pesiridis. (Year) Mild hybridization via electrification of the air system: electrically assisted and variable geometry turbocharging impact on off-road diesel engine. in ASME Turbo Expo. of Conference. San Antonio, Texas, USA.

(9) Tavčar, G., F. Bizjan & T. Katrašnik, (2011) Methods for improving transient response of diesel engines – influences of different electrically assisted turbocharging topologies. Proceedings of the Institution of Mechanical Engineers, Part D: Journal of Automobile Engineering. 225(9): p. 1167-1185.

(10) Bayomi, N.N. & R.M. Abd El-Maksoud, (2012) Two operating modes for turbocharger system. Energy Conversion and Management. 58(0): p. 59-65.

(11) Seichibarki., Y. Yamashta, K. Sumida, H. Ogita & Y. Ijnnai, (2006) Development of the "hybrid turbo", an electrically assisted turbocharger, in Mitsubishi Heavy Industries, Ltd, Technical Review.

(12) Yamashita, Y., S. Ibaraki & H. Ogita, Development of Electrically Assisted Turbocharger for Diesel Engine, L. Mitsubishi Heavy Indusstries, Tokyo, Japan, Editor.

(13) Balis, C., C. Middlemass & S.M. Shahed, (2003) Design & development of E-Turbo for SUV and light truck applications, in Diesel Engine Emissions Reduction Conference. Newport, RI.

(14) Costall, A.W., R. Ivanov & T.P.F. Langley. (Year) Electric turbo assist and an enabler for design downspeeding. in ASME Turbo Expo. of Conference. Copenhagen, Denmark.

(15) Suzuki, T., Y. Hirai & N. Ikeya. (Year) Electrically assisted turbocharger as as enabling technology for improved fuel economy in new European driving cycle operation. in 11th international conference on turbochargers and turbocharging. of Conference. The British Museum Woodhead.

(16) Heywood, J.B., (1988) Internal Combustion Engine Fundamentals. USA: McGraw-Hill, Inc.

(17) Chang, J., O. Güralp & Z. Filipi, (2004) New Heat Transfer Correlation for an HCCI Engine Derived from Measurements of Instantaneous Surface Heat Flux. SAE Technical paper series: p. 1-18.

(18) (2015) Engine Performance Application Manual. Gamma Technologies: 601 Oakmont Lane, Suite 220, Westmont, IL 60559, USA.

(19) Serrano, J.R., H. Climent, C. Guardiola & P. Piqueras, (2009) Methodology for characterisation and simulation of turbocharged diesel engines combustion during transient operation. Part 2: Phenomenological combustion simulation. Applied Thermal Engineering. 29(1): p. 150-158.

(20) (2014) Worldwide emission standard, Heavy duty and off-highway vehicles, in DELPHP Powertrain Systems.

Design of compressor for electrically decoupled turbocharger in downsized gasoline engine by 3D inverse design

I Ghazaly, M Zangeneh
Department of Mechanical Engineering, University College London, UK

ABSTRACT

Engine downsizing provides the most effective means of achieving the legislative requirements for reducing CO2 emissions from vehicles. However, it is well known that downsized engines when coupled with a conventional turbocharger can suffer from low engine torque at low rpm, which is major issue for drivability of downsized engines. One option is to use two stage turbocharging which has been shown to be an effective but also an expensive option. Another possibility is to couple an electrically driven supercharger with a conventional turbocharger. A third possibility is to use a completely decoupled electrically driven turbine and compressor together with an energy storage system. The compressor will be connected to an electric motor while the turbine will be connected to an electric generator. The electric motor and electric generator is then connected to an electric storage unit. This concept is advantageous as both compressor and turbine can be designed to operate in different operating speeds which is made possible by decoupling of the turboshaft. In addition, the use of wastegate can be circumvented as there is no risk of overspeeding.

In order to demonstrate the benefits of this option, a 2 litre naturally aspirated engine is first modelled in GT-Power. The performance of the engine is then compared to a 1 litre downsized engine coupled with a standard turbocharger. Also the downsized engine is then coupled with an electrically decoupled turbocharger and an estimate of the engine lug-line on the compressor map is obtained. This information is then used to design two different compressors for the electrically decoupled system. The new compressor impellers are designed by using commercially available 1D meanline code and a 3D inverse design method. The volutes are also designed by a 2D inverse design method. Finally, the resulting performance of the two stages are predicted by using 3D CFD code.

The results of the new map are then coupled with the GT-power model and the results confirm improvement in the low engine speed torque of the downsized engine coupled with the electrically decoupled turbocharger.

NOMENCLATURE

1-D	One dimensional
CFD	Computational Fluid Dynamics
PR	Pressure Ratio
NA	Naturally Aspirated
TDC	Decoupled Turbocharger
LE	Leading Edge
TE	Trailing Edge

1 INTRODUCTION

The current trend in gasoline engine technologies involves engine downsizing in order to satisfy the ever restricting emission requirements set by local legislative bodies. Engine downsizing is effective in that it offers better fuel consumption and lower frictional losses due to relatively smaller cylinder capacity [1]. Downsized engines are made possible through charging methods such as supercharging or turbocharging, the latter which is currently being explored by numerous researchers [1-3]. But when a turbocharger which is coupled to a downsized engine is compared to a NA engine with similar power and torque target requirement, downsized turbocharged engine suffers from low torque at low engine speed range [3]. This is due to the coupling of the compressor directly to the turbine which make the compressor speed thus power directly affected by the turbine power. At low engine speed the low mass flow of the exhaust gases of the engine will not be able to provide enough torque to the compressor for a higher pressure ratio which is needed for higher engine torque. Thus decoupling the compressor from the turbine electrically seems to be the most strategic option to be explored albeit at the cost of a introducing a complex motor-generator electrical system.

The work on electrification of superchargers and turbochargers are researched by several people such as Terdich et. al. [4], Godeke & Prevedel [5] and Bumby et. al.[6]. Terdich et. al. [4] researched on hybridization of the charging system where a diesel engine is coupled to an electrically assisted turbocharger and variable geometry turbine. Godeke & Prevedel [5] had studied on the transient behaviour of hybrid turbocharger while Bumby et. al. [6] had done a study on the implementation of assisted electric turbocharger on a city bus. Currently electrification of turbochargers is mainly focused on assisted electric turbocharging such as done by these researchers [4-6] which mainly couples a standard turbocharger to a bi-functional motor/generator unit. This approach is best when the engine space is limited and a compact unit is required but the downside to this approach is that the compressor is still connected directly to the turbine making each component dependant on each other and impossible to control without affecting the other component. This means that by using a single electric machine for the role of bifunctional motor/generator, it is hard to cater for both the compressor and turbine requirements. This is where the decoupling of the compressor and turbine which assigns dedicated motor to the compressor and dedicated generator to the turbine is essential to fully harness the electrification technologies of the turbocharging methods.

By decoupling the turbocharging process, meaning by separately coupling the compressor with the motor and turbine with a generator, both compressor and motor can be independently sized up according to engine requirement. This means that the compressor rotating speed can be sized to match the design motor requirement as electric motor have a lower speed range. In addition, the turbine can be sized to match the exhaust gas' maximum mass flow thus bypassing the need to use wastegate to control overspeeding when desired pressure ratio at compressor is reached. Traditionally, in a standard setup, usually this will lead to low performance during low engine speed due to sizing of a larger turbine, but because of the turbine is already decoupled from the compressor, at low engine speed the low mass flow will only create low electric power generation in turbine, the compressor can rely on the electrical energy stored in the battery unit to cater for the deficit. The compressor has no problem with this because it can easily be driven by stored battery power supply during the low engine rpm. At high engine rpm, excess high flow from the exhaust can be used to charge the battery using the generator the turbine is coupled with.

The aim of this project is to design compressor using 3D inverse design for a decoupled electric turbocharger in order to solve the low torque at low engine speed problem occurring in turbocharged downsized gasoline internal combustion engine.

2 METHODS OF RESEARCH

2.1 GT-Power modelling

This study uses the commercial 1-D engine simulation software, GT-Power to simulate spark ignition, port injected, 4-cylinder engine models in order to assess the performance of electrically decoupled turbocharger and then compare the performance to a conventional turbocharger setup as well as a baseline NA engine setup. The combustion rate was modelled with a Wiebe curve while heat transfer of the combustion chamber wall uses the Woshni heat transfer model. The setup of the GT-power model for a 1.5-Litre NA engine is shown in Fig. 1. The simulation for a 1.5 Litre NA engine had been run using the model and its results had shown good agreement with experimental results [7] in terms of brake power and brake torque. The same model was then used for the 2.0 L NA baseline engine for comparison purposes with the downsized turbocharged engine. All the runs in the study is full-load, steady state operations.

For the 1.0 L engine coupled to the standard turbocharger model (Fig. 2), a Garrett's GT1241 compressor [8] was used for the map inputs. The GT1241 is a wastegated turbocharger normally used for small displacement engines. Thus in the GT-power model a wastegate component will be included so that the maximum PR for the compressor can be set to 2. The decoupled turbocharger model uses the same base model as the standard turbocharger but with several modifications. The compressor map used initially for the decoupled model simulations was the GT1241 map but for the new compressors the map used was generated using CFD simulations which will be discussed in section 2.3. The compressor is connected to the electric motor while the turbine is connected to the electric generator. This can be seen in Fig. 3. The electrical power needed for the motor is supplied by the battery. The battery is charged by the electric generator which is driven by the turbine. For the decoupled model, the compressor speed is varied by using a controller unit which varies the torque produced by the motor for each engine speed. Another controller is then used to set the generator brake power for each engine speed based on the available turbine power and turbine speed at each engine speed. The electric motor model uses torque-speed curves map as well as efficiency map to calculate the conversion of brake power produced by the motor and the electrical power it required from the battery. In the model, the battery used has capacity of 20 A-h but in this study, capacity of the battery is less important since the main issue for the study is regarding the rate of electrical power used by motor and rate of electrical power generated by the generator in a steady state operation at specific engine speed.

The electric generator also uses a separate torque-speed map to calculate the power generated from the turbine as well as an efficiency map to calculate the amount of electrical power that can be used to charge the battery. Both compressor will be driven by an 8.5 kW, 120krpm maximum speed electric motor. The motors' torque-speed characteristic curve is used as inputs for the electric motor model in GT-power. In the case of the turbine, for both cases, the turbines used are identical and will be connected to a 14 kW electric generator. Since for the decoupled system, there will be no wastegate activating at specific compressor's pressure ratio, the turbine can utilize its maximum mass flow range. In the GT-power model, both electric motor and electric generator utilizes efficiency maps when calculating electrical power from mechanical power and vice versa.

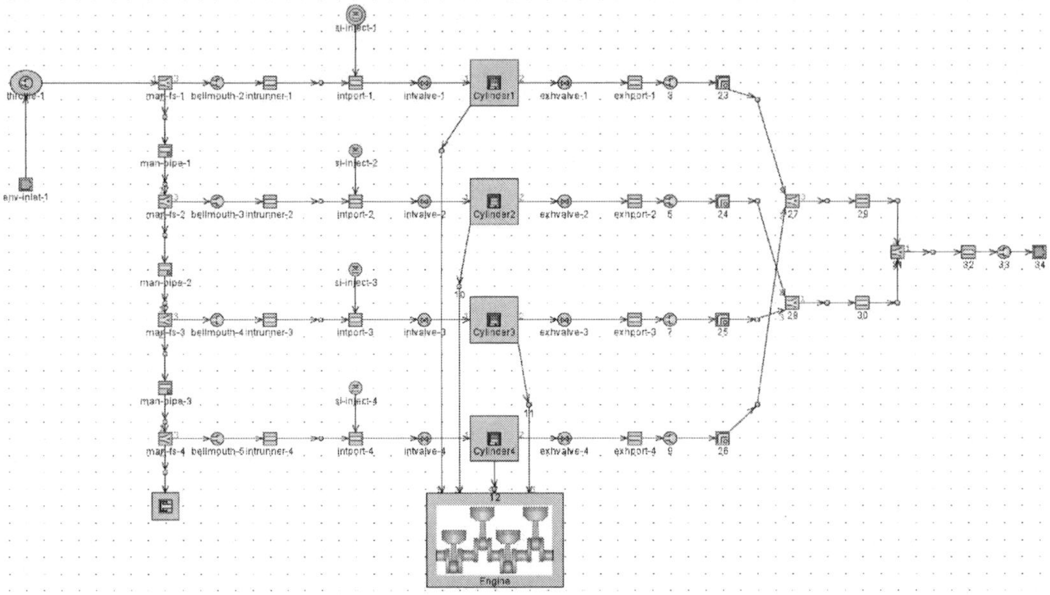

Figure 1: GT-Power model for NA engine

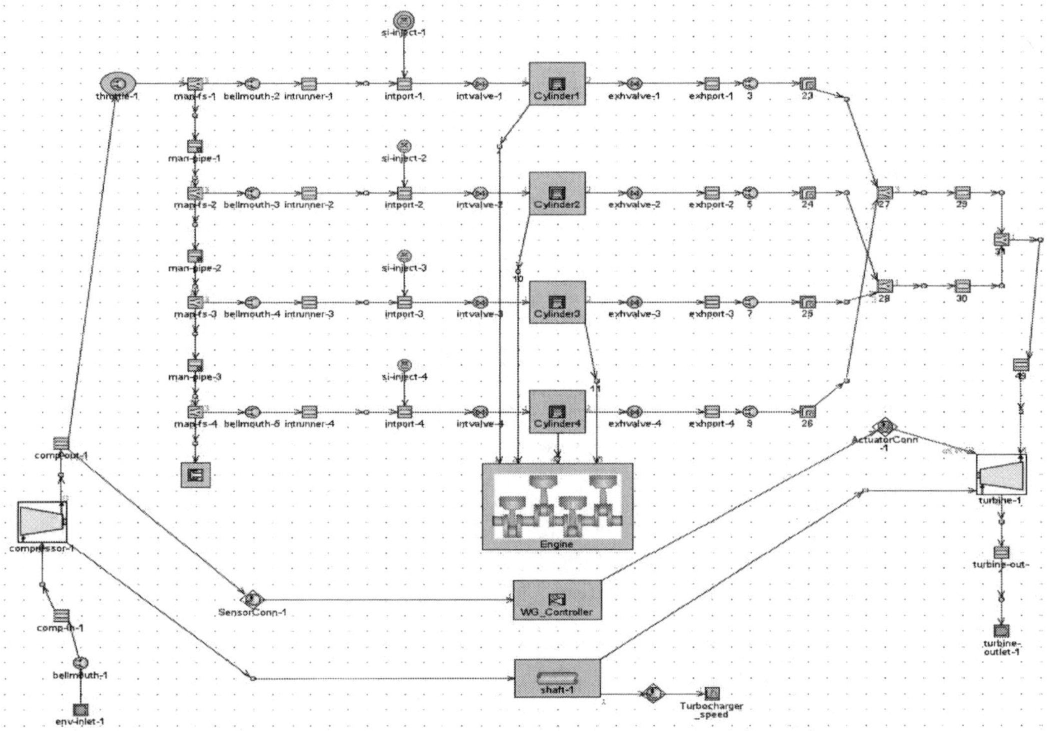

Figure 2: GT-Power model for downsized engine coupled with standard turbocharger

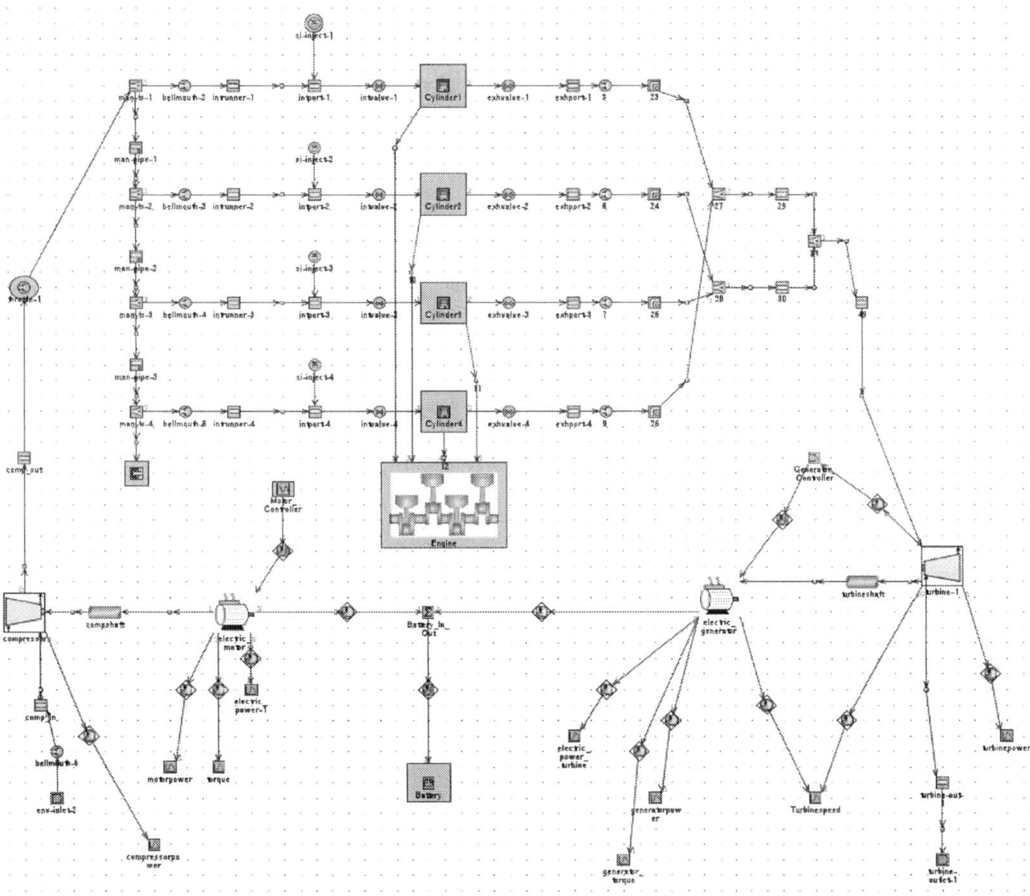

Figure 3: GT-Power model for downsized engine coupled with decoupled electric turbocharger

2.2 Compressor design and volute design

In this paper, the compressor will be designed by using the 3D inverse design method for turbomachinery developed by Zangeneh et. al. [9] and [10]. The method designs the compressor blade geometry for specified distribution of blade loading. It has been extensively applied to design of centrifugal compressors both in subsonic and transonic flow regimes, see [11], [12]. It has led to significant improvement in compressor performance by controlling secondary flows, e.g. [11] or reduction of shock strength in the inducer, e.g. [12]. The design conditions used is based on the required flow rate for the downsized engine and the boost pressure. However, the compressor rpm can have an effect on the motor costs and hence in this study two different compressor speeds were considered at the nominal design point. For Design A (110krpm) and Design B (90krpm) were used respectively.

Below are design point specifications of both compressors:

Table 1: Design specification for compressor A and compressor B

Parameters	**Design A**	**Design B**
Speed	110krpm	90krpm
Mass Flow	0.08 kg/s	0.08 kg/s
Stage Pressure Ratio	2	2

It is imperative to note that the design pressure ratio of the compressor was set to 2 in order to match the downsized engine power with a naturally aspirated 2-Litre engine. In addition, higher pressure ratio is undesirable as it may lead to engine knocking which is highly prevalent in spark ignition engines running on highly compressed air-fuel mixture. The design specifications (Table 1) with the additional input parameters in Table 2 are used to generate a meridional geometry from the meanline code TURBOdesign pre [13]. TURBOdesign Pre generates the meridional geometry of the compressor given the design conditions specified in Table 1 and additional information such as shaft diameter, impeller axial length over tip diameter ratio, impeller TE thickness and tip clearance. The additional conditions used to design the meridional shape for both Design A and Design B are shown in Table 2.

Table 2: Input parameters for meridional geometry generation

Parameters	Design A	Design B
Inlet Total Pressure	100 kPa	100 kPa
Inlet Total Temperature	300 K	300 K
Inlet hub diameter	0.01 m	0.01 m
Axial length ratio	0.325	0.325
TE blade thickness	0.001 m	0.001 m
Tip clearance	0.00025 m	0.00025 m
Tip Diameter	0.061 m	0.075 m
Impeller exit width	0.002997 m	0.002102 m
Diffuser width	0.002182 m	0.001894 m

Inlet hub diameter of 0.01 m was chosen because it gives the best efficiency prediction in TURBOdesign Pre compared to larger inlet hub diameters. The initial meridional geometry and additional data such as blade number provided by the meanline code is then used in the 3D inverse design code TURBOdesign1 to compute the 3D blade shape of the impeller. In the case of both Design A and B the impellers were designed with full blades only and no splitter was used. The blade loading used for the design of the impellers are shown in Fig. 4. For both Design A and Design B a strongly aft-loaded loading distribution is used at the hub. This tends to reduce meridional secondary flows and hence result in a more uniform exit flow. The loading at the shroud is compromise between a large stall margin and good efficiency at the design point.

Table 3: TD1 Input parameters

Parameters	Design A	Design B
Rotational speed	110 000 rpm	90 000 rpm
Blade number	13	12
Thickness at hub	1.2 mm	1.2 mm
Thickness at shroud	0.6 mm	0.6 mm
Inlet stag. density	1.16104 kg/m^3	1.16104 kg/m^3
Gas constant	287.1 J/kg·K	287.1 J/kg·K
Inlet stagnation temperature	300 K	300 K
Impeller inlet shroud diameter	32 mm	33.5 mm
Impeller tip diameter	61 mm	75 mm
Volute max diameter	124.8 mm	144.8 mm
Work Coefficient at Trailing edge	0.631	0.655

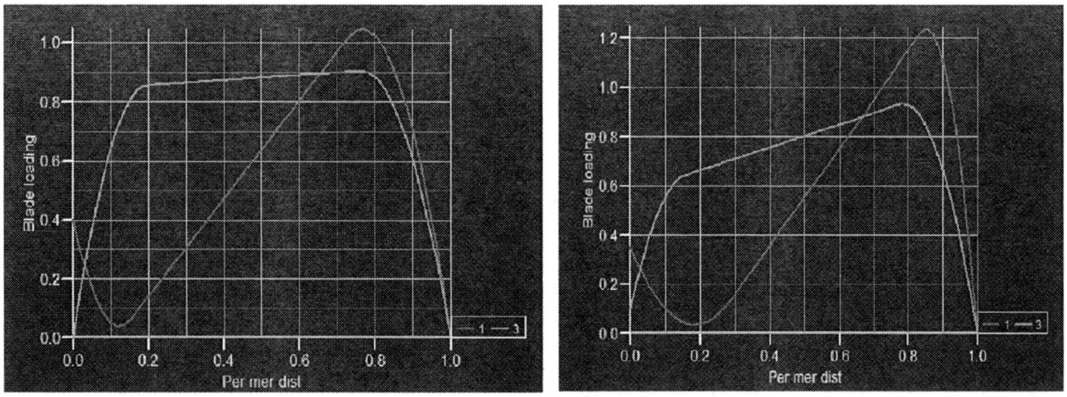

Figure 4: Blade Loading distribution at hub (red line) and blade tip (orange line) for design A (left) and design B (right)

The resulting impeller geometry for Design A and Design B is shown in Fig. 5, while the output blade angles for leading and trailing edges are shown in Table 4.

Table 4: TD1 Output parameters

Parameters	Design A	Design B
Impeller blade angle at LE, hub	20.49°	24.18°
Impeller blade angle at TE, hub	50.46°	44.92°
Impeller blade angle at LE, shroud	57.53°	57.39°
Impeller blade angle at TE, shroud	51.61°	45.31°

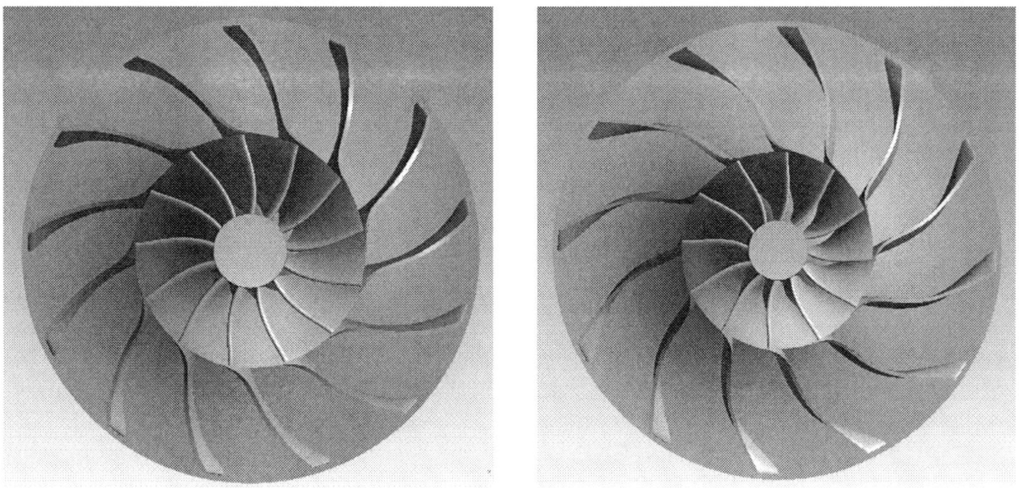

Figure 5: Impeller geometry obtained for design A (left) and design B (right)

The volute of the compressors were then generated using a 2-D inverse design code, TURBOdesign volute [14]. The code allows for the design of a 3D volute based on the circumferential flow variations.

The volute inlets' parameters such as static pressure, static temperature, circumferential velocity and radial velocity of the flow were taken from CFD simulations at design point using only the passage, inlet and diffuser domains. The design parameters of both volutes are given in Table 5:

Table 5: Volute Design parameters

Parameters	Design A	Design B
Volute Inlet Diameter	103.184 mm	123.46 mm
Volute Inlet Width	2.18 mm	1.89 mm
Outlet Tube exit Diameter	22 mm	22 mm
Outlet Tube exit length	60 mm	60 mm

2.3 CFD analysis and compressor map generation

Commercial software ANSYS CFX-14.5 is used for the CFD simulations. Four domains are modelled which are the inlet, passage, diffuser and volute domains. Inlet, passage and diffuser domains are modelled as single passage domains with periodic boundary conditions while the volute was modelled as the full volute. The inlet, passage and diffuser domains are meshed in structured multiblock meshes by using ANSYS Turbogrid, while boundary layer resolved unstructured meshes were used for the volute domain. All of the domains have Y+ value below 25 which indicates good mesh density near wall boundary surfaces. In Table 6, the number of meshes used for each cases is shown:

Table 6: Number of elements generated for each mesh at each domains

	No. of Elements	
Domain	**Design A**	**Design B**
Inlet	20160	22680
Passage	350082	436314
Diffuser	37044	38610
Volute	1579158	1367253

The stage frame change interface was used to connect the single passage diffuser outlet with the full volute inlet. The simulation utilized the SST k-omega turbulence model. Boundary conditions used for the inlet were total pressure and total temperature boundary conditions. For the outlet, mass flow and static pressure boundary conditions was used as boundary conditions. For each compressor speed, the simulations were done with pressure boundary conditions starting from choke region (very low static pressure boundary condition). The static pressure boundary conditions were then increased incrementally until design flow was reached. Next, the outlet boundary conditions were then switched to mass flow boundary condition until the flow enter the stall region. The monitoring parameters used to determine convergence were pressure ratio, torque and stage efficiency. The stall limit was estimated by incrementally decreasing the exit mass flow boundary condition until the monitoring parameters starts to fluctuate. Based on previous experience on various turbocharger compressors, this approach can provide a fairly accurate prediction of surge point as compared to experimental measurements. Our experience shows that this approach usually overpredicts the actual surge flow rate as compared to experiments and hence provides a rather conservative measure of the operating range at each speed.

For both compressor designs, the simulations were run at 70%, 80%, 90%, 100% and 110% design speed with varying outlet boundary conditions in order to generate the compressor map to be used in the GT-power simulations. In Fig. 6, the assembly of the domain with the computational mesh for the CFD simulations is shown.

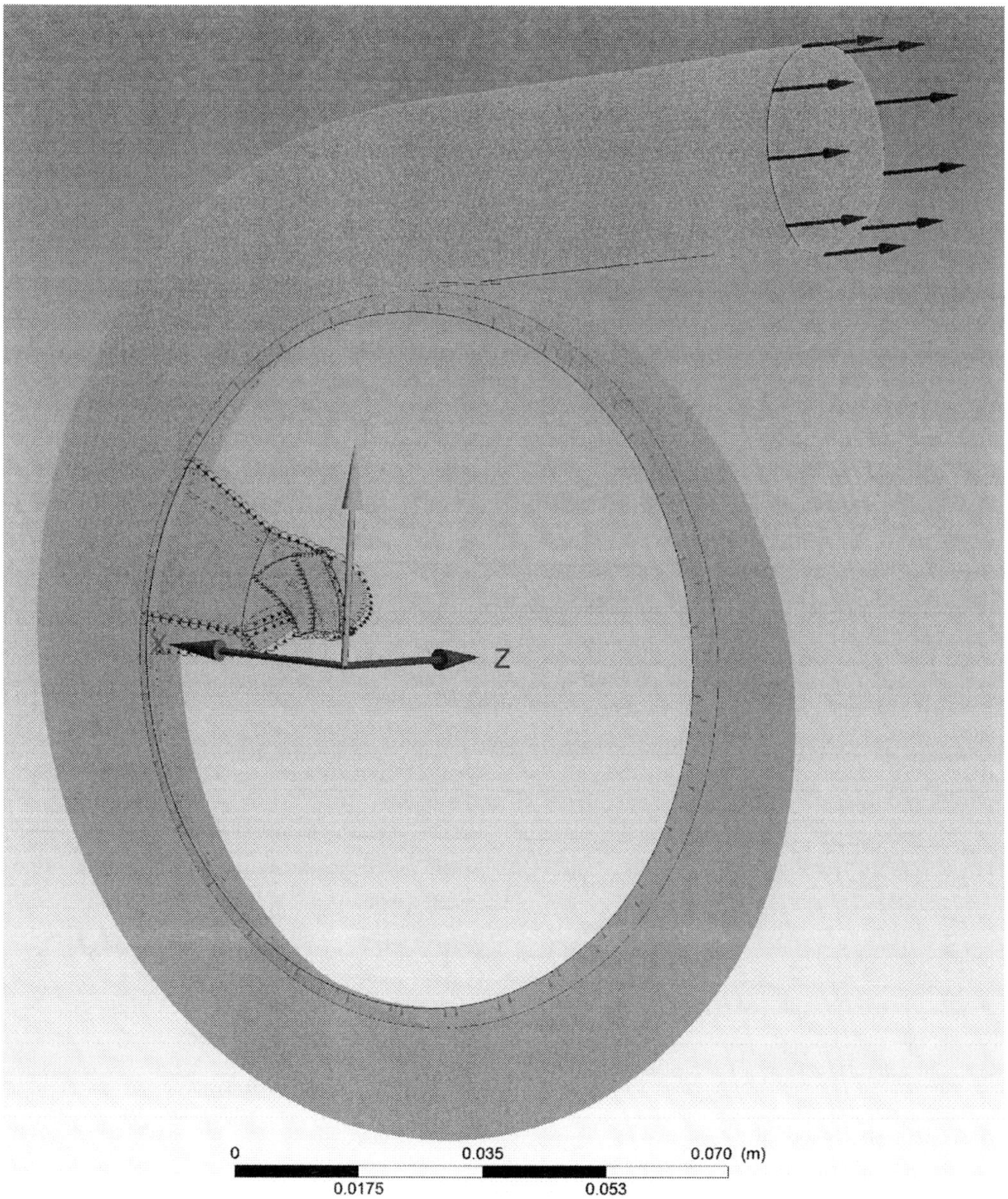

Figure 6: CFD computational mesh used in the simulation – Inlet, Impeller, Diffuser and Volute domains

3 RESULTS AND DISCUSSION

3.1 Brake torque comparison for decoupled system

Turbocharged engine has the issue of low torque at low engine speeds especially in downsized engine where the mass flow is reduced due to reduced total engine cylinder volume reduction. This trend can be seen in the simulation of the 1 Litre engine coupled with a standard turbocharger (GT1241) – see Fig. 7. It can be seen that from 1000 RPM to 2500 RPM, the torque of the engine is relatively low compared to the 2 Litre naturally aspirated engine. This can be attributed to turbine inability to operate efficiently at very low mass flow. This could be rectified by using a smaller turbine which will lead to the wastegate to be triggered at comparatively lower engine RPM, which means that more exhaust energy is wasted.

Compared to the standard turbocharger torque line, the decoupled turbochargers' engine torque line is comparatively higher at lower engine speed. This is due to the independent operation of the compressor and the turbine. The compressor can be operated at higher speeds and hence have higher boost pressure to the engine. The compressor map for the baseline compressor is shown in Fig. 8 together with engine lug-line for the downsized engine with standard turbocharger. According to this chart at 1500 rpm, the second point on the lug-line the compressor pressure ratio is 1.1. However, as can be seen from Figure 9 and 10, which show the compressor maps for Design A and Design B and the engine lug-line with the decoupled arrangement, the second point is actually operating at pressure ratio of around 1.6. This helps to explain the increase in engine torque from around 60Nm at 1500 rpm for the standard turbocharger to 105 Nm for the decoupled electrically driven turbocharger. However, by comparing the location of the lug-line in the electrically decoupled system as compared to the standard turbocharger it is clear that the new lug-line requires an increase in the compressor surge margin. So a key aspect of the design process for the new compressor was to achieve higher surge margins. Furthermore, the new compressors ran at higher speed implying usage of higher power, but that is not constrained anymore by the turbine due to the decoupling of the compressor and turbine, even though the turbine is experiencing low power due to low mass flow at low engine speeds. The required power at low engine speed is supplied by the energy storage of the system, in this case the battery. It is also interesting to note that the torque for the decoupled turbocharger with compressor A and B at 2000RPM and 2500RPM is able to match the torque of the baseline NA 2 Litre engine.

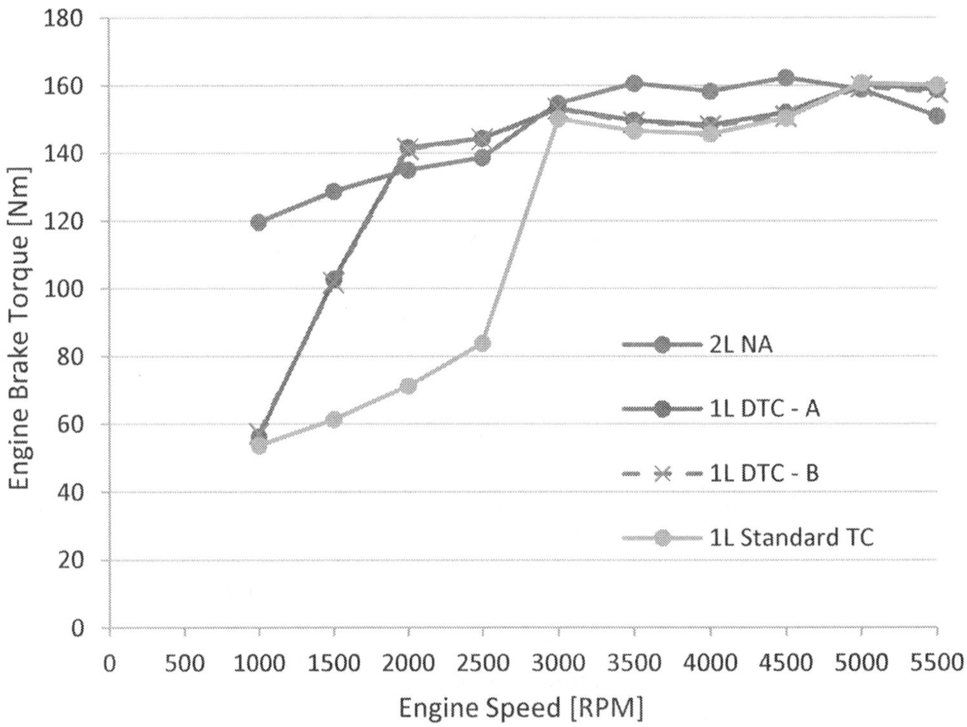

Figure 7: Comparison of brake torque

The torque at engine speeds of 3000 RPM and above are similar for both cases of decoupled and standard turbocharger due to the operations' boost pressure limit of 2 Bar. For the standard turbocharger, this is controlled by the wastegate while for the decoupled system, this was controlled by the motor controller which sets the speed of the motor to operate at the target pressure ratio.

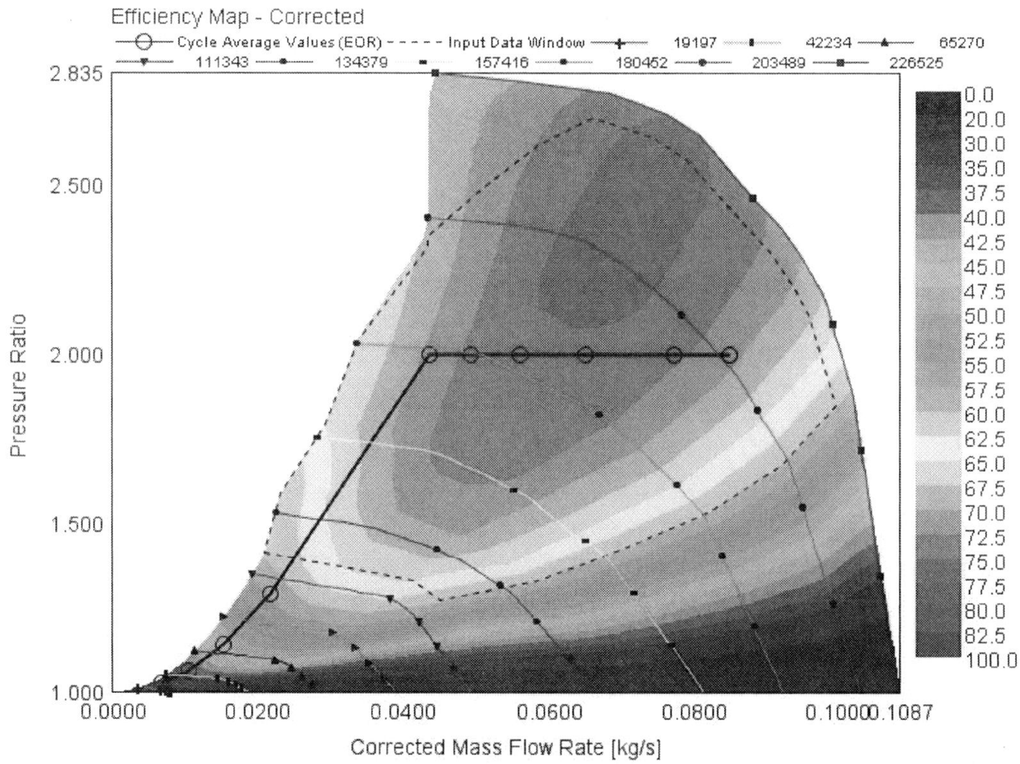

Figure 8: Operating points for standard 1-Litre engine using wastegated GT1241 compressor

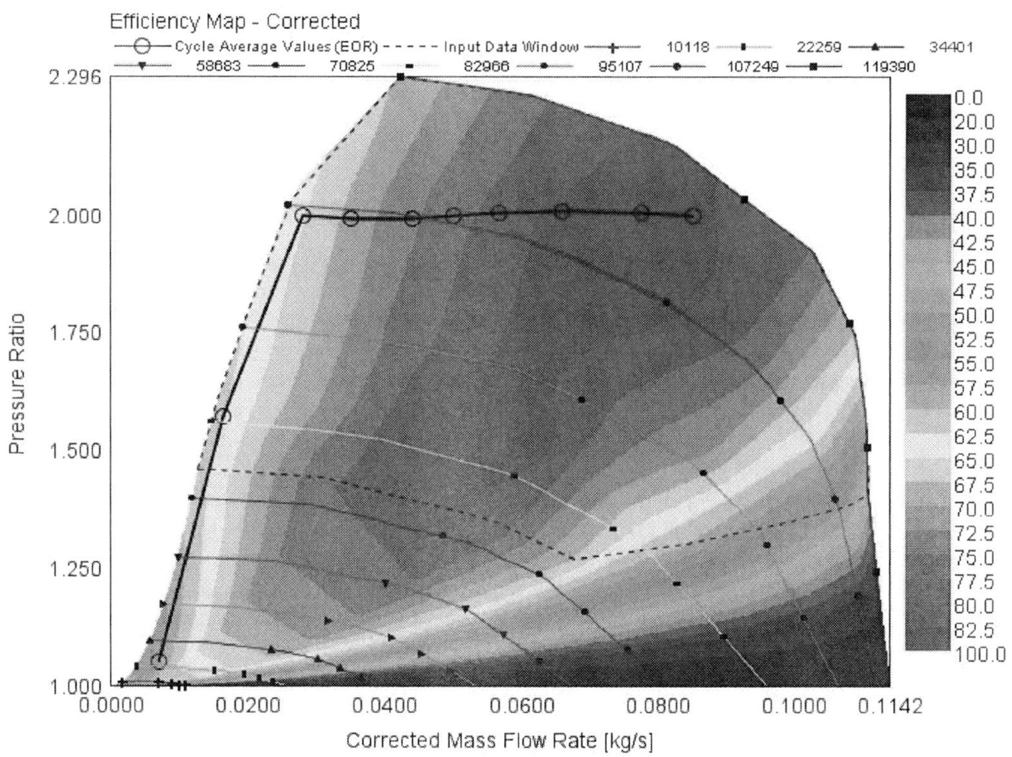

Figure 9: Operating points for 1-Litre engine coupled with electrically decoupled turbocharger – compressor A

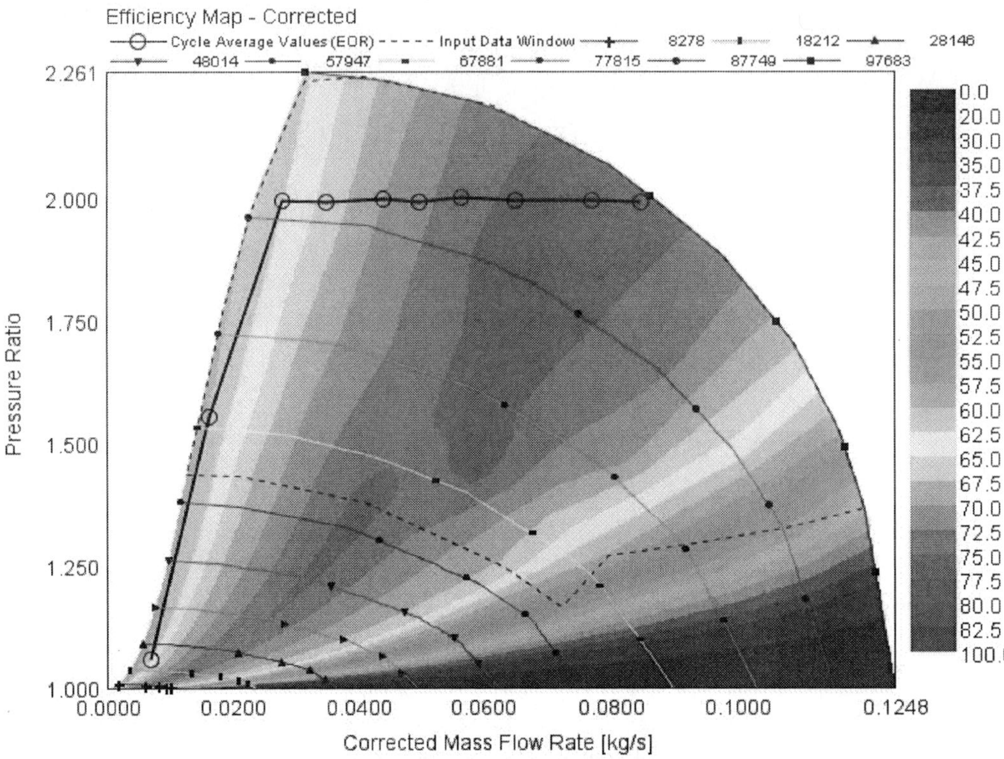

Figure 10: Operating points for 1-Litre engine coupled with electrically decoupled turbocharger – compressor B

3.2 Electrical power usage/generation comparison

Another important parameter to be observed in decoupled electric turbocharger system is the usage of electrical power of the battery. In this study, since the study is a steady-state simulation at different engine speeds, the parameter worthwhile to compare is the electrical power used by the electric motor to spin the compressor as well as the electrical power generated by the generator from being driven by the turbine. When the motor is using more electrical power than the generator is producing, the battery is in discharge mode and uses its storage of energy but when the generator is producing more electrical power than the motor uses, the battery is in charging mode. By comparing Fig. 12 & Fig. 13, the power consumption/generation is shown to be similar for both decoupled system using compressor A and B. The similar trend is due to the same engine capacity, target boost pressure and turbine setup. Note that at engine speed of 4000 RPM and below, the motor is using more power compared to the generation of electricity while at 4500 RPM and above, the electrical power generation exceeds that of the electrical power consumption of the motor. However, this could be partly because the turbine design has not been optimized for the decoupled operations. It may be possible to reduce the energy deficit at low to middle engine speed range by redesign of the turbine. In Fig. 11, it could be seen that the decoupled turbocharger was able to use the full capacity of the turbine compared to the standard turbocharger, which was capped at lower mass flow due to the wastegate actuation which makes part of the exhaust flow being wasted to avoid overspeeding of the compressor when the boost pressure had reached 2 Bar.

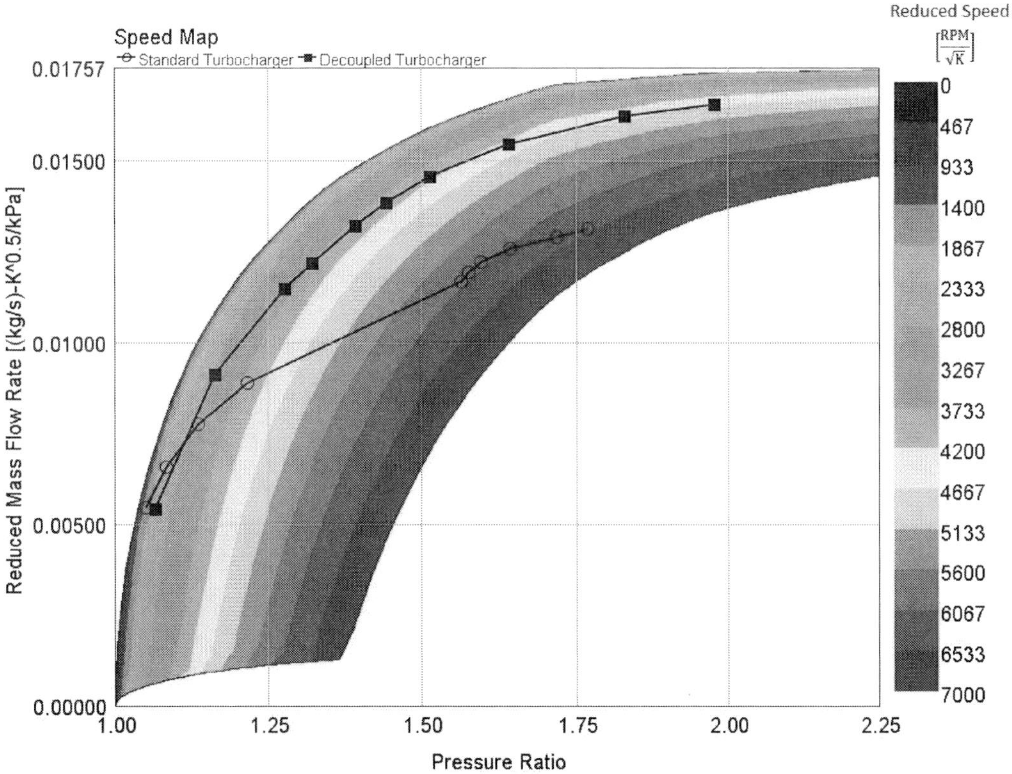

Figure 11: Comparison of turbine lug-line between decoupled turbocharger setup and standard turbocharger setup

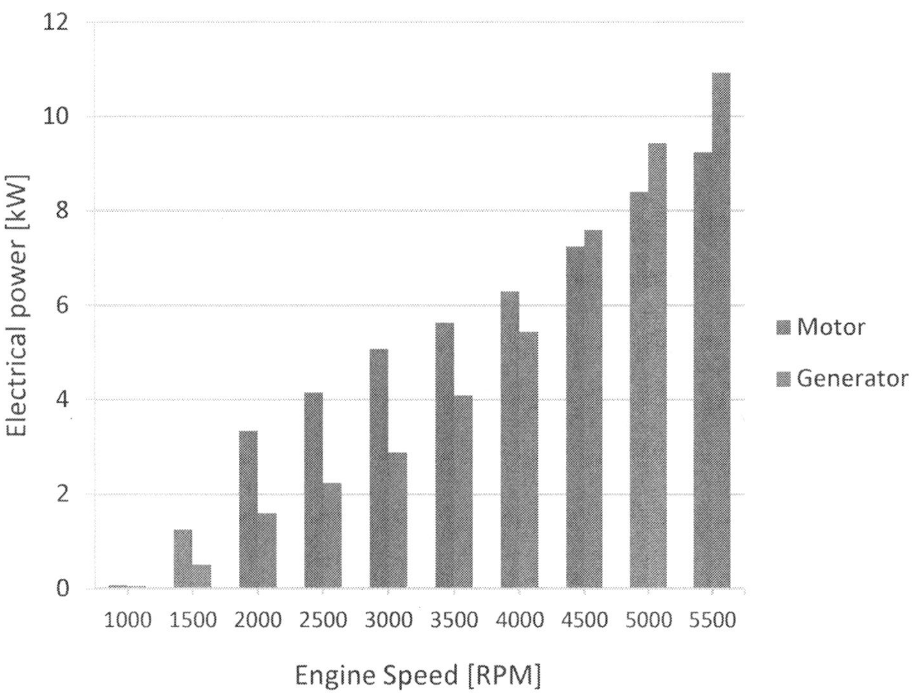

Figure 12: Electrical power consumed by the electric motor and generated by the power turbine for compressor A

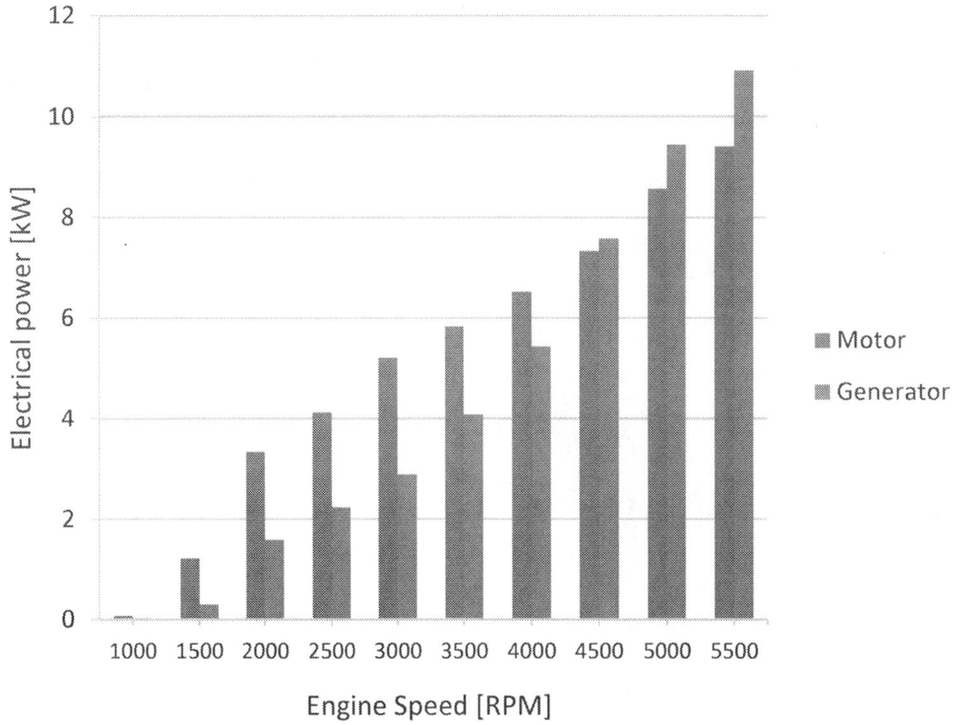

Figure 13: Electrical power consumed by the electric motor and generated by the power turbine for compressor B

4 CONCLUSIONS AND FUTURE WORK

There are several conclusions that can be made from this study:

1. Decoupling of the compressor from the turbine makes it possible to improve the low torque of turbocharged downsized engine at low engine speeds.
2. Wider compressor map is required in order for decoupled turbocharger to be efficiently used with turbocharged downsized engine.
3. Due to the absence of wastegate, the high energy from high mass flow at high engine speeds can be fully utilized as there is no risk of overspeeding the compressor when decoupling the turbine from the compressor. This also means that waste exhaust energy is used more efficiently instead of being bypassed from flowing through the turbine.

It is understood that the results in this paper was predominantly based on CFD calculations especially when generating the compressor maps and obtaining compressor efficiencies at very low RPM that was used in the 1-D engine simulations. Thus the next step of the project is to validate the compressor maps experimentally in order to justify the advantages of the new compressor designs and decoupled turbocharger setup that was discussed in this paper. In addition, transient simulations during key operating periods such as in-city driving and acceleration phases will be useful in identifying the electric decoupled turbocharger implementation to the real-driving conditions.

5 ACKNOWLEDGEMENT

The authors would like to thank Pierre-Julian Angelot of Valeo Power Systems Business Division for providing the motor and generator torque speed curves that were used in the GT power model. The authors would also like to acknowledge the Malaysian Government and National Defence University of Malaysia for their financial support of the first author during this research.

6 REFERENCE LIST

[1] Police, G., Diana, S., Giglio, V., Iorio, B. & Rispoli, N, 2006. "Downsizing of SI Engines by Turbocharging". *8th Biennial ASME Conference on Engineering Systems Design and Analysis*.

[2] Lake, T., Stokes, J., Murphy, R., Osborne, R. & Schamel, A., 2004. "Turbocharging Concepts for Downsized DI Gasoline Engines", SAE Technical Paper 2004-01-0036.

[3] Fraser, N., Blaxill, H., Lumsden, G. & Basset, M, 2009. "Challeges for Increased Efficiency through Gasoline Engine Downsizing", *SAE Int. J. Engines*. Vol 2(1).

[4] Terdich, N., Martinez-Botas, R.F., Romagnoli, A., Pesiridis, A., 2014. "Mild Hybridization via Electrification of the Air System: Electrically Assisted and Variable Geometry Turbocharging Impact on an Off-Road Diesel Engine", *Journal of Engineering for Gas Turbines and Power*, Vol 136(3).

[5] Godeke, H. & Prevedel, K., 2014. "Hybrid Turbocharger with Innovative Electric Motor", *MTZ Worldwide,* Vol 75(3), pp 26-31.

[6] Bumby, J., Crossland, S. & Carter, J., 2006. "Electrically Assisted Turbochargers: Their Potential For Energy Recovery". *Hybrid Vehicle Conference, IET, The Institution of Engineering and Technology,* pp. 43-52.

[7] I, Ghazaly, 2011. "Comparison of Performance and Emission between Gasoline and Compressed Natural Gas", Master Thesis, National University of Malaysia.

[8] Garret Products online catalogue. https://turbobygarrett.com/turbobygarrett/turbocharger#GT1241.

[9] Zangeneh, M., 1991. "A Compressible Three-dimensional Design Method for Radial and Mixed Flow Turbomachinery Blades". *International Journal for Numerical Methods in Fluids, 13*(5), pp. 599–624.

[10] Turbodesign suite version 5.2.5, Advanced Design Technology Ltd, www. Adtechnology.co.uk

[11] Zangeneh, M., Goto, A. and Harada, H., 1998. "On the design criteria for suppression of secondary flows in centrifugal and mixed flow impellers", *ASME Journal of Turbomachinery*, Vol 120, pp. 723-735.

[12] Zangeneh, M, Nikpour, B, Watanabe, H., 2010. "Development of a High Performance Centrifugal Compressor Using a 3D Inverse Design Technique", *IMechE 9th International Conference on Turbochargers and Turbocharging*, May 19th-20th 2010, London.

[13] TURBOdesign pre, Advanced Design technology Ltd, www.adtechnology.co.uk.

[14] TURBOdesign Volute, Advanced Design technology Ltd, www.adtechnology.co.uk.

A high-performance electric supercharger to improve low-end torque and transient response in a heavily downsized engine

B Richards, K Gray, H Tran, A Andah
Aeristech Limited, UK

M Bassett, J Hall
MAHLE Powertrain Limited, UK

ABSTRACT

Extreme downsizing is a modern solution for reducing the CO_2 emissions of internal combustion engines. Aeristech has developed a 48V electric supercharger with dual functions: steady state supercharger and transient booster, for application to heavily downsized engines. The electric supercharger has been designed in two variants: with electronics and compressor motor in a single integrated package and with these subsystems housed in separate, non-integrated packages. As much as 60% engine downsizing can be achieved when the electric supercharger is paired with a conventional mechanical turbocharger to create a two-stage boosting arrangement. This arrangement takes advantage of the rapid transient response of the electric supercharger and also of the continuous steady-state power delivery of which this electric supercharger is capable. This allows the electric supercharger to supplement a conventional boosting system and allows the electric supercharger with a single mechanical turbocharger to function as a multi-stage boosting system with superior transient response. The electric supercharger incorporates a new compressor design with high efficiency and large map width, benefiting from the unique high speed capability of Aeristech's electric supercharger, which in turn is derived from Aeristech's high-speed electric motor control technology.

ABBREVIATIONS

BMEP	Brake mean effective pressure
C	Compressor
C1	Name of compressor design 1
C2	Name of compressor design 2
CAN	Controller area network
CFD	Computational fluid dynamics
Di3	Direct injection, three cylinders
eS/eSupercharger	Electric supercharger
krpm	Kilo revolution per minute
L	Litre
M	Motor
RPM	Revolution per minute
T	Turbine
WCAC	Water charger air cooler

1 INTRODUCTION

1.1 Engine downsizing and CO_2 reduction

European CO_2 emissions of passenger vehicles is targeted at a 27% reduction from 130g/km in 2015 to 95g/km by 2020. The United States of America has proposed a 35% decrease in CO_2 emissions levels from 149g/km in 2015 to 97g/km in 2025. China's CO_2 emissions targets will drop 27% from 161g/km in 2015 to 117g/km in 2020 [1]. Moreover, the JLR Ultraboost engine has achieved a 35% fuel economy improvement (and hence CO_2 reduction) by downsizing from a 5.0L naturally aspirated configuration to a 2.0L supercharged and turbocharged layout. In the Ultraboost project, 23% CO_2 reduction is achieved via 60% engine downsizing, and 12% is achieved via reduced friction, improved thermal efficiency, and stop-start functionality [2]. MAHLE Powertrain have recently showcased their "dynamic downsizing engine", which employs 62.5% downsizing (from 3.2L naturally aspirated to 1.2L boosted) to achieve a 35% CO_2 reduction [3]. It is also postulated that the theoretical peak fuel economy improvement could be as high as 40% (depending upon vehicle application and drive-cycle) with 75% downsizing of the engine [3]. Engine downsizing, especially of gasoline engines which incur throttling losses, is likely to be a key approach adopted towards meeting legislated CO_2 levels in vehicle markets across the world.

1.2 Boosting system requirements

An earlier embodiment of the MAHLE downsizing demonstrator engine mentioned above achieved a 30% improvement to fuel efficiency via 50% engine downsizing (and the addition of start-stop), from a 2.4L naturally aspirated to a 1.2L single turbocharger layout [4]. This gasoline engine produced 120kW peak power and 100kW/L specific power. MAHLE characterized this engine as boost system limited, and it was clear to MAHLE at that time that further engine downsizing would require an improvement in the boosting system [3]. The next embodiment produced 144kW peak power and 120kW/L specific power with a two-stage turbocharger [5]. In a separate project at JLR, the 2.0L Ultraboost engine achieved 283kW peak power and 141.5kW/L with a turbocharger and supercharger [6]. Heavily downsized engines tend to require multi-stage boosting systems because of limitations of the turbocharger compressor map width and efficiency with a single stage boosting system. Furthermore, transient response and drivability can represent significant limiting characteristics of even a multi-stage system if mitigating technologies are not adopted. When moving to greater levels of engine downsizing and CO_2 reduction, MAHLE Powertrain adopted an electric supercharger on the basis of both transient characteristics and steady-state performance across a broad engine speed (and hence air mass flow) range.

1.3 The 48V electric supercharger

The 48V electric supercharger developed by Aeristech Ltd incorporates three power specifications: a steady state (continuous) power capability, intermittent, and transient. The steady-state capability defines the supercharger's ability to function as part of a multi-stage boosting system in a continuous steady-state engine operating condition, whereby the electric supercharger can be relied upon to contribute to the maximum torque capability of the engine. In addition to a high continuous power rating, the electric supercharger requires a wide compressor map (and therefore generally a high compressor speed) to be able to boost the engine across a broad engine speed range and allow the second boosting device (the mechanical turbocharger) to be well-matched at higher engine speeds, which can lead to the main turbocharger being unable to provide sufficient boost at lower engine speeds.

The second power rating of the electric supercharger, the intermittent rating, is an over-rating which the electric supercharger is capable of maintaining for a limited period. The purpose of this is to maintain a quasi-steady operating point while waiting for the conventional mechanical turbocharger to spool up (overcome turbo lag) and reach the required steady operating condition. This intermittent condition is generally less than five seconds in duration but the electric supercharger has been designed to support the turbocharger during much longer spool up events. The intermittent power rating of the electric supercharger, being higher than the steady state rating, allows increased air flow through the engine for improved tip-in response and quicker pick-up of the conventional mechanical turbocharger. When the mechanical turbocharger reaches operating speed and boosts in tandem with the electric supercharger, the power output of the electric supercharger drops to, or below, its steady-state limit.

The transient power rating of the electric supercharger is used only while the electric supercharger itself is accelerating, and it is a function of the rotating inertia of the electric supercharger and the tip-in response requested by the engine calibration. A typical requirement is that the electric supercharger should produce peak pressure from an idle condition within a time period usually less than 0.5 seconds in every load condition (ie following a compressor run-up line that represents maximum aerodynamic load on the compressor wheel at every intermediate speed during the electric supercharger's acceleration event).

The new capability offered by Aeristech's electric supercharger is to achieve high levels of boost in steady-state and intermittent conditions across a wide range of mass flow rates in the engine lower speed range. This allows the conventional mechanical turbocharger to be sized for higher RPM specific engine power than could be achieved with more conventional boosting systems (given the constraint on modern road cars to maintaining drivability and consistency of torque output across the full RPM engine speed range).

2 METHODOLOGY

2.1 The 1.2L Di3 engine's requirements

The MAHLE 1.2L Di3 engine is pictured in Figure 1. Aeristech's electric supercharger was designed in part to fulfill the gap in boosting system capability after this engine was specified with a large enough conventional mechanical turbocharger to allow the engine to achieve 161kW/L and 33 bar maximum BMEP. Table 1 provides further specification details of the Di3 engine.

Figure 1: MAHLE POWERTRAIN 1.2L Di3 engine

Table 1: 1.2L Di3 engine specification

Configuration	-	Inline 3 Cylinders
Displacement	cm^3	1200
Bore	mm	83
Stroke	mm	73.9
Compression ratio	-	9.3:1
Peak power	kW	193
Specific power	kW/L	161
Max BMEP	bar	33

Figure 2 shows the boosting system configuration for this engine. The 48V electric supercharger is located upstream of the turbocharger. A valve is integrated to bypass the electric supercharger at higher engine speeds, whereupon the turbocharger provides all of the air flow for the engine. A charger air cooler is placed between the 48V electric supercharger and the turbocharger.

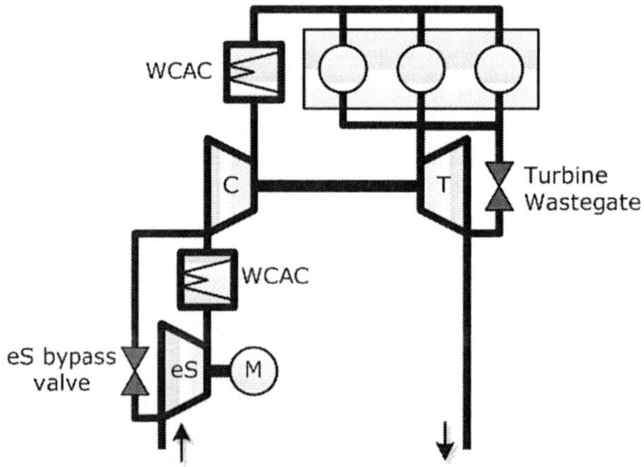

Figure 2: eSupercharger and turbocharger engine diagram

To understand the boost requirements of the electric supercharger in steady-state, intermittent, and transient conditions for this and other engines was a non-trivial task requiring engine simulation. Aeristech also made compromises in the design to allow one electric supercharger to satisfy a broad range of engine applications. This requirements specification exercise will be detailed separately in another paper.

2.2 Electric supercharger specification

Having established the requirements of the boosting system, Aeristech set about to design the electric supercharger starting with the compressor. The compressor design must provide a wide and efficient compressor map, which typically puts pressure on the compressor designer to demand a high operating speed. Higher compressor operating speed tends to allow a broader range of mass flow through a given geometry, as well as enabling a high flow rate in proportion to manufacturing clearances and gaps, improving the map width and efficiency.

From the compressor design, the requirements of the electric supercharger's motor can be specified and from there the requirements of the controller can be specified. In general, the motor is sized for thermal stability in the steady-state operating condition of the electric supercharger and the motor operates in a state of thermal disequilibrium in intermittent and transient operating conditions. In general, the controller's power electronics are sized for the maximum (transient) power requirement of the electric supercharger, with thermal equilibrium also achieved in intermittent operation. Thermal equilibrium of the power electronics is generally required in intermittent operation because the thermal timescales inside the controller are generally much shorter than inside the motor.

The above sequential design process must be caveated with various iterative loops from "downstream" controller design back to "upstream" compressor and motor design as well as across engineering disciplines from typically mechanical engineering derived thermal and packaging designs to typically electrical engineering derived circuit designs and isolation. The iterative design process enables cross-optimisation between the compressor, motor, and controller systems across the three performance requirements: steady-state, intermittent, and transient. However, engineering insight and a wealth of company experience in high-speed and high-power electric compressor design can streamline the process by adding assumptions and constraints to anticipate and avoid certain iteration paths in the design process.

3 COMPRESSOR DESIGN

3.1 First compressor: C1

The first compressor map of the 48V electric supercharger is shown in figure 3. The data, derived from CFD simulation, shows the compressor's predicted performance up to a maximum speed of 140krpm, and a region of operation with efficiency above 74%. This compressor will be hereby referred to as C1.

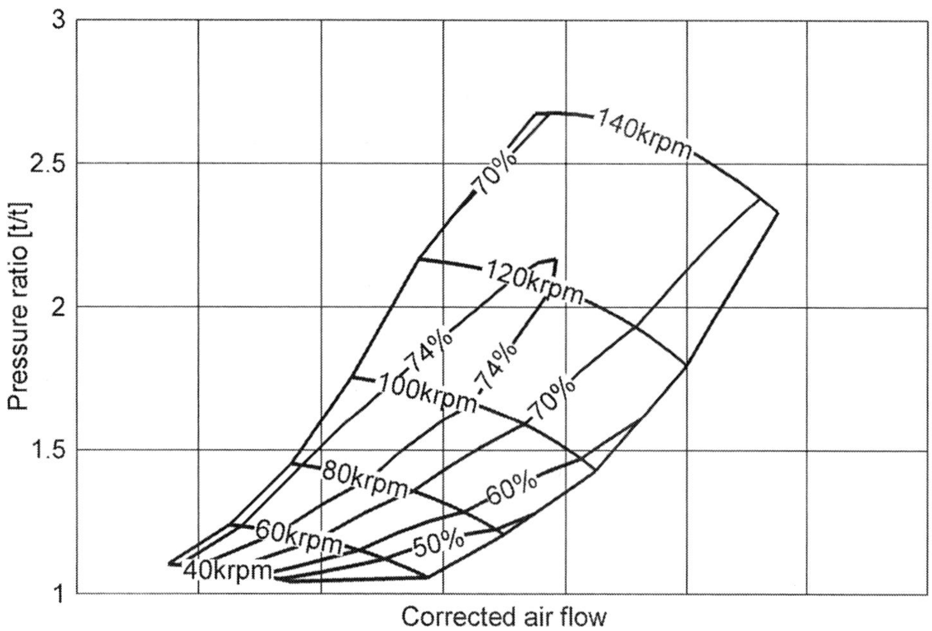

Figure 3: The compressor map of the first design

3.2 Second compressor: C2

The second compressor, hereby referred to as C2, improves the surge margin and the choke limit with the map being wider at every operating speed. C2 provides a region of peak operating efficiency above 78% which allows the power consumption of the electric supercharger to be lower with the C2 compressor than with the C1 compressor.

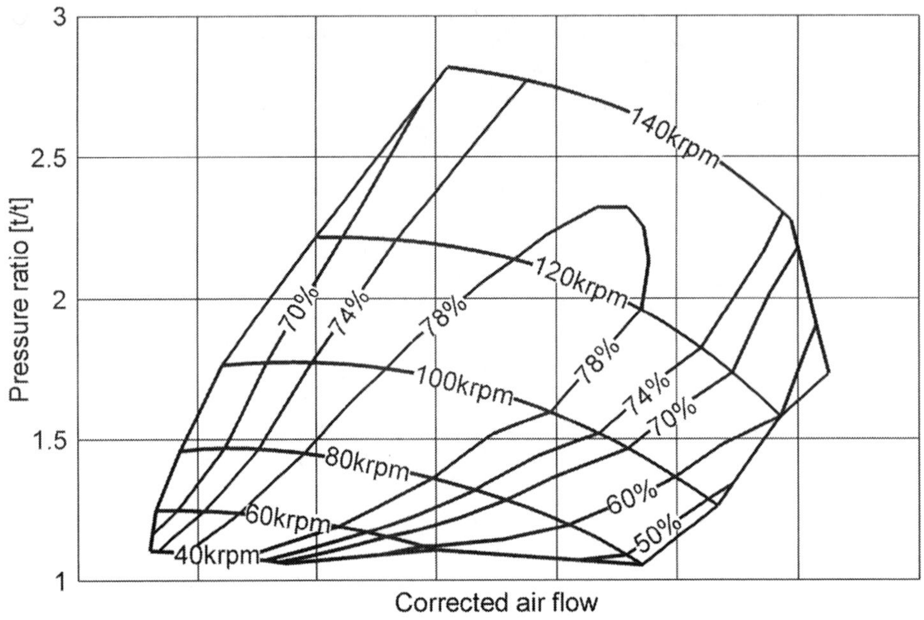

Figure 4: The compressor map of second design

3.3 Comparing the two compressor designs

The CFD simulation results of both compressors are shown in Figures 5 and 6. Figure 5 demonstrates the improvement of the new compressor design in terms of map width at various compressor speeds. The speed lines of both designs produce similar pressure ratios near surge, but C2 provides slightly higher pressure ratios at high mass flow. C2 provides a 50% better surge margin than C1 at 80krpm, 45% at 100krpm, 28% at 120krpm, and 17% at 140krpm. The choke flow limit of C2 is also 34% higher than C1 at 80krpm, 25% at 100krpm, 17% at 120krpm, and 9% at 140krpm.

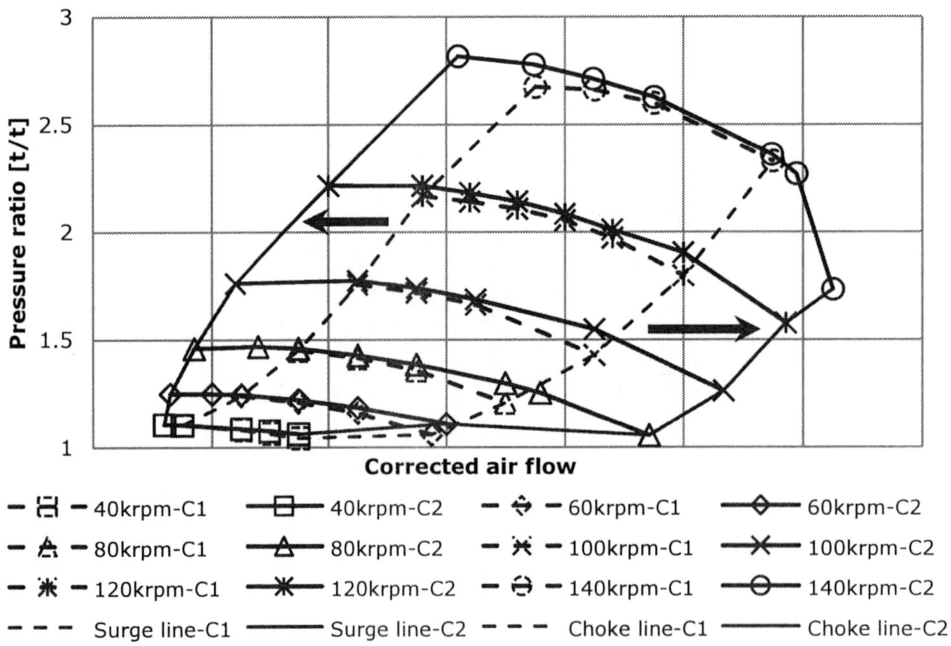

Figure 5: CFD result comparison of two compressor designs: Pressure ratio versus corrected air flow

Figure 6 presents the efficiency changes in particular speed lines. The compressor efficiency of C2 is higher at all speeds and peak values rise from approximately 74% for C1 to approximately 80% for C2. The surge and choke points of C2 also run more efficiently than C1. Regarding choke limit, C2 achieves 67.5% efficiency, compared to 43% for C1 at the 40krpm, and at 60krpm C2 achieves 55% efficiency compared to 26% for C1.

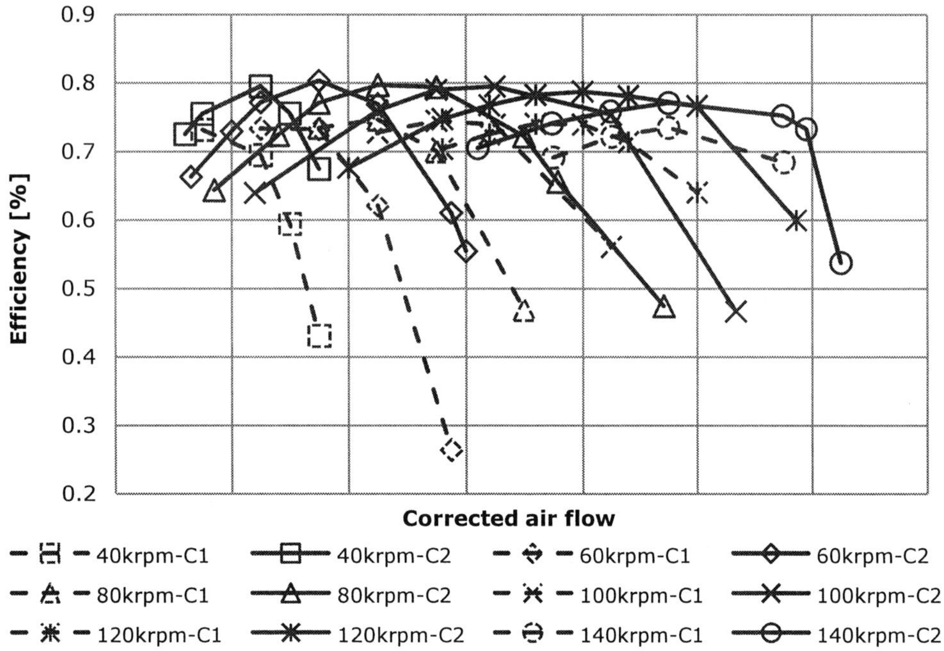

Figure 6: CFD result comparison of two compressor design: Isentropic efficiency versus corrected air flow

3.4 Compressor testing procedure

The experimental setup is shown in Figure 7. The inlet section contains a long radius (bellmouth) inlet nozzle for flow measurement and an inlet pipe containing pressure tappings and k-type thermocouples. The compressor is driven by Aeristech's electric motor. Precise speed control of the motor is achieved using Aeristech's controller which also provides the speed feedback signal used as the compressor speed measurement. Controller speed feedback is based on the measured back EMF from Aeristech's permanent magnet electric motor which, being a synchronous machine, provides a highly accurate measurement of compressor speed. Downstream of the compressor is the discharge pipe containing pressure tappings and k-type thermocouples. The compressor's load point (mass air flow rate and pressure ratio) is controlled by specifying rotational speed on the drive motor and by setting the opening of a butterfly valve in the compressor outlet pipe.

Pressure and temperature measurements are taken in the inlet and discharge pipe sections with the sensor positions in accordance with SAE J1723. These measurements are required for the calculation of parameters such as mass flow rate, pressure ratio and isentropic efficiency. The test involves recording a range of predetermined points, equally spaced from surge to choke, for several compressor operating speed lines. Before each data point is recorded, the stabilisation criteria as stated by SAE J1723 must be met.

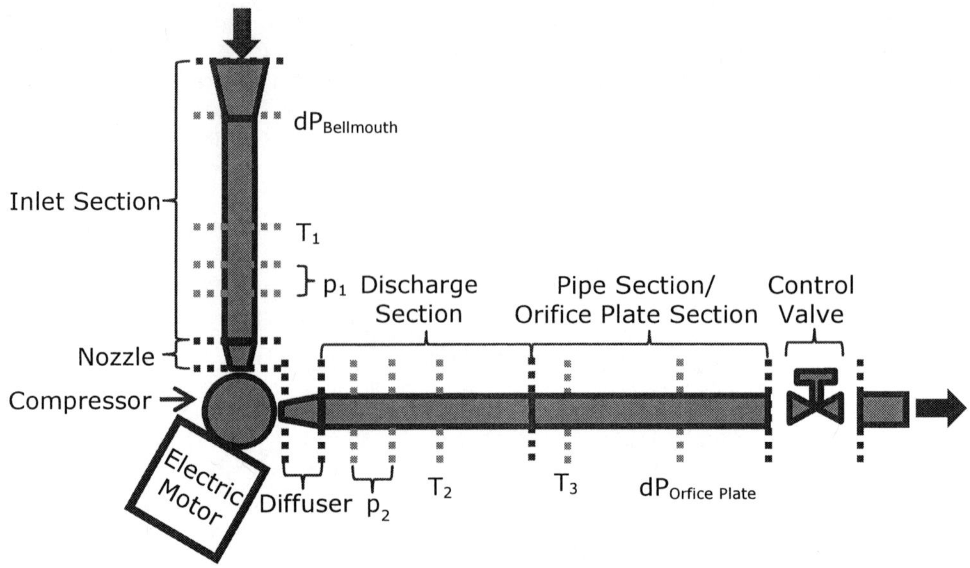

Figure 7: Experimental set-up

4 HARDWARE

The motor specification is based on the engine and compressor simulation results and the controller design is based on the motor simulation results (with iteration, insight and experience to allow cross-optimisation between components). Aeristech has existing designs for integrated power electronics (sharing one single housing with the motor) and non-integrated power electronics (separate from the motor housing).

The present designs are specified for 48V electrical input (assumed to be limited by vehicle architecture) although higher voltage variants have been developed for fuel cell and range extender engine applications. A 12V variant was considered but practical limits on the vehicle's electrical system, such as maximum permissible current draw, limit the usefulness of any 12V system even with the high levels of efficiency achieved by Aeristech's high-speed machine.

Motor speed in this application was limited to 120krpm, mainly due to bearing limitations. The impeller wheel is directly mounted to the motor shaft and therefore runs at the same speed.

Table 2: Electric supercharger specification

Air compressor	Centrifugal
Motor type	4-pole Permanent Magnet
Lubrication	Greased (for life)
Cooling	Liquid (water/glycol)
Input voltage	48V
Max speed	120,000 rev/min
Idle to max speed	< 0.5s (depends on compressor flow)

4.1 Non-integrated design

The controller and motor compressor units are presented as independent packages in Figure 8. The controller input is 48Vdc and its output voltage is a function of motor speed. The motor's electrical input is connected to the controller output. Both the controller and motor are liquid cooled and cooling water from the controller is connected to the motor housing in a series arrangement. CAN and other interface methods are provided on the controller unit.

Figure 8: Electric supercharger non-integrated design

4.2 Integrated design

The integrated design is shown in Figure 9. The motor and power electronics are combined into one package with an integrated cooling subsystem design. A CAN interface is provided on the housing. The housing covers incorporate ports and internal space for additional sensors and wiring for diagnostic purposes but these features are not intended to be used in engine and vehicle testing.

Figure 9: Electric supercharger integrated design

5 RESULTS

5.1 Electric supercharger test results

5.1.1 C2 compressor test result

The test results and CFD simulations of the C2 compressor are compared in Figures 10 and 11. Figure 10 shows that the pressure ratio performance (solid lines) was close to the simulated design (dotted lines) in the centre and low-flow regions of the map. The tested compressor provided a poorer surge margin than the simulated compressor. The compressor was tested beyond the CFD data in the choke region of the map with the results showing some drop-off in pressure ratio at the higher flow points compared to the simulation.

Figure 11 shows efficiency from the CFD and test data for the C2 compressor. The maximum efficiency of both the test and CFD results is nearly 80%. The efficiency of the test result is lower than the CFD simulation in the high airflow region at certain speeds. The reason for these differences is believed to be due, in part, to the actual test rig pipe design compared with the simulation setup. Another potential reason for the discrepancy at higher air flow rates is a high compressor volute outlet velocity; the CFD simulation treats high-velocity exit air as entirely efficient, whereas the compressor test rig will convert some of the velocity into pressure (with less than 100% efficiency) between the compressor outlet point and the position in the outlet pipe where the measurement is sampled.

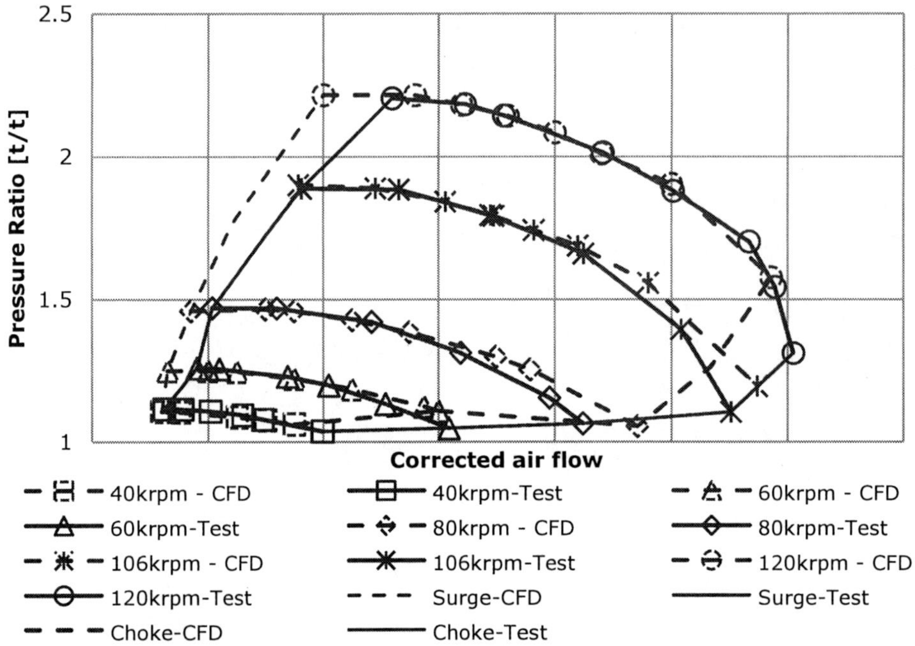

Figure 10: C2 CFD and test result comparison: Pressure ratio versus corrected air flow

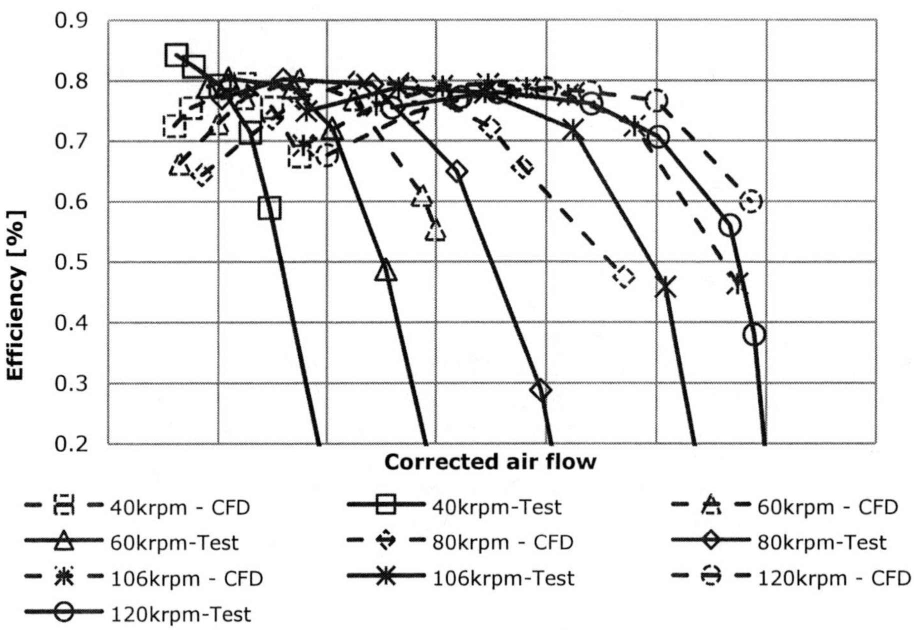

Figure 11: C2 CFD and test result comparison: Isentropic efficiency versus corrected air flow

5.1.2 C1 and C2 comparative results

Figures 12 and 13 show comparisons of the compressor test results, with the test result of C1 as the dotted line and C2 as the solid line. C1 was tested at 40krpm, 60krpm, 80krpm, 100krpm and 120krpm. C2 was tested at the same speeds with the exception that 106krpm was tested instead of 100krpm.

The comparison clearly demonstrates the improvement of the new (C2) design over C1. The map width (comparing surge and choke lines) of the new compressor is significantly better than C1 for each speed line. C2 also generally achieves a higher pressure ratio compared to C1 across the whole map, this is particularly evident along the 120krpm speed line. The surge margin of C2 shows a 26% improvement at 40krpm over C1, 35% at 60krpm, 46% at 80krpm and 6% at 120krpm. The choke limit of C2 is increased from that of C1 by 13% at 40krpm, 9% at 60krpm, 12% at 80krpm, and 6% at 120krpm.

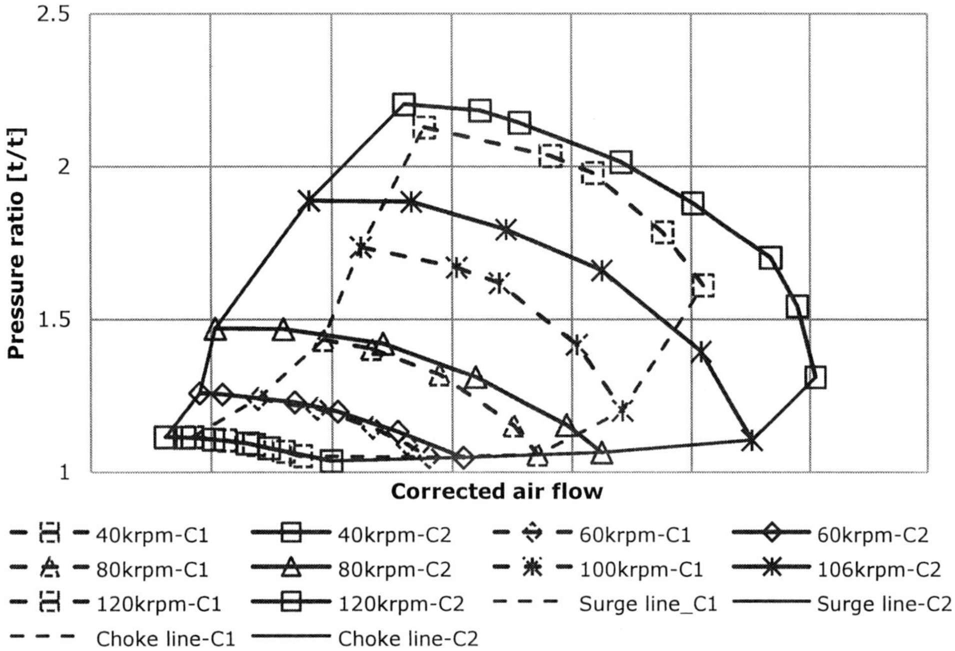

Figure 12: C1 and C2 comparative compressor maps

Figure 13 shows an improvement in compressor efficiency of C2 relative to C1 at every point sampled.

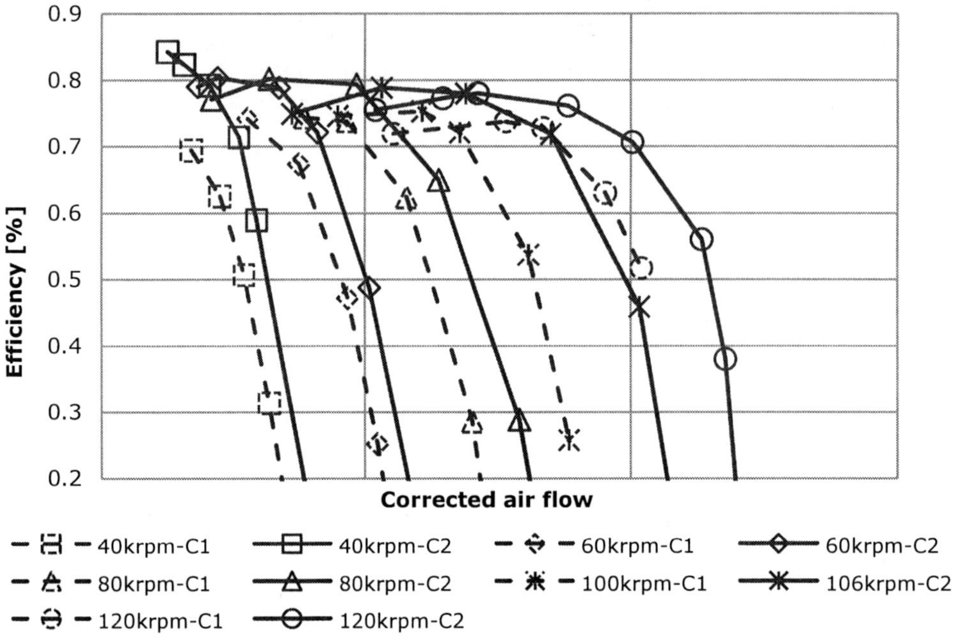

Figure 13: C1 and C2 comparative efficiency

5.1.3 Transient results

The electric supercharger transient response is presented in Figure 14. The MAHLE 1.2L Di3 engine has been tested over a transient load step at a fixed engine speed of 1500 rev/min. The electric supercharger has taken 0.8 seconds to accelerate from an idle speed of 10krpm up to 90% of the target speed of 120 krpm (106krpm). The turbocharger speed is plotted on the same graph, and demonstrates slower response than that of the electric supercharger.

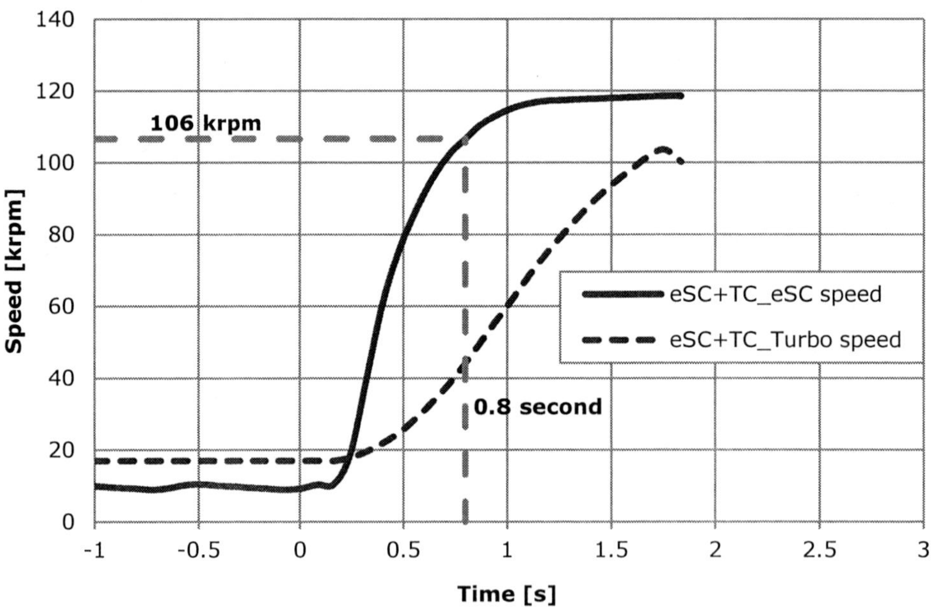

Figure 14: Transient response of the electric supercharger during a transient engine load step at 1500 rev/min

5.2 Engine test results

The MAHLE 1.2L Di3 engine has been tested both in steady-state and transient load conditions. Test results are shown in Figures 15 and 16, where it can be seen that the engine demonstrates extremely high specific power with the electric supercharger and turbocharger.

5.2.1 Steady state engine test results

The steady state test results are shown in Figure 15, incorporating two steady state modes: single turbo (dashed line) and eSupercharged + Turbo (solid line). The single turbo line represents the original version of the MAHLE engine and it can be seen that this achieves maximum torque of 30 bar BMEP at 1600 rev/min and maintains this until almost 4000 rev/min. The maximum power achieved is 100 kW/L. The eSupercharged + Turbo line sits above the original torque curve across the entire speed range of the engine. Maximum torque of 33 bar BMEP is achieved at 1500 rev/min and this is maintained until 4500 rev/min. The maximum power achieved is 161 kW/L. The eSupercharged version of the engine has been mapped at 35 bar BMEP from 1500 to 4000 rev/min. This excess torque capacity is reserved for powering the alternator, in order to enable the eSupercharger to be driven indefinitely, irrespective of the state of charge of the vehicle battery, without any compromise to the net torque output of the system.

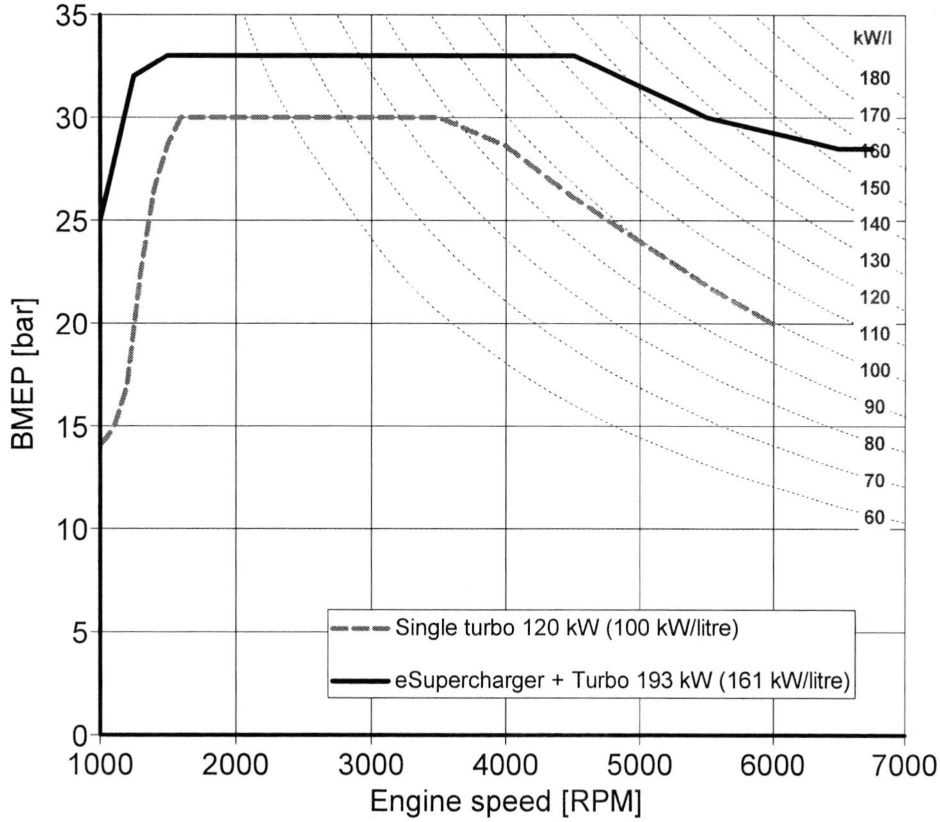

Figure 15: Steady state engine test result of 1.2L Di3 engine by combination of eSupercharger and turbocharger

5.2.2 Transient engine test results

The engine transient response, during a 1500 rev/min fixed speed load step from 4 bar BMEP, is shown in Figure 16. Several tests have been undertaken: electric supercharger plus turbocharger, electric supercharger alone, turbocharger alone and naturally aspirated 1.2L Di3 engine response. The responses of these different configurations are shown at 1500rev/min constant engine speed. The engine achieves 35 bar BMEP at this speed (over torque) with both the electric supercharger and the conventional mechanical turbocharger contributing to the total boost pressure. The engine achieves 90% of this BMEP after 1.2 seconds. The engine also produces 27 bar BMEP with the electric supercharger alone, which is an achievement that takes advantage of the intermittent power rating of the electric supercharger to provide good tip-in torque response before the conventional turbocharger reaches operating speed. The turbocharger alone produces very little boost pressure at 1500rev/min, so the engine generates 14.5 bar peak BMEP. The 1.2L naturally aspirated response is approximately 11 bar peak BMEP.

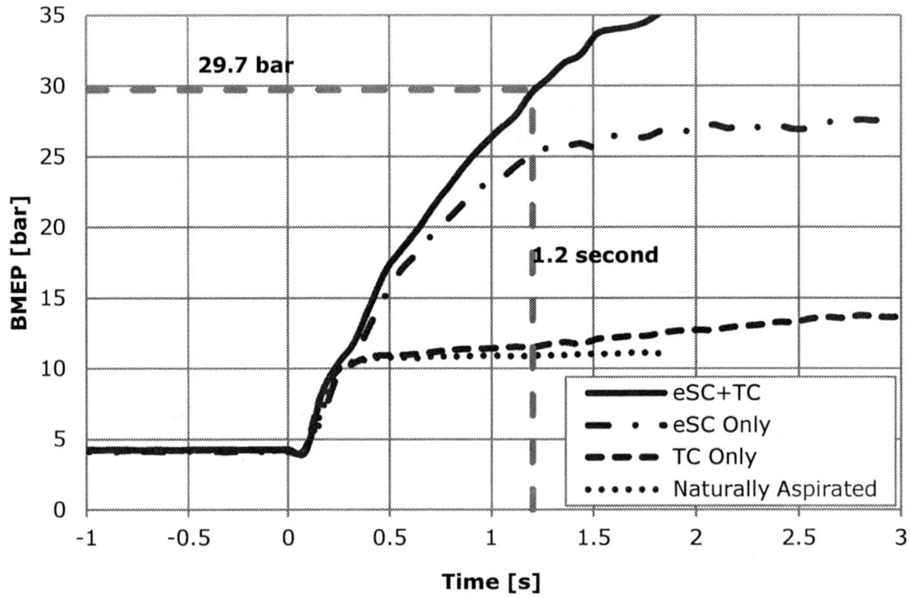

Figure 16: Transient test result at 1500rev/min of 1.2L Di3 engine by combination of eSupercharger and turbocharger

6 DISCUSSION

6.1 Electric supercharger test results

The new compressor design (C2) demonstrates significant improvements in map width and isentropic efficiency compared to the C1 design. Both maps necessitate the use of a high-speed motor directly coupled to the compressor but the significantly wider C2 map and the subsequent engine tests demonstrate the possible advantages to engine downsizing of a high speed electric supercharger with high continuous power rating and even higher intermittent rating for tip-in response.

The design process yielded integrated and non-integrated designs that met, or exceeded design targets. Further design requirements for future machines also evolved during the course of the project related to packaging, cost and vehicle integration.

6.2 Engine test results

The 1.2L Di3 engine produces 193kW peak power and 161kW/L specific power by combining the electric supercharger with a conventional mechanical turbocharger. Baseline configurations available for this engine include a 120kW peak power and 100kW/L specific power variant using a single mechanical turbocharger boosting arrangement [4] and a 144kW peak power 120kW/L specific power variant with a two-stage mechanical turbocharger boosting arrangement [5]. The engine reaches peak BMEP of 35 bar with the electric supercharger and mechanical turbocharger compared to 30 bar peak in baseline configurations [5].

The transient response of the engine during a 1500 rev/min fixed speed load step from 4 bar BMEP was 1.2 seconds to 90% of maximum torque in the electrically supercharged 193kW variant compared to a baseline value of 3.3 seconds in the 120kW peak power single stage mechanical turbocharger variant. This result is even more impressive when considering that the target torque output of the 193kW engine was higher (reflecting 33 bar BMEP) than the baseline engine (reflecting 30 bar BMEP) and the time to torque was lower. In a separate study, the electric supercharger was simulated on a 2.0L gasoline engine, which achieved time to torque of 0.64 seconds at 1200rev/min and 0.78 seconds at 1500rev/min, in comparison to 1.71 seconds at 1200rev/min and 1.44 seconds at 1500rev/min for the baseline engine simulation [7].

By demonstrating that the electric supercharger enables greater specific power and a greater degree of engine downsizing, the authors provide a strong indication of reduced CO_2 emissions in-line with international targets. The link between these factors is widely discussed and well-understood. Actual results will depend on the drive cycle and vehicle to which the engine is applied, and the authors intend to publish results from a C-class demonstration vehicle to be fitted with the electric supercharged Di3 engine.

7 CONCLUSIONS

The electric supercharger achieves dual functionality of steady state supercharger operation to improve BMEP and to allow a better turbocharger match and transient booster operation to improve time to torque. The combination of the electric supercharger and mechanical turbocharger demonstrated represents a relatively simple and easily tuned boosting system, compared to a multi-stage conventional system or a multi-stage conventional system with a transient-only electric booster.

To provide the necessary functionality for MAHLE's extreme downsizing Di3 application Aeristech specified the electric supercharger with three power levels: steady-state, intermittent, and transient, representing engine steady-state operation, the turbo spool up (turbo lag) event, and the tip-in event, respectively.

The new (C2) compressor design clearly shows the improvement of compressor performance in terms of map width and efficiency compared with the C1 design. The benefit of wide map width was demonstrated on the 1.2L MAHLE Di3 engine, where the conventional mechanical turbocharger could be enlarged and made more efficient given the ability of the electric supercharger to satisfy steady-state boosting requirements at lower engine speeds. This enabled greater specific power than would have been possible using an electric supercharger either without a steady-state capability or with a narrower compressor map. The turbocharger matching and engine tuning will be discussed in future publications, as well as CO_2 emissions results following integration of the electric supercharged Di3 engine with a C-segment demonstration vehicle.

ACKNOWLEDGEMENTS

The authors acknowledge the contributions of their colleagues to this work at Aeristech Ltd and MAHLE Powertrain UK Ltd. This research was supported by a grant from the Technology Strategy Board / Innovate UK and a grant from the Niche Vehicle Network.

REFERENCES

[1] The international council on clean transportation, "CO2 emissions from new passenger cars in the EU: car manufacturer's performance in 2014", (2015). http://www.theicct.org, accessed on Oct, 2015

[2] R. Patel, L. Jeffcoat, J. Turner, "Ultraboost- improving fuel economy by 35% of a spark ignition engine through downsizing", Panel presentation at Engine expo 2013, (2013).

[3] Bassett, M., Hibberd, B., Hall, J., Gray, K., Richards, B. "Dynamic Downsizing for Gasoline Engines", 24th Aachen Colloquium Automobile and Engine Technology, (2015)

[4] Hancock, D., Fraser, N., Jeremy, M., Sykes, R. *et al.*, "A New 3 Cylinder 1.2l Advanced Downsizing Technology Demonstrator Engine," SAE Technical Paper 2008-01-0611, (2008)

[5] Fraser, N., Blaxill, H., Lumsden, G., Bassett, M., "Challenges for Increased Efficiency through Gasoline Engine Downsizing," SAE Int. J. Engines 2(1):991-1008, (2009)

[6] Turner, J., Popplewell, A., Patel, R., Johnson, T. *et al.*, "Ultra Boost for Economy: Extending the Limits of Extreme Engine Downsizing," SAE Int. J. Engines 7(1), (2014)

[7] B. Richards, H. Tran, O. Porter, M. Parrett, P. Matajsz "A 110kW/L, 200Nm/L Downsized Gasoline Engine with a High-Speed, Permanent Magnet Electric Supercharger", 4th International Conference on Energy Efficient Vehicle, (2015)

Performance testing of an electrically assisted turbocharger on a heavy duty diesel engine

E Winward [(1)], **J Rutledge** [(2)], **J Carter** [(3)], **A Costall** [(4)], **R Stobart** [(1)], **D Zhao** [(1)], **Z Yang** [(1)]

[(1)] Dept. Aeronautical and Automotive Engineering, Loughborough University, UK
[(2)] Research and Advanced Engineering, Caterpillar Inc., USA
[(3)] BorgWarner Turbo Systems, UK
[(4)] Imperial College London, UK

ABSTRACT

An electrically assisted turbocharger was designed, built and tested by a consortium consisting of Caterpillar Inc., BorgWarner Turbo Systems, Imperial College London and Loughborough University. The Electric Turbocharger Assist (ETA) device was based on a BorgWarner BV63 variable turbine geometry (VTG™) turbocharger with a bearing housing that was extended to accommodate a switched reluctance (SR) electrical machine. The ETA device was evaluated over a range of steady state and transient engine conditions with the ETA providing electric assist or electric regeneration. The response of the single stage ETA turbocharger matched a production 2-stage turbocharger on the same engine.

1. INTRODUCTION

Electric Turbocharger Assist (ETA) describes a turbocharger device in which the turbocharger operation is assisted by means of an electric motor/generator which is mechanically coupled to the turbocharger shaft. Such devices are also referred to as a hybrid turbocharger or turbocharger motor/generator. Electrification of the turbocharger can be applied to both standard fixed or variable geometry turbines.

There have been a number of research programs which have investigated the integration of an electric machine with an exhaust gas turbocharger. In many concepts the electric machine is built into the turbocharger bearing housing (1-5). There is at least one example where the electrical machine is overhung on the compressor side (6).

An ETA device such as that described in this paper has two operating modes: 1) Electrical assist mode (herein referred to as motoring) in which the electric motor provides additional steady state boost pressure or accelerates the rotor during transients to give faster boost pressure rise rate and improved engine transient response (1,4,6-10). 2) In electrical regenerative mode (herein referred to as generating) the ETA can be used to recover part of the otherwise waste heat of the exhaust by generating electricity from excess turbine power (2,11,12). A battery can be used to supply the electrical power to the ETA when motoring and store energy when the ETA is generating.

Electrically assisted turbochargers provide an additional degree of freedom for engine air path control and have the potential to improve the engine-turbocharger matching and resizing of the turbine for optimal efficiency (maintaining transient performance via motoring assist) (12,13). These benefits can potentially give thermal efficiency gains of 10% (6). The ETA as a technology can be an enabler for engine down-sizing, engine down-speeding, improving power density and reducing parasitic losses.

This paper describes the ETA hardware which was developed and tested by a consortium consisting of Caterpillar Inc., BorgWarner Turbo Systems, Imperial College London and Loughborough University and which was co-funded by Innovate UK (formerly the Technology Strategy Board UK). The paper presents results from engine steady state and transient testing of the ETA on a 7 litre off-highway diesel engine.

2. DESCRIPTION OF ETA HARDWARE

The turbocharger used for this project was a modified BV63 from BorgWarner Turbo Systems which has a variable turbine geometry (VTG™). The electric machine was a switched reluctance topology supplied by Magnomatics. The integration of the electric machine with the turbocharger was carried out by BorgWarner. The core assembly can be seen in Figure 1. In order to accommodate the electric machine, the rotor shaft length is increased significantly compared to the standard turbocharger. The electric rotor has a larger diameter than the bearings and therefore requires the bearing housing to be split into two halves. It was found that a vertical split line as seen in Figure 1 was preferable to a horizontal split line. Nevertheless, careful control of the concentricity of the two halves is required to maintain rotor shaft motion within limits.

The switched reluctance machine has the advantage of a simple rotor structure without any magnetic material, which results a high rotor temperature capability and no concern about the high cost of rare-earth materials. However, compared to permanent-magnet machines, the SR machine typically has lower power density and greater noise signature.

Figure 1: Integration of electric machine with turbocharger

There are a considerable number of different switched reluctance motor topologies which can be defined by the number of phases and the associated numbers of stator and rotor poles. The choice of topology depends on many factors including the starting capability, torque ripple, iron core losses, converter cost, power switching losses and commutation frequency.

Single phase and two phase machines generally have poor starting capability but they have an advantage in reduced switching frequency compared to designs with more phases. They also require fewer components in the power converter and have a lower assembly cost due to fewer machine coils.

A commonly used SR topology is 3-phase machine with 6 stator poles and 4 rotor poles (e.g. 6/4 type) and this was selected for the ETA, Figure 2. This machine is capable of starting from any position in either direction. This was selected based on the balance of requirements spanning torque ripple, iron core losses, converter cost, power switching losses and commutation frequency. The SR motor was designed to give a peak electrical power in excess of 5kW in both motoring and generating. This enabled evaluation of both modes of operation of the prototype ETA on the test engine over a useful electrical power range.

The electric machine's rotor is clamped between turbine wheel and compressor wheel. This has the advantage of simple assembly, not requiring press- or shrink-fit of the electrical rotor to the shaft. Lubrication oil was used to cool the stator of the electrical machine. This eliminates the risk of oil and water mixing through casting porosity and reduces the hoses required.

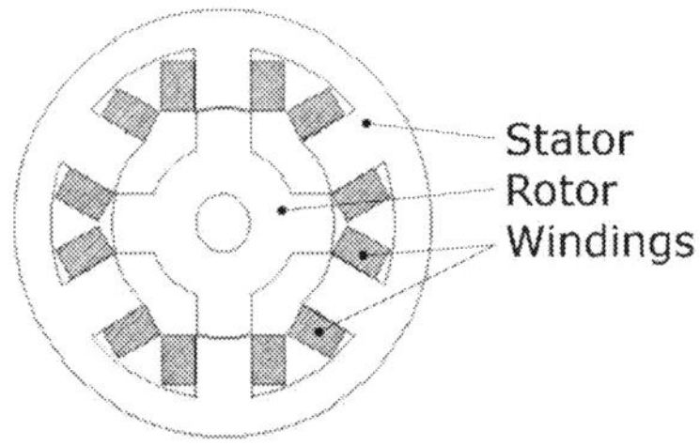

Figure 2: 6/4 type switched reluctance electrical machine

A position sensor-less control method was devised to control the switched reluctance electrical machine, this method avoided the requirement to package a motor angular position sensor within the ETA device. This sensor-less control enabled electrical operation of the machine between 40kRPM and 130kRPM (14). Figure 3 summarises the electrical operating envelope of the ETA with sensor-less control which could be practically explored on the test engine at Loughborough University.

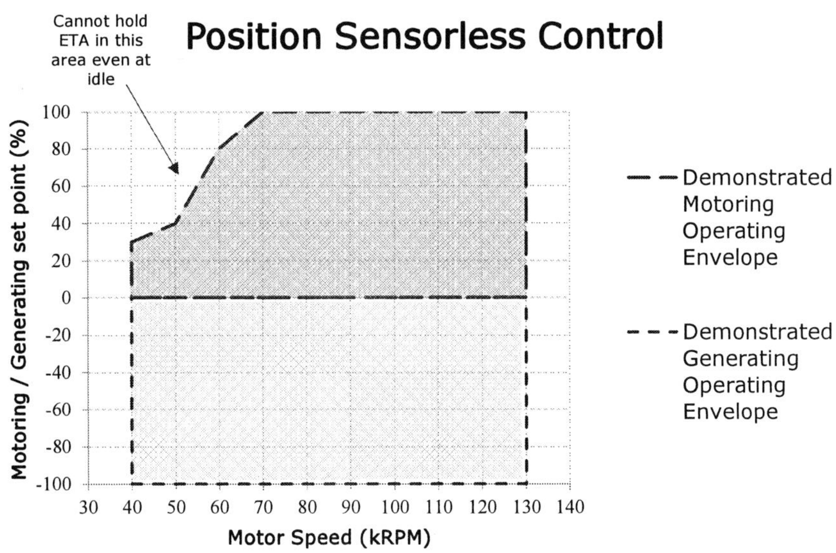

Figure 3: ETA position sensor-less control operating envelope

In Figure 3 the vertical axis represents the % of the maximum achievable torque of the SR electrical machine in both motoring (+ve) and generating (-ve). The ETA was controlled by setting a % of the maximum achievable torque. The speed of the ETA was therefore a balance between the ETA % torque demand and engine conditions (speed, load, air system settings, etc.). It was found that the ETA could not be operated at high motoring electrical power demand set points at lower turbocharger speeds (40 to 70kRPM) as the motor torque exceeded compressor needs at those speeds. The other points within the operating envelope could be reached under steady state conditions using a combination of varying engine speed, load, and air system settings.

3. GAS STAND TESTING

A 'gimballed' version of the ETA device bearing housing that allowed bearing housing reaction torque to be measured was manufactured to be used for gas stand testing at Imperial College London. This gas stand testing allowed the electrical machine to be characterised both electrically and in terms of machine efficiency. This electrical characterisation was an important part of developing the position sensor-less control method mentioned above. The electrical machine also allowed the turbine performance to be mapped at much lower flow rates than is possible on a conventional gas stand (points at which turbine power is insufficient to overcome bearing and other losses). This allowed extension of the turbine performance maps using measured data, rather than extrapolation, into areas of operation seen during engine transient load shedding. More information on the testing arrangement can be found in the literature (15).

4. ETA INSTALLATION

The engine testing of the device was carried out both at BorgWarner in the US (for standard turbocharger hardware sign off testing) and at Loughborough University in the UK (for controls development and performance evaluation). The testing at Loughborough was carried out by fitting the device to a 7 litre displacement tier 4 off-highway diesel engine in place of the production 2-stage turbochargers. This engine is equipped with high pressure loop cooled EGR and is used in such applications as excavators, wheel loaders and gen sets. The ETA was mounted remotely from the exhaust manifold for ease of installation of the prototype turbocharger, Figure 4. This mounting solution led to some heat loss between the exhaust manifold and the turbine which reduced the available energy in the exhaust but also reduced the thermal load into the ETA and avoided excessive mechanical load on the engine exhaust manifold. This installation solution also allowed initial testing with a conventional (non-electrified) BorgWarner BV63.

The ETA was equipped with several K-type thermocouples which measured the temperature at several locations inside the electrical machine stator. The oil supply (bearing lubrication and stator cooling) to the ETA was independent of the engine with a specially designed oil cart used, Figure 4. This allowed precise control of the oil temperature and pressure. The test cell was equipped with a motoring dyno that was capable of rapid load application to allow assessment of engine transient response. Engine fuel consumption, air mass flow and exhaust emissions were all measured.

Figure 4: ETA fitted to test engine in Loughborough University test cell

5. ENGINE STEADY STATE ETA RESULTS

5.1 ETA winding temperatures (non-electrified)

Initial testing of the ETA prototype device focussed on resolving some issues with winding wiring, winding insulation, and device sealing. Once these issues were overcome tests were run to understand the sensitivity of electrical machine winding temperatures to engine parameters such as load and exhaust temperatures without any electrical assistance. Figure 5 presents results of the winding temperature measurements compared to related parameters.

The winding temperature shown in Figure 5 is the measured temperature at one of the end winding positions. The end windings are furthest from the cooling oil and showed higher temperatures than other measurement points. Figure 5 shows a generally reasonably linear relationship between end winding temperature and engine load/exhaust temperature at the mid region but departs from this at the higher and lower ends. It is believed that the lower end departure from the linear trend is due to the windings being heated by the leak off from the bearing oil circuit.

The intermittent departures from the linear temperature at the top end were believed to be due to exhaust gas blow-by from the turbine leaking across either the horizontally split bearing housing or the turbine bearings. Higher levels of exhaust gas blow-by into the turbocharger oil return line were seen with the ETA device than with production turbochargers.

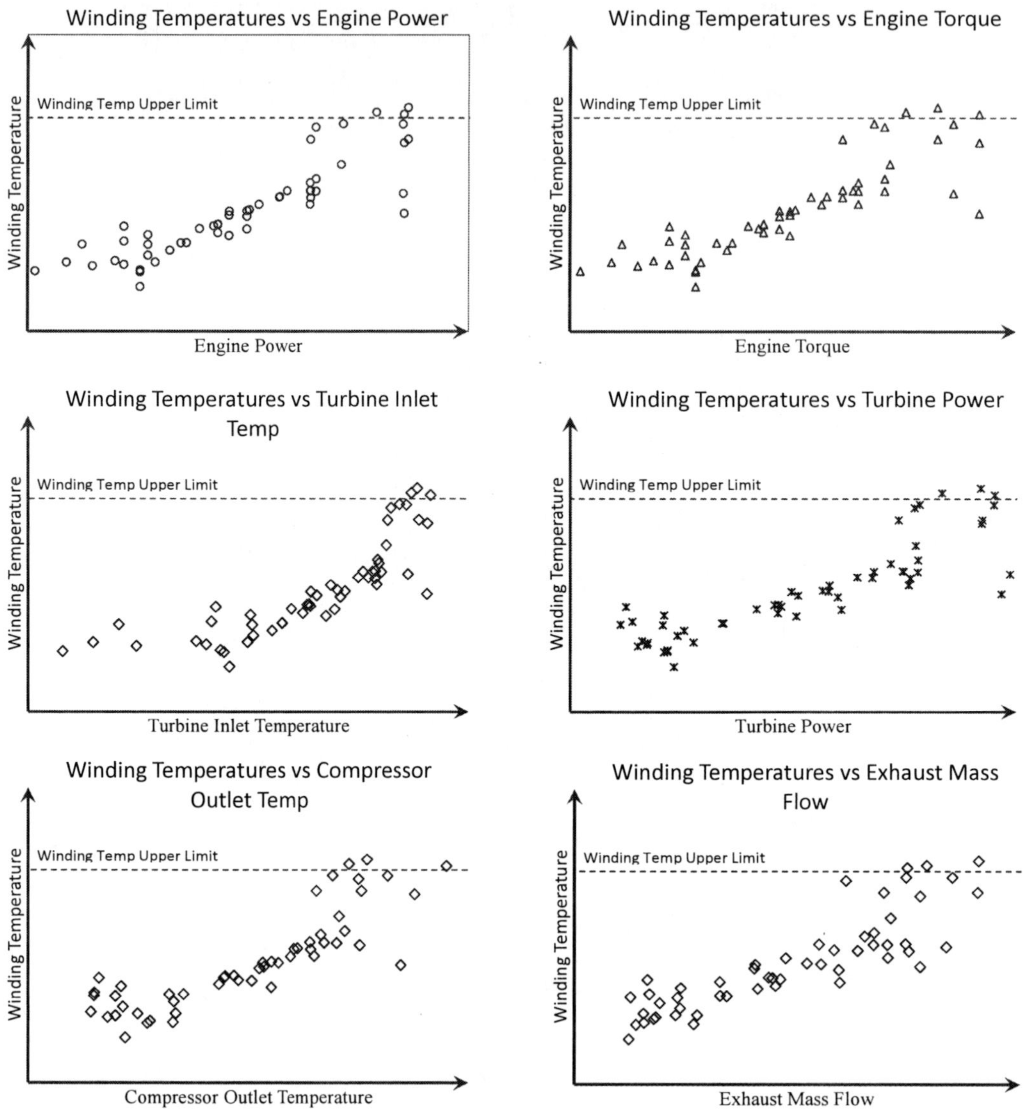

Figure 5: ETA winding temperature sensitivity (1800 RPM)

An example of a sudden increase in winding temperature at higher loads can be seen in Figure 6. In this case the end winding temperature starts to increase very quickly once high load has been reached. The sudden increase in the rate of end winding temperature does not coincide with the preceding increase in load, instead it is delayed by nearly one minute. This suggests that the higher temperatures seen at high loads is not solely due to the higher exhaust temps but some other effect, such as the increased blow-by rates detailed earlier.

A correlation analysis was carried out to understand which factors affect the winding temperature of the non-electrified ETA, Figure 7. Many of the parameters that correlated with end winding temperature were not unexpected (engine load, exhaust temps etc.) but it was also seen that increased bearing oil and cooling oil flow rates led to increased end winding temperatures while increased bearing and stator oil supply pressures led to reduced winding temperatures. The higher winding temps from higher cooling flow rates may be due to the effect of greater heating of the windings from increased splash from heated bearing oil (end windings are close to drainage side of bearings). The unexpected correlations is a clear indication of the complexity of the ETA as a system to cool.

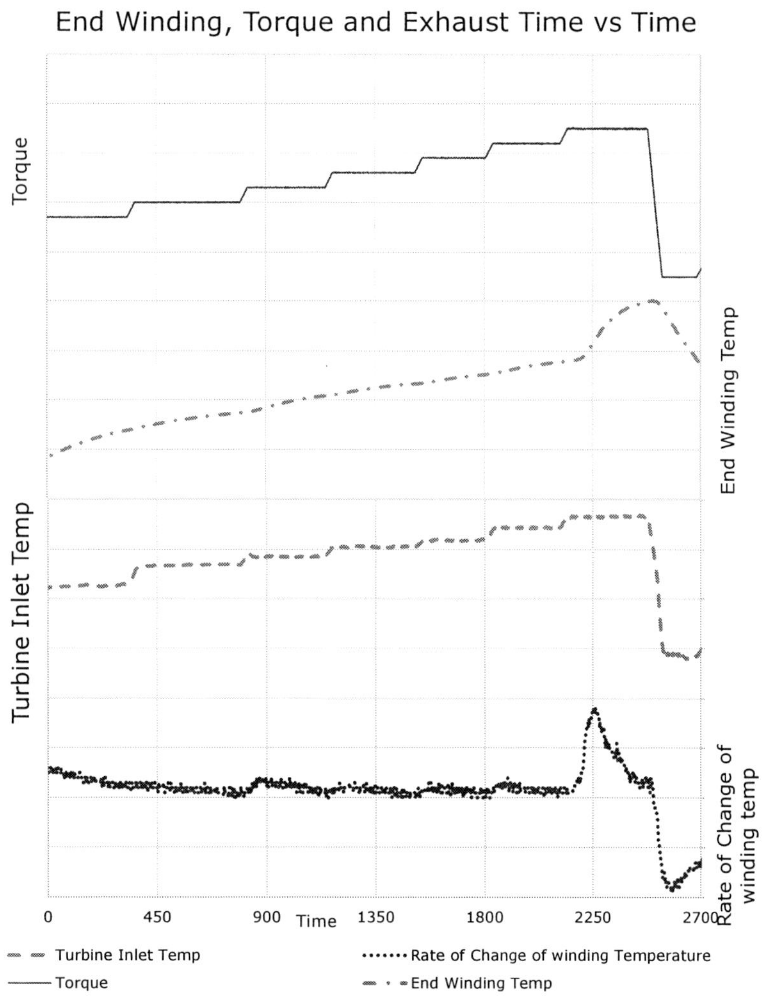

Figure 6: Load independent winding temperature increase (1800 RPM)

	No.	Channel
	1	Engine Speed
	2	Engine Torque
	3	ETA Speed
	4	Stator Temp 1
	5	Stator Temp 2
	6	Winding Temp 1
	7	Winding Temp 2
	8	Winding Temp 3
↓	***9***	***Oil Cart Bearing Pressure***
↑	10	Oil Cart bearing flow rate
↑	11	Oil Cart bearing oil supply temp
↓	***12***	***Oil Cart Stator Pressure***
↑	13	Oil Cart Stator Flow Rate
	14	Oil Cart Stator oil supply temp
	15	ETA Oil Return Temp
	16	Exh Manifold Temp
	17	Exh Temp Turbine Out
	18	Air Mass Flow
	19	Fuel mass Flow
	20	Ambient Air temp
	21	Compressor inlet temp
	22	Compressor outlet temp
	23	Exhaust Mass Flow

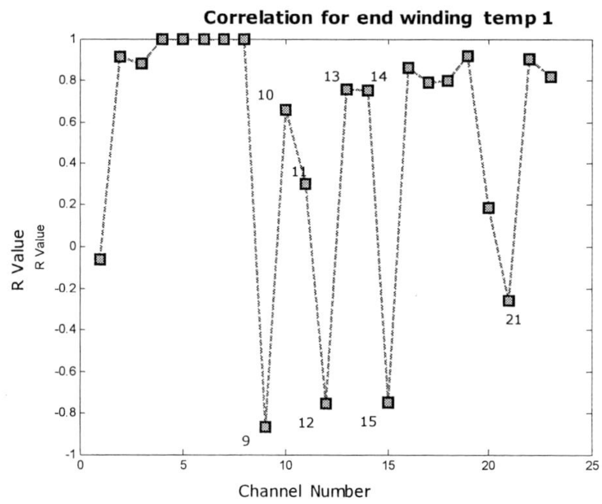

Figure 7: End winding 1 temperature correlation

Once the non-electrified temperature limits of the device had been established, work was done to implement the ETA sensor-less control method throughout the operating speed range of the machine. The position-sensor-less controls were demonstrated under steady state conditions over the area shown in Figure 3.

5.2 ETA winding temperatures (electrified)

The effects of ETA electrical assist on winding temperatures was also investigated. A series of sensitivity tests were carried out to assess rates of heating and cooling. An 1800 RPM example of this is shown in Figure 8. This shows a series of tests where the electrical machine was continuously motored at a constant low % torque demand at different engine torque levels. The motoring continued until a predetermined temperature was reached.

It was seen that at higher starting loads the winding temperature was already at a higher temperature (as per the trend in Figure 5) and that it increased at a higher rate from this point (i.e. the 300 Nm case took 40 seconds less time to reach the upper temperature limit than the 100 Nm case took for the same temperature rise). This faster rate of electrical heating is believed to be due to a combination of the higher torque points having higher electrical power for the same motor torque setting (due to the higher turbocharger speed) and increased winding resistance from higher starting temperatures. The rate of cooling was also seen to be slower at the higher exhaust temperatures. This clearly suggests that prolonged use of the ETA at higher loads may lead to temperature issues.

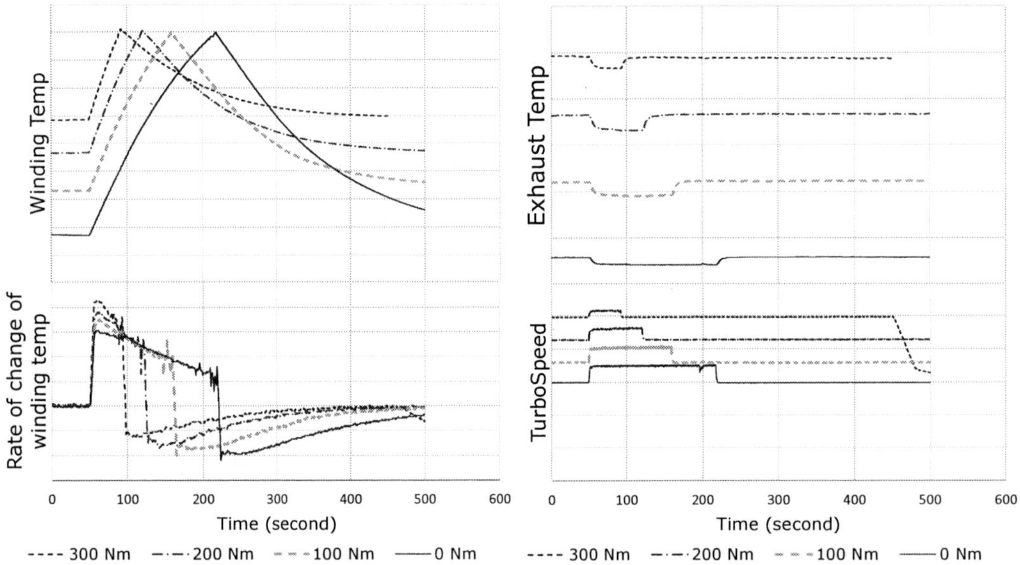

Figure 8: ETA electrical heating rate comparison at different loads

The winding temperatures were also monitored during engine performance testing. An example is shown in Figure 9 where it can be seen that over a series of 6 block load tests (load quickly increased at a constant desired engine speed) that the effect of a few seconds of motoring assistance accumulates over each assistance event leading to a noticeable divergence in winding temperature from an identical case ran without the electrical machine active. This suggests that more intermittent and shorter duration use of the device may lead to less temperature problems.

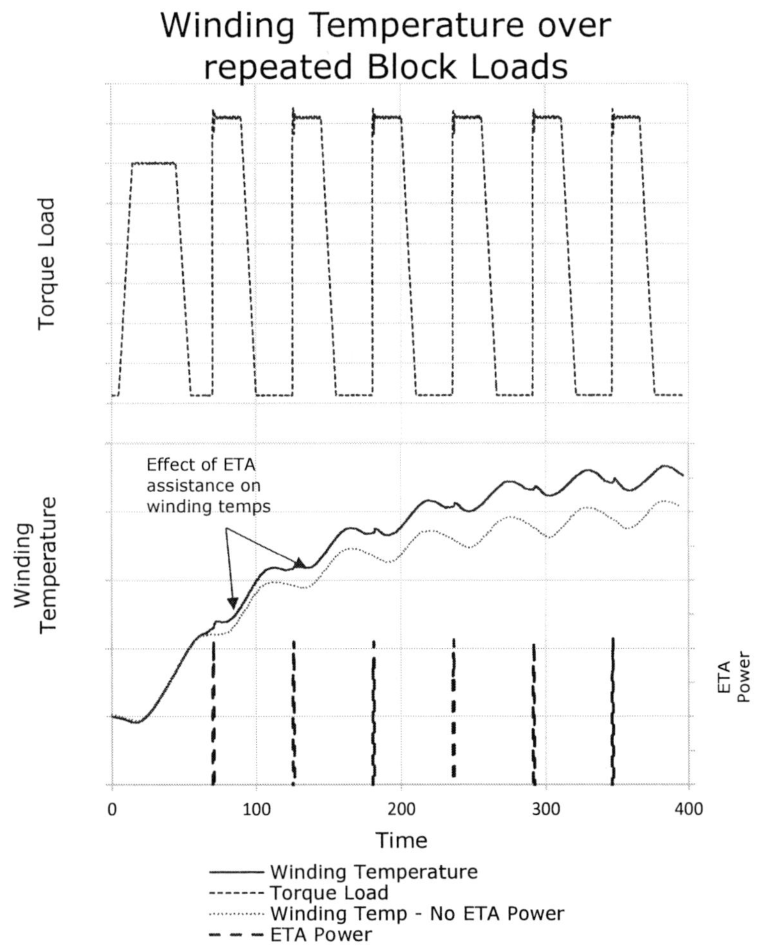

Figure 9: Effect of ETA block load assistance on winding temps

5.3 Effect of ETA motoring and generating on engine AFR and EGR

In order to assess the effect of the ETA on the steady state operation of the engine, a range of ETA power swings (running the device at a series of motoring and generating % torque demands) were carried out. These were run with the other air system actuators either locked at a fixed position or attempting to control to a set-point. The addition or subtraction of extra torque to the turbo-shaft had a significant effect on the engine air system. The effect of a limited amount of ETA motoring and generating power on the engine Air Fuel Ratio (AFR) and Exhaust Gas Recirculation (EGR) achievable 'space' (with EGR and VTG vane position actuators fixed) is illustrated in Figure 10. This shows that motoring the ETA will tend to increase engine AFR while generating will reduce both AFR and EGR. The engine particulate and NOx emissions will obviously also be affected by any changes in AFR and EGR rate that the ETA drives.

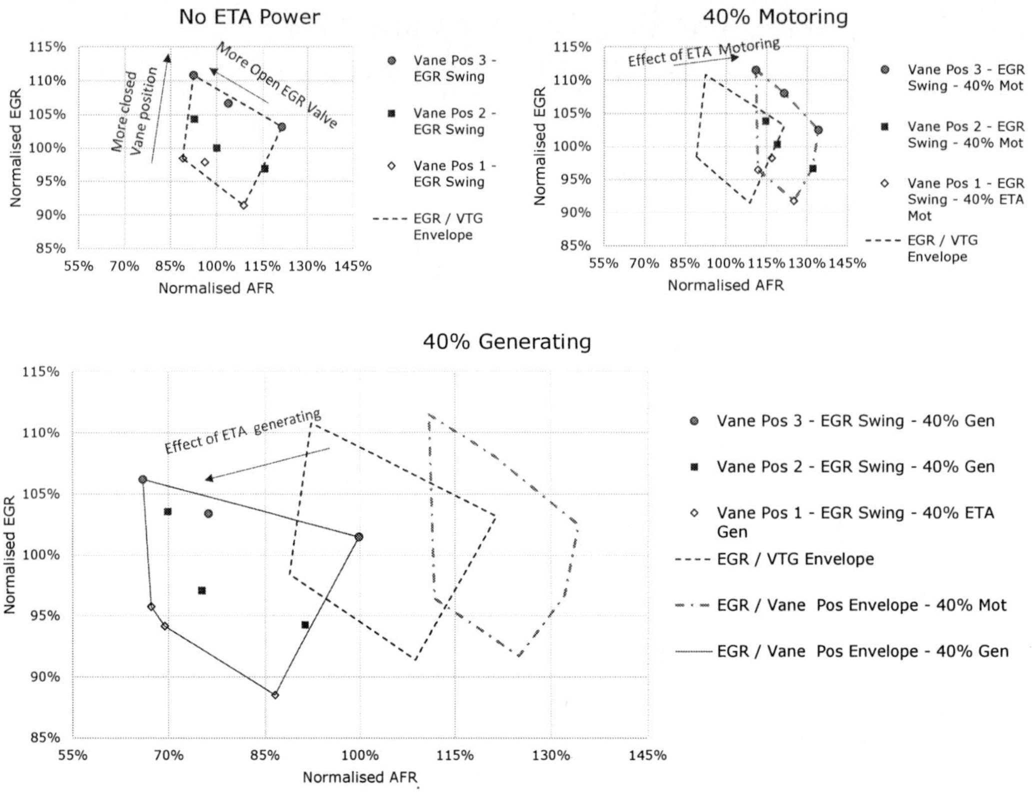

Figure 10: Effect of ETA motoring and generating on engine AFR and EGR (1800 RPM 40% load)

5.4 Effect of ETA electrical power on engine pumping efficiency

The other benefit of motoring is that it will increase the inlet manifold pressure with no requirement for more turbine power (i.e. waste gate closure) that could lead to an increase in exhaust pressure (although the increased air flow will increase turbine flow leading to a possible increase in turbine expansion ratio and exhaust manifold pressure).

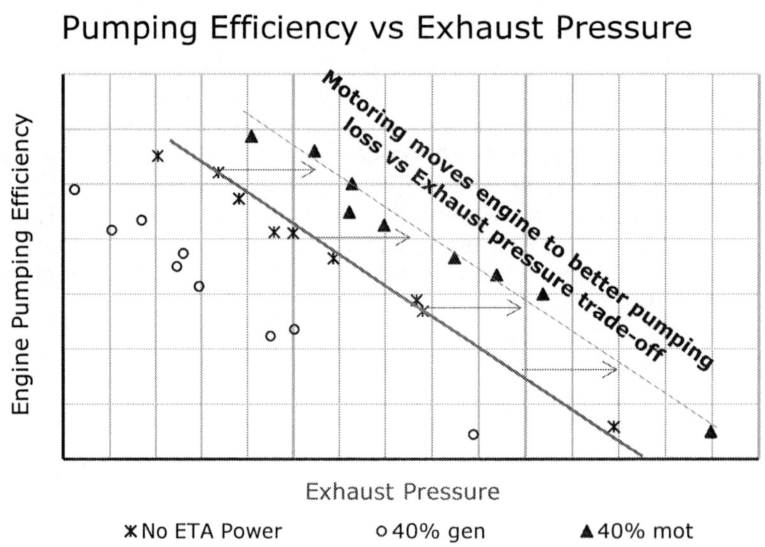

Figure 11: Effect of ETA power on engine pumping efficiency (1800 RPM 40% load)

An increase in inlet manifold pressure relative to exhaust manifold pressure will lead to a reduction in engine pumping losses. Pumping losses can be defined as the 'pumping efficiency' which is the ratio of the 720 deg. Indicated Mean Effective Pressure (IMEP) to the 360 deg IMEP for a 4 stroke engine. This is plotted in Figure 11 against exhaust pressure for the data from Figure 10. The improved pumping efficiency for a given exhaust pressure when the ETA is motoring can clearly be seen. However, there is a reduction in pumping efficiency associated with generating. This illustrates that the ETA is capable of decoupling exhaust pressure from inlet manifold pressure which can be used to reduce engine pumping losses.

5.5 Sweeps of ETA power at different VTG and EGR valve positions

The ETA has also been seen to affect the engine efficiency in others ways. Motoring has been seen to increase the engine friction due to the increased cylinder pressure from increased air mass in the cylinder. Generating has been seen to lower AFR which then leads to slower combustion, which reduces the engine closed cycle efficiency.

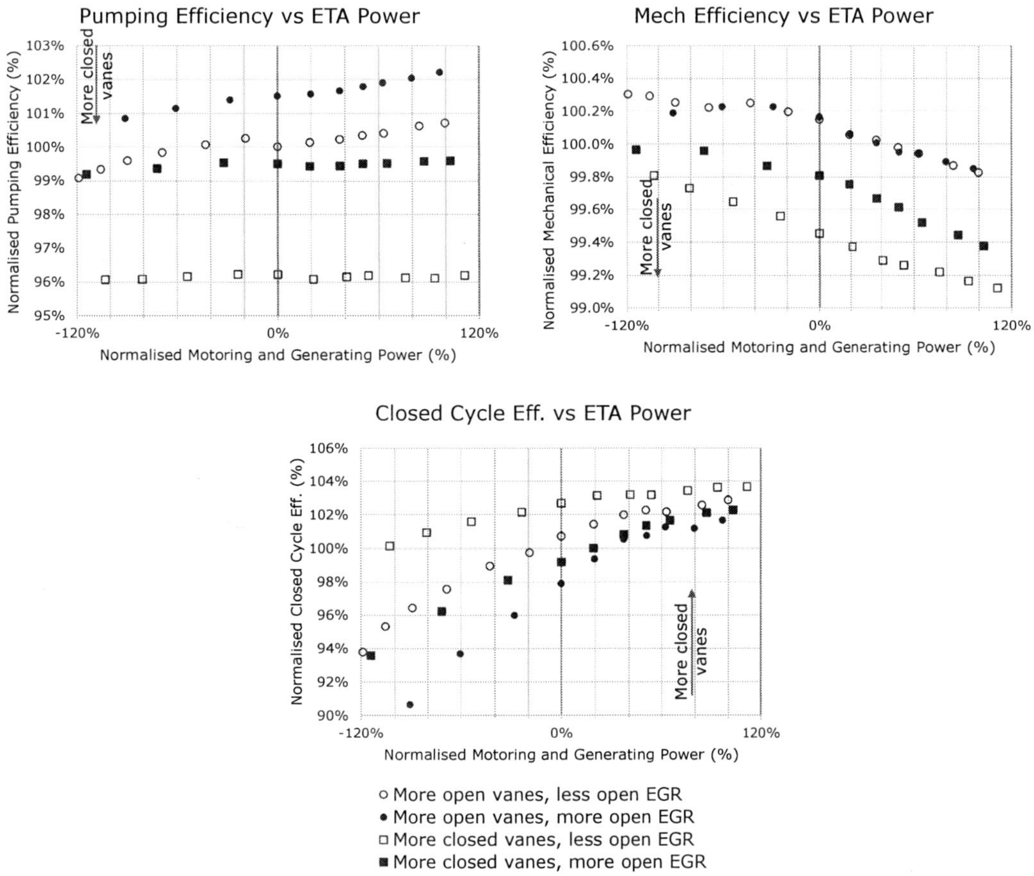

Figure 12: Normalised engine pumping eff., mechanical eff., closed cycle eff. vs ETA power (1800 RPM 40% load)

Some of the ETA effects can be counter-acted through adjustments to other air system actuators. Closing the VTG vanes will help maintain a higher AFR during generating, reducing the gross indicated efficiency penalty. This will however be at the cost of increased pumping losses. Figure 12 and Figure 13 show normalised plots of engine brake, mechanical, pumping, closed cycle efficiency as well as turbo speed and AFR against normalised ETA electrical power (-ve: generating, +ve: motoring, 100% being the maximum ETA power achieved at a mid VTG and EGR point). The data was taken at 1800 RPM 40% load. The data is plotted at two vane positions and two EGR valve positions. The effect of the different engine air system actuator settings on the engine's response to ETA motoring and generating can be seen.

From Figure 12 and Figure 13 it can be observed that there is a clear pumping and mechanical efficiency penalty from having a more closed VTG vane position but it can be seen that there is significant AFR and closed cycle efficiency benefit from the more closed vanes, especially under higher levels of generating.

The brake efficiency plots in Figure 13 show the combined effects of ETA power on engine mechanical, pumping and indicated efficiencies. It can be seen that a more closed vane position gives lower brake efficiencies under non-motoring or motoring cases (due to a combination of lower pumping efficiency from more restrictive turbine and slightly lower mechanical efficiency from higher cylinder pressure).

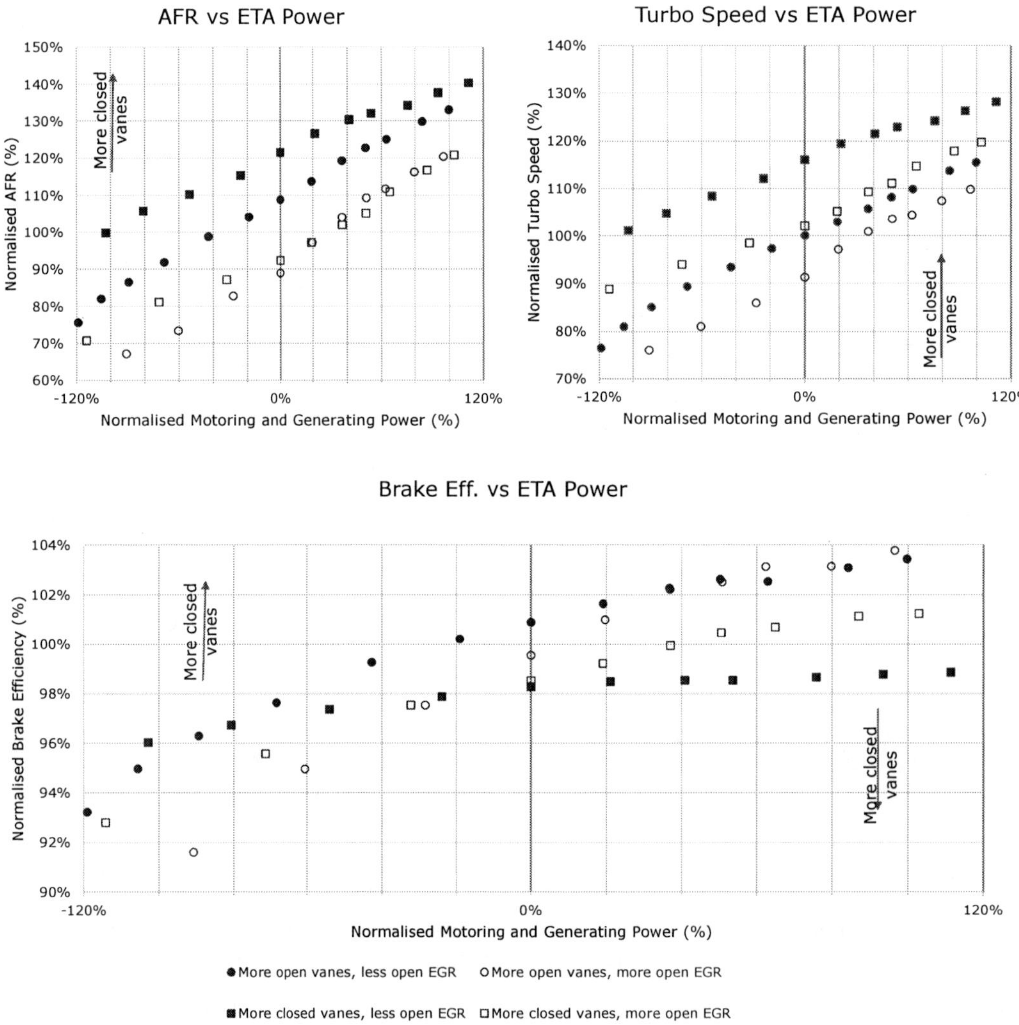

Figure 13: Normalised engine AFR, turbo speed and brake Eff. vs ETA power (1800 RPM 40% load)

As the ETA generates there is a drop in AFR. The more closed vane case sees a higher minimum in AFR due to:

(i) The higher initial AFR.
(ii) A smaller proportional reduction in compressor power due to the electrical power extracted from the turbine being a smaller portion of the overall higher turbine power (more closed vanes gives a greater expansion ratio).
(iii) The higher turbo speeds also appear to be of benefit as the turbocharger stays in a more efficient part of the compressor map.

The higher minimum AFR leads to higher closed cycle efficiency which outweighs the lower mechanical and pumping efficiencies of the more closed vane case. This leads, under high generating power levels, to equivalent or better brake efficiencies compared to the more open case. The particulate emissions penalty associated with running to such low AFRs with the more open vane cases is also significant.

The results presented in Figure 12 and Figure 13 illustrate that careful selection of the ETA's turbocharger is required to find one that can operate well both in motoring and generating mode.

6. ENGINE TRANSIENT ETA RESULTS

A key potential benefit of ETA technology is the potential to improve transient response. This can be used to either:

(i) Improve engine load acceptance to improve productivity.
(ii) Improve engine load acceptance to allow engine down-sizing or down-speeding for improved fuel consumption.
(iii) Allow an engine to use a turbocharger matched for Brake Specific Fuel Consumption (BSFC) rather than response and meet the transient requirements using the ETA.

A range of transient tests were carried out on the ETA to quantify the potential response benefits. These included:

- Actuator step response tests.
- Engine block load acceptance tests.
- Simulated application working cycles.

The first two tests were relatively simple discrete transient events and could be run without having the ETA and engine controls fully integrated. The final one required a fully developed control system for the ETA, VTG and EGR.

6.1 Actuator step response tests

The actuator response tests involved carrying out comparable turbo speed step changes using the ETA motor and the VTG vanes. The response of turbo speed and other parameters was compared. The turbo speed results are shown in Figure 14.

From Figure 14 it can be seen that the 5-95% turbo speed response time achieved by ETA motoring assist was on average 1.2 seconds faster than that achieved by an equivalent VTG vane step change. The 5-50% response time was faster by an average of 0.2 sec and therefore the ETA advantage was mostly over the second half of the step change. It can also be seen that the benefit was lower at lower turbo speeds. This was thought to be due to the lower available motor electrical power at those speeds.

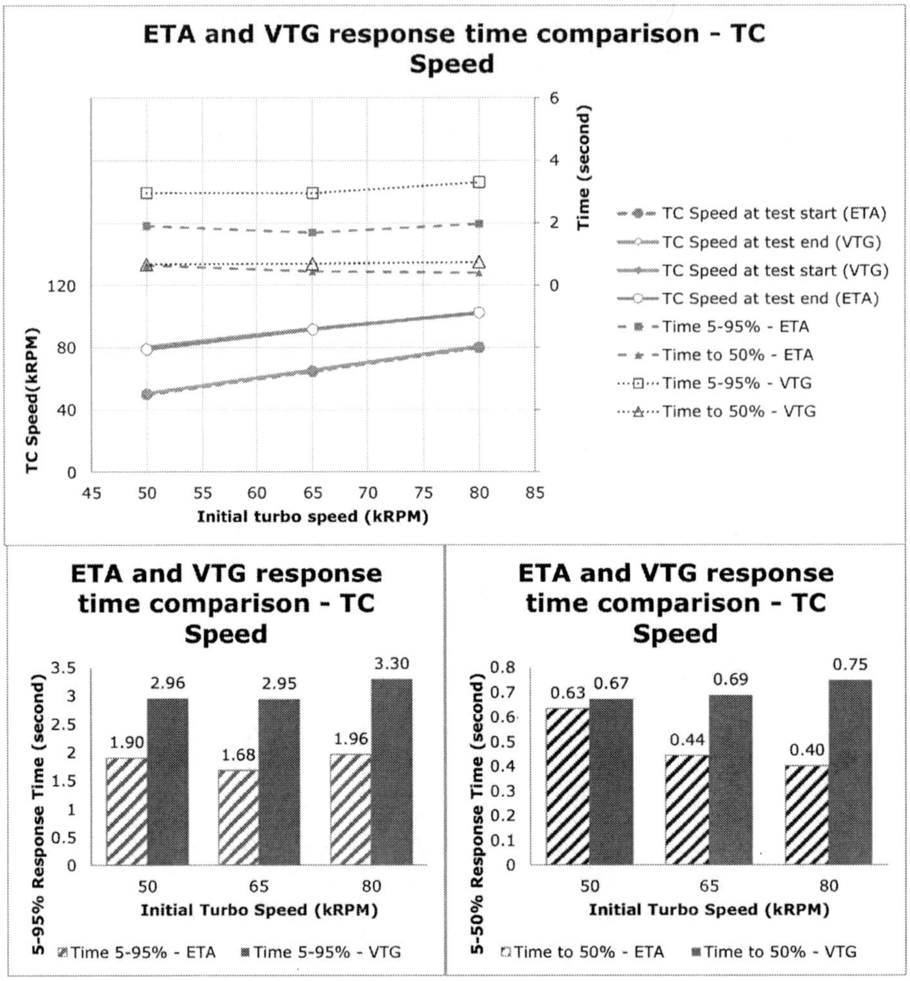

Figure 14: ETA vs VTG vane position turbo speed response test

An example of one of the tests (65 kRPM starting speed) is presented in Figure 15 and shows the ETA power, vane position and exhaust manifold pressure. Exhaust manifold pressure increased significantly with closing the vanes while less so with the ETA motoring assist. Thus, for the same step change in turbo speed, VTG will result in increasing pumping losses (Figure 12). The reduced pumping loss penalty on acceleration is therefore an additional benefit of the ETA.

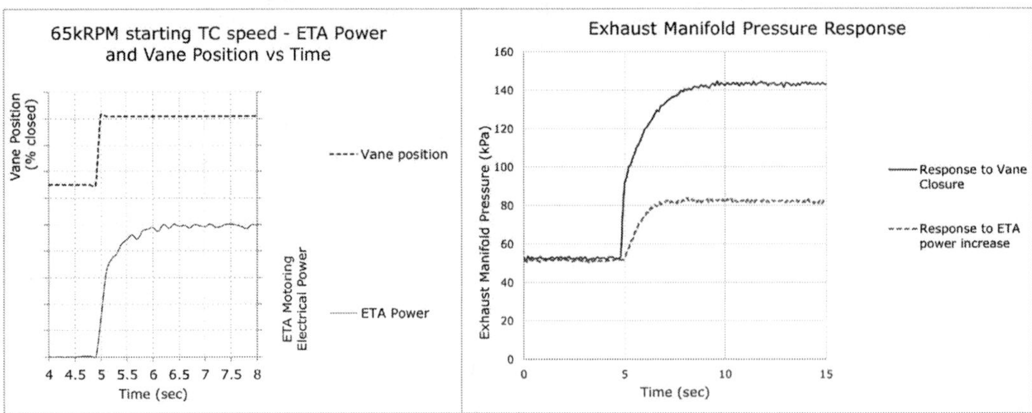

Figure 15: 65 kRPM actuator response test

6.2 Engine block load test results

Engine block load tests were also carried out involving running the engine at low load at a fixed speed and then increasing the load on the engine to a pre-determined level as quickly as the dyno will apply it. The amount by which the engine speed drops upon load acceptance (speed drop) and the time taken for it to recover to its prior speed (recovery time) are measures of an engine's transient capability.

The performance of the baseline engine with its production 2-stage turbocharger was set as an initial target for the ETA. It was found that the BorgWarner BV63 single stage turbocharger with VTG of the same aero specification as the ETA was not capable of matching the 2-stage response. The non-electrified single stage had worse speed drop and recovery time. This can be seen in Figure 16 which also shows that the 2-stage case had slightly greater initial boost, which would increase engine response. Despite lower initial boost the ETA was capable of matching the 2-stage recovery time but did have slightly worse engine speed drop (which is more affected by initial boost than rate of boost rise).

Block load tests were run over a range of other speeds and loads looking at the effects of different ETA power levels, different assist durations, different initial conditions and other factors. An example of some of the results is shown in Figure 17 where the effects of ETA electrical power on block load performance at 1400 RPM are shown.

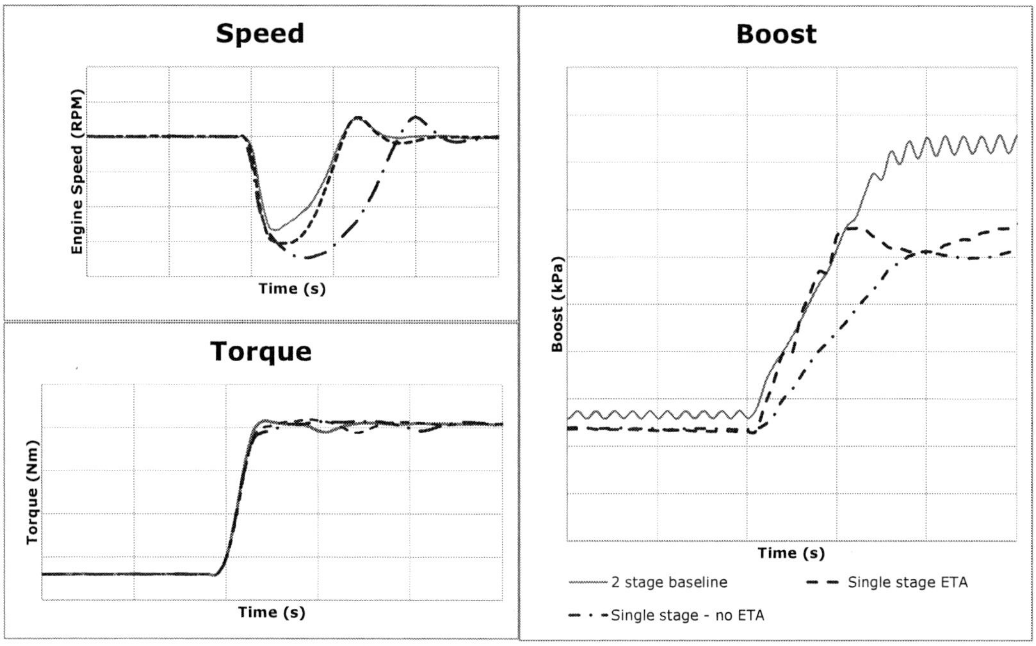

Figure 16: ETA block load response vs 2-stage engine response

Figure 17 shows that the ETA is more effective at reducing recovery time than speed drop and that the incremental response benefit from increased ETA power levels starts to decrease (35% ETA power gives large benefit compared to no assist case but 70% power does not double this). This is due to the engine speed recovery becoming less air system constrained with increased levels of ETA power, making any further boost increase with higher ETA power less effective.

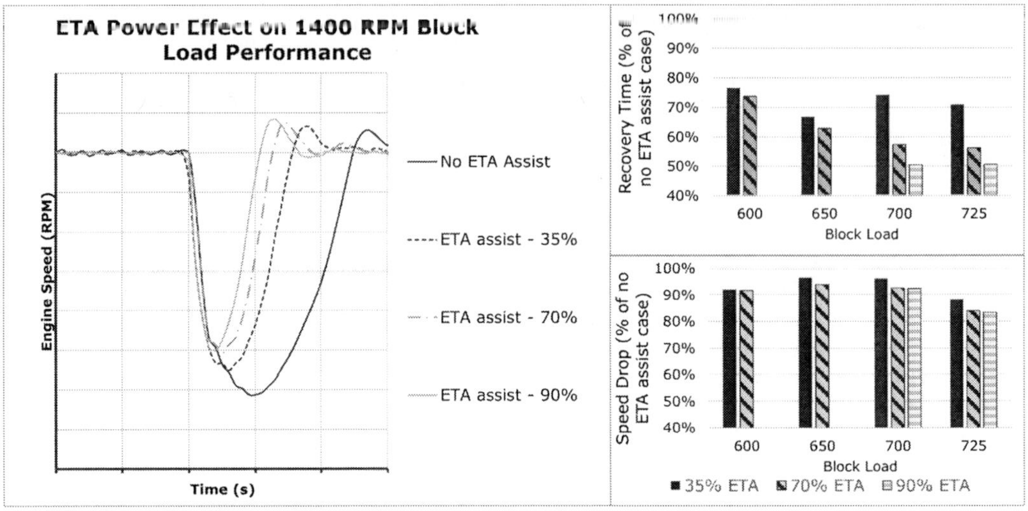

Figure 17: 1400 RPM block load sensitivity to ETA power

6.3 Engine working cycle test results

Working cycles were run once a control system for the ETA had been integrated with the engine controller. In the simplest control system implementation, ETA transient motoring assist was triggered based on inlet manifold conditions and other engine related parameters. The control system also had a battery state of charge (SoC) maintenance algorithm which regulated ETA motoring and generating in steady state conditions. This was required to assess effects on cycle efficiency of maintaining sufficient battery state of charge to allow transient assistance.

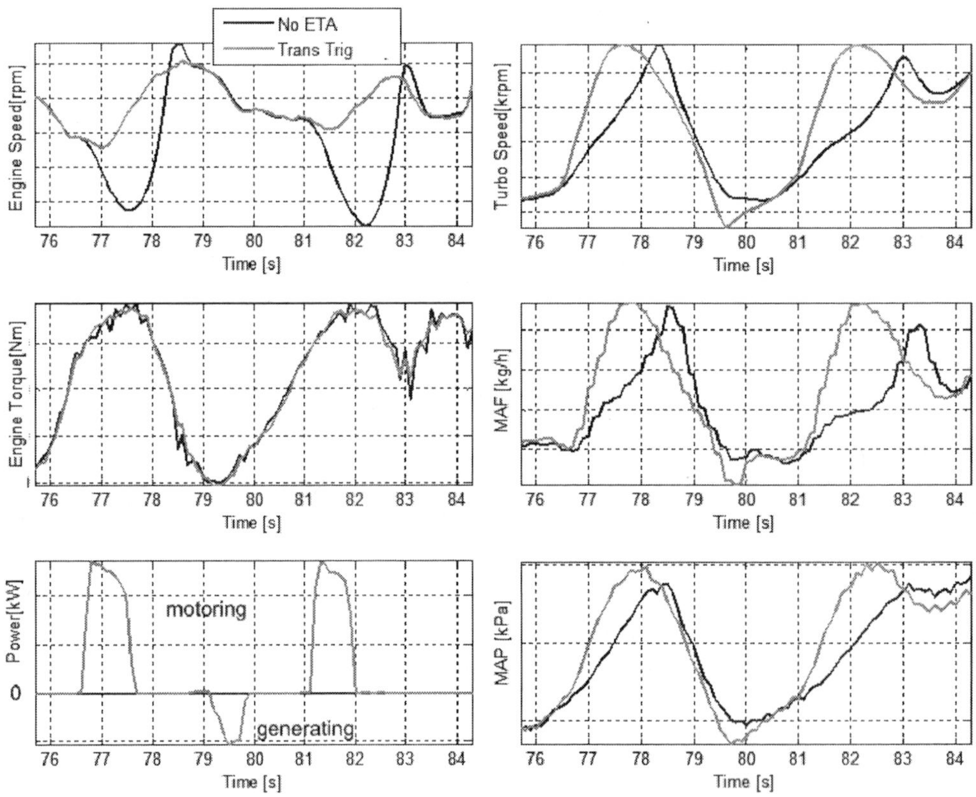

Figure 18: Trenching test working cycle (with and without ETA assist at an engine speed demand =1800 RPM)

One of the working cycles used to evaluate the ETA was a so called trenching test in which the engine speed demand is fixed whilst the engine torque is constantly changing rapidly over the course of several minutes. Figure 18 shows a range of engine responses (engine speed, ETA electrical power, turbo speed, intake Mass Air Flow (MAF) and the intake Manifold Air Pressure (MAP)) for two trenching tests, one run with ETA assistance the other without.

The results presented in Figure 18 show very clearly how ETA motoring assist improved the engine transient response relative to the no ETA assist case. In this example the ETA is providing motoring assist during rapid load acceptance and then generating briefly between. The increased acceleration of the turbocharger with motoring assist and the increased MAP rise rate are clear and these contribute to the improved engine speed response.

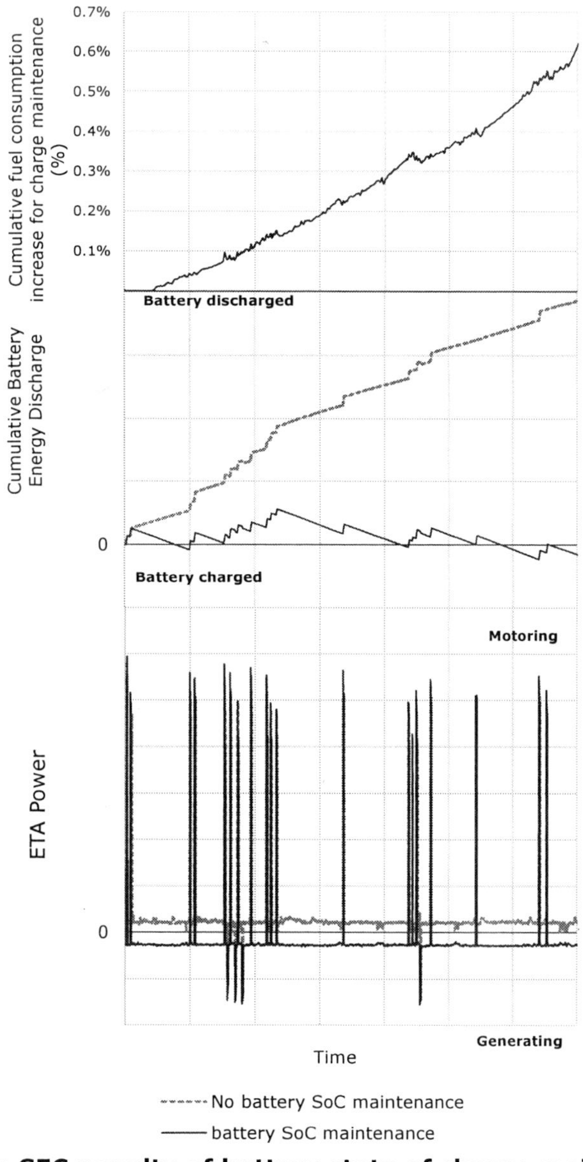

Figure 19: SFC penalty of battery state of charge maintenance

Utilisation of the ETA over a working cycle requires the battery SoC be maintained. Figure 19 compares two tests that employ the same transient ETA control method as illustrated in Figure 18; one cycle with battery SoC maintained (using a controls strategy) and one where the battery is allowed to discharge. This shows that battery SoC can be maintained but that it is at the cost of a 0.6% fuel penalty over the working cycle.

Several different control systems were developed in simulation and tested on the ETA equipped engine at Loughborough University (16-21). This work highlighted that there exists very strong interaction between the ETA electrical power, VTG and EGR that requires careful control system design and system energy management.

7. CONCLUSIONS

The presented work has demonstrated that operation of the ETA offers an additional degree of freedom for air system control and can be used to influence the engine pumping losses. Steady state motoring operation of the ETA will lead to improved SFC through pumping benefits while steady state generating operation of the ETA can lead to an SFC penalty through a combination of increased pumping losses and the effects of reduced air fuel ratio on combustion. Prolonged or high engine load operation of the ETA device is currently limited by stator winding insulation temperature limitations. These cooling issues are attributed to the use of oil for cooling, the design of the ETA cooling system and excessive blow-by at high loads.

It has also been shown that the ETA device is capable of enabling improved engine transient response (>60% reduction in block load recovery time) compared to a variable geometry turbocharger through the ability to more quickly increase inlet manifold pressure (allowing a faster increase in engine fuelling). Diminishing benefits from increased ETA power were observed and illustrate that the there are other controlling factors on engine transient response beyond the rate of boost increase (i.e. initial boost level at start of transient, system inertia etc.) that start to become the limiting factors. The ETA device allowed the block load recovery time performance of a 2-stage turbo to be matched with a single stage turbocharger.

The ETA device has shown response benefits during on engine testing of a simulated working cycle (reduced speed drop and faster recovery time). A fuel penalty of 0.6% is seen when the battery state of charge is maintained over the working cycle compared to allowing the battery to discharge.

This work illustrates the potential of ETA technology but highlights both the challenges of cooling the electrical machine and the fuel penalty from maintaining battery state of charge over a working cycle.

8. FUTURE WORK

BorgWarner has used the experience gained with the ETA prototype described in Section 2 to develop an improved electric turbocharger concept, the BorgWarner eTurbo™, Figure 20. This is a permanent-magnet machine with a higher power density. This design incorporates a new rotor structure which allows improved tolerances and optimal stiffness for best rotordynamic behaviour. The stator is cooled by engine coolant (rather than oil) which allows greater power output whilst maintaining temperatures within the limit of the winding material.

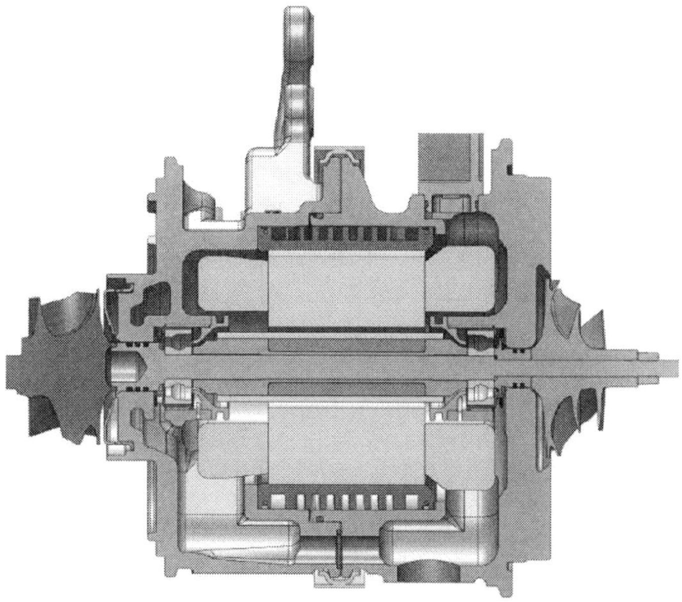

Figure 20: Core assembly of eTurbo™

ACKNOWLEDGEMENTS

This work was co-funded by Innovate UK (formerly the Technology Strategy Board UK), under a grant for the Low Carbon Vehicle IDP4 Programme (TP14/LCV/6/I/BG011L). Innovate UK is an executive body established by the United Kingdom Government to drive innovation. It promotes and invests in research, development and the exploitation of science, technology and new ideas for the benefit of business - increasing sustainable economic growth in the UK and improving quality of life.

REFERENCE LIST

[1] Balis, C., Middlemass, C., and Shahed, S. M., "Design and Development of e-Turbo for SUV and Light Truck Applications". In 9th Diesel Engine Emissions Reduction (DEER) Conference, Technical Session 4: Waste Heat Utilization, 2003.

[2] Hopmann, U., and Algrain, M. C., "Diesel Engine Electric Turbo Compound Technology". In Proceedings of the SAE Future Transportation Technology Conference, paper 2003-01-2294, 2003.

[3] Arnold, S., Balis, C., Barthelet, P., Poix, E., Samad, T., Hampson, G., and Shahed, S. M., Garrett Electric Boosting Systems (EBS) Program. Final report, grant DEFC05-00OR22809, US Department of Energy, 2005.

[4] Ibaraki, S., Yamashita, Y., Sumida, K., Ogita, H., and Jinnai, Y., "Development of the hybrid turbo, an electrically assisted turbocharger". Mitsubishi Heavy Industries, Ltd. Technical Review, 43(3), pp. 1–5, 2006.

[5] Ryder, O., Sutter, H., and Jaeger, L., "The Design And Testing Of An Electrically Assisted Turbocharger For Heavy Duty Diesel Engines". In Proceedings of the IMechE 8th International Conference on Turbochargers and Turbocharging, pp. 157–166, 2006.

[6] Panting, J., Pullen, K. R., and Martinez-Botas, R. F., "Turbocharger motor-generator for improvement of transient performance in an internal combustion engine". Proceedings of the IMechE, Part D: Journal of Automotive Engineering, 215(3), pp. 369–383, 2001.

[7] Kolmanovsky, I., Stefanopoulous, A. G., Powell, B. K., "Improving Turbocharged Diesel Engine Operation With Turbo Power Assist System," in IEEE International Conference on Control Applications, vol. 1, pp. 454–459, 1999.
[8] Katrašnik, T., Rodman, S., Trenc, F., Hribernik, A., Medica, V., "Improvement of the Dynamic Characteristic of an Automotive Engine by a Turbocharger Assisted by an Electric Motor," Journal of Engineering for Gas Turbines and Power, vol. 125, no. 2, p. 590, 2003.
[9] Terdich, N., Martinez-Botas, R. F., Howey, D. A., Copeland, C. D., "Off-Road Diesel Engine Transient Response Improvement by Electrically Assisted Turbocharging", 10th International Conference on Engines & Vehicles, 2011.
[10] Terdich, N., Martinez-Botas, R. F., Romagnoli, A., Pesiridis, A., "Mild Hybridization via Electrification of the Air System: Electrically Assisted and Variable Geometry Turbocharging Impact on an Off-Road Diesel Engine," Proceedings of ASME Turbo Expo 2013, 2013.
[11] Millo, F., Mallamo, F., Pautasso, E., Mego, G. G., "The Potential of Electric Exhaust Gas Turbocharging for HD Diesel Engines," SAE Technical Paper 2006-01-0437, 2006.
[12] Costall, A. W., Radoslav, I., Langley, T. P. F., "Electric Turbo Assist: Efficient Rapid Boost for Heavy-Duty Diesel Engines". THIESEL Conference on Thermo and Fluid Dynamic Processes in Direct Injection Engines, 2012.
[13] Costall, A. W., Radoslav, I., Langley, T. P. F., "Electric Turbo Assist as An Enabler for Engine Downspeeding" Proceedings of ASME Turbo Expo, June 11-15, Copenhagen, Denmark, 2012.
[14] Watkins, S., Winward, E., "Modelling and Control of an Ultra-high-speed Switched Reluctance Machine", 8th IET International Conference on Power Electronics Machines and Drives, 19-21 April, Glasgow, Scotland, 2016.
[15] Terdich, N. and Martinez-Botas, R. F, "Experimental Efficiency Characterization of an Electrically Assisted Turbocharger", SAE Technical Paper 2013-24-0122, 2013.
[16] Zhao, D., Stobart, R., Dong, G., Winward, E., "Real-Time Energy Management for Diesel Heavy Duty Hybrid Electric Vehicles", IEEE Transactions on Control Systems Technology, Vol. 23, No. 3, May 2015.
[17] Zhao, D., Liu, C., Stobart, R., Deng, J., Winward, E., Dong, G., "An Explicit Model Predictive Control Framework for Turbocharged Diesel Engines", IEEE Transactions on Industrial Electronics, Vol. 61, No. 7, July 2014.
[18] Zhao, D., Liu, C., Stobart, R., Deng, J., Winward, E., "Explicit Model Predictive Control on the Air Path of Turbocharged Diesel Engines', American Control Conference, 2013.
[19] Zhao, D., Stobart, R., "Real-Time Optimal Energy Management of Heavy Duty Hybrid Electric Vehicles", SAE International Journal of Alternative Power. 2(2):369-378, 2013.
[20] Zhao, D, Winward, E., Yang, Z., Rutledge, J., Stobart, R., "Control-Oriented Dynamics Analysis for Electrified Turbocharged Diesel Engines", SAE Technical Paper 2016-01-0617, 2016.
[21] Zhao, D., Winward, E., Yang, Z., Stobart, R., "Decoupling Control of Electrified Turbocharged Diesel Engines", American Control Conference, 2016.

Design of a turbine-generator-unit for commercial vehicle ORC applications

H S Kunte, J R Seume
Institute of Turbomachinery and Fluid Dynamics,
Leibniz University Hannover, Germany

ABSTRACT

This paper introduces a supersonic impulse turbine which was designed for an ORC for a commercial vehicle running with ethanol. The application is characterized by challenging conditions for a turbine, such as a very high pressure ratio of up to 49 and a low mass flow. This results in a very high rotational speed of up to 110,000 min^{-1}. The developed expansion machine consists of the turbine and a directly coupled high-speed generator suitable for these speeds. The paper presents the mechanical design of the machine and shows the rotordynamics of the system. Finally, experimental results are presented for selected operating points.

NOMENCLATURE

D	m	diameter
h	m	blade height
$\dot{m}$	$kg\ s^{-1}$	mass flow
N	-	number of blades/passages
n	min^{-1}	rotational speed
P	kW	power
p	Pa	absolute pressure
s	$kJ\ kg^{-1}\ K^{-1}$	entropy
T	K	temperature
δ	m	tip clearance
ε	-	degree of partial admission

Subscripts

Design	design condition
in	inlet
max	maximum
out	outlet
q	heat flow
R	turbine rotor
S	turbine stator
Shroud	shroud
t	total

Abbreviations

bw	backward
fw	forward
OP	operating point
ORC	organic Rankine cycle

1 INTRODUCTION

Vehicle manufacturers are continuously looking for new systems to reduce the fuel consumption and decrease the emissions of their combustion engines. A thermodynamic analysis of the losses of these combustion engines showed that a significant part of the chemical energy is lost as thermal energy in the exhaust gas [1]. One approach for recovering some of that thermal energy is an organic Rankine cycle (ORC).

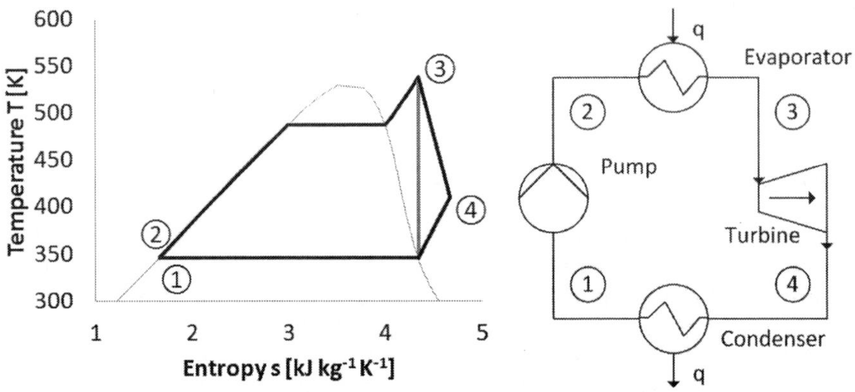

Figure 1: Basic organic Rankine Cycle with ethanol

The basic ORC is shown in Figure 1. It consists of 4 main components. A pump raises the pressure of the liquid working fluid from ①→②. Subsequently, the working fluid is heated, evaporated and superheated in an evaporator (②→③). The gaseous working fluid enters an expansion machine and is expanded with a simultaneous delivery of power (③→④). Finally, the fluid is condensed in a condenser and returns to the pump. The power output of this cycle is significantly dependent on the used working fluid and the chosen boundary conditions.

Ethanol works well as working fluid for the basic ORC without recuperation and a moderate condensation temperature, which is provided by the existing motor cooling [2] [3] [4]. Contrary, refrigerants, like R1233zd(e), are dry fluids and have a much higher expander outlet temperature. This requires the application of a recuperator for an efficient operation. Additionally, many refrigerants have a lower saturation temperature at ambient pressure. This requires a lower condensation temperature and thereby the application of an additional air cooled condensator [3] [5].

Additionally, the overall efficiency of the ORC is significantly dependent on the efficiency of the expansion machine. Possible expansion machines are scroll expanders [7] [8] [9], screw expanders [10], piston expanders [11] and turbines [11] [12]. Until now, no expansion machine is established for this application and further investigations are necessary.

The authors compared the Pelton turbine, the Heron turbine, the radial inflow turbine, the axial reaction turbine and the impulse turbine regarding their performance characteristics [13] [14]. They identified the supersonic impulse turbine as most convenient for this application. For this reason, the authors developed a supersonic impulse turbine for this application. This turbine was developed for the operation with ethanol. Special blade shapes are used to enable an efficient operation at very high pressure ratios. However, the mass flow in supersonic turbines is only dependent on the inlet conditions. For this reason a variable partial admission was applied to gain an additional setting parameter. This variable partial admission enables to drive the turbine at low mass flow with high inlet pressures, which is beneficial for the efficiency of the thermodynamic cycle. The turbine is directly coupled with a high-speed generator, which is designed for the high rotational speeds, and both form the expansion machine [15]. Within the paper, the mechanical design of this expansion machine is presented. This includes the bearing, the rotordynamics and the sealing. Finally, the operation of the expansion machine is demonstrated by means of two operating points.

2 OPERATING CONDITIONS

The boundary conditions were chosen on the basis of a truck with a 12.8 l diesel engine and a power of 375 kW. The investigation was focused on the high temperature heat contained in the exhaust gas, which is easier to recover than low temperature heat sources like engine cooling.

The selection of the proper working fluid depends on the boundary conditions and cycle limitations. Kunte and Seume [14] and Span et al. [2] performed a thermodynamic analysis to determine an adequate working fluid. Within these studies, the inlet pressure was limited to a maximum of 4 MPa and the outlet pressure to a minimum of 0.05 MPa at a condensation temperature of 343 K. They determined, that ethanol promises the highest power output for the given boundary conditions. However, the absolute outlet pressure was raised to 0.081 MPa, because of the condensation temperature and an assumed pressure loss between turbine and condensator. The boundary conditions are listed in Table 1.

Table 1: Operating conditions of the ORC

Parameter	Units	Min	Max
Fluid	Ethanol (95%mass), water (5%mass)		
$\dot{m}_{S,in}$	kg s^{-1}	0.006	0.091
$T_{S,in}$	K	458	539
$p_{t,S,in}$	MPa	1	4
$p_{R,out}$	MPa	0.081	0.081

3 DESIGN OF THE SUPERSONIC IMPULSE TURBINE

The development of a turbine for these boundary conditions is quite ambitious. Firstly, the high pressure ratio results in high supersonic flow velocities if expanded in a single stage turbine. Secondly, the volume flow, provided by the ORC, is very low and leads to small turbine diameters or low blade heights. The small dimensions of the turbine promise higher secondary losses, such as tip gap losses which decrease the efficiency. Additionally, a small turbine diameter combined with high flow velocities result in a high rotational speed, which makes the bearing and coupling of the turbine more challenging.

An axial impulse turbine was developed with consideration to these boundary conditions. The impulse turbine has two major benefits for this application. Firstly, impulse turbines have a lower rotational speed compared with reaction type turbines. Secondly, it is possible to operate impulse turbines with a partial admission. That means that just a part of the circumference is provided with flow and the working fluids only flows through this part of the turbine regularly. The rest of the circumference merely ventilates. On the one hand, the blade heights can be increased, therefore decreasing tip gap losses. On the other hand, the ventilation lowers the efficiency of the turbine significantly. However, the developed turbine features a variable partial admission, which can be changed between 20%, 40%,

60% and 80% of the entire circumference. Through this, the performance map of the Turbine is significantly extended to cover all operating points of the truck.

Verneau [16] showed that impulse turbines, operated at very high pressure ratios, can have good efficiencies, although these turbines must have special blade shapes. The passages of the turbine stator are formed like Laval-Nozzles to enable an efficient acceleration of the fluid to supersonic velocities. The stator of the developed turbine is shown in Figure 2 (l). It possesses 8 Laval-nozzles with a quadrangular cross section. However, the nozzle throats have a small gap of 0.46 mm making manufacturing difficult. For this reason, the stator is assembled with multiple elements. The main component of the stator is a stator blisk, where every second blade is missing. Instead, a notch is manufactured. The segments consist of a stator blade and a blade root, which fits into the notch. The segments are form closure connected to the blisk with bolts. Additionally, the gaps between the segments and the blisk are filled with glue to avoid leakage. Finally, the shroud is heat shrunk. The final stator is shown in Figure 2 (r).

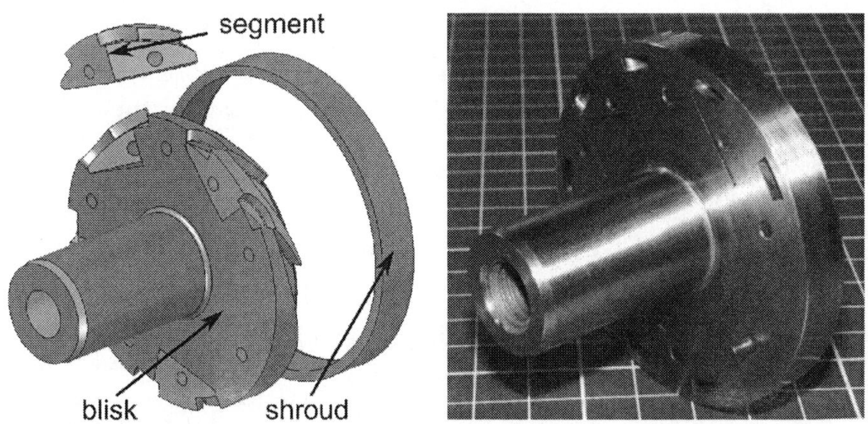

Figure 2: Turbine stator, disassembled (l), assembled (r)

The turbine wheel is shown in Figure 3. The blades are formed like normal impulse blades, which just redirect the flow without altering the flow velocity. The blisk was made from titanium via milling. Titanium was chosen as it is lighter compared to a steel blisk, which is beneficial for the rotordynamics. The radii at the leading and trailing edges are 0.2 mm. Smaller radii would be better for the efficiency of the turbine, but mechanical integrity would not be guaranteed.

Figure 3: Turbine rotor

Table 2: Parameters of the investigated Turbine

Parameter	Unit	Value
D_{Shroud}	m	0.0631
h	m	0.00343
δ	m	0.00013
$n_{Operation}$	min^{-1}	60,000...110,000
n_{Design}	min^{-1}	100,000
N_s		8
ε_{Design}		0.4
ε		0.2, 0.4, 0.6, 0.8
N_R		33

The design parameters of the turbine are listed in Table 2. The design speed of the turbine is 100,000 min^{-1} at a partial admission of 0.4. The predicted power is 8.0 kW for this operating point. However, it is possible to increase the partial admission on 0.8 to cover the whole performance map of the combustion engine. In this case, the power increases on 18.3 kW at a maximum rotational speed of 110,000 min^{-1}.

4 DESIGN OF THE PROTOTYPE

Figure 4 shows the design of the turbine-generator-unit. It consists of two main components; the single stage turbine (①, ②) and the generator (③-⑨). The turbine rotor ② is directly mounted on one shaft with the rotor of the generator. A special high speed generator is used ⑦, which is designed to withstand the centrifugal forces up to a rotational speed of 110,000 min^{-1} [17]. It is a synchronous generator with permanent magnets on the shaft with a maximum efficiency of up to 97.5 %. The shaft is mounted with four ball bearings arranged in pairs⑥. The pair arrangement offers a much higher stiffness of the mounting than single bearings and has a favorable effect on the rotordynamics and on the position of harmful natural frequencies in the Campbell diagram. The bearings are lubricated for life with grease. However, this grease is not resistant to ethanol and could be washed out, if contact with ethanol is not prevented. Additionally, the ethanol vapor has a low pressure, which could cause a contamination with ambient air and makes the sealing challenging. For this reason, an aerodynamically lubricated seal is implemented, which was originally developed for turbocharger applications [18]. It is able to seal the vacuum of 0.081 MPa at these high rotational speeds and prevent a contamination of the ethanol cycle. However, the seal loses permanently a small amount of ethanol, due to its functioning. This ethanol is flushed away with gas ⑤ to prohibit remains of ethanol in the machine. The concept of sealing is discussed later in more detail.

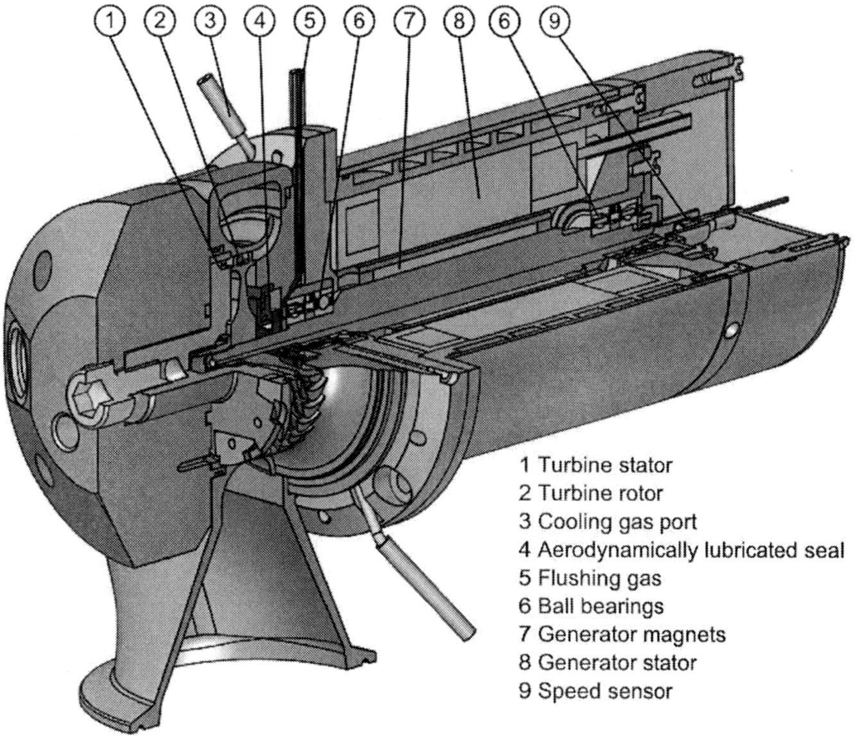

Figure 4: Turbine-Generator unit

The generator stator is cooled by a water cooling jacket. However, the cooling of the generator rotor is reasonable too, because the maximum temperature of the neodymium-iron-boron magnets is limited. For this reason, the interior of the generator is flowed through by a cooling gas (inlet port: ③). Although, the power electronic cares for a constant rotational speed, the expansion machine is equipped with an additional optical speed sensor ⑨. This sensor scans a surface at the shaft end and detects a mark.

4.1 Rotor assembly and rotordynamics

A main challenge in the design of high speed generators is the rotordynamics. The operation of the machine near a harmful natural frequency would result in a damage or destruction. The part of the machine, investigated within the rotordynamics analysis, is depicted in Figure 5. It consists of a shaft, on which the other components are mounted. On the left side is one pair of bearings mounted, which are separated by a spacer. The spacer offers the space for the spring, which is positioned at the outer ring and preloads the bearings. The used bearings are high precision spindle ball bearings of Schaeffler, which are able to be operated at the desired rotational speed (HCB7000-C-2RSD-T-P4S). Next to the bearing is another spacer, which separates the inner ring of the aerodynamically lubricated seal and the bearing. The turbine blisk is then mounted on the shaft. The torque of the blisk is transmitted by static friction through the previous components to the shaft. A nut cares for the required pre-load, which prevents torsion being exacted on the components. The right pair of ball bearings is tightened by a nut, as well. The left pair of ball bearings is the fixed bearing and absorbs the axial forces. The right pair of ball bearings has no axial fixing to enable the compensation of a thermal expansion of the shaft.

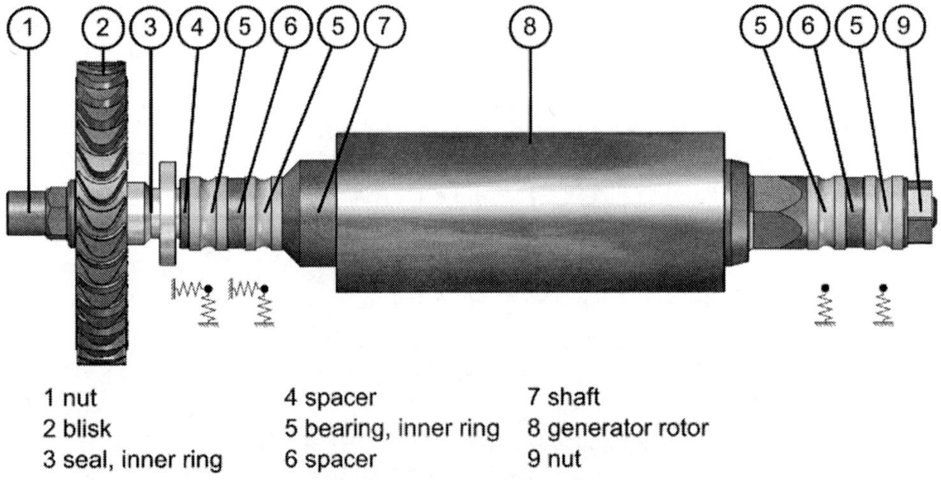

Figure 5: Rotating assembly considered in the rotordynamics

4.2 Analysis of the rotordynamics

Madyn 2000 (v3.6) was used for the investigation of the rotordynamics. It is not possible to load a CFD model directly into Madyn. Instead, the system was composed by 36 sections. Some of them were superimposed, e.g. the shaft consists of steel, but the superimposed magnets are from neodymium-iron-boron. On this way, each of the elements can be linked with the corresponding material properties. Exclusively, the turbine rotor was modeled by a normal mass with a disk inertia. The ball bearings were modeled by a general spring (GSP), which allows specifying the stiffness and damping coefficients in all three global directions. These stiffness coefficients are not rotational speed dependent, however it was tested, that the effect on the results is tolerable. Additionally, ball bearings exhibit low damping features, thus the damping coefficients were set to zero.

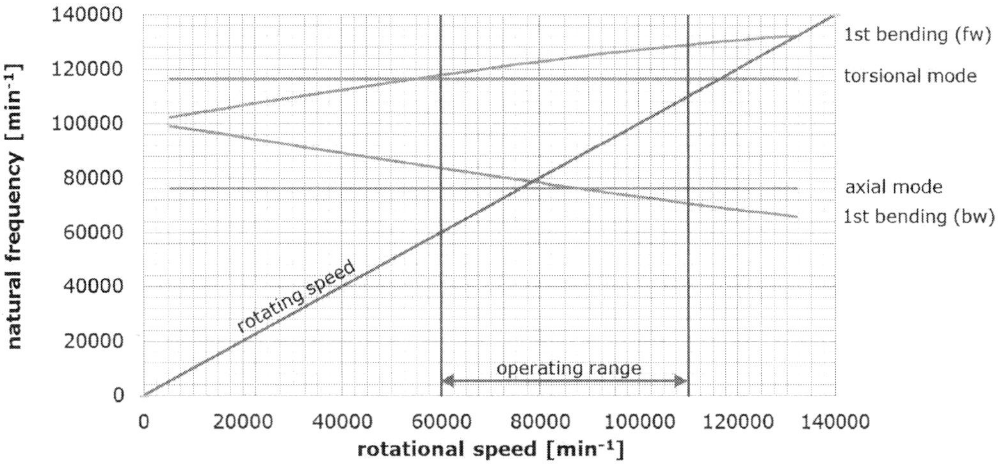

Figure 6: Campbell diagram of the investigated assembly

The Campbell diagram of the rotating assembly is shown in Figure 6. The operation range of the turbine is between 60,000...110,000 min^{-1}. Three eigenfrequencies exist in the observed speed range. The first one is an axial mode at a speed of 76,339 min^{-1}. This mode results from the axial stiffness of the fixed bearings and the total mass of the rotating part. It is assumed, that no axial excitation exists with the rotational frequency; thus this natural frequency is not harmful.

The second natural frequency is the first bending mode. The bending mode splits into a forward bending mode and a backward bending mode. Both bending modes show an obvious dependency on the rotational speed. The backward bending mode starts at a frequency of 101,900 min^{-1} and meets the rotational speed at a frequency of 78,750 min^{-1}. However, backward bending modes are not excited by imbalances in the rotor, thus this natural frequency is not harmful [19]. The forward bending mode increases with the rotational speed. It starts at a frequency of 101,900 min^{-1}, too, and meets the rotational speed at 132,000 min^{-1}. Forward bending modes are excited by imbalances and are harmful. However, this bending mode is outside of the operating range, which ends at 110,000 min^{-1}.

The third natural frequency is a torsional mode of the shaft. This mode is not dependent on the rotational speed and is constant at 116,500 min^{-1}. No torsional incitement is expected so this mode is not harmful.

The existence of a natural frequency within the operating range could not be prevented with this system. However, the only harmful natural frequency, the first forward bending mode, is at a high frequency of 132,000 min^{-1} and outside of the speed range. This was achieved by the high stiffness of the pair wise arranged ball bearings. Other natural speeds, an axial mode and the backward bending mode are within the operating range, but these modes are harmless and tolerable.

4.3 Sealing of the ethanol cycle

The sealing of the ethanol cycle is challenging. Firstly, a low absolute pressure of 0.081 MPa exists in the ethanol cycle and it can be contaminated by ambient air. Secondly, ethanol is a solvent and many elastomers are not resistant to it, especially at higher temperatures of up to 423 K. Additionally, the high rotational speed prohibits the application of touching seals, like rotary shaft seals. In contrast, labyrinth seals are not touching, but an additional sealing gas flow from ethanol is necessary, to avoid a contamination of the ethanol cycle.

An aerodynamically lubricated seal (Figure 8) is able to avoid this contamination without additional sealing ethanol gas flow. The used seal was developed to minimize the oil loss in turbochargers [18]. This seals consist of a stationary slide ring and a rotating counter ring (Figure 7). If the seal is not rotating, a spring presses the parts axially together. Grooves in the counter ring, cause a layer of vapor between the two parts, if the counter ring is rotating. This gas flows out of the ethanol cycle into a clearance. The clearance has channels to purge the ethanol. This purging is supported by an additional nitrogen flow, which secures, that no ethanol gets to the bearings. The sealing arrangement was already tested in an ethanol environment. It successfully sealed an absolute pressure of 0.081 MPa against the nitrogen at ambient pressure. Additionally, the seal was able to seal a vacuum of 0.02 MPa when the counter ring was not rotating.

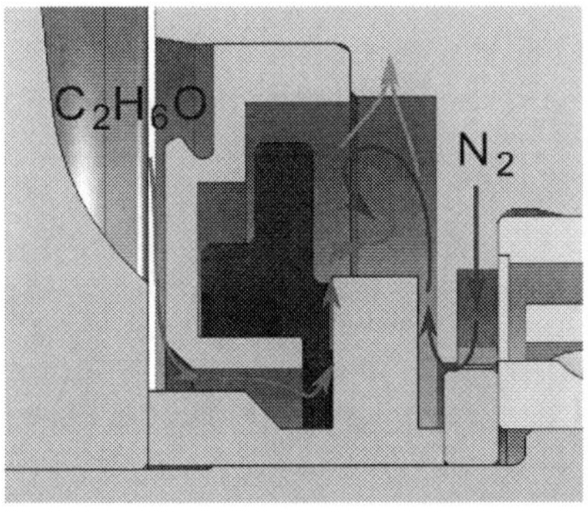

Figure 7: Fluid flows in the seal

Figure 8: Aerodynamically lubricated seal

4.4 Control device for the variable partial admission

One feature of this expansion machine is the ability to change its degree of partial admission under operating conditions. The regulation device for this variable partial admission is shown in Figure 9. It consists of a junction, which splits the flow into three pipes. One of these pipes is always open and does not possess a valve. The two other pipes are equipped with a ball valve, which can switch the pipes on. The valves are driven pneumatically, thus the partial admission can be safely changed under operating conditions. Afterwards, the fluid enters a flow divider and is splitted up and guided to the stator passages. Pipe one and two respectively provides two turbine passages; pipe three provides four turbine passages. Four degrees of partial admission are possible with this arrangement; 20%, 40%, 60% and 80%. However, the design of the flow divider was not possible without internal canals. This prohibits the manufacturing by milling. Additionally, the channel heights are too low, to be manufactured by precision casting. For this reason, the flow divider was made by selective laser melting, which allows manufacturing the internal channels. The pipes are corrugated hoses able to withstand the pressure. This provides flexibility in the regulation device to compensate temperature differences. Additionally, all pipes were isolated to minimize the thermal losses.

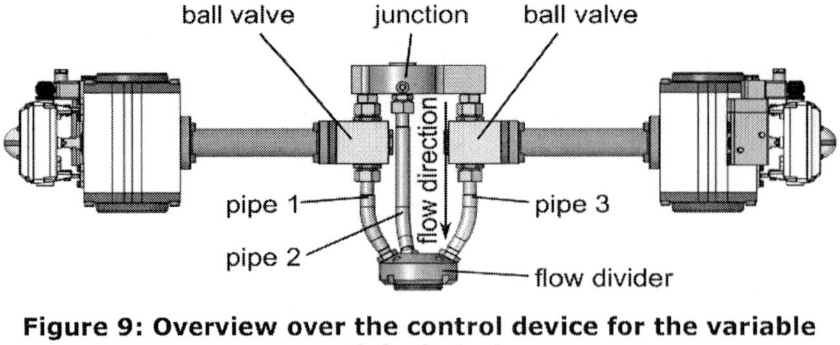

Figure 9: Overview over the control device for the variable partial admission

5 EXPERIMENTAL RESULTS

The experiments were performed on an ORC-test bench of the University of Hannover. This test bench consists of a turbocharger test bench with a burning chamber, which provides the required thermal energy. The outlet of this chamber is connected with the ORC-module. This ORC-module is able to provide the required ethanol mass flow.

5.1 Measuring instrumentation for the performance determination

One of the major tasks of the test bench is the performance estimation of expansion machines for ORCs (a simplified scheme of the test bench is shown in Figure 10). For this reason, instrumented tubes were positioned at the inlet ① and outlet ③ connection of the expansion machine. The instrumented tubes are able to measure the temperature and the pressure of the fluid. Additionally, the test bench has a mass flow meter in the liquid phase ⑤ and a flow meter in the vapor phase ④. The mass flow meter measures the whole mass flow provided by the pump, which flows through the evaporator. Behind the evaporator, the flow is split into the turbine mass flow and a bypass mass flow. The flow meter is positioned in the turbine line section and only measures the turbine flow. During the measurement, the bypass is completely closed and both meters measure the turbine flow. With this equipment, a thermodynamic analysis of the expansion machine is possible. However, it is known from investigations of turbochargers, that thermal losses significantly affect the efficiency, if it is determined by the inlet and outlet conditions of the turbine. For this reason, an additional pressure and temperature probe was implemented directly in front of the turbine stator ②, to minimize the effect. Additionally, the electric power of the generator was measured with a power measuring device ⑥. This device is positioned between the generator and the power electronics. This allows the determination of the isentropic efficiency of the turbine, if the mechanical efficiency of the expansion machine is known.

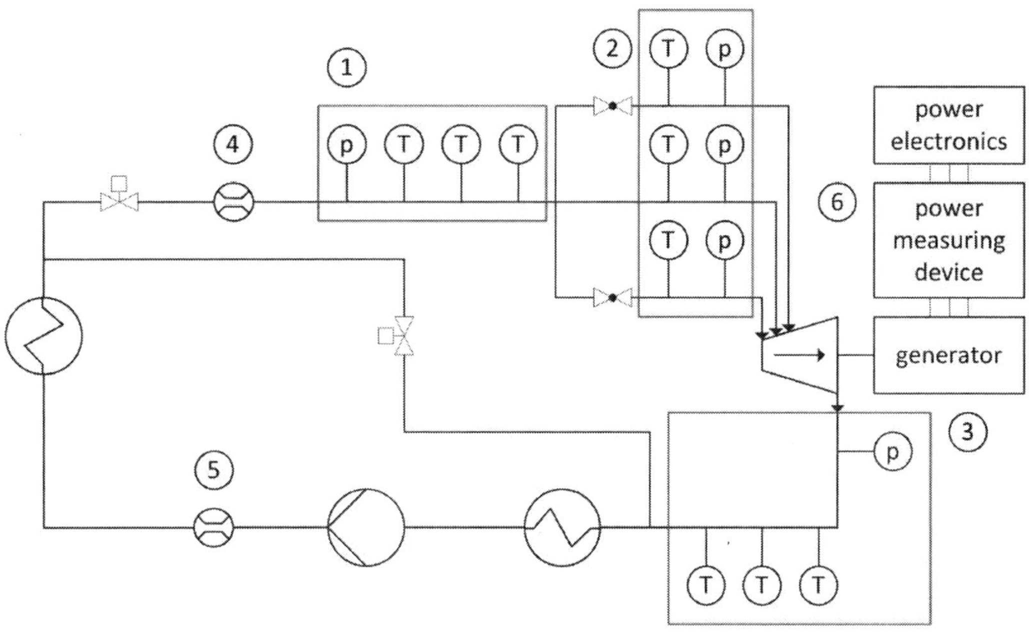

Figure 10: Simplified scheme of the test bench presenting the measuring points for the performance determination

5.2 Electrical power output of the expansion machine at selected operating points

Two operating points (OP) were selected for the presentation in this paper (Table 3). The OP have an inlet pressure of 2 MPa and an inlet temperature of 507 K. However, the degree of partial admission was changed; it is 20% for OP 1 and 40% for OP 2. The outlet pressure was adjusted by a balancing valve positioned between the outlet of the turbine and the condenser. The boundary conditions are listed in table 3.

Table 3: Selected operating points

Parameter:	Units	OP 1	OP 2
ε:		0.2	0.4
p_{in}:	MPa	2.00	1.99
T_{in}:	K	507.8	506.2
p_{out}:	MPa	0.0815	0.0811
P_{max}:	kW	1.54	3.85

The results of this investigation are depicted in Figure 11. Shown is the electrical power of the expansion machine depending on its rotational speed. The turbine was investigated up to a rotational speed of 80,000 min^{-1} during this measuring series. A maximum power of 1.54 kW is achieved for OP 1 at a rotational speed of 70,000 min^{-1}. A deviation of the rotational speed from 70,000 min^{-1} decreases the electrical power. The power decreases about 7.4%, if the rotational speed is reduced to 45,000 min^{-1} and about 5.8% if it is increased to 80,000 min^{-1}. Two additional stator passages are opened in OP 2. The electric power increases disproportionately high from 1.54 kW to 3.85 kW. This means a power increase of 250% while the mass flow is only doubled. This is caused by the lower partial admission losses of the turbine at higher degrees of partial admission, which results in a higher turbine efficiency. This effect is particularly notable, if the degree of partial admission is low. Additionally, the optimal rotational speed increases to 80,000 min^{-1}. The electrical power decreases significantly with a decreasing rotational speed. It decreases about 19.6% if the rotational speed is reduced to 45,000 min^{-1}.

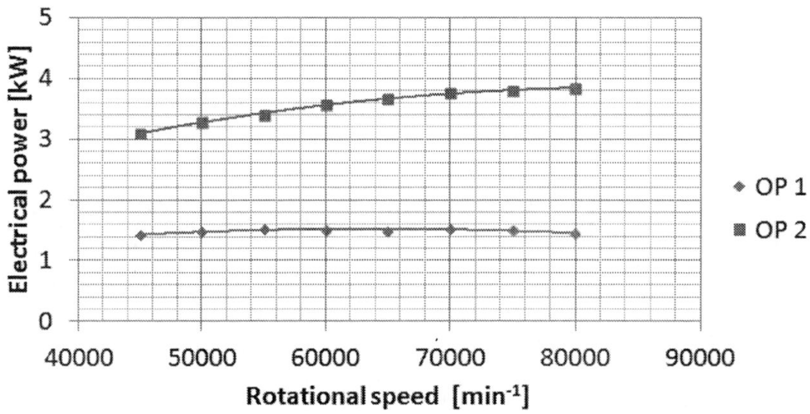

Figure 11: Electrical power output of the expansion machine

6 CONCLUSIONS

Prior investigations have shown the potential of ORCs for improving the efficiency of combustion engines in automobiles and commercial vehicles. The overall efficiency of these ORCs is heavily dependent on the efficiency of the expansion machine. However, currently no expansion machines are established for this application.

For this reason, a supersonic impulse turbine was developed at the Institute of Turbomachinery and Fluid Dynamics Hannover (TFD). This turbine is characterized by a high pressure ratio of up to 49 and a maximum rotational speed of up to 110,000 min^{-1}. Additionally, the turbine provides a variable partial admission, which extends the performance map and enables the expansion machine to cover all requested operating points of the considered truck engine.

A prototype of this turbine was manufactured at the TFD for an experimental investigation. The high rotational speed and the ethanol are challenging for designing a prototype. However, high-speed generators are available for the required rotational speeds, enabling the direct coupling of the turbine with the generator on one shaft. The shaft is mounted on two pairs of spindle ball bearings. This pair arrangement provides the system with high stiffness shifting the critical forward bending mode out of the operating range. Two more modes are in the speed range, the backward bending mode and an axial mode. These modes are not harmful and thus tolerable.

The turbine showed its operability within a speed range between 45,000 min^{-1} and 80,000 min^{-1}. Two operating points were presented with an inlet pressure of 2 MPa and a degree of partial admission of 20% (OP 1) and 40% (OP 2). The maximal electrical power gained was 3.85 kW for OP 2, which is 2.5 times higher than the power gained at OP 1 while the mass flow is only doubled. This can be explained by a much higher turbine efficiency, caused by the higher degree of partial admission. However, the power output is heavily dependent on the adequate rotational speed, which has to be adjusted to maximize the turbine efficiency. The absolute outlet pressure was 0.081 MPa at these operating conditions. The aerodynamically lubricated seal showed the ability to seal this vacuum and avoid a contamination of the ethanol cycle with ambient gas. Further experiments will explore the ORC turbine performance at higher loads.

ACKNOWLEDGEMENTS

This investigation was funded by industry fees in the context of the Sonderforschungsprogramm CO_2 of the Forschungsvereinigung Verbrennungskraftmaschinen e.V. (FVV) and the Forschungsvereinigung Antriebstechnik e.V. (FVA) which is gratefully acknowledged.

Substantial support was provided by Antriebstechnik und Entwicklungs GmbH, Leutkirch im Allgaeu/Germany, who contributed the rotor and stator of the generator, by Schaeffler, Herzogenaurauch/Germany, and its British subsidiary Barden Corp, Plymouth/UK, who contributed the design of the bearings and the bearing hardware, by Dr.-Ing. Martin Busch of Schaeffler for his support with the analysis and design suggestions regarding the rotor dynamics, and by Sieb & Meyer, Lueneburg/Germany, who furnished the power electronics. The authors also gratefully acknowledge the guidance of the industrial partners of the FVV Expansion Machine committee under the guidance of Dr.-Ing. Thomas Streule of Daimler and the valuable advice of Prof. Dr.-Ing. Bernd Ponick's team at Leibniz Universitaet Hannover Dept. of Electrical Engineering which helped to overcome challenges of the electrical generator.

The ORC Test Facility was partly financed by the European Community under an EFRE grand and co-financed by the Institute of Power Plant Technology at the Leibniz University Hannover, headed by Prof. Dr.-Ing. R. Scharf. The ORC test facility is located at the EFZN, an energy research center of Lower Saxony at Goslar/Germany.

REFERENCE LIST

[1] Bourhis G., Leduc P.; Energy and Exergy Balances for Modern Diesel and Gasoline Engines; Oil and Gas Science and Technology; Vol. 65; 2010

[2] Span R.; Eifler W.; Struzyna R.; Nutzung der Motorwärme durch Kreisprozesse; presented at the FVV Informationstagung Motoren/Turbomaschinen; Bad Neuenahr; Germany; 2011

[3] Nelson C.; Application of Refrigerant working fluids for mobile organic rankine Cycles; ASME-ORC; Bruessels; Belgium; 2015

[4] Howell T.; Gibble J.; Tun C.; Development of an ORC system to improve HD truck fuel efficiency; DEER 2011 Conference; 2011

[5] Rosset K.; Mounier V.; Guenat E.; Pajot O.; Schiffmann J.; Potential of Small-Scale Turbomachinery for waste heat Recovery on Automotive Internal Combustion Engines; ASME-ORC; Bruessels; Belgium; 2015

[6] Bockelhoff E.; Rosenberger T.; Energie zu Nulltarif; Magazin Lastauto Omnibus; page 24-25; 5/2013

[7] Meyer D.; Wong C.; Frithjof E.; Krumdieck S.; Design and Build of a 1 kilowatt Organic Rankine Cycle Power Generator; 35th New Zealand Geothermal Workshop; Rotorua, New Zealand; 2013

[8] Lemort V.; Quoilin S.; Cuevas C.; Lebrun J.; Testing and modeling a scroll expander integrated into an Organic Rankine Cycle; Applied Thermal Engineering; 2009

[9] Declaye S.; Quoilin S.; Guillaume L.; Lemort V.; Experimental study on an open-drive scroll expander integrated into an ORC system with R245fa as working fluid; Energy; 2013

[10] Smith I.K.; Stosic N.; Kovacevic A.; Langson R.; Cost effective small scale ORC systems for Power Recovery from Low Enthalpy Geothermal Resources; Geothermal Resources Council Transactions (Vol. 31); Davis; California, 2007

[11] Seher D.; Lengenfelder T.; Gerhardt J.; Eisenmenger N.; Hackner M.; Krinn I.; Waste Heat Recovery for Commercial Vehicles with a Rankine Process; 21st Aachener Colloquium Automobile and Engine Technology; 2012

[12] Pei G.; Li J.; Wang D.; Ji J.; Construction and Dynamic test of a small-scale organic rankine cycle; Energy; 2011

[13] Eilts P.; Seume J.; Brümmer A.; Sourell R.; Kunte H.; Hütker J.; Nikolov A.; Koch T.; Expansionsmaschine (Vorstudie); presented at the FVV Informationstagung Motoren/Turbomaschinen; Bad Neuenahr; Germany; 2012

[14] Kunte H.S.; Seume J.R.; Partial Admission Impulse Turbine for Automotive ORC Application; SAE-International Conference on Engines; Paper-No. 2013-24-0092; Capri; Italy; 2013

[15] Kunte H.; Seume J.; Experimental Setup of a small supersonic turbine for an automotive ORC application running with ethanol; Paper-ID 172; 3rd International Seminar on ORC Power Systems; October 12-14; Bruessels; Belgium; 2015

[16] Verneau A.; Supersonic Turbines for Organic Fluid Rankine Cycles from 3 to 1300 kW; von Karman Institute for Fluid Dynamics, Lecture Series 1987-07, Bertin, France; 1987

[17] ATE, Antriebstechnik und Entwicklungs GmbH, technical data sheet of the generator, No. 582 0001 100, 2012-07-13

[18] Simon C., Lang K., Feigl K., Bock E.; Turboladerdichtung für Null Ölverbrauch und minimiertes Blow-By; MTZ; 2010-04

[19] Swanson E.; Powell C.D.; Weissman; A Practical Review of Rotating Machinery Critical Speeds and Modes; Sound and Vibration; Acoustical Publications; 2005

SESSION 5B

TURBINE: UNSTEADY FLOW

The next generation of pulse optimized multi-scroll turbines for automotive applications – DualVolute and TwinScroll

M Gugau, F Winkler, S Weiske, T Heuer, R Sauerstein
BorgWarner Turbo Systems, Germany

ABSTRACT

Customer and legislation fuel economy requirements drive car manufacturers to develop downsized engines. Turbocharging is a key technology for higher power output with reduced displacement. Excellent transient performance for good drivability is a strong focus with downsizing. Single-scroll turbocharger turbines are an economic solution, albeit variable turbines extend the turbine performance range. Accordingly, for engines with relatively low exhaust gas temperatures, variable turbine guide vanes are standard solutions. For gasoline engines with high exhaust gas temperatures, the multi-scroll turbine regulated with waste-gate is a very attractive method to improve transient response.

Multi-scroll turbines use two or more scrolls feeding one turbine wheel making effective usage of pulse charging and enabling lower exhaust backpressure. The well-known design is a TwinScroll turbine, established in many passenger car gasoline applications. Recently, high attention has been paid to the DualVolute concept, allowing even more effective scroll separation. The two concepts are compared with respect to their performance characteristics, the benefits and challenges in engine interaction, and the specific design challenges and optimization parameters. Also, questions regarding new engine concepts, specific validation requirements, and validation results for DualVolute are discussed.

In addition, the latest improvements of thermodynamic application and matching methods for multi-scroll turbines are presented. Enhanced single admission efficiency outperforms high equal admission efficiency. The strong effect of the separation level on the engine control strategy is a significant benefit for a DualVolute turbine, although an additional scroll connection device might be necessary to reduce high load turbine back pressure.

NOMENCLATURE

A	Area	[m^2]
DH	Divider Wall Distance to Turbine Wheel	[m]
NW	Nozzle Width	[m]
m	Mass flow	[$kg \cdot s^{-1}$]
m*	Flow capacity	[$kg \cdot K^{0.5} \cdot s^{-1} \cdot bar^{-1}$]
p	Pressure	[bar]
R	Radius	[m]
T	Temperature	[°C]

Greek Symbols

β	Scroll Pressure Ratio	[-]
Δ	Tongue to Wheel Distance	[m]
λ	Air-Fuel-Ratio	[-]
η	Efficiency	[-]
π	Expansion Ratio	[-]

Subscripts

1, 2	Volute Channel 1, 2
2s	Cylinder Inlet
3	Turbine Inlet
4	Turbine Exit
DL	Drop Level
min	Minimum Value
max	Maximum Value
PL	Peak Level

Abbreviations

BMEP	Brake Mean Effective Pressure
BSFC	Brake Specific Fuel Consumption
CA	Crank Angle
CFD	Computational Fluid Dynamics
COHR	Center Of Heat Release
EA	Equal Admission
ER	Expansion Ratio
EVC	Exhaust Valve Closing
FSL	Flow Separation Level
HCF	High Cycle Fatigue
IVO	Intake Valve Opening
MFR	Mass Flow Ratio
OEM	Original Equipment Manufacturer
PMEP	Pumping Mean Effective Pressure
RGF	Residual Gas Fraction
SA	Single Admission
SO	Scroll Overflow
SSL	Scroll Separation Level
SYMF	Symmetry Level Flow
TMF	Thermo-Mechanical-Fatigue
TW	Turbine Wheel
UA	Unequal Admission
WLTC	Worldwide Harmonized Light Vehicles Test Cycle
WLTP	Worldwide Harmonized Light Vehicles Test Procedures

1. INTRODUCTION AND LITERATURE REVIEW

Pulse charging and better engine gas exchange are the main drivers for multi-scroll turbines in high temperature applications. These turbines consist of one wheel and multiple scrolls, separated in different ways, the most common of which are the TwinScoll and the DualVolute entry designs. The definition will be addressed later in this paper. Multi-scroll turbines have a long history in commercial turbocharging. Transient effects of pulse charging have been discussed in numerous publications, a comprehensive review was provided by Baines [1]. TwinScroll turbines are widely used in commercial vehicle turbochargers, Daimler for instance uses an asymmetric TwinScroll turbine to drive exhaust gas recirculation [2], [3]. The launch of TwinScroll turbines in gasoline passenger car engines started later, because the increased exhaust gas temperature drives material requirements and cost. However, TwinScroll turbines, notably in combination with variable valve trains, significantly improve transient response [4]. Consequently, turbocharger manufacturers have developed materials and simulation methods to design cost-effective TwinScroll solutions and largely offer appropriate solutions today. The TwinScroll became a standard option for 4- and 6-cylinder gasoline engines with displacement larger than 1.6-2.0l, introduced by various car manufacturers, for example BMW [5], [6], Figure 1, Volvo [7], or AMG, repeatedly awarded engine of the year [8]. Although the benefits of DualVolute turbines have been long known, and even used in niche applications like the Porsche 917 in racing, see figure 1, there has not yet been a market launch for a high volume pass car gasoline engine.

Figure 1: TwinScroll- and DualVolute applications

Interest in the DualVolute concept and the extensive usage of pulse energy with a 4-scroll segment turbine system continues, as for example proposed by Schenkel et al./Daimler [9], including prototype testing. As reported by Sauerstein et al. [10], BorgWarner has investigated the DualVolute in combination with a variable vane ring turbine, finding high potential for residual gas reduction and turbine downsizing. However, a thorough comparison to the existing TwinScroll concepts is needed in order to ultimately justify the replacement.

During the last years, many researchers have worked on a deeper understanding of both the flow behavior in multi-scroll turbines, mainly characterized by partial admission conditions, and the interaction with an internal combustion engine. Important scientific work has been published by the research teams of Martinez-Botas and Copeland. They experimentally studied a double-entry (dual) turbine, comparing transient pressure traces in the steady-state interpolation map [11] and proposed a possible method for a cycle-averaged efficiency approach for transient pulse admission [12]. Additional activities have concentrated on numerical map calculation, on understanding the typical loss-mechanisms [13], [14], on performance comparisons between twin and dual in steady-state admission by Romagnoli et al. [15], on finding substantial differences for single admission flow, and on the need for map extrapolation of unequal and partial (single) admission published by Newton [16]. Simultaneously to this work, which yielded a deeper understanding of the operation modes of twin and dual regarded as a turbomachine, the desire to substitute the TwinScroll by a DualVolute concept for passenger car turbochargers was recognized and pushed.

In 2011, Brinkert et al. released an analysis of the twin in gasoline engine operation and derived the importance of single admission condition for essential engine targets [18]. The consequential next step was the development and integration of a new turbine model in engine process simulation, taking into account separation and scroll interaction [19]. Hence, the assessment of twin and dual turbines applied to gasoline engines was made possible. It requires both, the knowledge of how these turbines should be characterized and the integration of adequate modelling of the interaction of the turbine and the engine. With this methodology, Uhlmann et al [20] or Schorn [21] have shown the thermodynamic benefits in conjunction with downsizing. Besides the possible benefits of dual turbines, Lückmann et al. [22], Brinkert et al.[23], and others have pointed out that a scroll connection device might be necessary at high scroll separation level. Most recently, Walkingshaw et al. [17] presented a comprehensive work comparing single, twin and dual turbine concepts.

BorgWarner is committed to deliver application specific thermodynamically optimized DualVolute turbochargers with respective, thorough validation of their components. Main aspects of BorgWarner's specific approaches are presented here.

2. AERO DESIGN OF MULTI-SCROLL TURBINES

Multi-scroll turbines improve engine steady state and transient torque by separating exhaust gas flows from the manifold down to the turbine wheel entry. Each of the two manifold/volute scrolls has a significantly lower volume compared to a single scroll. The two volutes and the manifold are designed in such a way, that the respective firing cylinder is always connected to a manifold/volute pipe that has not been loaded by the previous exhaust stroke.

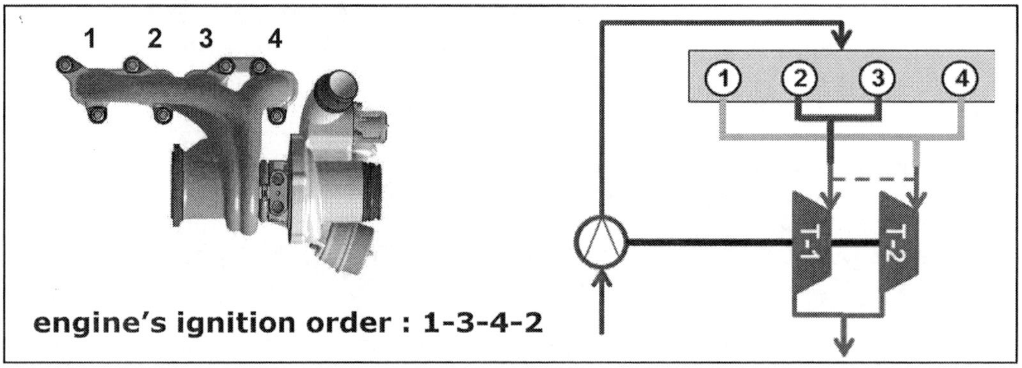

Figure 2: TwinScroll application at 4-cylinder engine

A possible solution is shown in Figure 2 for a 4-cylinder engine. Typical engine types suitable for multi-scroll turbines are 4, 6 and 8-cylinder, the latter often with double TwinScroll charging.

This type of turbine application enables improved gas exchange and turbine efficiency by amplifying pulse charging. Hence, the main performance benefits of multi-scroll turbines are related to improved system efficiency engine/turbine and high pulse utilization of the turbine. Consequently, designing, optimizing, and matching a multi-scroll turbine for a specific engine requires an understanding of turbine design for ideal pulse utilization in transient operation, as well as the understanding of engine interaction.

2.1 Performance characteristics of multi-scroll turbines

Multi-scroll turbines have a single turbine wheel that is fed by two or more volute scrolls. For ideal separation, each cylinder is equipped with its own exhaust flow duct. The potential of this approach has been demonstrated by Daimler [9] for a 4 cylinder medium duty diesel engine. Possible solutions with two scrolls are shown in figure 3. Both are currently being further developed for automotive applications. TwinScroll turbines have a scroll layout with a circumferential divider wall, thus loading the turbine wheel on the full inflow diameter but only half of the inlet width, scroll-1 towards the bearing housing, scroll2 towards the exhaust system side. DualVolute turbines load the wheel on the full inlet width but only half of the circumference. The inner scroll-1 with the smaller volume from wheel to flange and the outer scroll-2 with the larger volume. The literature uses various names, e.g. twin-entry and double-entry [15] or segment-turbine [22].

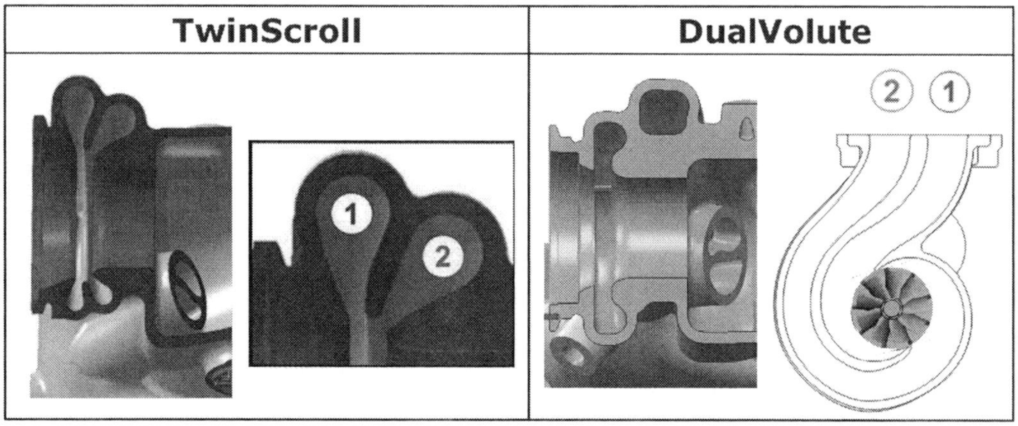

Figure 3: TwinScroll and DualVolute turbines / scroll definitions

For simplicity, in engine exhaust gas modelling, the two types of turbines are typically modelled as two single-scroll turbines, however the two scrolls are somehow connected. Perfect scroll separation is structurally not possible and flow exchange between the scrolls always occurs when the pressure level in the two scrolls is not equal. Test methods to account for this effect and correctly characterize dual and twin for engine simulation have been developed.

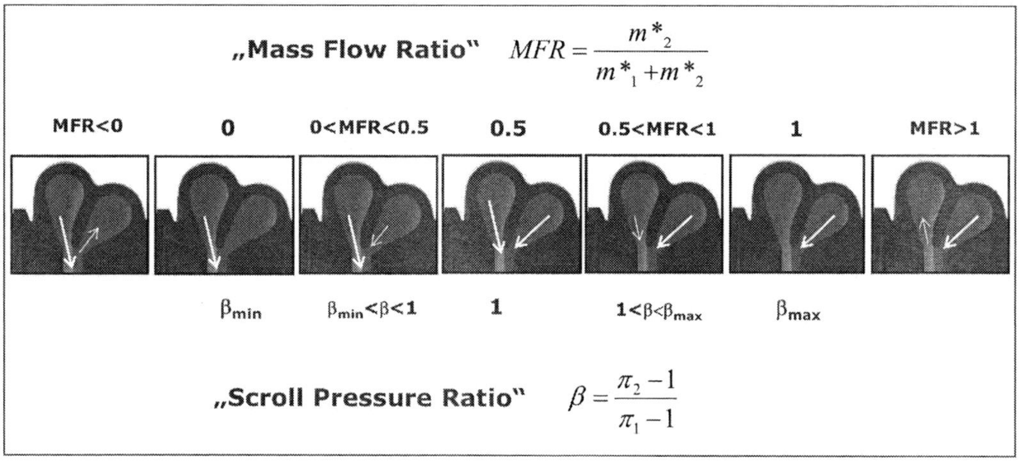

Figure 4: Possible flow conditions in TwinScroll turbines

The importance of single admission maps has been clearly demonstrated. Eventually, these allow for comparing twin and dual turbines in engine simulation. Additionally, the deeper understanding helps to derive performance relevant aero characteristics for both TwinScroll and DualVolute turbines.

The operating conditions of multi-scroll turbines are represented by flow, efficiency, and the pressure ratio between the scrolls. The possible flow conditions in real engine operation can best be specified by two characteristic numbers, the mass flow parameter MFR and the modified scroll pressure ratio β. The MFR was introduced by Uhlmann et al [20], the scroll pressure ratio by Brinkert et al. [18]. Both numbers help in identifying the turbine's operating point in engine simulation and in assigning the respective turbine flow and efficiency based on extended maps. All this is shown exemplarily for a TwinScroll turbine in figure 4.

The full set of extended turbine maps for multi scroll turbines needed for enhanced engine simulation methods and the naming used below are listed in table 1.

Table 1: Multi-Scroll Turbine Maps

Type	Flow Condition	Name	MFR	β
1	Equal Admission	EA	0.5	1
2	Single Admission Scroll 1	SA-1	0	β_{min}
2	Single Admission Scroll 2	SA-2	1	β_{max}
3	Unequal Admission	UA	0<MFR<0.5 0.5<MFR<1	$\beta_{min} < \beta < 1$ $1 < \beta < \beta_{max}$
4	Scroll Overflow	SO	MFR<0 MFR>1	

The map types 1 and 2 can be measured on a standard test rig, whereas map type 3 needs a special test rig including two separate gas burners. In addition, map type 4 requires a special backflow measurement device. The scroll separation level β_{min} and β_{max} of the turbine can also be measured together with single admission using a standard test rig. Obviously, the full information about aero performance of the turbine includes the knowledge of unequal admission and overflow prediction, both of which can be measured or numerically derived.

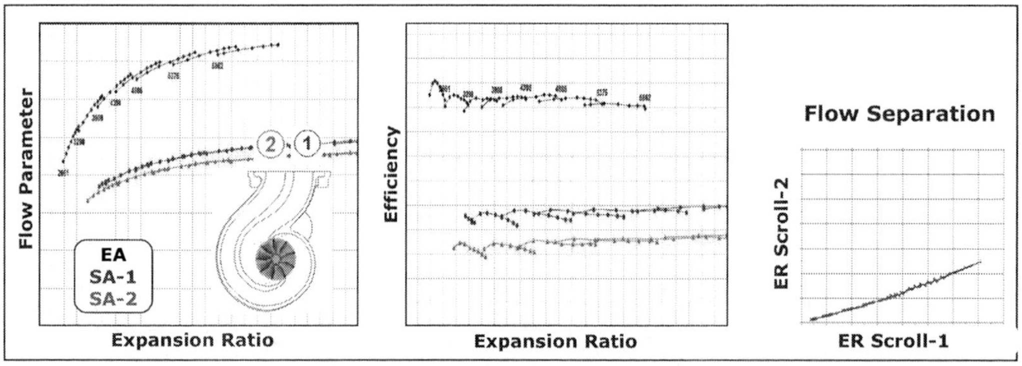

Figure 5: Standard hot gas tests for DualVolute turbine

For simplicity, figure 5 shows results only for equal and single admission of a dual. Both, flow capacity m* and efficiency η are much lower for single admission compared to equal admission. The difference in flow and efficiency for the two single admission tests is a measure for the symmetry of the design.

As will be shown in chapter 3, the turbine's most important operating condition is single admission. Assessing this system information, the essential performance parameters of the turbine can be derived from the maps in figure 5. The following parameters allow for aerodynamic performance comparison of multi-scroll turbines and an estimation of their performance on engine:

- Scroll separation level – a measure for the separation capability

$$SSL_1 = 1 - \frac{1}{\beta_{max}} \qquad SSL_2 = 1 - \beta_{min}$$

- Efficiency peak and drop level – a measure for max efficiency in equal admission and the drop for single admission. Efficiency is evaluated for turbine tip speeds 280 m/s (EA) and 340 m/s (SA)

$$\eta_{PL} = \eta_{max}(EA) \qquad \eta_{PL,1} = \eta_{max}(SA,1) \qquad \eta_{PL,2} = \eta_{max}(SA,2) \qquad \eta_{DL} = \frac{\eta_{PL,1} + \eta_{PL,2}}{2 \cdot \eta_{PL,EA}}$$

- Flow separation level – a measure for separation capability

$$FSL = 2 \cdot \left(1 - \frac{\overline{m^*(SA)}}{m^*(EA)}\right)$$

- Symmetry Level for Flow and Efficiency – a measure for design symmetry

$$SYMF = \frac{m^*(SA_1)}{m^*(SA_2)} \quad \{m^*(SA_1) < m^*(SA_2)\}$$

All parameters are defined in such a way that a value closer to 1 means better performance. These parameters are component-related quality criteria. Figure 6 shows some of the above listed performance parameters for some selected TwinScroll (12 parts) and DualVolute (3 parts) turbines from BorgWarner's portfolio. The parts are listed in the same sequence (1-15) for each parameter. These values show that an almost symmetrical scroll behavior can be designed. Both, twin and dual, offer almost identical flow in the scrolls, while symmetrical efficiency is much harder to achieve. Evidently, the dual offers much better separation capabilities than the twin with SSL around 0.9. Simultaneously, the flow separation level FSL (not shown in figure 6) is higher for dual, providing lower flow capacity in single admission. Peak efficiency of dual in equal admission is higher than of twin, even comparable to single-scroll turbines. In equal admission operation, the dual generates a flow field at the turbine wheel inlet that is almost similar compared to a single-scroll. However, the twin efficiency in equal admission suffers somehow from the flow being accelerated in the two nozzles first and then decelerated and mixed just upstream of the wheel. The efficiency drop level of dual is lower than twin, meaning a lower efficiency in single admission. The general efficiency characteristics of twin versus dual are also shown in figure 6 for a fixed expansion ratio. To summarize, the dual is superior with respect to equal admission efficiency and separation level, while the twin offers higher efficiencies for single admission but also worse separation behavior. As a single turbomachine, the two types of multi-scroll turbines perform differently and they offer different options to the engine.

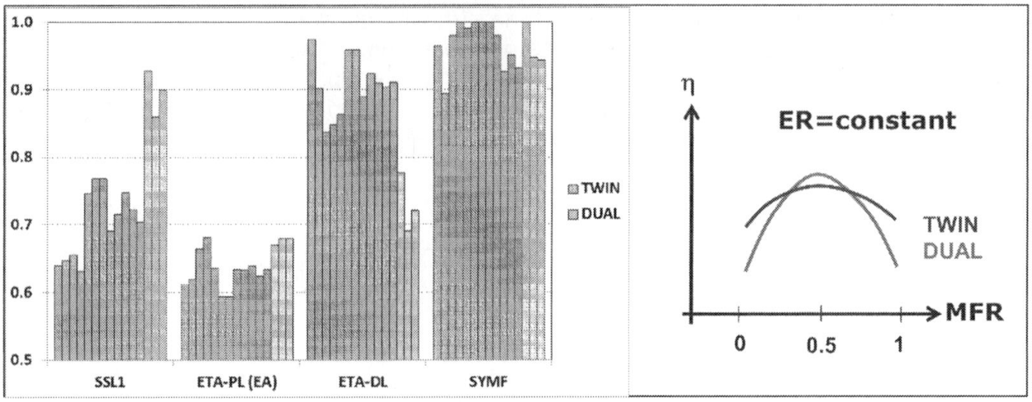

Figure 6: Performance parameters and characteristic multi-scroll

Although the correct matching of multi-scroll turbines is essential, a highly performing turbine stage is as important to reach good performance. The BorgWarner strategies for designing and optimizing twin and dual turbines will be discussed below.

2.2 Design parameters and -strategies

The challenge of turbine design usually means finding the best compromise between mechanical integrity for lifetime requirements and aerodynamic performance. This also applies to multi-scroll turbines, yet the aerodynamic performance optimization is a compromise on its own. Contrary design targets exist, since the turbine must fulfil engine targets at operating points, which always represent part load conditions for the turbine. For the engine torque knee point, the turbine operates at low mean mass flow, medium rotational speed, but high pressure amplitudes and thus works in an extreme pulse charging mode. On the other hand, for engine rated power, the multi-scroll turbine operates under high mean mass flow at high rotational speed and rather works in pressure charging mode.

The aerodynamic main design targets for multi-scroll turbines are:

- Maximized and equalized turbine wheel power output for each scroll under pulsating turbine inflow conditions.
- Optimized scroll separation capability for improving engine performance in low-end torque operation.
- Equalized scroll backpressure for engine rated power (waste-gate opened) and low-end torque operation (waste-gate closed) assuming pulsating inflow conditions.

Hence, the engine requires a turbine with both scrolls performing equally, producing high level equal power output for the torque knee and low, equal back pressure at rated power operation. This clearly means the turbine design and optimization process includes all parts, namely wheels, scrolls, manifold pipes, and waste-gate channels. Ideally, all aero targets are checked and optimized with real pulsating mass flow as engine boundary condition.

2.2.1. TwinScroll design

For steady-state inflow conditions, as measured on the hot gas stand and shown in figure 5, the TwinScroll acts like a high degree of reaction turbine for equal admission and like a low degree of reaction turbine for single admission. According to the chosen combination of wheel flow capacity and volute sizing, this tendency can be amplified or diminished. Having the wheel defined, the volute sizing can be varied within reasonable limits. Nevertheless, the main design feature of the twin controlling and influencing the turbine aero performance is the nozzle area at the exit of the scrolls. This is a challenging design task, since the nozzle area is extremely sensitive to casting and thermomechanical strength requirements and therefore limited in parameter variation.

Figure 7 shows the parameters defining the nozzle: The divider wall distance to the turbine wheel, the nozzle widths of both scrolls, and the inclination angles of the scrolls. A symmetric nozzle design is usually not possible for twins in gasoline engine applications, because the engine compartment package requirements and the proximity to the bearing housing extremely limit the available space.

First, the base sizing of the turbine, namely wheel flow characteristics and volute size, is defined. The volute size is determined by the throat area A/R ratio. Then, depending on the base sizing and the customer's emphasis on certain design targets, the nozzle area is optimized application specific by CFD. Figure 7 shows results for two possible nozzle designs with the same base sizing, providing different wheel torque. The turbine wheel inlet width is increased for design B, which is necessary for a certain nozzle width. The smallest nozzle width is determined by casting requirements and the divider wall thickness is controlled by casting and thermomechanical constraints. A comprehensive study on the effect of nozzle parameter variations on performance was done by Reich [24].

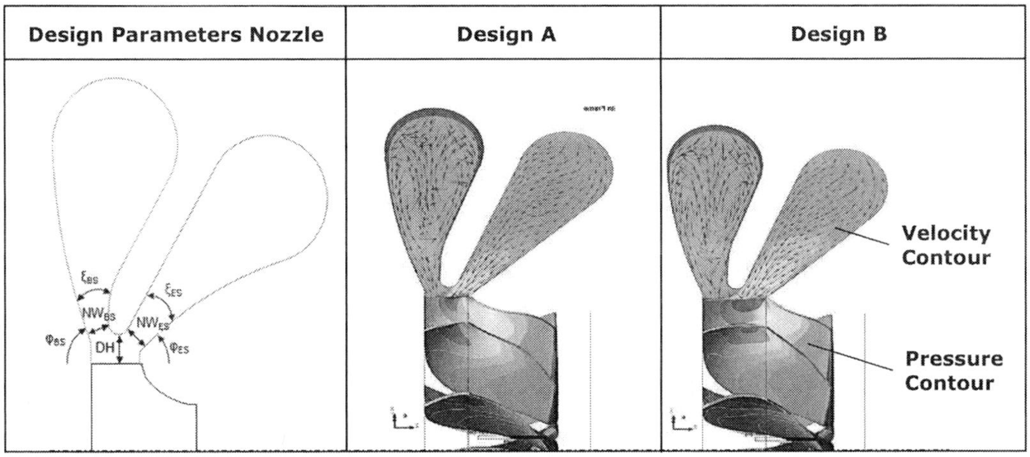

Figure 7: Nozzle design TwinScroll

The CFD optimization of the nozzle area is performed together with the manifold design. The design optimization requires steady-state and transient simulation. In particular, the wheel torque and the turbine back pressure should be evaluated with transient inflow CFD to account for the filling and emptying behavior of the pipes. Due to the fact that turbine scrolls need to be connected by manifold pipes to the engine's exhaust ports within very limited space available, the volumes of the two scrolls are never completely equal. Thus, one design target is to minimize and to equalize the two volumes of scroll and manifold as much as possible, which will help to fulfil the aero targets. The remaining difference in volume and the effect on aero targets can be compensated for by nozzle design and volute area sizing.

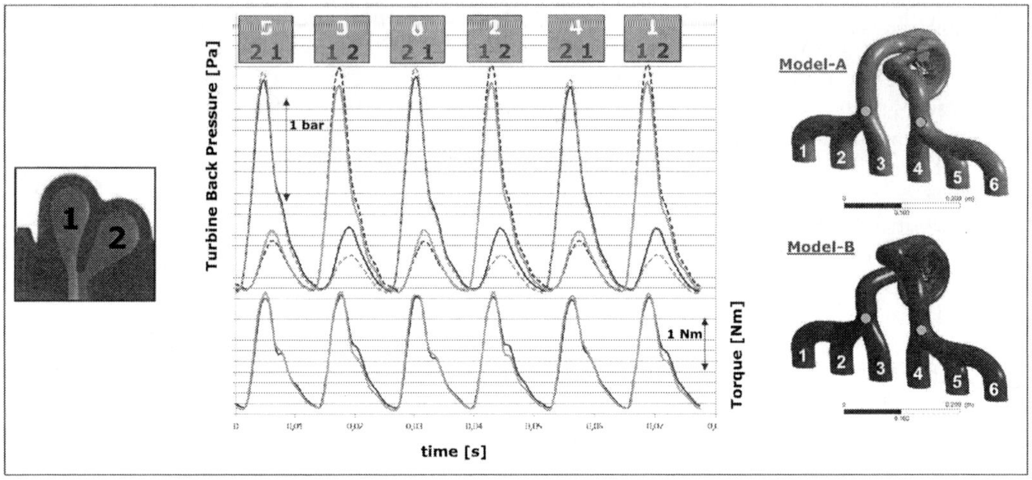

Figure 8: Effect of unsteady inflow including manifold

As an example, the results of a CFD simulation including the full turbine and manifold are presented in figure 8. A 6-cylinder engine has been equipped with a twin in two different arrangements: the connection of cylinders 1-3 and 4-6 with the respective volute scroll has been changed from model A to model B. Turbine back pressure and wheel torque are plotted versus time in blue for model A and in brown for model B. All of the 6-cylinder exhaust gas pulses are shown (ignition order 1-5-3-6-2-4), marked in white numbers in the squares. The blue and brown numbers are the admitted scrolls for both model cases.

It can be observed for both cases that the turbine back pressure is higher if scroll 2 is admitted. Overall, the pressure pulses are more equalized for model B. Also, the integrated torque output is slightly higher, which yet cannot be figured out from the plot. The important message of figure 8 is the fact that manifold and volute should be simultaneously optimized by transient CFD. This applies for both, TwinScroll and DualVolute turbines. The nozzle area is a key feature and the challenge for adjusting and optimizing performance of a twin.

2.2.2. DualVolute design

The DualVolute turbine offers less degrees of freedom for design than the TwinScroll. The scroll design is rather simple, using two half-circumference volute bends. The asymmetric volume distribution of a dual for an inline engine offers the possibility to arrange the scrolls with the respective manifold pipes targeting equalized scroll volume. The main design parameters controlling the aero performance are the turbine wheel, the degree of reaction, and the gap between the volute tongue and the turbine wheel's leading edge. This gap, also called tongue-to-wheel distance Δ, is the key design feature for scroll separation. A small gap will help improve SSL as well as turbine torque in single admission, but has to be matched with dynamic blade excitation and the resulting durability requirements.

Figure 9 shows the effect of tongue-to-wheel distance Δ and the turbine wheel type on wheel torque by means of CFD. The simulation is performed for single admission, since this operating condition has been determined as dominant for engine low-end-torque. The pressure in the admitted pipe is near 3 bar, the pressure in the empty pipe is almost ambient. Overflow occurs, as shown in figure 9, right side, but it is a function of Δ: A small gap reduces overflow, improves separation and wheel torque. The torque output with Δ=30% is 5% higher than the same wheel with Δ=100%. Turbine wheel B with adapted blade shape increases torque by another 0.5%. For accuracy, CFD simulations for dual in single admission should be run transiently with rotating wheel.

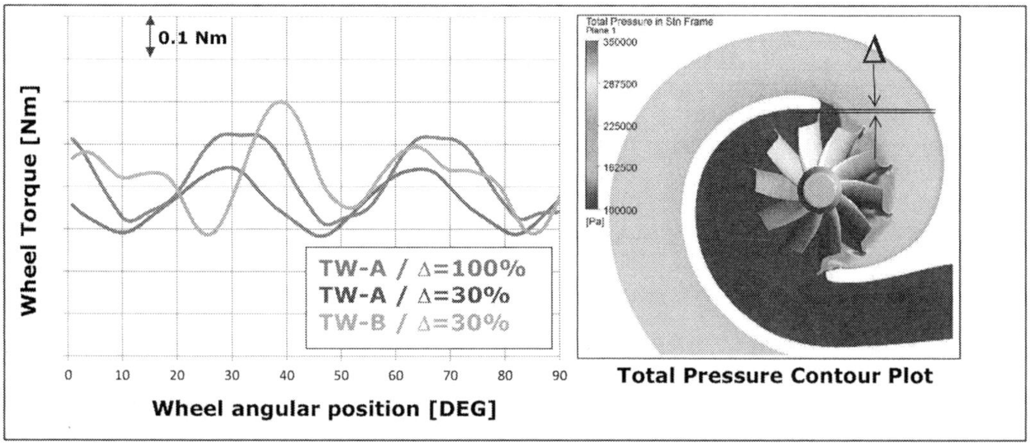

Figure 9: DualVolute wheel torque simulation

Evidently, the tongue design is the essential design feature of the dual. The tongue-to-wheel distance Δ should be as small as possible for aero performance. One limitation is given by the casting tolerance unless the tongue area is machined. The second limitation, rather a challenge since Δ is being pushed to the limit for high SSL and wheel torque, is the turbine wheel high-cycle-fatigue (HCF) resistance. The wheel suffers from the high pressure amplitudes and frequencies of harmonic blade excitation when passing the two tongues. Indeed, the demand for lifetime durability of the turbine wheel in the dual is the most critical part of the design. The approach BorgWarner has chosen to meet this challenge for dual will be discussed in chapter 4. Figure 10 shows a recent DualVolute prototype as an example of an application.

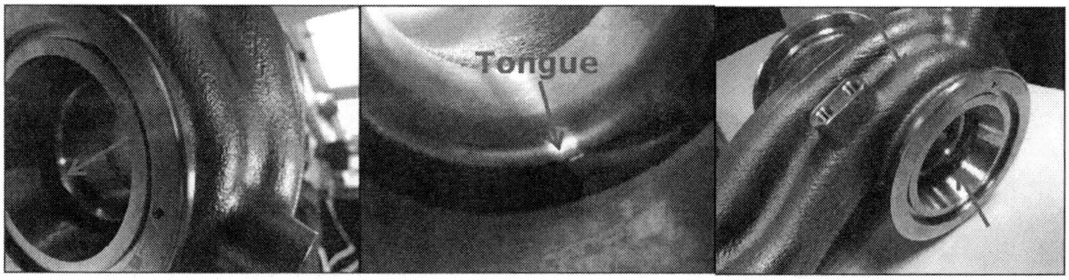

Figure 10: DualVolute prototype turbine housing for 4-cylinder gasoline engine

The design strategy for DualVolute is to optimize the single admission flow. Besides the gap Δ, turbine wheel and volute sizing are critical. Since only half of the circumference is pressurized in single admission, blade loading is beneficial for wheels with higher number of blades, even splitter wheel design is a promising option. The specific flow effects generating high entropy in the gap area and in the empty scroll leading to rather low efficiencies in single admission condition have been fundamentally investigated and analyzed by Copeland et al. [13], [14]. Wheel inertia plays a minor role in multi-scroll turbines compared to its importance for single-scroll turbines. Transient response is mainly driven by high pulse charging abilities and by scroll separation. Yet, this high level of separation, which is desired for low-end-torque, could become a drawback of the dual concept at engine rated power. In high specific power applications, turbine back pressure might rise undesirably, leading to the need of an additional scroll connection device.

Existing design parameters for the aero layout of TwinScroll and DualVolute have been shown above. The degree of reaction is the fundamental choice - a function of the sizing of wheel and the two volutes. Regarding turbine operating range with respect to blade speed ratio, this decision is primarily influenced by the compressor wheel diameter. With the turbine aero design, either engine low-end-torque or rated power can be emphasized. Modifying some appropriate geometrical design features of turbine wheel and turbine volute causes amplifying the efficiency either for single or for equal admission (see figure 5), each at the expense of the other. Absolute efficiency will in most cases be highest for equal admission operation. The best strategy is sometimes application specific, while design for single admission generally seems to be the right way. The next chapter will prove evidence for this strategy.

3. THE MULTISCROLL TURBINE IN ENGINE OPERATION

Engine knock is one of the challenges in gasoline engines. Inline 4 and 6 cylinder engines suffer from exhaust pulse interference of neighboring cylinders, mainly determined by exhaust valve duration. The residual gas fraction RGF in the cylinder at the beginning of the compression stroke limits the amount of air. Minimum RGF and low temperatures for optimal spark timing are key targets to optimize combustion processes and engine performance. This, as well as limitations on the use of scavenging will further stretch multi-scroll turbo performance and time to torque targets.

Within the scope of a technology benchmark, different turbo technologies have been measured on a modern DI I4 gasoline engine. To match the different turbochargers, a specific power of 90 kW/l and a specific torque of 180 Nm/l were targeted. Three different concepts were evaluated. The baseline was defined by a single-scroll turbocharger, which has been rated against a TwinScroll and a DualVolute turbocharger.

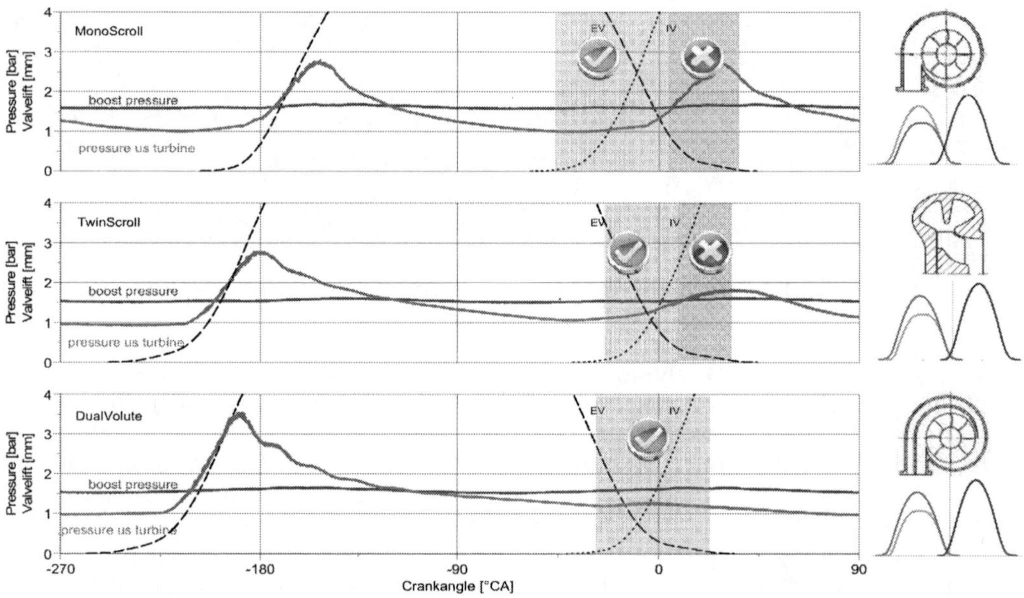

Figure 11: Pressure indication during charge exchange at 1250rpm, constant BMEP for single (top), twin (middle) and dual (bottom)

Figure 11 shows the comparison of upstream turbine pressure (p_3, red line) in relation to boost pressure (p_{2s}, blue line) during a charge exchange period at 1250 rpm. The excess of p_3 compared to p_{2s} indicates backflow of hot burned gas into the cylinder. This gas dynamical effect requires a compromise in exhaust valve event length. At low engine speed, a short valve opening is beneficial for low residual gas fraction, while at high engine speed, long opening durations help to reduce charge exchange losses. In modern mid-power engines these contradicting targets usually lead to exhaust valve event lengths of 195° CA to 205° CA at a reference lift of 1mm for single-scroll applications. Due to better pulse separation during gas exchange, multi-scroll turbines enable a stretch in event length of about 8-12% depending on scroll separation level. Comparing single-scroll, TwinScroll and DualVolute, due to its almost perfect scroll separation, the DualVolute is able to deliver complete positive dynamical scavenging (p_{2s}-p_3) during gas exchange solely. Even by reducing valve event length for single-scroll to 180° CA, it is not possible to eliminate negative dynamical scavenging effect.

Figure 12 shows the engine operation parameters at low engine speed and maximum torque comparing single, twin and dual. At low engine speed, the single-scroll turbine operates with a short exhaust cam of 180° CA profile. The exhaust event lengths for the multi-scroll turbines have been extended as described above.

The low end torque evaluation has been done at identical scavenging behavior represented by the O_2 fraction upstream of the catalyst and identical cylinder lambda, represented by the CO fraction upstream of the catalyst.

All turbochargers deliver an identical torque level. Both multi-scroll turbines show a 100-150 mbar higher positive scavenging ratio (p_{2s}-p_3) than the single-scroll reference. Up to 1500rpm, the dual shows slight advantages compared to the twin. The enhanced combustion boundary conditions translate into 2-3° CA earlier center of combustion and a fuel consumption benefit of 5-6%.

At rated power, the smaller cross section areas per scroll for the multi-scroll turbines lead to an increase in charge exchange losses. This negative effect could be reduced by adapting exhaust valve event length. The superior scroll separation behavior of the DualVolute requires an additional stretch in event length. As tradeoff between reduction in charge exchange losses and rise in expansion losses, an increase of 2-4% compared to TwinScroll seems to be possible. A standard exhaust cam was used for the single-scroll evaluation.

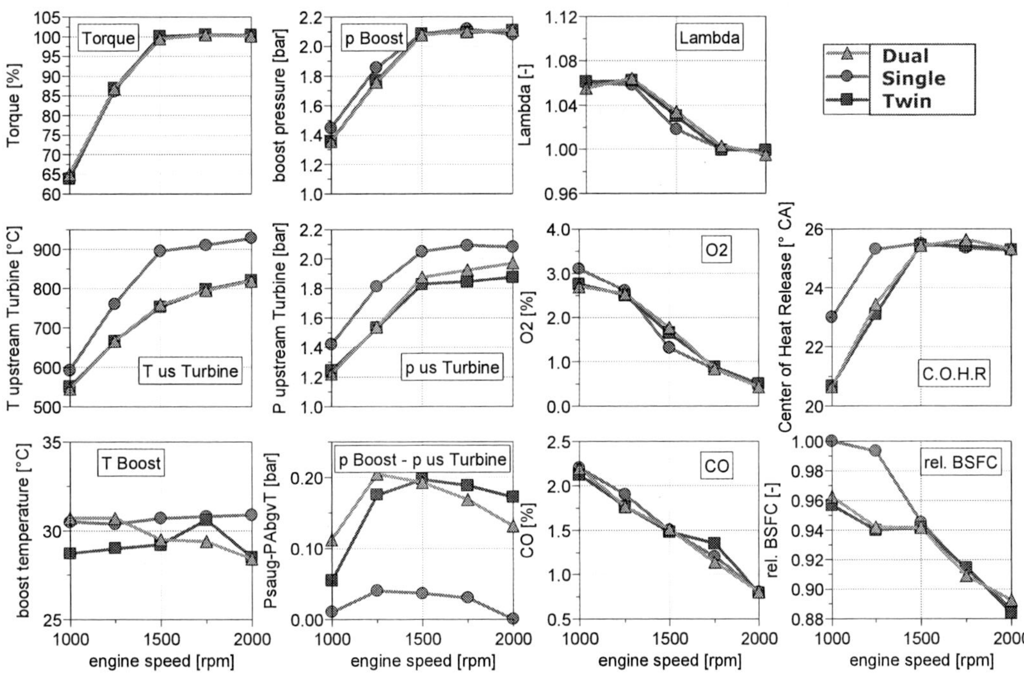

Figure 12: Engine operation in scavenging restricted low end torque area for single (red), twin (blue) and dual (green)

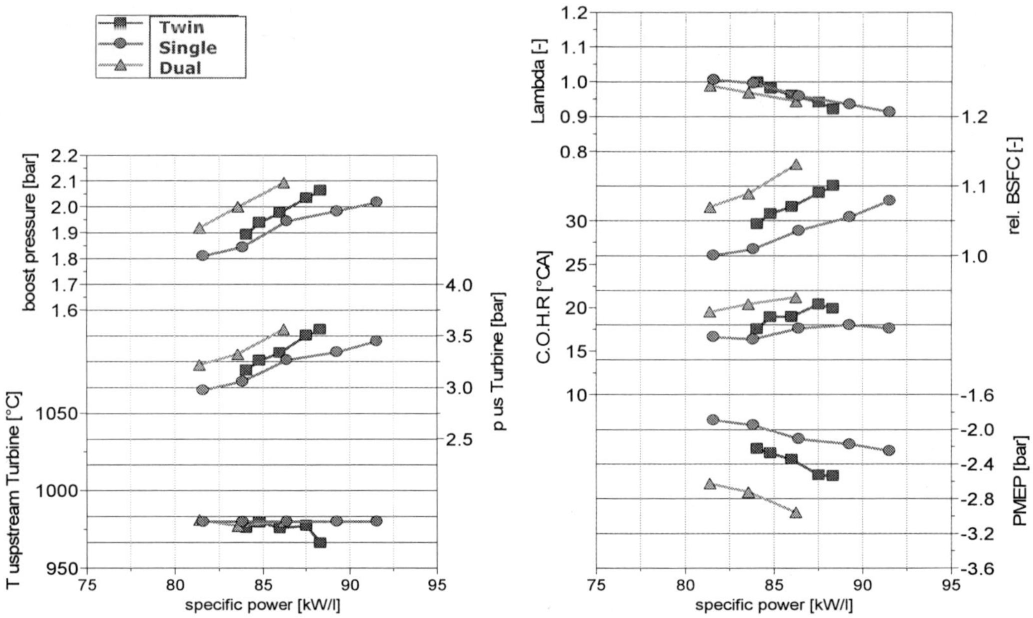

Figure 13: Boost pressure sweep at 5500rpm for single (red), twin (blue) and dual (green)

Figure 13 shows a boost pressure sweep at 5500 rpm. Valve timing has been optimized for all turbochargers. Corresponding to the increase in charge exchange losses, the indicated operation point as well as boost pressure demand and fuel consumption of dual and twin rise. Despite adapting the exhaust cam in the analyzed scenario, PMEP remains 700mbar higher than single-scroll and approximately 500mbar higher than TwinScroll.

A power evaluation is done on maximum upstream turbine pressure level (p_3) of 3.5bar. The increase in charge exchange losses of the multi-scroll turbines result in a deviation of 3-4% in max power for twin and ~6% for dual.

After the steady state evaluation, load step tests were performed. All rotors have comparable rotor inertia. The transient tests were started at a BMEP of 2bar stepping to full load while engine speed was kept constant. The procedure at 1500rpm is exemplarily shown in Figure 14. Valve timings (EVC and IVO) in transient operation were set identical for all variants. The single-scroll tests were performed with 180° CA exhaust valve event to ensure maximum dynamic performance.

Both multi-scroll turbines showed improved transient performance than the single-scroll with a short exhaust cam by 27-35%. Among both multi-scroll turbines, the DualVolute technology shows 11% better transient response at the analyzed speed point. At engine speeds below 1500rpm the DualVolute turbine has been performing even better. Further improvement, especially for the DualVolute is expected if the base engine would be adapted to have the best system layout and ECU setting.

The multi-scroll turbine technology seems to show the best tradeoff between high power and good transient response. The disadvantage at rated power can be addressed with a scroll connection. For such a concept, the DualVolute is better suited than the TwinScroll. Matching exhaust valve event length and pressure drop of that switching device seem to be key factors for dual performance.

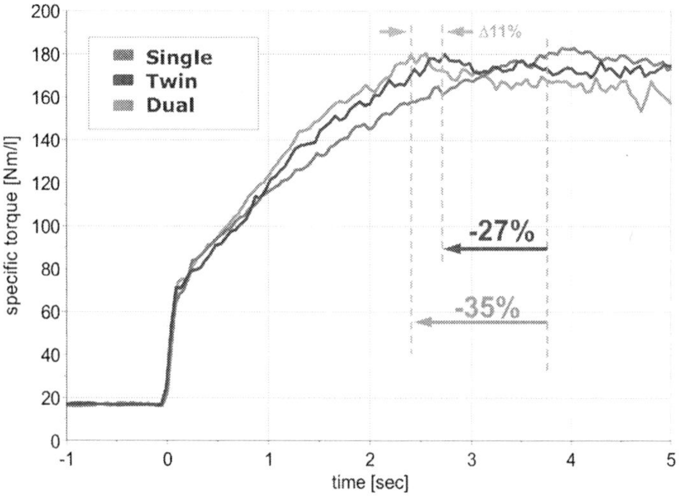

Figure 14: Load step from 2 bar bmep to full load at 1500 rpm for single (red), twin (blue) and dual (green)

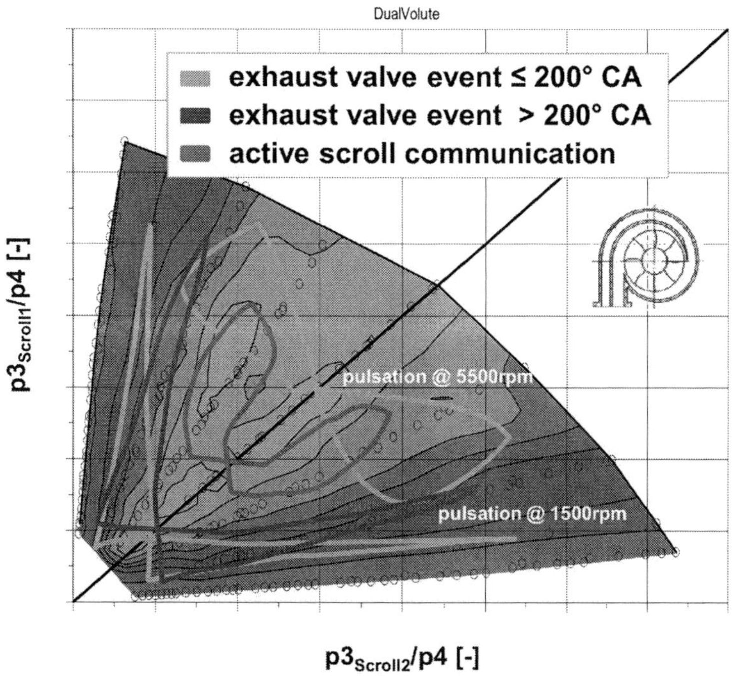

Figure 15: Turbine operation ranges for different pulsation setups

Figure 15 characterizes the turbine's operation during different exhaust cam timing strategies. The diagram shows the turbine efficiency characteristic as contour color of dual as function of pressure ratio per scroll and mass flow ratio of both scrolls according to Brinkert et al. [23]. Similar pressure traces in pulse admission of a dual concept (double entry) had also been shown by Copeland et al.[11], [12]. Especially in low engine speed operation with high exhaust pressure amplitudes, a stretching in valve event length leads to a shift in turbine efficiency. Due to its efficiency characteristics and scroll separation level as shown in figure 6, the DualVolute shows much higher peaks of pressure traces than the TwinScroll. At rated power operation, the elimination of scroll separation ("active scroll communication", meaning a control device that enables the scroll connection and separation on demand) and the resulting shift towards equal admission (MFR 0.5) indicates an additional benefit of the dual. The pressure drop minimization of the switching device in this context is prerequisite.

4. VALIDATION OF MULTI-SCROLL TURBINE COMPONENTS

The durability challenges for new turbine stages have remained the same for past decades: turbine housing thermo-mechanical fatigue (TMF), turbine wheel high-cycle-fatigue (HCF) and waste gate system pulse load. Given the nature of strong pulse loads in multi-scroll turbine stages in gasoline application, the latter two are of critical concern for any TwinScroll or DualVolute turbine stage, while the first topic of turbine housing TMF has proven a particular challenge for modern high temperature gasoline engines.

Based on extensive experience in multi-scroll turbine development and production over the past decades in both commercial vehicle as well as high power and temperature gasoline passenger car applications, the early focus on these topics has been an integral part of the development of the DualVolute turbine the very beginning.

To make sure the new DualVolute turbine stage is well equipped regarding the challenges resulting from its specific high load due to nearly perfectly pulse separated turbine design, BorgWarner has decided not only to simulate the waste gate pulse load for the two common turbine designs single-scroll and TwinScroll in comparison with the DualVolute, but also to build hardware and measure actual waste gate pulse load on engine. These results show a clear increase in pulse load going from single-scroll constant pressure charged turbine stages to the pulse charged multi-scroll turbines of twin and dual (see figure 16).

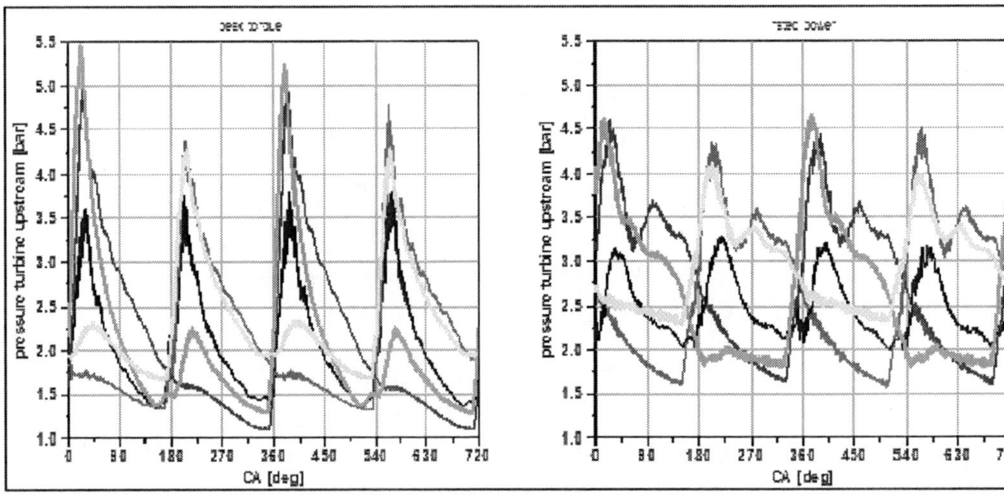

Figure 16: Measured pulses on the same engine for single (black), twin (green/yellow), and dual (red/blue)

As a result, all BorgWarner gasoline waste gate components include the application of the DualVolute turbine stage in their base specification requirements.

Specific focus has been set on both, wear conditions, and the cost effective, robust, and low noise solutions for all waste gate turbochargers, all implemented and available for DualVolute and TwinScroll turbines.

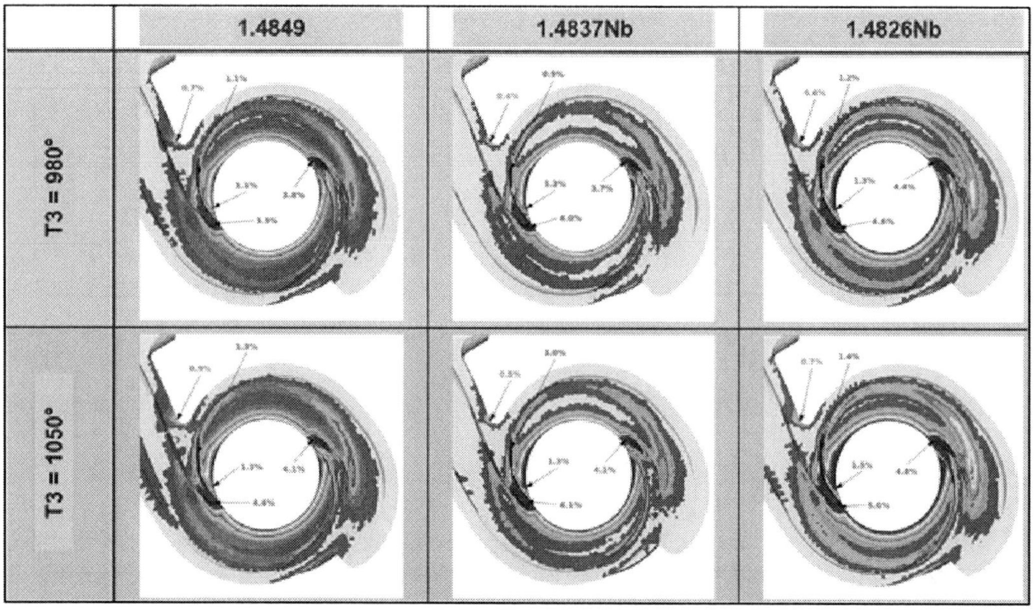

Figure 17: DualVolute turbine housing TMF simulation matrix

The typically tight package requirements drive turbine housing designs to a great extent. Therefore BorgWarner decided to pick challenging customer packages for the turbine housing design and validation process. A setup of various version of fully and partly integrated manifolds have been designed, simulated, built, and tested to prove the superior robustness of the DualVolute design, leading to good cost effectiveness through lower material requirements.

To prove better TMF durability, a matrix of different turbine housing materials and temperatures was simulated. Three austenitic steels 1.4849, 1.4837Nb, and 1.4826Nb, in combination with 980°C and 1050°C turbine inlet peak temperature T_3 have been used. No changes were made to the simulation processes, boundary conditions, and cycle times, as well as acceptance criteria compared to the standard TMF analysis processes. For the less challenging application of 980°C, a comparison TwinScroll turbine housing was simulated in the highest material grade 1.4849.

Results showed a clear advantage of the DualVolute turbine housing , with fewer and lower plastic deformation hot spots. Moreover, significantly lower plastic deformation at the very high stress area tongue were found, with a delta of around 3% to 4%.

The temperature and materials matrix has confirmed the superior TMF performance of the DualVolute turbine design, including the potential to operate at 1050°C with existing materials, e.g. 1.4849 or the use of 1.4837Nb at current 980°C turbine inlet peak temperature applications, see results matrix in figure 17.

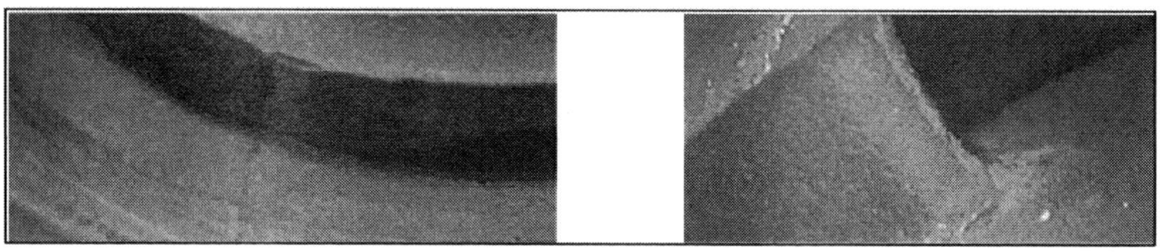

Figure 18: Samples of both DualVolute turbine housing volute tongues after durability testing up to 1050°C

The material 1.4849 was ultimately chosen for a lifetime customer release test on engine. The full rapid prototype turbocharger successfully passed the accelerated duty cycle of a 1050°C T_3 peak engine application. The tear down analysis showed no failures of the DualVolute turbocharger. Most notably, the tongues, typically the turbine housing's high stress area and the initial cracking location, came out without any indication of cracking. Figure 18 shows both tongue areas of the prototype after durability testing.

Turbine wheel HCF is of particular interest in the DualVolute turbine stage development, because the pulsation from two separate volutes will cause higher than normal HCF stresses on the turbine wheel.

Figure 19: Turbine wheel designed for twin used in dual (left) and designed and optimized for dual (right)

In figure 19 the left picture shows a turbine wheel failure of not "dual-tailored" wheel design used in a DualVolute turbine. This wheel design had been used for years in TwinScroll applications without any issues. However, it was not strong enough to withstand the DualVolute's loads. This underlines the requirement for turbine wheel HCF robustness in a DualVolute turbine. The right turbine wheel in figure 19 is a new wheel, specifically designed for the new DualVolute turbine stage. The photo shows the situation after a first successful engine endurance test. The wheel design incorporates several robustness measures in wheel properties like natural frequencies and wheel design features. Also, turbine scroll to wheel geometric features were achieved with specific manufacturing processes, and specific scroll features were implemented, contributing to robustness of the wheel. Obviously, the two tongues do shift the critical speeds to half of the typical values of single-scroll turbines, which is beneficial.

To summarize, from extensive test and validation exercises, BorgWarner designed a DualVolute turbine wheel and turbine scrolls with excellent robustness. For validation, comprehensive hardware testing of parts was performed with advanced Agilis light probe testing methodology. Contrary to traditional frequency testing of a static shaft & wheel, this method allows to define the model frequencies and n-mode weaknesses of each turbine wheel blade in operation. By doing so, critical operation points are identified, which eventually allows to accelerate the subsequent tests to failure or fatigue strength. In addition to modifications of the scrolls and the wheel, new waste-gate designs and validation tests have been developed to ensure meeting next generation engine NVH criteria [25].

5. SUMMARY AND CONCLUSION

Pulse-separating multi-scroll turbines enable the engine to improve performance, in particular low-end-torque and transient response. Three types of turbines have been compared with respect to basic performance on engine, as well as design features and challenges regarding durability.

The aerodynamic characteristics and the potentials of single, twin and dual turbines are well understood today, especially with respect to the benefits and challenges in their interaction with the engine. For the aero component design of the three types of turbines, state of the art steady-state and transient numerical simulation approaches, but also enhanced methods, have been used. The key design features controlling the aero performance of TwinScroll and DualVolute turbines as well as corresponding CFD design and optimization methodologies have been presented. Regarding engine gas exchange simulation, remarkable progress can be confirmed to the modelling techniques of multi-scroll turbines, also driving for improved steady-state input data from testing or simulation. Consequently, suitable tools for the ideal matching of a turbine for a certain engine application are available today. A comparison of the thermodynamic impact of turbine technology on the engine performance has been shown though the results of an engine test series. The DualVolute turbine technology offers the highest potential for the best tradeoff between high power output and transient response targets. The elimination of scroll separation on demand for optimizing charge exchange in rated power operation will even expand the potential of the DualVolute technology.

Nevertheless, the required validation of the turbine does not only include tailored testing facilities for aero performance like interaction and overflow maps , but also comprehensive component testing. Multi-scroll turbines used in high temperature gasoline applications with high pressure pulse amplitudes produce particularly high demands on all turbine components, including the volute, the waste-gate flap, and especially the turbine wheel. To meet these requirements and the challenges of mechanical integrity with high effective lifetime, BorgWarner has successfully proven thermomechanical fatigue strength for volutes and high cycle fatigue strength for turbine wheels as well as the durability of other DualVolute components. Today, BorgWarner offers various solutions for TwinScroll turbines in a wide range of engine displacements in series production. Design and up-to-date matching methodology for multi-scroll turbines are available, while validation work for the successful application of a robust and high-performing DualVolute turbine, including an optional variable scroll connection device, are largely accomplished.

REFERENCE LIST

[1] Baines, N.C., 2010, "Turbocharger Turbine Pulse Flow and Modelling - 25 Years on", Turbochargers and Turbocharging, Institution of Mechanical Engineers, pp. 347-362

[2] Müller, M., Streule, T., Sumser, S., Hertweck, G., Nolte, A., Schmid, W., 2008, "The asymmetric twin scroll turbine for exhaust gas turbochargers", Proceedings of ASME Turbo Expo, GT2008-50614

[3] Brinkert, N., Sumser, S., Weber, S., Fieweger, K., Bauer, H.J., 2010, "The asymmetric Twin-Scroll turbine under Engine operating conditions", THIESEL - Conference on Thermo- and Fluid Dynamic Processes in Diesel Engines

[4] Björnsson, H., Ottosson, A., Rydquist, J.-E., Späder, U., Schorn, N., 2005, "Optimizing the SI-Engine turbo systems for maximum transient response. Methods, factors, findings.", 10. Aufladetechnische Konferenz Dresden

[5] Klauer, N., Klüting, M., Steinparzer, F., Kretschmer, J., 2009, "Innovative Turbocharging, Variable Valvetrains and 8-speed Transmissions - A new Generation of Drivetrains", 18th Aachen Colloquium Automobile and Engine Technology

[6] Steinparzer, F., Klauer, N., Kannenberg, D., Unger, H., 2011, „Der neue aufgeladene 2.0-l-Vierzylinder-Ottomotor von BMW", in MTZ 72, Nr. 12

[7] Björnsson, H., Collinder, M., Daunius, O., 2007, "Si6T - A new generation of 6 cylinder turbo engines from Volvo Car Corporation", 12. Aufladetechnische Konferenz Dresden

[8] Gindele, J., Ramsteiner, T., Fischer, J., Tschamon, B., 2013, „Der neue 2.0-l-Hochleistungs-Vierzylindermotor von Mercedes-AMG", MTZ 09/2013

[9] Schenkel, S., Sumser, S., Fledersbacher, P., Daudel, H., Finger, H., 2002, "4-fach Stoßturbine am Vierzylinder-Reihenmotor Nutzfahrzeuge", 8. Aufladetechnische Konferenz Dresden

[10] Sauerstein, R., Dabrowski, R., Becker, M., Schmalzl, H.-P., Christmann, R., 2009, "The Dual-Volute-VTG from BorgWarner – A new Boosting Concept for DI-SI-Engines", 14. Aufladetechnische Konferenz Dresden

[11] Copeland, C., Martinez-Botas, R., Seiler, M., 2008, "Unsteady Performance of a double entry turbocharger turbine with a comparison to steady flow conditions", Proceedings of ASME Turbo Expo, GT2008-50827

[12] Copeland, C., Martinez-Botas, R., Seiler, M., 2009, "Comparison between steady and unsteady double-entry turbine performance using the quasi-steady assumption", Proceedings of ASME Turbo Expo, GT2009-59290

[13] Copeland, C., Newton, P., Martinez-Botas, R., Seiler, M., 2010, "The effect of unequal admission on the performance and loss generation in a double-entry turbocharger turbine", Proceedings of ASME Turbo Expo, GT2010-22212

[14] Copeland, C.D., Newton, P., Martinez-Botas, R., Seiler, M., 2012, "The effect of unequal admission on the performance and loss generation in a double entry turbocharger turbine", Journal of Turbomachinery, Vol 134, 021004, ASME

[15] Romagnoli, A., Copeland, C., Martinez-Botas, R., Rajoo, S., Seiler, M., Costall, A., 2011, "Comparison between the steady performance of double-entry and twin-entry turbocharger turbines", Proceedings of ASME Turbo Expo, GT2011-45525

[16] Newton, P., Romagnoli, A., Martinez-Botas, A., Copeland, C., Seiler, M., 2013, "A method of map extrapolation for unequal and partial admission in a double entry turbine", Proceedings of ASME Turbo Expo, GT2013-95815

[17] Walkingshaw, J., Iosifidis, G., Scheuermann, T., Filsinger, D., Ikeya, N., 2015, "A comparison of a mono, twin and double scroll turbine for automotive applications", Proceedings of ASME Turbo Expo, GT2015-43240

[18] Brinkert, N., Sumser, S., Schulz, A., Weber, S., Fieweger, K., Bauer, H.J., 2011, "Understanding the Twin-Scroll Turbine – Flow Similarity", Proceedings of ASME Turbo Expo, GT2011-46820

[19] Brinkert, N., Kuhn, T., Sumser, S., Weber, S., Fieweger, K., Bauer, H., 2011, "Modellierung der zweiflutigen Turbine in der Motorprozesssimulation", in Motorprozesssimulation und Aufladung 3, Berlin

[20] Uhlmann, T., Lückmann, D., Aymanns, R., Höpke, B., Scassa, M., Schorn, N., Kindl, H., 2013, "Development and Matching of Double Entry Turbines for the next Generation of highly boosted Gasoline Engines", 34. Internationales Wiener Motorensymposium

[21] Schorn, A., 2013, "Potentialabschätzung einer Doppelstromturbine für kleine Ottomotoren", Studienarbeit TU Kaiserslautern, SAM

[22] Lückmann, D., Stadermann, M., Pischinger, S., 2015, "The segment turbine – a future technology for highly boosted gasoline engines ?", 24th Aachen Colloquium Automobile and Engine Technology

[23] Brinkert, N., Rott, M., Friedrich, J., Weber, S., Freisinger, N., Karl, G., 2014, "Further development of pulse / constant pressure turbocharging on 4-cylinder gasoline engines", 19. Aufladetechnische Konferenz Dresden

[24] Reich. T., 2015, "Design guideline for the nozzle dimensioning of a twin-scroll turbine with pulsating inflow", Master Thesis TU Kaiserslautern, SAM and BorgWarner Turbo Systems GmbH

[25] Sauerstein, R., Cempirkova, D., Hadas, R., Ruh, M., Stilgenbauer, M., Zahoransky, R., Takabatake, L., 2015, "A new generation of Turbochargers for gasoline engines for E100 fuel capability and high pulsation requirements for Wastegate control", 20. Aufladetechnische Konferenz Dresden

Analyses of flow structure through radial turbine twin-entry volute under pulsatile flows

M Cerdoun [1], **A Ghenaiet** [2]
[1] Laboratory of Thermal Power Systems, Applied Mechanics,
Ecole Militaire Polytechnique, Algeria
[2] Faculty of Mechanical Engineering, University of Sciences and Technology USTHB, Algeria

ABSTRACT

In turbocharging applications, twin-entry radial turbines will better use the energy of pulsating exhaust gas flows by avoiding expansion losses into exhaust manifolds and interference between exhaust pulses. In the present study, the flow structure through a twin-entry volute is investigated, by means of code ANSYS-CFX, under the engine-like conditions to have a better insight into the unsteady flow behavior and effects of pulsating gas flows. The unsteady boundary conditions imposed at the inlet and outlet of this radial turbine are related to the engine real operating environment, which were combined from average and instantaneous experimental data. The obtained results show that the flow in the spiral part of twin-entry volute is subject to centrifugal forces which induce a secondary flow in the form of counter-rotating vortices. Also, there is a typical dissymmetrical behaviour characterized by high non-uniformities at the rotor inlet in both spanwise and circumferential directions. The flow incidence at rotor inlet is highly affected circumferentially around the tongue due to interactions between the main flow and that infiltrating to feed the rotor, hence generating extra mixing losses. On the other hand, in the spanwise direction, the flow is affected by the Dean vortices formed within the hub side. The swirling flow motion is only seen at the hub entry side and convected from the outer wall to the rotor entry, and then leaves the volute before a circumferential position of 180 deg from the tongue.

Keywords: Radial inflow turbine, Twin-entry volute, Pulsatile flow, CFD, Dean vortices

NOTATION

N_b	number of blades	
C_{mean}	mean chord	(m)
h_{in}	height of blade at inlet	(m)
P	pressure	(Pa)
P_{ave}	average outlet pressure	(Pa)
Pt_0	inlet total pressure	(Pa)
Pt_{0_ave}	average total pressure	(Pa)
R	radial coordinate	(m)
R_{in}	radius a rotor inlet	(m)

R_{out}	radius a rotor exit	(m)
Re	Reynolds Number	
De	Dean number	
t	time	(s)
Tt0	inlet total temperature	(K)
Tt_{0_ave}	average inlet total temperature	(K)
U	peripheral speed	($m.s^{-1}$)
X	axial coordinate	(m)
X_{rot}	axial rotor length	(m)
θ	circumferential coordinate	(rad)
Ω_{rot}	speed of rotation of rotor	($rad.\ s^{-1}$)

1. INTRODUCTION

Twin-entry volute of a radial turbine converts the static enthalpy of gases into kinetic energy and directs the flow into the rotor at a prescribed flow angle with minimum losses. In contrast to a single-entry volute, a twin-entry volute better uses the pulsating energy of an engine exhaust gases, and by the way the expansion losses into the exhaust manifolds are avoided and the interference between exhaust pulses are reduced (1). Moreover, when using a twin-entry volute, the turbine response at low rotational speed is increased (2). Pischinger and Wunsche (3) studied the interactions between the inlet gas flows from a comparable double and twin-entry volute under the steady conditions, and concluded that both turbines have a better efficiency at full admission, but that of a twin-entry is better with respect to the partial admission. Dale and Watson (4) measured the performance of a symmetrical twin-entry turbine over a wide range of partial admission conditions. They have indicated that the best efficiency is not under full admission conditions, but the point of maximum efficiency occurs when the mass flow at shroud-side is more than hub-side for a ratio $\dot{m}_s/\dot{m}_h$=1.66 to 2, which was later confirmed by Capobianco and Gambarotta (5) ($\dot{m}_s/\dot{m}_h = 7.14$). A large difference between obtained values (7.14 and 1.66-2) is noted and it may be attributed to different shapes of used volutes. Yeo and Baines (6) used laser two-focus velocimetry techniques to examine the flow inside a turbine, and showed (7) that for different operating points and equal admission the incidence angle is close to −30°. Romagnoli et al.(8) compared the performance of a twin-entry and a double-entry of mixed flow turbine and showed that under partial admission, the interaction between the two entries at the interspaces is significant as a consequence of flow leakage from the flowing entry into the non-flowing entry. Also, Srithar Rajoo et al.(9) concluded that at the full admission, the efficiency is not fully representative of the out-of-phase flow condition in the turbine, which at each instant in time is more likely to act as in partial admission conditions. The performances of a twin-entry radial turbine were predicted analytically and experimentally by Aghaali and Hajilouy (1) and Hajilouy et al. (10), under the steady state for the full and partial admission conditions. The obtained results reveal that when the ratio of mass flows between the shroud and hub side is very high, the efficiency becomes lower than that of full admission. Recently, Costall et al. (11) by using a dynamic code developed a 1-D model to evaluate the pulse flow performance of a twin-entry turbine under full and unequal admissions. The results revealed that a twin-entry turbine can be modelled as a simplified single constant diameter pipe with an equivalent volume for the full admission operating conditions, while a more complex model was required for unequal admission condition. Based on the same approach, Chiong et al. (12) developed five 1-D models to investigate the geometrical effects on flow performance under a full admission. Their findings show that the use of variable area ducts had a tendency to magnify secondary fluctuations in mass flow. On the other hand, the model that treats the shroud and hub entry separately until the rotor inlet, gave the best compromise of results.

In the open literature, there is an interest in studying the flow structure through a twin-entry radial turbine under pulsatile flows. Indeed, the pulsatile flows through curved tube lead to complicated secondary flows with flow reversals which are difficult to analyze, in addition to phenomena not yet fully understood, according to Hamakiotes and Berger (13), Kundu et al. (14). Adding to that, the fact that a twin-entry volute splits the exhaust flow arriving from separate groups of engine cylinders until reaching the inlet of rotor which is sensitive to velocity flow angle, and consequently the turbine efficiency is strongly affected (7).

The present paper is another contribution in the study of pulsatile flows through a twin-entry volute considering the engine-like operating conditions, in order to get a better insight into the pulsatile flow effects and the phenomena of volute/rotor interaction. The unsteady three-dimensional flow through the full geometry of the turbine was obtained by using ANSYS-CFX code under engine like conditions. The flow field in terms of streamlines through the volute is analyzed and the relative flow at volute exit and rotor entry show strong interaction effects.

2. GEOMETRY AND GRIDS

Table 1 summarizes the most important parameters for this model of a twin-entry radial turbine, which consists of two spiral-shaped flow ducts separated by a divider into two zones: the hub side and shroud side which are brought together at the interspace. The considered turbine has a non-symmetric twin-entry volute (Figure 1) and was divided into two domains: an inlet domain and a spiral domain. At the inlet domain, there is a surface that remains fixed from the inlet section to the horizontal section, but the angle between the two slots varies. In fact, the inlet domain of casing is twisted. Following the horizontal section the geometry of volute is a spiral. At this domain the angle between the two planes remain constant and only the dimensions of the sides vary according to the spiral shape.

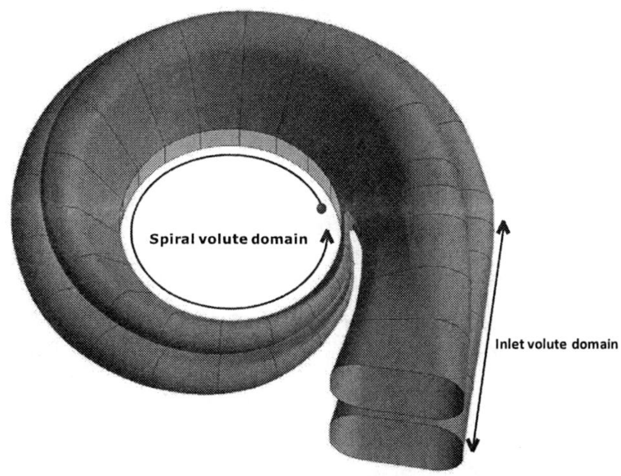

Figure 1: The twin-entry volute

Table 1: Geometry parameters of the twin-entry radial turbine

	Parameter	Designation	Value
Rotor	Number of blades	N_b	12
	Inlet diameter	D_2	96 mm
	External outlet diameter	D_{s3}	86.3 mm
	Internal outlet diameter	D_{h3}	30.5 mm
	Blade hub angle	β_{3h}	54°
	Blade shroud angle	β_{3s}	37°
Volute	Volute inlet area	As/Ah	2 x 2.263 10^{-3}m^2
	Volute outlet diameter	D_v	96.8 mm

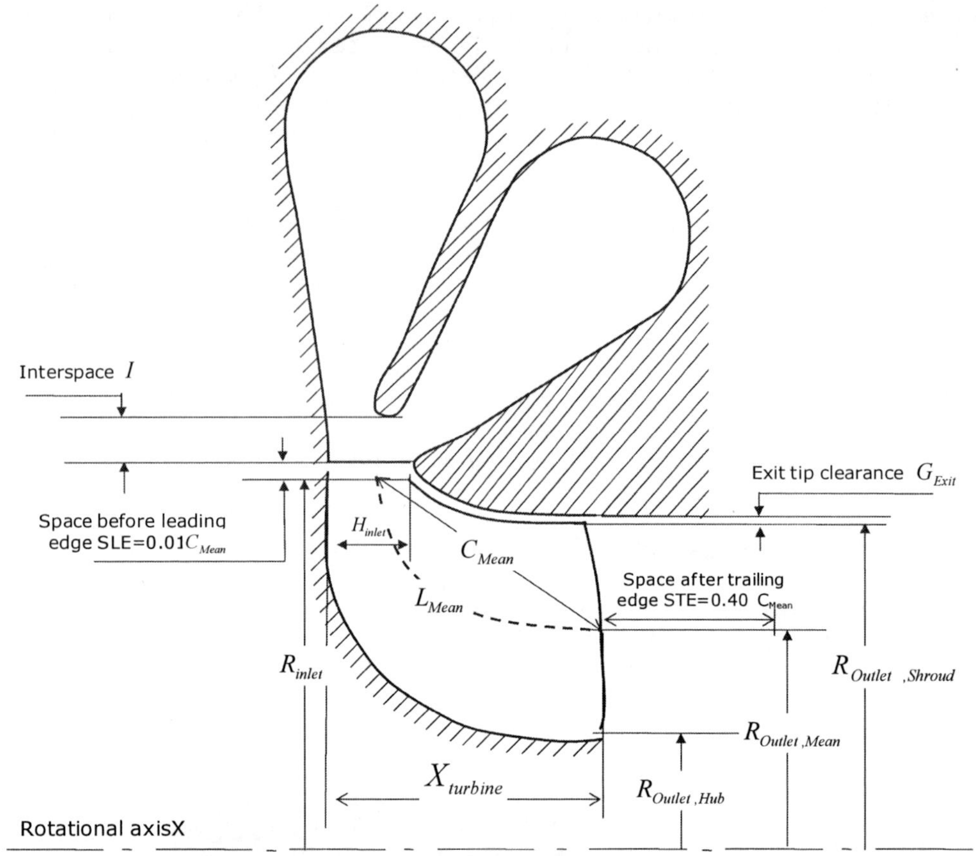

Figure 2: The rotor and volute cross sections

To reproduce the complex geometry of the rotor, 3D optical measurement technique accessed this geometry with a sufficient accuracy. The space before leading edge SLE is related to the clearance between volute exit and rotor blade leading edge (about 0.4 mm), while the space after trailing edge STE is chosen to cover the volume delimited by the rotor axis and casing. Considering the definition of mean chord length $C_{Mean} = \sqrt{(X_{Turbine} - \frac{1}{2}H_{inlet})^2 + (R_{inlet} - R_{Oulet,Mean})^2}$, the spaces before leading edge SLE and after trailing edge STE are equal to 1.25% and 39.9% of C_{Mean}, respectively (Fig.2).

For the twin-entry volute an hexahedron mesh of 1063249 nodes was generated and refined in areas of high curvatures and expected high gradients, particularly around tongue, lip and interface with rotor. In order to achieve a better prediction of boundary layers, ten prism layers were used near the walls with an expansion ratio of 1.15. An H-grid topology was adopted with an O-grid around the rotor blades to minimize the skew angles. In order to guarantee a better boundary layer resolution around the blade, especially at trailing edge and leading edge, ten prisms were generated with an expansion ratio of 1.1. The tip gap height of the present radial inflow turbine is about 0.11 mm, where twelve uniform cells were used in the clearance gap between the tip of rotor and shroud to capture the tip leakage flow. The total number of nodes used to mesh the full blades of rotor is 1599684 nodes.

The good quality of the rotor grid allowed producing a minimum skew angle exceeding 16.58 deg and an aspect ratio less than 100. A mesh size of 265006 nodes (figure 2) was used for the diffuser.

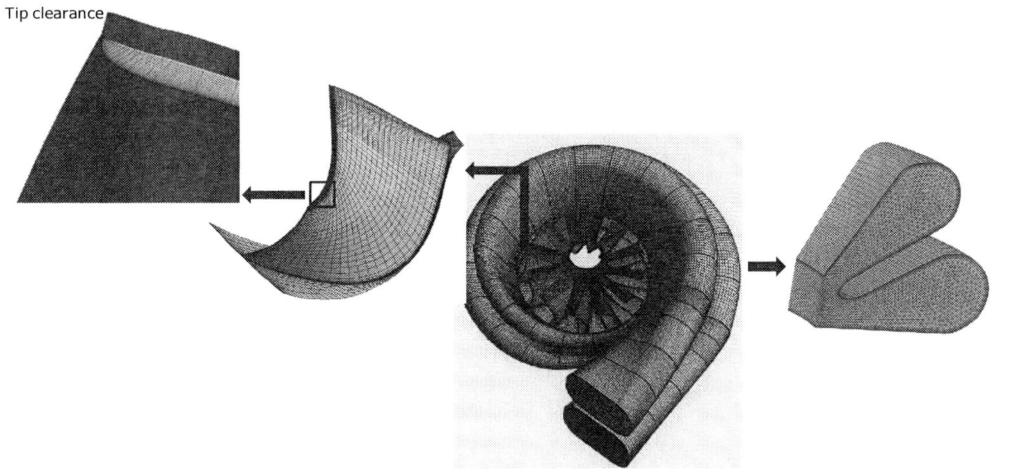

Figure 3: Meshes for volute and rotor

3. CFD MODELLING

3D RANS equations are solved by means of CFX-Solver with mass-averaged velocity and time-averaged density, expressed in fully conservative form is as follows:

$$\frac{\partial \rho}{\partial t}+\frac{\partial}{\partial x_j}\left(\rho u_j\right)=0 \tag{1.a}$$

$$\frac{\partial}{\partial t}\left(\rho u_i\right)+\frac{\partial}{\partial x_j}\left(\rho u_j u_i\right)=-\frac{\partial P}{\partial x_i}+\frac{\partial}{\partial x_j}\left[\mu_{eff}\left(\frac{\partial u_i}{\partial x_j}+\frac{\partial u_j}{\partial x_i}\right)\right]+S_{ui}$$

$$\vec{S}_u=-2\vec{\Omega}\times\vec{U}-\vec{\Omega}\times\left(\vec{\Omega}\times\vec{r}\right) \tag{1.b}$$

$$\frac{\partial}{\partial t}(\rho H)-\frac{\partial P}{\partial t}+\frac{\partial}{\partial x_j}(\rho u_j H)=-\frac{\partial}{\partial x_j}\left(\lambda\frac{\partial T}{\partial x_j}+\frac{\mu_t}{Pr_t}\frac{\partial h}{\partial x_j}\right)+\frac{\partial}{\partial x_j}\left\{u_i\left[\mu_{eff}\left(\frac{\partial u_i}{\partial x_j}+\frac{\partial u_j}{\partial x_i}\right)-\frac{2}{3}\mu_{eff}\frac{\partial u_l}{\partial x_l}\delta_{ij}\right]+\mu\frac{\partial k}{\partial x_j}\right\}+S_E$$

$$H=h+\frac{u_i u_i}{2}+k \tag{1.c}$$

The integration used the finite volume technique using a second order upwind scheme. The first order upwind scheme is generally used to obtain an approximate solution. To account for the frame of reference transfer and pitch change between the volute exit and rotor inlet, the transient rotor/stator interface was selected. The flow was solved for the volute in a stationary frame and in a rotating frame for the rotor. In the present flow simulations, the k-ω SST turbulence model with an automatic wall function was considered. The minimum and maximum values of y+ on the volute and rotor surfaces were in between 0 - 92. When y+ exceeds 2, the solver automatically uses the wall function, whereas the low Reynolds method is used for y+ less than 2. The used time-step for computations was 4.67 micro-second corresponding to a rotation of approximately 1.5 deg which represents 240 time step for one full rotor rotation. In order to reduce the time integration errors in unsteady-state simulations, a time step $\Delta t \sim \frac{\Delta x}{U}$, recommended where Δx is grid spacing and U the front speed. To reduce computer run time and facilitate post-processing in the present study this time step was set equal to $\Delta t = 3\frac{\Delta x}{U}$.

At the inlet domain of volute, the flow is assumed subsonic and pulsatile (the flow is composed of steady and periodic components), therefore the total pressure and temperature in the stationary frame of reference as well as flow direction were imposed. These inlet pressure and temperature were scaled from the experiments by Ehrlich et al. (15)(Fig 4.a-b). At the turbine outlet, where the flow is subsonic, a scaled static pressure (Fig 4.c) is imposed. The frequencies of these signals correspond to the engine working with an open throttle. A constant rotating speed of 60000 rpm was set for this rotor which corresponds to the design speed (16).

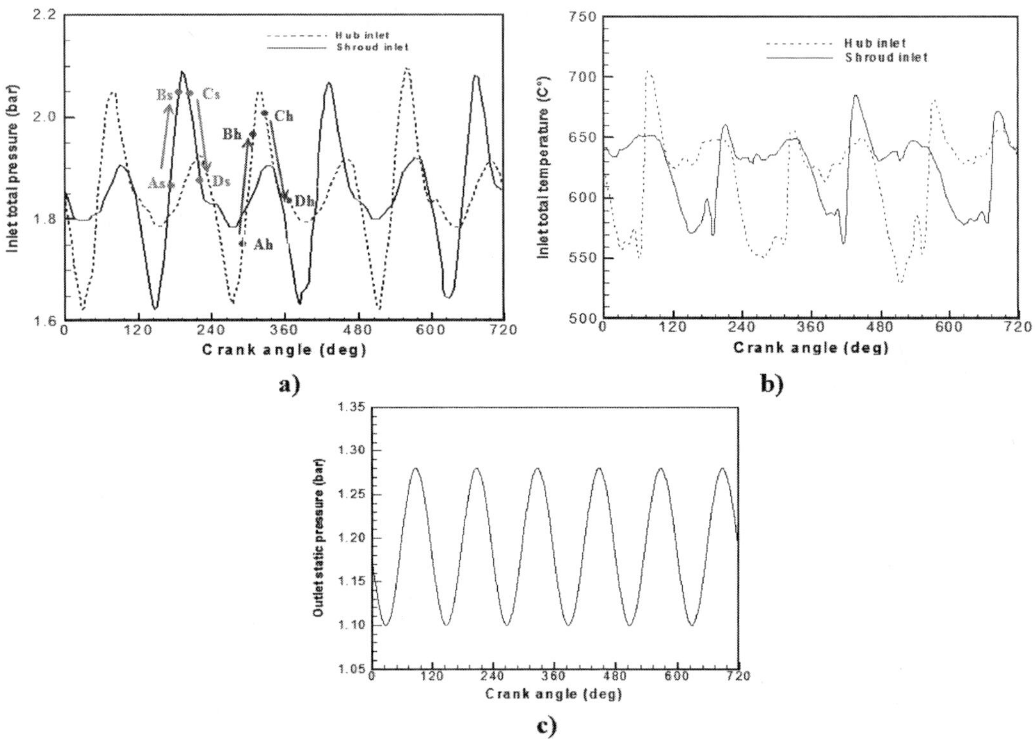

Figure 4: Scaled boundary condition: a) inlet total pressure, b) inlet total temperature, c) outlet static pressure

The pressure boundary condition results in the reflection of the outgoing waves, as a consequence, the interior domain will contain spurious wave reflections. Non-reflecting boundary conditions provide a special treatment to the domain boundaries to control these spurious wave reflections (17) which suggests infinitely long pipes upstream and downstream of rotor. It was shown by Fatsis (18) that the solution was not affected by the upstream location of the boundary. Unfortunately, the non-reflecting boundary conditions are not available for the pressure-based solver implemented in ANSYS-CFX, as consequence an induct pipe and a diffuser of length $2D_r$ and $3D_r$ (D_r represent inlet rotor) were added to the inlet and outlet of twin-entry radial turbine, to avoid any interference produced by the boundaries.

4. FLOW FIELD THROUGH VOLUTE

When the flow enters the bend, the fluid is accelerated near the inner wall and simultaneously, the fluid near the outer wall is decelerated due to adverse pressure gradient which induces a secondary flow in the transverse plane of volute. Furthermore, at downstream of the bend the centrifugal force induces a second flow in the central part of cross section from inner wall to outer wall and forms two counter-rotating vortices. This flow is referred to as Dean vortex flow, and the accompanying pairs of streamwise oriented vortices are referred to Dean vortices (19). A non-dimensional term called Dean number (19) $\mathrm{De} = \mathrm{Re}\sqrt{\frac{\mathrm{D_h}}{\mathrm{Rc}}}$ is used to describe the secondary flow behaviour. Where Reynolds number Re equals UD_h/v, U is the mean (bulk) streamwise velocity, D_h is the aerodynamic diameter, and R_c is the radius of the convex curved surface. The strength of secondary flow is dependent on the dimensions of channel, radius of curvature and flow velocity.

The flow through the twin-entry volute is fully turbulent and pulsatile characterized by a high Reynolds number and the fluid could be expected as a flow through a bend pipe with 180 deg in curvature, except a portion of main flow leaving the volute at each circumferential angle to feed the rotor. Also, the outer radii of the two entries are not constant and take the shape of a spiral. In order to have an estimation of the Reynolds number Re used to determine the Dean number De, two mean planes are plotted in the two entries. For each time step (corresponding to 1.5 deg of the rotor rotation) and for each entry, the average area velocity is evaluated, hence the Reynolds number is calculated.

The outer radius of two entries of volute is variable whereas the inner radius of two exits is constant. Therefore, the radius of the centroid line of the two entries is chosen to represent the radius of the curved surface. The radius of curved surface of hub side is taken as the distance between the centroid line and the rotational axis of rotor and the radius of curved surface of shroud side is taken as the distance between the centroid line and the rotational axis of shroud entry cross section (see Fig.5). The radius of the curved surface is variable; hence, a mean value for each entry side is taken as follows:

$$\bar{R}_{c-h} = \frac{1}{\pi}\int_0^{\pi} R_{c-h,\theta}\, d\theta \qquad 2.a$$

$$\bar{R}_{c-s} = \frac{1}{\pi}\int_0^{\pi} R_{c-s,\theta}\, d\theta \qquad 2.b$$

In the present study, the two aerodynamic diameters D_s et D_h are not constant and decrease with the decrease of the two entries cross sections, thus a mean value is taken to represent the D_s et D_h as follows:

$$\bar{D}_s = \frac{1}{\pi}\int_0^{\pi} \frac{4S_{h,\theta}}{Per_{h,\theta}}\, d\theta \qquad 3.a$$

$$\bar{D}_s = \frac{1}{\pi}\int_0^{\pi} \frac{4S_{s,\theta}}{Per_{s,\theta}}\, d\theta \qquad 3.b$$

Where, *S* and *Per* represent the cross section and its perimeter for each circumferential angle θ.

By analogy to the dean number for a bend pipe and by considering the above parameters, the Dean number for the two entries related to this model of volute is calculated as following

$$\mathrm{De\_h} = \mathrm{Re_h}\sqrt{\frac{\bar{\mathrm{D}}_\mathrm{h}}{\bar{R}_{c-h}}} \qquad \mathrm{De\_s} = \mathrm{Re_s}\sqrt{\frac{\bar{\mathrm{D}}_\mathrm{s}}{\bar{R}_{c-s}}} \qquad 4$$

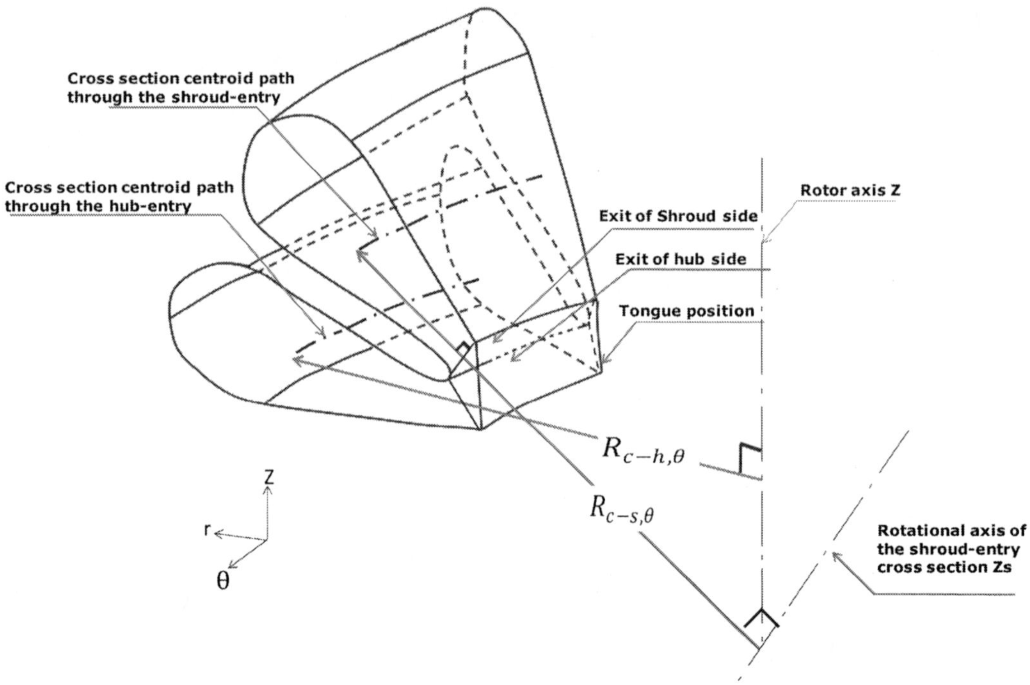

Figure 5: Qualitative length representations for shroud entry and hub entry

Figure 6 represents the variation of the Dean number *De* through the two entries versus the crank angle. At the beginning of simulations, a strong behaviour was experienced. However, the effect of pulses within the turbine implied a dynamic process; thus, the instantaneous developed power was not a simple function of the current boundary condition. Consequently, a lapse of time corresponding to 45 deg of crank angle equivalent to a half of the complete period of pulses was required for these simulations to ensure that such details history is attained and a periodic solution is reached. The obtained Dean number evaluated through the hub entry is shown to be bigger than that of shroud entry since the radius of curved surface of hub entry is the lowest. The mean value of Re and the aerodynamic diameter through the two entries are shown to be approximately the same. The hub entry is susceptible to more instabilities than the shroud entry. In the open literature, the critical Dean number corresponding to the first formation of the two contra-rotating vortices for this particular geometry is not available. However, in the present study, the mean value of Dean number through the hub entry and shroud entry are equal to 50130 and 36150 (Figure 5), respectively. As seen from Fig. 6, the formation of the swirling motion is shown to occur only at the hub entry. Therefore, the critical value of Dean number for this model of volute could be clearly greater than a value of 36000.

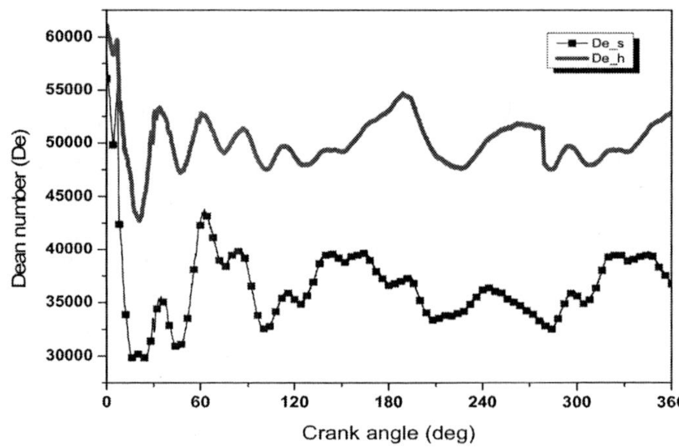

Figure 6: Instantaneous Dean number through the two entries

To identify the flow structure through volute, the streamlines (Fig.7) superimposed to static pressure are plotted for different cross sections at an instant corresponding to a high total pressure within the hub entry (instant Bh Fig.4). The flow in the inter-space near the tongue depicts a clear flow mixing near the divider and the streamlines show a form of billow (position 10 deg and 15 deg) created between the main flow from the two incoming entries and the infiltrated flow due to high differences in velocities between the incoming flow through the tongue and the main flow. Thus, the flow structures in the inter-space between the shroud side and hub side is seen to affect the flow incidence at rotor entry to cause high losses. This billow is revealed to extend approximately to 60 deg after the tongue.

When the flow enters the spiral part of twin-entry turbine, the mechanism of secondary flow is initialized near the outer wall (at positions 10 deg and 15 deg). The flow is subject to inertial forces in the circumferential direction and to centrifugal forces in radial direction. The centrifugal force induced by the curvature is proportional to the square of bulk velocity acting strongly on the fluid close to the volute exit more than the two outer walls, since the highest fluid velocity is at the rotor entry due to converging shape of volute. At the same time, the fluid is moving from the straight part of twin-entry volute to the spiral part and this curvature causes the fluid particles to change their main direction. A sharp velocity gradient develops near the outer wall between the particles positioned in the bulk flow and those located near the outer wall, where the flow velocity is nearly zero. The decrease in the centrifugal force on the fluid near the outer wall leads to a stronger pressure gradient between the outer wall of twin entry volute and that of exit. The flow velocity near the outer wall of twin-entry volute is not significant enough to provide a complete balance of pressure gradient, thus causing a rise to a secondary flow motion superimposed on the primary flow due to an imbalance between the centrifugal force and viscous force. This effect is shown at both entries but is more marked at the hub as the radius of curvature is shorter. The secondary flow started by creating a first vortex near the outer wall of hub entry and the divider at a cross section located at 70 deg from the tongue and this evolution is quickly convected by the main flow to occupy the centre of the cross section of hub entry. On the other side, in the shroud entry cross section, the secondary flow tends to create a swirling motion, and because the centrifugal force is not sufficient to imbalance the cross-stream pressure gradient, the formation of the vortices is not initiated. The fluid in the centre of the cross section is being swept towards the outer wall of the volute and the fluid near the exit volute returns towards the inside of the hub duct curvature, thus, two contra rotating vortices are completely created around 90 deg from the tongue. These two vortices are shown (Fig. 7) to be spherical and not symmetrical, however their main axes are parallel to the low surface and the divider of the twin-entry volute. These two vortices as located at the mean radius of the cross section (90° from the tongue) are stopped before entering the interspaces region. At an angular position of 110 deg from the tongue, the intensity of vortices and their sizes decrease and the streamlines tend to form an "S" shape. The perturbation created by the secondary flow in the shroud entry tends to be dumped and the streamlines become parallel to the divider. At cross sections located between 130 deg and 170 deg, the volume occupied by the secondary flows is reduced and subsequently, the vortices could move from the centre of section (180 deg from the tongue) to the interspaces region. As a result, the flow incidence at rotor entry induces high losses around 180 deg from the tongue.

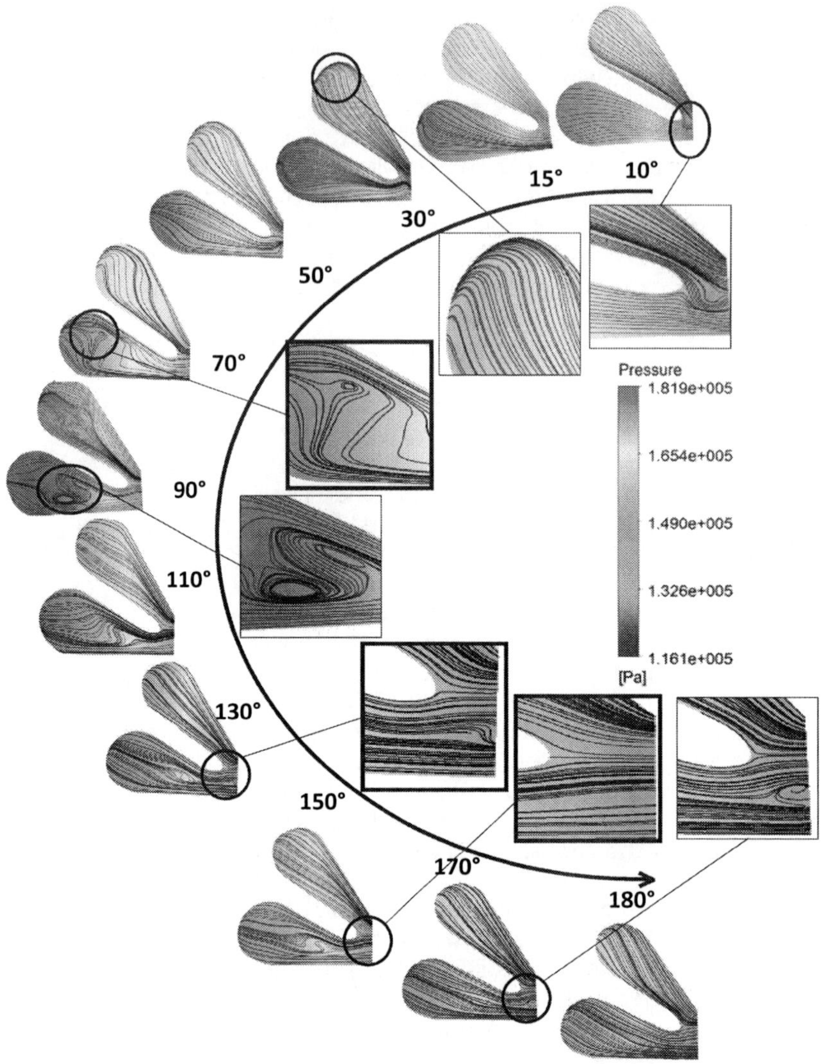

Figure 7: Streamlines superimposed to static pressure at different cross sections of twin-entry volute

In order to expect the effect of rapid accelerations and decelerations inherent in such waveforms, the pulsating flow is divided in two cases according to the pulse state within the two entries, such as a rise time and a descent time (see Fig.4). Table 2 presents the different studied instants and the terminology used in the remaining section.

Table 2: Terminology used in studying the pulse effects

	Through the shroud entry	Through the hub entry
Rise time	$A_s \rightarrow B_s$	$A_h \rightarrow B_h$
Descent time	$C_s \rightarrow D_s$	$C_h \rightarrow D_h$

Figure 8 represents streamlines superimposed to static pressure at selected cross sections of twin-entry volute at different instants which correspond to a pulse through the shroud entry. The selected cross sections (10 deg, 60 deg, 90 deg and 160 deg) which correspond to positions of main changes in the flow structure are chosen to better understand the effect of the pulses. Near the tongue, the form of the billow created due to differences in velocities between the incoming flow through the tongue and main flow is shown to lose its intensity in the rise time of the pulses through the shroud side, whereas in the descent time the form of billow keeps relatively constant. When the pulse is applied to shroud side entry, the form of streamlines through the cross sections for the same position and different instants are shown to be unvarying. These observations are expected since at any time the calculated Dean number reaches a critical number, so, the swirling motion do not initiate.

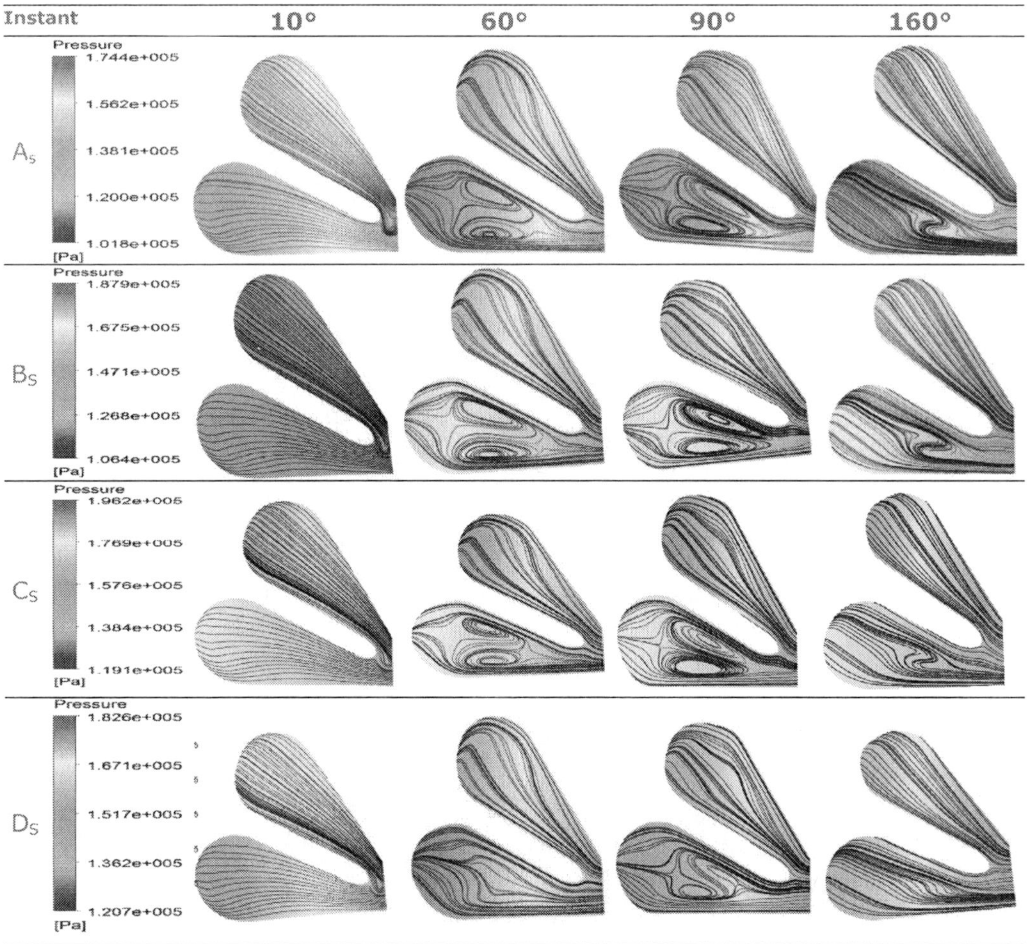

Figure 8: Streamlines superimposed to static pressure at cross sections of twin-entry volute at different instants which correspond to a pulse through the shroud side entry

Figure 9 represents the streamlines at selected cross sections of twin-entry volute at different instants, which correspond to a pulse through the hub entry. The qualitative comparison reveals that the rise time delays and pushes back the formation of Dean vortices to a cross section about 90 deg from the tongue. Inversely, the descent time of the pulse anticipate their formation around a position of 60 deg from the tongue. Also, the streamlines show that the rise time of the pulse accelerates the departure of two miniature vortices from the interspace to the rotor, which are convected before the position of 180 deg from the tongue.

Consequently, the rise time of the pulse decreases the period of residence of the swirling motion through the twin-entry volute. This behaviour may happen since the turbine has insufficient time to evacuate the gas mass within the working pipe before the next pulse cycle. Therefore, due to the pressure wave; an accumulation leads to an increase in the maximum mass flow rate through the twin-entry volute, which may maintain a balance between the centrifugal force and inertial force, thus delaying the initiation of Dean instabilities. Therefore, the rotor evacuates quickly the storage mass characterized by a high pressure within the volume of volute, so, the miniature vortices are convected to the rotor entry before reaching the position of 180 deg.

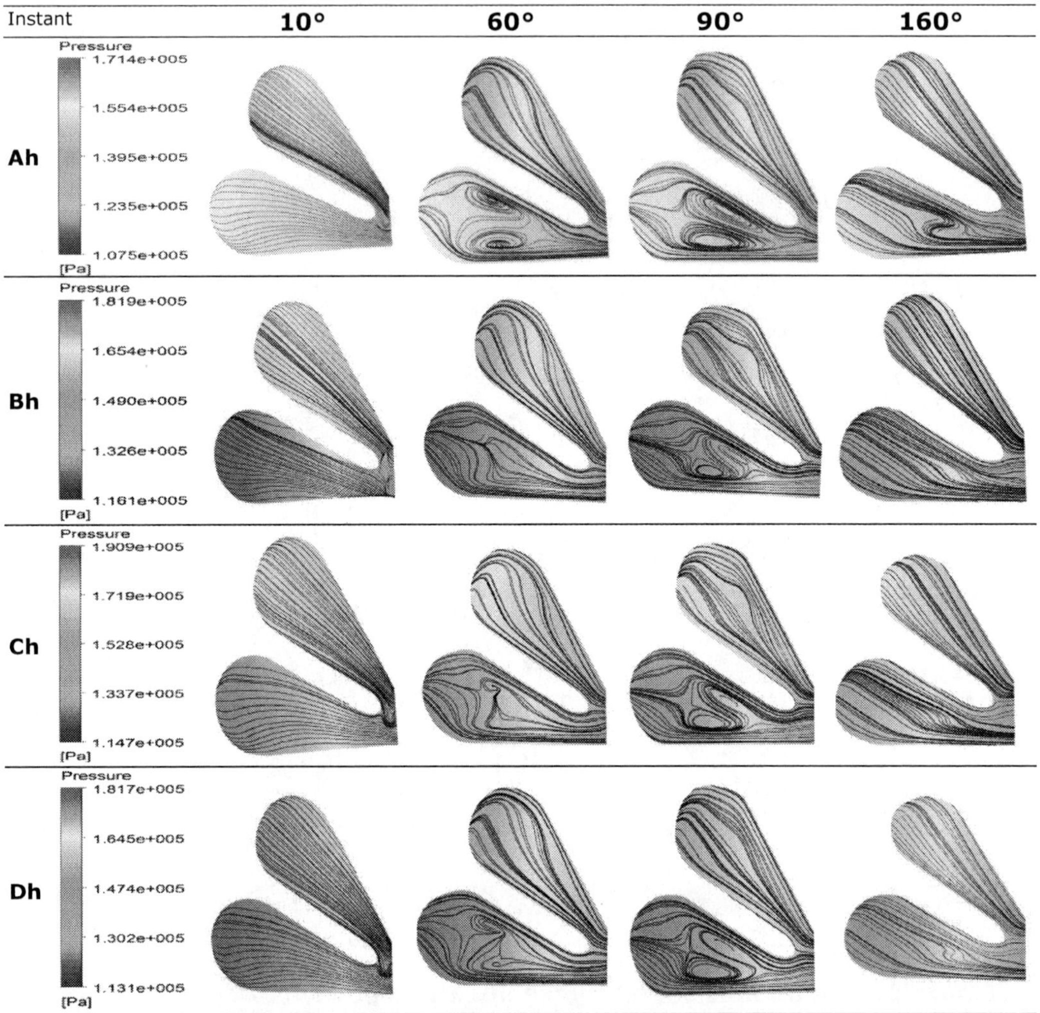

Figure 9: Streamlines superimposed to static pressure at different cross sections of twin-entry volute and at different instants corresponding to a pulse through the hub side entry

5. FLOW FIELD AT EXIT VOLUTE

To investigate the effects of incoming flow of each entry on the flow behaviour at rotor inlet, three circumferential lines are placed at different spans 25%, 50% and 75% of rotor inlet. The geometric blade angle β is set equal zero aligned with radius to the entrance of rotor blade and is measured relatively to this direction. Also, the position 0 deg of circumferential direction is set at the tongue. Figure 10 depicts the relative flow angle at the rotor entry corresponding to the rise time and descent time through the hub side entry. For the different circumferential lines, the relative flow angle is shown to fluctuate between maxima and minima due to reversed flow along the pressure side and because the flow is strongly affected both forward and backward of the tongue. The recording plane was too close to the turbine blade (about 0.4 mm) to detect this effect. The high flow angle around the tongue might be attributed to an inversion in the relative flow direction due to interaction effect between the main flow and that infiltrating through the tongue to feed again the rotor and also due to interaction between the incoming flows through the two entries when entering the spiral part of volute. The flow takes high relative angle around the tongue which is highly influenced by the tongue since at upstream the flow is turned radially, whereas downstream the tongue it coincides with the mixing region, which affects the inlet flow conditions to rotor (Cerdoun et Ghenaiet (20), Miller et al.(21)). Previous investigations related to volute have revealed high levels of non-uniformities (Benisek (24)), as well as significant departures from intended design values of rotor inlet flow angle (Hussain (22)). The relative flow angle has always been known to have a major effect on the radial turbine performance. Baines et al. (7) showed that the optimal flow angle is between 20 and -40 deg, which is in agreement with the measurements of Baines and Yeo (8). Japikse and Baines (26) suggested an incidence angle value of 20 deg to 30 deg as the optimum to suppress the flow separation on both suction and pressure sides of rotor blade, thus minimizing cross passage flow and losses. These different optimum values (7,8,26) may be related to the form of volute used in such studies.

Figures 10-11 show the relative flow angle behaviour as it is strongly related to the pulse state and also to the span position. Figures 10 depict the relative flow angle corresponding to the rise time and descent time through the hub entry. At 25% of span, the relative flow angle moves from a strong fluctuation corresponding to a low total pressure (instant Ah) at the hub entry to a weak fluctuation (instant B_h). The axisymmetric distribution of the flow around the rotor is far from occurring but at the instant B_h (corresponding to high total pressure) the relative flow angle is relatively adjusted. Inversely, through the descent time of the pulse, the relative flow angle tends to recover the high fluctuations. Also, it is noticed that the relative flow angle becomes strongly non-uniform when the pulse passes by its maximum. In addition, at this position the average value of flow angle decreases for both times. At 50% of span, the relative flow angle is affected by the wake downstream the divider and the boundary layer separations from the two sides of divider. The relative flow angle shows approximately the same behaviour except at the rise time where the fluctuations of relative flow angle are largely dumped. At 75% of span, close to the shroud exit of volute, the flow angles are mainly affected by the shroud boundary condition. This latter happens in the rise time, whereas in the descent time, the average value is relatively out of interval. In this zone, it is also noted that the rise phase increases the relative flow angle, but in the descent time it decreases it. Consequently, the twin-entry volute presents a dissymmetrical behaviour as depicted by the high non-uniformities in terms of amplitude at the rotor inlet in both spanwise and circumferential directions. Also, the relative flow angle reveals the Dean vortices effect around the position of 180 deg from the tongue where the incidence is highly influenced by the migration of the two vortices from the hub side entry to the rotor entry.

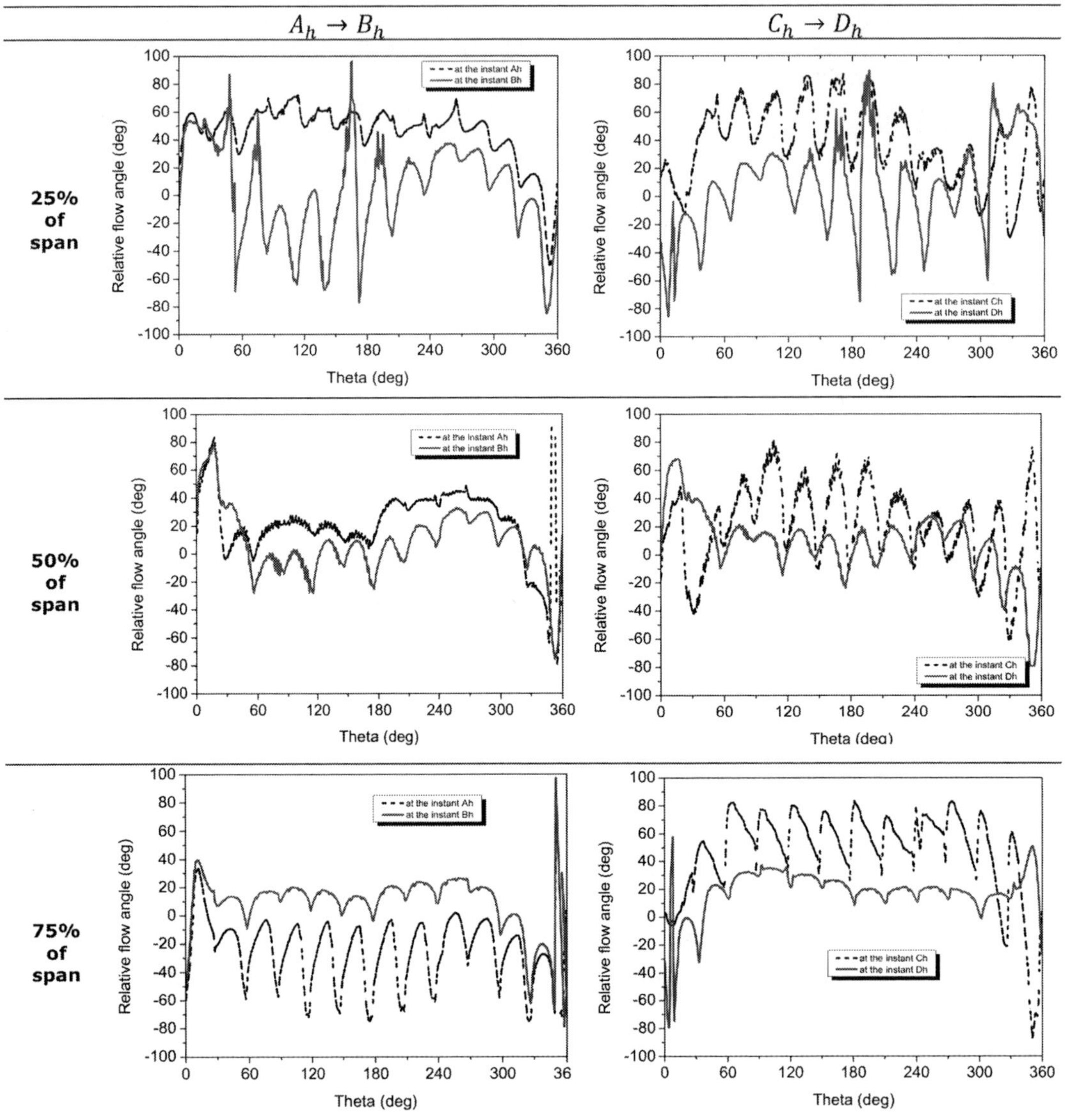

a) Rise time of the pulse through the hub entry Ah-->Bh

b) Descent time of the pulse through the hub entry Ch-->Dh

Figure 10: Relative flow angle at the rotor entry corresponding to a) Rise time and b) Descent time through the hub entry

Figures 11 exhibit the relative flow angle corresponding to the rise time and descent time through the shroud side entry. At 25% of span, the relative flow angle shows a high variation through the rise time which is dumped in the descent time. At the mid of span, the relative flow angle is comprised between -30 deg and 40 deg except around the tongue where it reaches higher values. Near the shroud (75% of span), the high inclined shroud side entry of this asymmetrical twin-entry radial turbine allows the flow entering directly to the inter-space region and then to the rotor, and subsequently destroying the wake generated by the divider. As a result, the flow angle presents a high fluctuation behaviour compared to the case when the twin-entry volute is fed from the hub entry, which maybe results from the interaction between the rotor and the flow emerging from the shroud side.

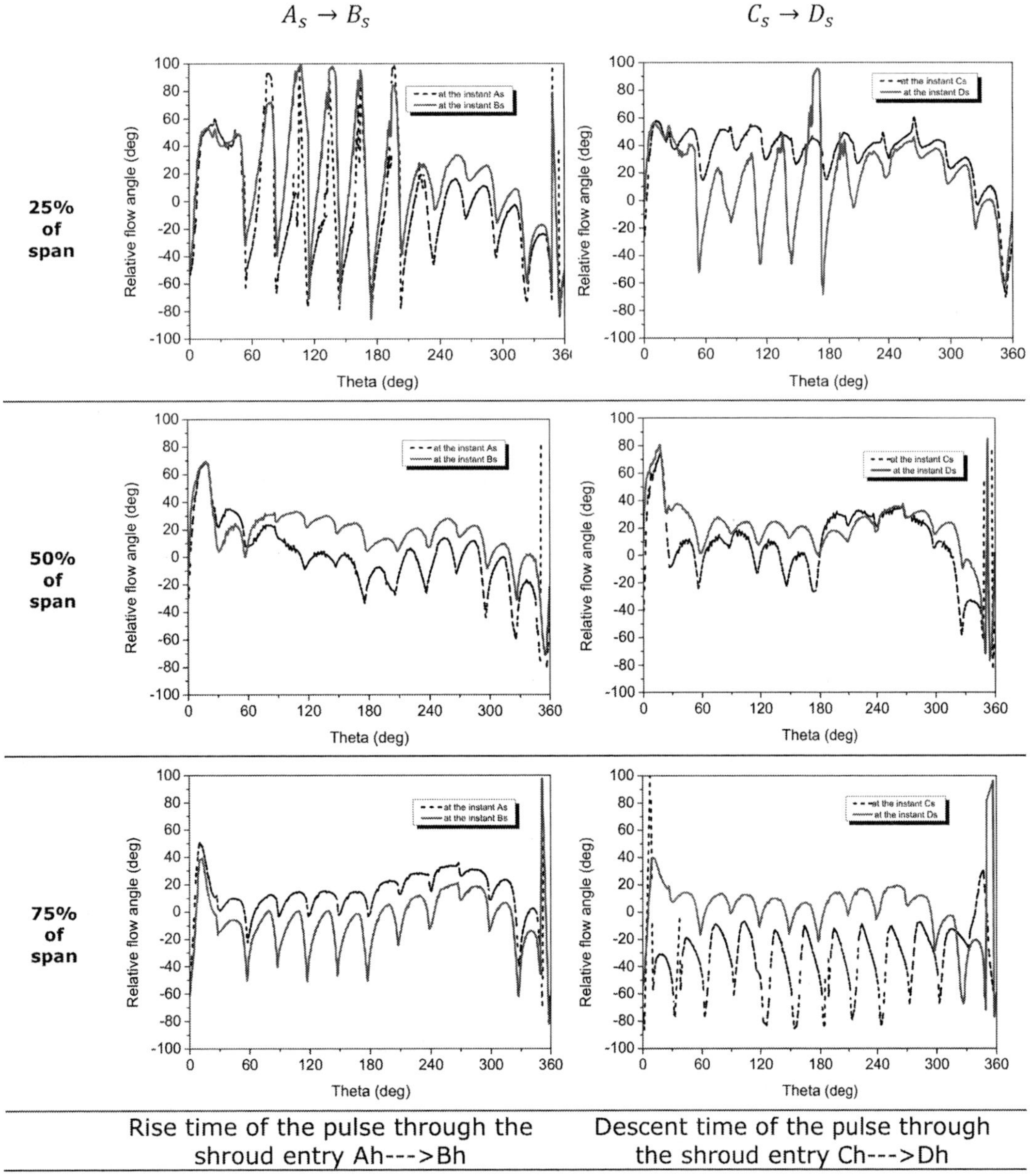

Figure 11: Relative flow angle at rotor entry corresponding to: a) Rise time and b) Descent time through the shroud entry

CONCLUSION

The flow instabilities occurring mainly through the hub entry are shown to be related to the inlet conditions. In order to assess the effect of pulses, they were subdivided into characteristic periods, namely the rise time and the descent time. As revealed the formation of the vortices anticipated around 60 deg from the tongue or delayed around 90 deg, depend on the inlet conditions and the state of the pulse such as rise time and descent time. The secondary flow migrates from the inner wall of volute to rotor entry, leaving the interspace zone before the position 180 deg from the tongue. The variation in relative flow angle reveals that the twin-entry volute presents a dissymmetrical behaviour depicted by high flow non-uniformities at the rotor inlet in both spanwise and circumferential directions. In addition the flow incidence is highly affected around the tongue due infiltrated flow and around 180 deg after the tongue near the hub side due to dean vortices. The behaviour of relative flow angle is strongly related to the pulse state; therefore the incidence angle in the descent time tends to recover high fluctuations than at rise time. In addition, the formation of Dean vortices occurs around a position 60 deg from the tongue through the hub entry, during the descent time of pulse. In contrary, in the rise time of pulses, these phenomenon is shown be delayed to a position of 90 deg from the tongue. These results bring out the effects of secondary flows during the pulses on flow structures of a twin-entry volute. This detailed flow depiction help improving the understanding of unsteady behaviour to reduce undesired volute/rotor interactions and instabilities during the turbine operation.

REFERENCES

1. Aghaali, H. and Hajilouy-Benisi, A., Experimental modelling of twin-entry radial turbine. Iranian Journal of Science & Technology, Transaction B, Engineering, Vol. 32, No. B6, pp 571-584, 2008.
2. Watson N. and Janota M.S., Turbocharging the internal combustion engine, Longman Scientic and Technical Publishing Company, 1982.
3. Pischinger F and Wunsche A., The characteristic behaviour of radial turbines and its influence on the turbocharging process. In: CIMAC Conference, Tokyo, 1977, pp.545–568.
4. Dale A. and Watson N., Vaneless radial turbocharger turbine performance. IMechE, 1986; C110/86: 65–76.
5. Capobianco, M. and Gambarotta, A., Performance of a twin-entry automotive turbocharger turbine, ASME paper 93-ICE-2, 1993.
6. Yeo, J.H and Baines, N.C., Pulsating flow behaviour in a twin-entry vaneless radial-inflow turbine, IMechE paper no. C405/004/90, 1990.
7. Baines, N.C and Yeo, J.H., Flow in a radial turbine under equal and partial admission conditions. IMechE paper no. C423/002/91, 1991.
8. Romagnoli A, Colin D Copeland, Ricardo Martinez-Botas, Martin Seiler, Srithar Rajoo and Aaron Costall, Comparison Between the Steady Performance of Double-Entry and Twin-Entry Turbocharger Turbines, Journal of Turbomachinery ASME January 2013, Vol. 135 / 011042-1.
9. Srithar Rajoo, Alessandro Romagnoli and Ricardo F. Martinez-Botas "Unsteady performance analysis of a twin-entry variable geometry turbocharger turbine" journal of Energy 38 176e189, 2012.
10. Hajilouy-Benisi A, Rad M and Shahhosseini M. R., Modelling of twin-entry radial turbine performance characteristics based on experimental investigation under full and partial admission conditions.Mech Eng; 16: 281–290. 2009.
11. Costall A.W, McDavid R.M, Martinez-Botas R.F, Baines N. C., Pulse performance modeling of a twin-entry turbocharger turbine under full and unequal admission. J Turbomach 2011; 133(2). pp. 021005-1–9.
12. M.S. Chiong, S. Rajoo, R.F. Martinez-Botas, A.W. Costal, Engine turbocharger performance prediction: One-dimensional modelling of a twin-entry turbine, Journal of Energy Conversion and Management 57 68–78(2012).
13. Kundu, K. P., Cohen, M. I. & Dowling, R. D. 2012 Fluid mechanics. Elsevier Inc., 5th ed.
14. Hamakiotes, C.C. & Berger, S.A. 1988. Fully developed pulsatile flow in a curved pipe. J. Fluid Mech. 195, 23–55.
15. Ehrlich, D.A., Characterisation of Unsteady on-Engine Turbocharger Turbine Performance, Doctoral Thesis, Purdue University 1998.
16 Cerdoun M. and Ghenaiet A., CFD analyses of a radial inflow turbine, 8th International Conference on Compressor and their Systems, pp635-646, 2013.
17. Givoli, D., Numerical method for problems of infinite domain, Elsevier, 1992.

18. Fatsis. A., Three dimensional unsteady flow calculations in radial components, Von Karman Institute Lecture Series, 'Spacecraft Propulsion', 1993-01.
19. Dean, W. R. 1928 The stream-line motion of fluid in a curved pipe. Phil. Mag. 5, 671–695.
20. Cerdoun, M., Ghenaiet, A., Analyses of steady and unsteady flows in a turbocharger's radial turbine, Journal of process Mechanical Engineering. vol.229(2), pp130-145, 2015.
21. Miller, E. C., L'Ecuyer, M. R., and Benisek, E. F., 1988, "Flow Field Surveys at the Rotor Inlet of a Radial Inflow Turbine," ASME J. Turbomach., 110, pp. 552–561.
22. Benisek, E.F., and Spraker, W. A., 1987, "Laser Velocimeter Measurements at the Rotor Inlet of a Turbocharger Radial Inflow Turbine," ASME Third International Symposium on Laser Anemometry, Vol. 55.
23. Hussain, M., and Bhinder, F.S., 1984, "Experimental Study of the Performance of a Nozzle-Less Volute Casing for Turbocharger Turbines," SAE Paper No. 840571.
24. Japikse D, Baines N.C., Introduction to turbomachinery. 1st ed. Concepts ETI Inc. Oxford University Press, Vermont, 05055, USA; 1994.

FOR PUBLICATION ONLY

Evaluation method of the exhaust system of the turbocharged diesel engine based on exergy and enthalpy analyses

C C Ma[1], L W Sun[1, 2], N Fang[3], M X Qi[1]
[1] School of Mechanical Engineering, Beijing Institute of Technology, China
[2] Automotive Engineering Institute, Dezhou University, China
[3] Technical Center of Dongfeng Automobile Company, China

ABSTRACT

In this paper, thermodynamic analysis of the engine exhaust process is carried out on a turbocharged inter-cooled diesel engine. The exergy analysis method is proposed to evaluate the energy quality transfer efficiency of the exhaust from the inner side of exhaust valve of diesel engine to turbine inlet. On the other hand, enthalpy analysis method is adopted to evaluate the conversion process of thermal energy to power in the turbine. Several evaluation analysis cases of a turbocharged engine are carried out by means of simulation. The results show that the design of engine exhaust system has influences on exhaust gas flow process in exhaust system and in turbine in different ways. In order to obtain most power from the exhaust gas of engine by the turbine under certain condition, the impacts of both exergy loss in exhaust system and power conversion efficiency have to be considered at the same time. That means a trade-off point has to be found.

NOMENCLATURE

e_x	Specific exergy of the exhaust [kJ/kg]
h_x	Specific enthalpy of the exhaust [kJ/kg]
h_0	Specific enthalpy of the exhaust in the environment condition[kJ/kg]
T_0	Environment temperature [K]
s_x	Specific entropy of the exhaust[kJ/kg·K^{-1}]
s_0	Specific entropy of the exhaust in the environment condition[kJ/kg·K-1]
m	Total exhaust flow of one operating cycle of the engine[kg]
Δh	Specific enthalpy drop [kJ/kg]
c_p	Specific heat capacity at constant pressure [kJ/kg·K^{-1}]
ΔT	Gas temperature difference [K]
c_{p1}	Specific heat at constant pressure and the temperature of T_1 [kJ/kg·K^{-1}]
c_{p2}	Specific heat at constant pressure and the temperature of T_2 [kJ/kg·K^{-1}]
E_{xt}	Exergy of the exhaust gas at turbine inlet [kJ]
E_{xc}	Exergy of the exhaust gas in cylinder [kJ]
$H_{t,in}$	Enthalpy of the exhaust gas at turbine inlet [kJ]
ΔH	Enthalpy drop across the Turbine [kJ]

1 INTRODUCTION

Turbochargers convert thermal energy of engine exhaust gas to power, which flows through the exhaust system. Therefore, the performance of turbocharged engine depends a great extent on the design of the engine exhaust system, besides the design of turbocharger[1].And a major concern in the design of exhaust system is the conversion ratio from the exhaust available energy to the turbine's power. This is closely related to the turbine's flow area, the timing of the exhaust valve, the size of exhaust manifold's diameter, length and so on. The proportion of exhaust energy's utilization by turbocharger reflects the design level of the exhaust system. The exhaust system refers to the pipelines from the engine's exhaust valve to the turbine. In recent years, the turbocharged engine's exhaust energy has been given more and more attention [2-7]. However a universal evaluation method has not been put forward for the optimal design of exhaust system. In this paper, a turbocharged six-cylinder diesel engine is taken as the research object, and the exergy-enthalpy analysis method is proposed to evaluate the exhaust energy utilization of the turbocharged diesel engine. In detail, the exergy analysis is applied to evaluate the energy transfer efficiency of the exhaust from the exhaust valve of diesel engine to the turbine inlet, and, the enthalpy analysis to evaluate the exhaust energy transfer efficiency respectively. From the simulation results of the engine by GT-power under some working condition, the datum of exergy and enthalpy at different position of the exhaust system and turbine are obtained by software REFPROP. Through analysis of the datum, some relationships between exergy, enthalpy, and the size of exhaust system has been got, which may give a direction of exhaust system design.

2 ANALYSIS OF GAS ENERGY IN EXHAUST SYSTEM

Since there is no power converted from the thermal energy contained in exhaust gas flowing through the exhaust pipes, the major concern is the loss of the power capability. So the exergy analysis was carried out in the exhaust pipes. In the turbine system, the only power component is the turbine. And the only way for the gas to work is doing work by expanding to the environment pressure, whereas at the same time, the temperature may not be equal to that of the environment. This makes the real power ability of the exhaust in the turbine is not equal to the exergy. That is to say, the exergy change does not represent the turbine's heat-power conversion. The decrease of exergy might be due to the exhaust working, also might be due to the entropy production caused by the irreversible process. And the turbine's adiabatic efficiency varies with the change of the exhaust state at the turbine's inlet, and the gas state at the turbine's outlet does not balance with the environment. Therefore, the exergy is not suitable to evaluate the turbine's energy utilization. As a result, the exergy analysis method is proposed to evaluate the energy transfer efficiency of the exhaust from the exhaust valve of diesel engine to the turbine inlet, and the enthalpy analysis method was adopted to calculate the turbine's heat-power conversion efficiency from the turbine inlet to outlet.

2.1 The exhaust gas's exergy analysis in the exhaust pipes

In this paper, the turbocharged six-cylinder diesel engine was the research object, and its exhaust process was analyzed based on the thermodynamics. The environment temperature T_0 was set to be 298.15K, and the environment pressure P_0 was 0.1013MPa. In the process of the exhaust energy's calculation, the exhaust energy in the engine at the time of the exhaust valve opening was set to be the base point. The method of exergy analysis calculated the population distribution of the exhaust energy loss in the exhaust system from the exhaust valve, to design the exhaust pipes and to evaluate the design.

Based on a dynamic model with the assumption of one dimensional fluid mechanics, Numerical simulations in GT-power are implemented by means of Finite Volume Method (FVM). Not only all engine modules but also some detailed modules for engine key working conditions are included in GT-power. In this simulation environment, different engine models could be built to simulate the variations of engine performance parameters and solve many problems in the engine development and improvement processes.

The engine's exhaust is the products of the fuel combustion reaction, and is the gas mixture. At the condition of ideal combustion, if the excess air coefficient is greater than one, the main components in the fuel gas are CO_2、H_2O、O_2、N_2, and other trace products in the actual combustion are exclusive. The exhaust gas is assumed as the steady, high-temperature and mixed flow to calculate the available exhaust energy.

In the general analysis of the engine cycle, the exergy analysis is generally necessary, and the exergy is the evaluation parameters of energy utilization. Under certain environment condition, the greatest proportion of energy which can be converted into useful work is the exergy. Perhaps when the thermodynamical system reversibly changes from any status to the status which is balanced with the environment, and only interacts with the environment, the greatest useful work of the thermodynamical system is called to be the exergy of the system. Here the concept of exergy was used in the calculation of the exhaust exergy[8], that is:

$$e_x = (h_x - h_0) - T_0(s_x - s_0) \tag{1}$$

Eq. (1) represents the exergy of unit mass exhaust. The exergy represents the ability of converting the thermal energy into power and the degree of availability technically, and evaluates the quality of the energy[9].Therefore, the method of exergy analysis may be adopted in the energy analysis of the exhaust system.

As the value of exergy at any position of the engine fluctuates with time or crankshaft angle, a whole working circulation of four stroke of the engine is taken into account to calculate exergy, that is 720° of crankshaft angle, as shown in formula (2).

$$E = m\int_0^{720} e_x(\phi)d\phi \tag{2}$$

2.2 The exhaust energy's enthalpy analysis in the turbine

The enthalpy is a state function representing the material energy of the system in the thermodynamics, and is usually expressed by H, as shown in Eq. (3).

$$H = U + pV \tag{3}$$

One kilogram working medium of the enthalpy is called specific enthalpy, and that is expressed by h, as shown in Eq.(4).

$$h = u + pv \tag{4}$$

The exhaust's enthalpy represents the power capability of the exhaust. Enthalpy drop signifies the decrease of the exhaust's power ability, and is basically equal to the turbine's work. In the turbine, the exhaust flows into the turbine from the inlet, and then applies the expanding power. In the system, the relative change of the enthalpy is necessary to be considered, and it is called the enthalpy drop (ΔH).

In this paper, with the high temperature and low pressure, the exhaust gas is reasonable to be regarded as the ideal gas.

For the ideal gas, the intermolecular forces and the internal potential energy do not exist, and the thermodynamic energy only includes the internal kinetic energy depending on the temperature. Therefore, the thermodynamic energy of the ideal gas is the single value function of temperature. From Eq. (5), in the ideal condition, the enthalpy value is also a single value function of temperature, and is not related with the pressure.

$$\Delta h = c_p \Delta T \tag{5}$$

In this paper, the exhaust's enthalpy drop of an operating cycle in the turbine was calculated, and the calculation formula was shown in Eq. (6).

$$\Delta H = m\left(c_{p1}T_1 - c_{p2}T_2\right) \tag{6}$$

3 THE ANALYSIS OF THE EXHAUST ENERGY TRANSFER PROCESS

3.1 The exergy analysis on the exhaust energy of three kinds of exhaust system

The composition of exhaust gas was set by the software REFPROP: the mass fractions of CO_2、H_2O、N_2、O_2 respectively are 9.29%、3.3%、74.68% and 12.73%. The values of temperature and pressure are obtained by simulation, and the density, enthalpy and entropy in this condition are able to be calculated by REFPROP automatically. Therefore, the thermodynamic parameters of exhaust were calculated by this method in this paper.

The thermodynamic model diagrams of the three kinds of exhaust system in turbocharged engines are shown in Fig.1, Fig.2 and Fig.3 respectively.

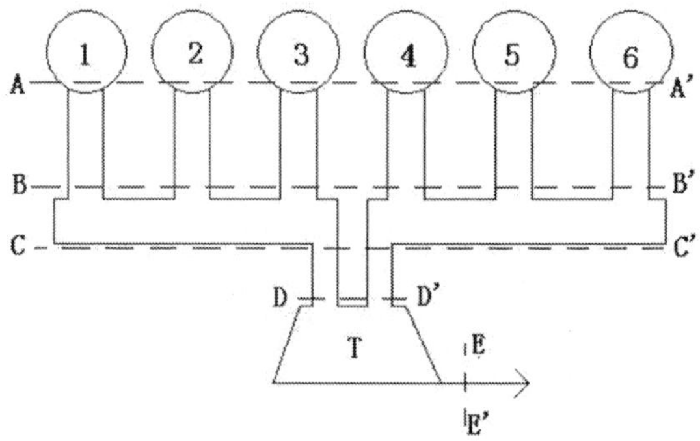

A-A': the ahead cross section of exhaust valve;
B-B': the cross section of exhaust manifold;
C-C': the cross section of exhaust main;
D-D': the cross section of the turbine inlet;
E-E': the cross section of the turbine outlet

Fig.1: Thermodynamic model diagram of original pulse exhaust system

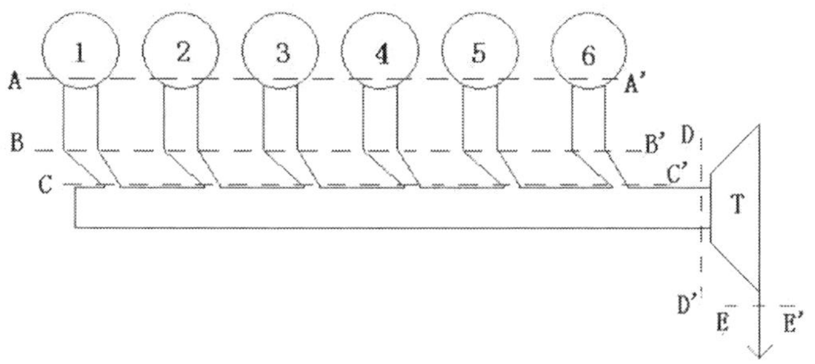

A-A': the ahead cross section of exhaust valve;
B-B': the cross section of pulse nozzle inlet of exhaust manifold
C-C': the cross section of pulse nozzle outlet of exhaust manifold and entering the main;
D-D': the cross section of the turbine inlet;
E-E': the cross section of the turbine outlet

Fig.2: The thermodynamic model diagram of the MPC exhaust system

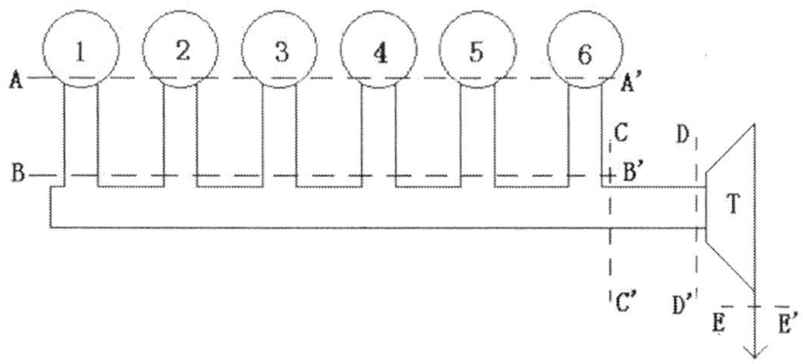

A-A': the ahead cross section of exhaust valve;
B-B': the cross section of exhaust manifold;
C-C': the cross section of exhaust main inlet entering from Cylinder 6;
D-D': the cross section of the turbine inlet;
E-E': the cross section of the turbine outlet

Fig.3: The thermodynamic model diagram of the constant pressure exhaust system

The engine simulation model is usually divided into several are both independent and interrelated modules. Without regard to the air filter and muffler, the original engine model established in GT-power is mainly composed of some basic modules such as cylinder, crankshaft, valve, injector, compressor, turbine and pipelines. For MPC exhaust system, the original engine exhaust system is modified with no change in the structural parameters of both intake system and turbocharger. The ejector nozzles of MPC exhaust system are represented by conical nozzles with different diameters of both ends. For constant pressure exhaust system, the size of exhaust manifold's entrance is equal to the size of cylinder's exhaust outlet. Since the cylinder design has been confirmed, from the perspective of effective utilization of exhaust gas, shorter exhaust pipe should be taken to reduce the exhaust resistance.

The hypothesis is that the flow in the exhaust pipes was quasi steady flow, the gas state in the exhaust pipes was uniform, and no gas leakage. The energy calculation of the system was carried out by the instantaneous integrals in the exhaust process. The operating conditions of calculation were the rated speed (2100r/min) and the speed at maximum torque (1500r/min) separately. By taking the exergy of the exhaust in an operating condition, the thermodynamic calculation results of the three models were obtained.

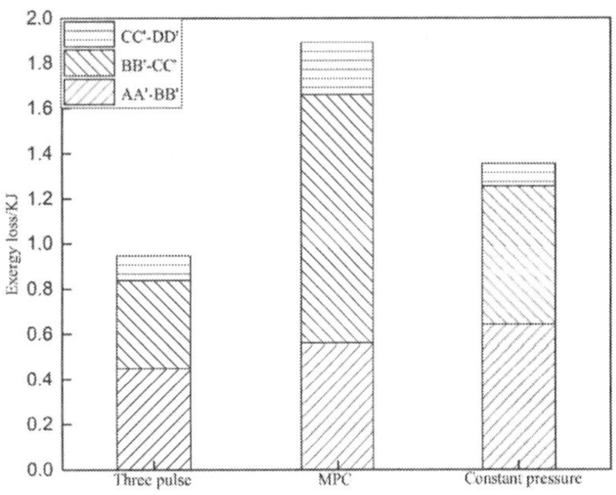

(a) Engine speed 1500r/min

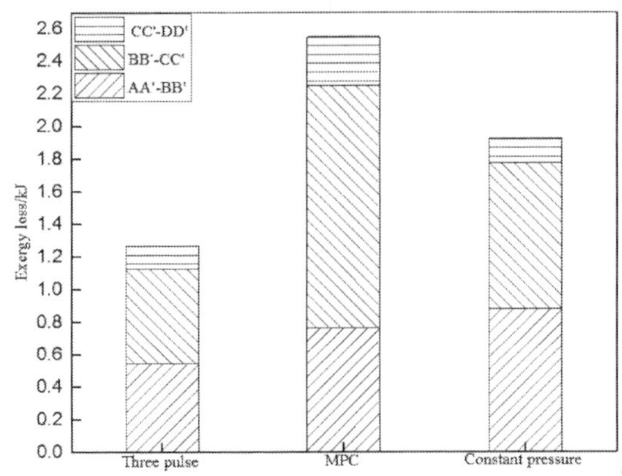

(b) Engine speed 2100r/min

Fig.4: Exergy loss calculation results

Fig.4 indicates the variation conditions of the exhaust's exergy loss of different exhaust systems at two different engine speeds. Judging from the entire calculation results of the exergy loss, the exergy loss of three pulse system in each stage are basically consistent.

For constant pressure system, the exhaust loss at section AA'-BB' was mainly the exergy loss at the exhaust valve and flowing into the manifold, and the loss of constant pressure was the highest at this section. At the opening of exhaust valve, the high pressure difference of both sides led to large throttling loss, therefore, most energy lost at this section. The constant pressure system's loss at section BB'-CC' was mainly the loss produced in the process from exhaust manifold to exhaust main, and was relatively small. In constant pressure system, six exhaust manifolds were connected to a main, and the exhaust entered into the main through the exhaust manifolds. Therefore, the exhaust's interference effect between each cylinder was strong. That is the main reason for the loss in constant pressure system larger than that in three pulse system.

For three pulse system, at section AA'-BB', the small diameter of exhaust pipes enabled the pressure difference in the exhaust pipes and cylinder not to be large at the moment of the exhaust valve opening. Therefore, the exergy loss of the three pulse system at this section was the lowest of the three kinds of exhaust systems. For three pulse system's loss at section BB'-CC', the exhaust pipes were divided into two parts, so the exhaust interference, energy loss and the energy interference of the exhaust pulse were all small. Therefore, the loss in this system was the smallest. Generally speaking, the total exergy loss of three pulse system from section AA' to section DD' was the smallest.

In MPC system, at section AA'-BB', the throttling effect of the pulse converter enabled the pressure in exhaust pipes to build up instantly ,and the pressure difference before and after exhaust valve was not high, therefore, the exergy loss was relatively low. The energy loss of MPC system at section BB'-CC' was mainly from the pulse nozzle. In MPC system, the exhaust accelerated through the pulse converter between manifolds and main. The pulse converter was equivalent to a convergent nozzle, and the pressure difference of both ends was large, so the throttling loss at this section was large. Therefore, the exergy loss of MPC system was the largest at this section. Section CC'-DD' was from the exhaust main to the turbine. The energy loss of the exhaust in this section was small. The flow impact and flow loss have been performed in the previous section, and the loss of friction and heat transfer generated in this section. In MPC system, the acceleration for the flow performed at the previous section, so the exhaust velocity at this section was the largest, and the loss of friction and heat transfer was relatively large. Therefore, in MPC system, the exergy loss at this section was the largest.

3.2 The enthalpy analysis on the exhaust energy of three kinds of exhaust system

The operating conditions of calculation were the rated speed (2100r/min) and the speed at maximum torque (1500r/min) separately, and the thermodynamic calculation results of the three models were obtained.

Fig. 5 indicates the exhaust enthalpy drop of three kinds of exhaust systems at different engine speed. The enthalpy drop in constant pressure was highest, pulse exhaust system second, and the MPC system the lowest. The value of enthalpy drop represents the ability of turbine power. The larger enthalpy drop indicates the more turbine power, and the more transition from exhaust's available energy to the useful power for turbocharging.

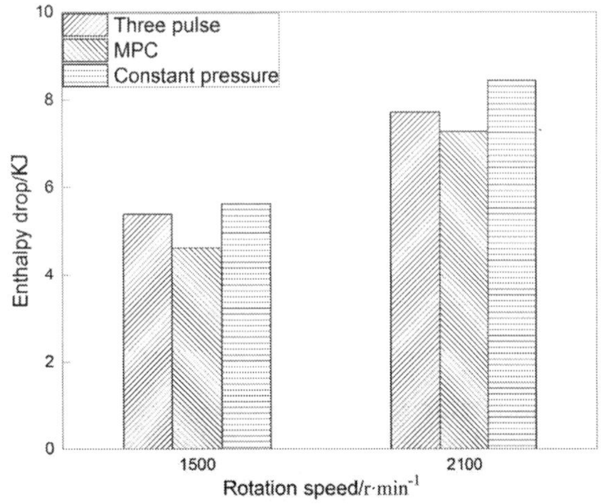

Fig.5: Enthalpy drop calculation results

Fig.6 and Fig.7 indicates the fluctuation of turbine inlet pressure and turbine efficiency of three kinds of exhaust systems at different engine speeds respectively. For constant pressor exhaust system and MPC exhaust system, the turbocharger models are in single scroll. And for pulse exhaust system, the turbocharger model is in twin scroll. In constant pressure system, the slight fluctuation of exhaust pressure entering the turbine enabled the turbine keeping high operating efficiency. In three pulse system and MPC system, the more severe fluctuation enabled the turbine efficiency more fluctuant.

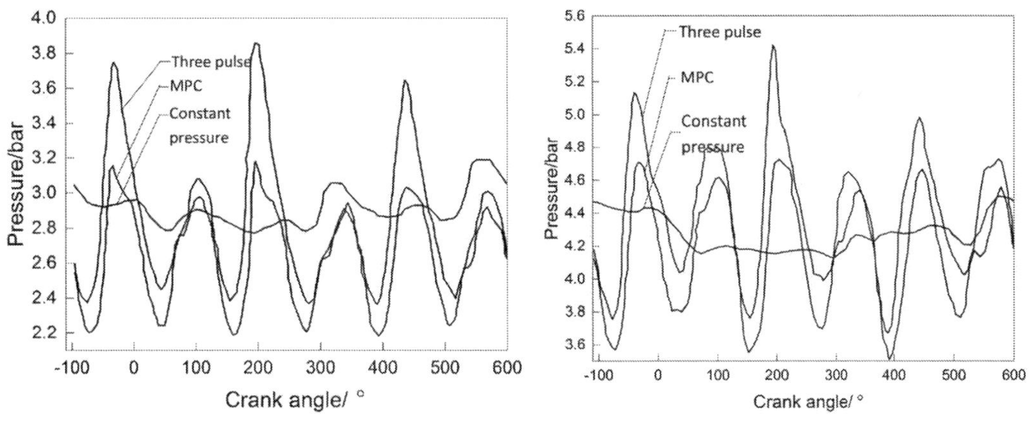

(a) Engine speed 1500r/min (b) Engine speed 2100r/min

Fig.6: Exhaust pressure at turbine inlet

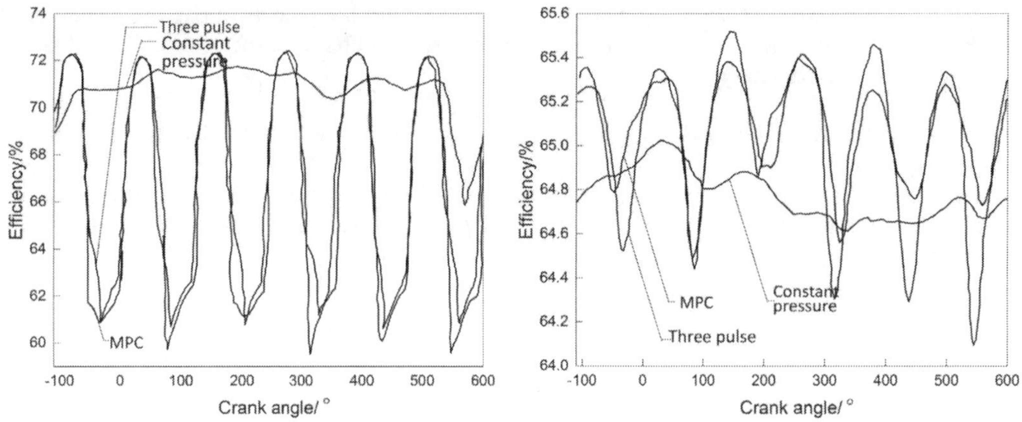

(a) Engine speed 1500r/min (b) Engine speed 2100r/min

Fig.7: Turbine efficiency

4 EVALUATION INDEX AND EVALUATION METHOD OF THE EXHAUST SYSTEM

4.1 Evaluation index

An available exergy coefficient is defined to evaluate the design of the exhaust system composed of the exhaust valve and the exhaust pipe, as shown in Eq.(7).

$$\Phi_e = \frac{E_{xt}}{E_{xc}} \tag{7}$$

Table 1 shows the available exergy coefficients of three different exhaust systems. A larger exergy coefficient does mean less energy dissipation of the exhaust system, more energy transferred to the turbine inlet and a higher transfer efficiency. Nevertheless, influences of the gas pulsation of the turbine inlet exhaust gas on the turbine efficiency should be considered to determine the best exhaust system.

Table 1: Exhaust system exergy coefficient

GT simulation model	Three pulse system		MPC system		Constant pressure system	
Rotational speed(r/min)	1500	2100	1500	2100	1500	2100
Available exergy Coefficient(-)	0.8740	0.8965	0.7674	0.8081	0.8376	0.8598

An turbine enthalpy drop coefficient is proposed to indicate the energy converted to the compressed air power and evaluate the turbine design, as shown in Eq.(8).

$$\Phi_t = \frac{\Delta H}{H_{t,in}} \tag{8}$$

Table 2 shows the turbine enthalpy drop coefficients of three exhaust systems. A larger enthalpy drop coefficient means a higher utilization ratio of the exhaust gas. Three pulse system is most suitable for this six-cylinder turbocharged diesel engine.

In conclusion, not only a high utilization ratio of the turbine available energy but also a high transfer efficiency of the exhaust valve and exhaust pipe system could achieve a high efficiency of the exhaust system.

Table 2: Exhaust system enthalpy drop coefficient

GT simulation model	Three pulse system		MPC system		Constant pressure system	
Rotational speed(r/min)	1500	2100	1500	2100	1500	2100
Turbine enthalpy drop Coefficient (-)	0.2054	0.2762	0.1807	0.2606	0.2071	0.2037

4.2 Evaluation method and process

The exhaust gas utilization analysis method combining exergy analysis with enthalpy analysis takes advantages of pressure and temperature. And the method is easy to operate, and is practical.

The evaluation method and process are described as following:

(1) Test position parameters, including pressure, temperature and mass flow rate, could be obtained by means of experimental or simulation method for different engine speeds. The higher available exergy coefficient and turbine enthalpy coefficient mean the exhaust gas energy can be better utilized. The proposed energy evaluation method could help to select the suitable exhaust system and has a guide for the optimal design of the exhaust system.

(2) The exhaust gas exergy of the selected cross-sections can be calculated from Eq. (1) and Eq. (2). The exhaust pipe position that has the largest available energy loss.

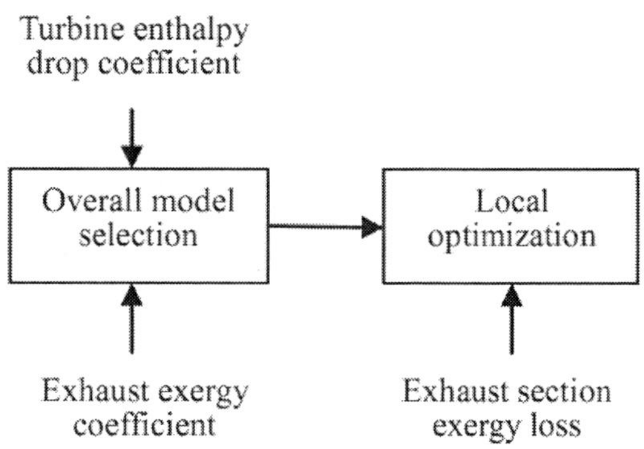

Fig.8: Evaluation process of the exhaust system

5 CONCLUSIONS

(1) From the calculation results of the exergy and exergy loss, the exhaust energy of pulse exhaust system are less than other two kinds of exhaust systems', and the exhaust energy of constant pressure system is the highest. Judging from the trend of the entire curve, the curve of the pulse system's exergy is relatively smooth, and the exergy losses in each stage are basically consistent.

(2) In constant pressure exhaust system, the turbine's enthalpy drop is the highest, this indicates that the transition from exhaust's available energy to the useful power for turbocharging is the most. In constant pressure system, the slight fluctuation of exhaust pressure entering the turbine enables the turbine keeping high operating efficiency. In three pulse system and MPC system, the more severe fluctuation enables the turbine efficiency more fluctuant.

(3) By calculating for the whole and part of exhaust system using the method of combining with exergy and enthalpy, the evaluation criterion of the exhaust system, the available exergy coefficient and the turbine enthalpy drop coefficient, are proposed. The evaluation method of exhaust system is developed and hoped to provide an effective guide for overall selection and local optimization design of exhaust system.

ACKNOWLEDGMENT

This research was supported by National Natural Science Foundation of China (NSFC, 51276018).

REFERENCE LIST

[1] WANG Shang-xue,LIU Sheng, HUANG Shu-he. Research on the energy state of exhaust system of turbocharged diesel engine based on exergy analysis[J]. Railway locomotive&car. 2011, 31: 202-4.
[2] Zhuge W, Zhang Y, Zheng X, et al. Development of an advanced turbocharger simulation method for cycle simulation of turbocharged internal combustion engines[J]. Proceedings of the Institution of Mechanical Engineers, Part D: Journal of Automobile Engineering. 2009, 223: 661-72.
[3] Huang KD, Quang KV, Tseng K-T. Study of recycling exhaust gas energy of hybrid pneumatic power system with CFD[J]. Energy Conversion and Management. 2009, 50(5): 1271–8.
[4] Hung T-C. Waste Heat Recovery Of Organic Rankine Cycle Using Dry Fluids[J]. Energy Conversion and Management. 2001, 42: 539-53.
[5] LIU Jing-ping,FU Jian-qin, FENG Kang, et al. Characteristic of engine exhaust gas energy flow[J]. Jouunal of central south university(Science and technology): 2011, 42: 3370-6.
[6] GAO Wen-zhi, LV Deng-ke, FENG Li-ming, et al. The analysis of engine exhaust heat recovery utilization[C]. Chinese society for internal combustion engine: 2009.
[7] LI Bao-guo. The exergy analysis of engine's exhaust system[C]. Chinese Society of Engineering Thermophysics: 2000.
[8] SHEN Wei-dao,TONG Jun-geng. Engineering thermodynamics [M]. Higer education press, 2007.
[9] FU Qin-sheng. Thermodynamic analysis method of energy system [M]. Xi'an jiaotong university press, 2005.

AUTHOR INDEX